183 Period Domains over Finite and p-adic Fields

Period Domains over Finite and p-adic Fields

JEAN-FRANÇOIS DAT
Université de Paris VI

SASCHA ORLIK
Universität Wuppertal

MICHAEL RAPOPORT
Universität Bonn

CAMBRIDGE
UNIVERSITY PRESS

CAMBRIDGE UNIVERSITY PRESS
Cambridge, New York, Melbourne, Madrid, Cape Town, Singapore,
São Paulo, Delhi, Dubai, Tokyo, Mexico City

Cambridge University Press
The Edinburgh Building, Cambridge CB2 8RU, UK

Published in the United States of America by Cambridge University Press, New York

www.cambridge.org
Information on this title: www.cambridge.org/9780521197694

First published 2010

A catalogue record for this publication is available from the British Library

ISBN 978-0-521-19769-4 Hardback

Contents

Preface

This monograph is a systematic treatise on *period domains* over finite and over p-adic fields. The theory we present here has developed over the past fifteen years. Part of it has already appeared in various research articles or announcements, sometimes without detailed proofs. Our goal here is to present the theory as a whole and to provide complete proofs of the basics of the theory, so that these research articles can be accessed more easily. As it turned out, when working out the details, we had to change the very foundations of the theory quite a bit in some places, especially to accomodate isocrystals over non-algebraically closed fields, and also isocrystals with G-structure. Our hope is that our book can serve as the basis of future research in this exciting area.

Period domains over p-adic fields arose historically at the confluence of two theories: on the one hand, of Fontaine's theory [80] of the "mysterious functor" conjectured by Grothendieck, which relates p-adic Galois representations of p-adic local fields and filtered isocrystals; on the other hand, of the theory of formal moduli spaces of p-divisible groups and their associated period maps [183]. Via the latter theory, they are naturally related to local Langlands correspondences between ℓ-adic representations of the Galois groups of p-adic fields and smooth representations of p-adic Lie groups. In recent times, it became apparent [34, 176] that, via the former theory, period domains also show up in connection with the conjectural p-adic Langlands program relating p-adic representations of the Galois groups of p-adic fields, and p-adic representations of p-adic Lie groups.

There are at least three possible motivations for investigating the period domains of the title. First of all, they are somehow a natural analogue of the Griffiths period domains in Hodge theory. This is true in two respects. Conceptually, a Griffiths period domain is a moduli space for Hodge structures of a certain type; similarly, a p-adic period domain is a moduli space for "weakly admissible filtered isocrystals," which are the p-adic analogues of Hodge structures in Fontaine's theory. Technically, p-adic period domains are defined by a semi-stability formalism; the same is true for Griffiths period domains, although their historical definition is different.

The second motivation is of an arithmetic nature and follows from the relations between Galois representations and p-adic Hodge structures. More precisely, the universal p-adic Hodge structure above a p-adic period domain should conjecturally provide us with a "universal" relative crystalline Galois representation (of a certain type). However, making precise this hope is a very difficult problem of current interest, which lies beyond the scope of this monograph (see the last chapter).

The third motivation was our guide in this monograph. It comes from the formalism used in this theory and its analogies with other topics in algebraic geometry. As already mentioned, the "weakly admissible filtered isocrystals" may be seen as semi-stable objects in the category of filtered isocrystals, once the latter is endowed with a suitable "slope function." Hence there is a direct analogy with the category of vector bundles on a Riemann surface, when endowed with the usual slope function. This analogy has been very fruitful, since many of the "classical" concepts of the theory of vector bundles, such as the Harder–Narasimhan filtration or the GIT criteria for semi-stability, turn out to have natural analogues in the context of filtered isocrystals. Even the theory of G-bundles, i.e., torsors over a reductive group G, has an analogue in the context of filtered isocrystals. This comes from the fact that the category of isocrystals is tannakian, and that the semi-stability condition is compatible with tensor products. In fact, we could have replaced the category of isocrystals by any tannakian category. Choosing the easiest one, namely that of vector spaces over a field, we get rid of the p-adic nature of isocrystals and obtain a theory over any abstract field. It turns out that when the field is finite, there is a moduli space for "semi-stable filtered vector spaces" of a fixed type, and such moduli spaces are the "period domains over finite fields" of the title. Although they are no longer related to any type of Hodge theory, their study remains a good way of approaching that of their p-adic brothers, avoiding most of the intricate technicalities of the p-adic case. For example, when studying period domains over finite fields, only basic algebraic geometry is needed, while in the p-adic case, rigid-analytic geometry is required. Therefore, we like to consider the first two parts of this book as a pedagogical introduction to the formalism of slopes, semi-stability, and related concepts, in the most elementary context where it appears. These first two parts should be accessible to any graduate student with a basic background in algebraic geometry and algebraic group theory.

In the Introduction following this Preface, we give more details on the background of this book and also give a brief description of its contents. Throughout the text, there are numerous examples (which one may also regard as exercises). At the end of each section there are remarks on open questions, on history, and directions to the literature.

We are happy to acknowledge the important contributions of G. Faltings, J.-M. Fontaine, U. Hartl, R. Kottwitz, B. Totaro and Th. Zink to the theory presented here. We also thank M. Cabanes, U. Görtz, M. Harris, X. He, R. Huber, L. Lafforgue, G. Laumon, P. Schneider and A. Strohmaier for helpful discussions. We gave courses on this material at the universities of Wuppertal, Bonn, Leipzig and Paris. We are grateful for the feedback from the audiences of these courses.

Introduction

In this Introduction we give a brief description of the background of this monograph, then explain its scope, and finally give an overview of its contents.

Background 1: Classical Hodge theory The concept of a period domain was created by Griffiths in his work on periods of integrals on algebraic varieties over the field of complex numbers [97]. Let us first explain Griffiths' construction, cf. [96].

Let $H_{\mathbb{R}}$ be a finite-dimensional $\mathbb{R}$-vector space and let $H_{\mathbb{C}} = H_{\mathbb{R}} \otimes \mathbb{C}$. We also fix an integer n and a collection of non-negative integers $\{h^{pq}\}_{p,q}$ which satisfy $h^{pq} = h^{qp}$ and $h^{pq} \neq 0$ only if $p+q = n$, and such that $\sum h^{pq} = \dim H_{\mathbb{R}}$. There is a natural structure of a complex manifold on the set of all Hodge structures of weight n on $H_{\mathbb{R}}$, with h^{pq} as its Hodge numbers. This comes about as follows.

Let $\mathcal{F}$ be the set of all decreasing filtrations of subspaces

$$\cdots \subset \mathcal{F}^{p+1} \subset \mathcal{F}^{p} \subset \mathcal{F}^{p-1} \subset \cdots \subset H_{\mathbb{C}}\,,$$

such that $\dim \mathcal{F}^p = \sum_{i \geq p} h^{i,n-i}$. Then $\mathcal{F}$ forms in the obvious way a partial flag variety, and as such has the structure of a smooth projective algebraic variety. The group $\mathrm{GL}(H_{\mathbb{C}})$ acts algebraically and transitively on $\mathcal{F}$. Consider the subset $\mathcal{F}^{\circ}$ of filtrations $\mathcal{F}^{\bullet}$ which satisfy

$$H_{\mathbb{C}} = \mathcal{F}^p \oplus \overline{\mathcal{F}^{n-p+1}}\,, \text{ for every } p\,.$$

Then $\mathcal{F}^{\circ}$ is an open subset of $\mathcal{F}$, and is the parameter space of Hodge structures on $H_{\mathbb{R}}$ of the given type. Indeed, a Hodge structure

$$H_{\mathbb{C}} = \bigoplus_{p,q} H^{p,q}\,,\ \overline{H^{p,q}} = H^{q,p} \tag{0.1}$$

defines the filtration $\mathcal{F}^{\bullet}$ in $\mathcal{F}^{\circ}$ given by

$$\mathcal{F}^p = \bigoplus_{i \geq p} H^{i,n-i}\,.$$

Conversely, if $\mathcal{F}^\bullet$ is a point of $\mathcal{F}^\circ$, then $\mathcal{F}^\bullet$ corresponds to the Hodge structure (0.1) with

$$H^{pq} = \mathcal{F}^p \cap \overline{\mathcal{F}^q} .$$

For technical reasons, one pays special attention to *polarized* Hodge structures. Let Ψ be a non-degenerate bilinear form on $H_\mathbb{R}$ which is symmetric when n is even, and skew-symmetric when n is odd. Let $\mathcal{F}_\Psi$ be the subset of all those filtrations $\mathcal{F}^\bullet$ in $\mathcal{F}$ which satisfy

$$\Psi(\mathcal{F}^p, \mathcal{F}^{n-p+1}) = 0 \,, \text{ for every } p \,.$$

Then $\mathcal{F}_\Psi$ is a closed subvariety of $\mathcal{F}$. Let

$$G = \mathrm{Aut}(H_\mathbb{R}, \Psi)$$

be the automorphism group of the form Ψ, i.e., either the orthogonal or the symplectic group. The complex Lie group $G(\mathbb{C})$ acts transitively on $\mathcal{F}_\Psi$. In particular, $\mathcal{F}_\Psi$ is a smooth projective variety. Consider the subset $\mathcal{F}_\Psi^\circ$ of filtrations $\mathcal{F}^\bullet$ in $\mathcal{F}^\circ \cap \mathcal{F}_\Psi$ which satisfy

$$\Psi(Cv, \overline{v}) > 0 \text{ for } v \in H_\mathbb{C} \,, \; v \neq 0 \,.$$

Here $C = C_{\mathcal{F}^\bullet}$ is the Weil operator $C : H_\mathbb{C} \longrightarrow H_\mathbb{C}$ defined in terms of the Hodge structure corresponding to $\mathcal{F}^\bullet$ by

$$Cv = i^{p-q} v \,, \; v \in H^{p,q} \,.$$

Then $\mathcal{F}_\Psi^\circ$ parametrizes all Hodge structures on $H_\mathbb{R}$ which are polarized by Ψ and have the h^{pq} as Hodge numbers. The subset $\mathcal{F}_\Psi^\circ$ is open in $\mathcal{F}_\Psi$, and is acted on transitively by $G(\mathbb{R})$. Fixing a base point in $\mathcal{F}_\Psi^\circ$ we therefore have identifications

$$\begin{array}{ccc} \mathcal{F}_\Psi^\circ & \subset & \mathcal{F}_\Psi \\ \| & & \| \\ G(\mathbb{R})/V & \subset & G(\mathbb{C})/P(\mathbb{C}) \,, \end{array}$$

where P is a parabolic subgroup of $G_\mathbb{C}$, and where the subgroup $V = P(\mathbb{C}) \cap G(\mathbb{R})$ turns out to be compact.

It is $\mathcal{F}_\Psi^\circ$ that is the prototype of a classical period domain, i.e., a period domain in the sense of Griffiths. The name arises from the connection with families of Hodge structures defined by families of algebraic varieties. Let $f : X \longrightarrow S$ be a polarized smooth family of projective algebraic varieties parametrized by a complex variety S. Then for each n, the nth *primitive* cohomology groups of the fibers of f form a local system $PR^n f_*(\mathbb{R})$. Over the universal covering $\tilde{S}$ of S, this local system can be trivialized. Choosing such a trivialization, and associating to a point $\tilde{s} \in \tilde{S}$ with image $s \in S$ the Hodge structure on $PH^n(X_s, \mathbb{R})$, we obtain the *period morphism*

$$\varphi : \tilde{S} \longrightarrow \mathcal{F}^\circ. \tag{0.2}$$

Here $\mathcal{F}^\circ = \mathcal{F}^\circ_\Psi$ is the period space relative to the choice of the polarization, of n, of the appropriate Hodge numbers $h^{p,q}$, and of the trivialization of $PR^n f_* \mathbb{R}$ over $\tilde{S}$.

Griffiths and Schmid made a deep study of the differential-geometric properties of period domains, cf. [97]. For any $\mathcal{F}^\bullet \in \mathcal{F}^\circ$, the Lie algebra $\mathfrak{g}$ of G inherits a real Hodge structure of weight 0 from $\mathrm{End}(H_\mathbb{R})$, and the Lie algebra $\mathfrak{p}$ of the parabolic stabilizer P of $\mathcal{F}^\bullet$ is the 0th step of the associated Hodge filtration. The holomorphic tangent space of $\mathcal{F}^\circ$ at the point corresponding to $\mathcal{F}^\bullet$ is naturally isomorphic to $\mathfrak{g}/\mathfrak{p}$. The subspace $\mathfrak{p} \oplus \mathfrak{g}^{-1,1}/\mathfrak{p}$ of $\mathfrak{g}/\mathfrak{p}$ is the fiber at $\mathcal{F}^\bullet$ of a $G(\mathbb{C})$-invariant holomorphic subbundle T_h of the holomorphic tangent bundle T, the *horizontal tangent subbundle* [194]. It is a fundamental fact that this subbundle has negative holomorphic sectional curvature bounded away from zero, for a suitable $G(\mathbb{C})$-invariant hermitian form on T. This circumstance allows the application of a version of the Schwartz Lemma. The relevance of this result comes from the fact that any period map φ as in (0.2) is *horizontal*, i.e., $d\varphi(T_{\tilde{S}}) \subset T_h$. These facts have important implications for the local systems $PR^n f_* \mathbb{R}$ defined by families of algebraic varieties (e.g. the proof of Borel [52] of the monodromy theorem).

Also, Schmid [195] studied the L^2-cohomology of the restriction of homogeneous line bundles to period domains, and identified the representations of $G(\mathbb{R})$ afforded by them. This gives a cohomological realization of discrete series representations of $G(\mathbb{R})$.

A striking special case of Griffiths' construction arises from real Hodge structures of type $\{(0,-1),(-1,0)\}$. These can be identified with complex structures on $H_\mathbb{R}$. Choosing a polarization Ψ, the corresponding period domain $\mathcal{F}^\circ_\Psi$ can be identified with the union of the upper and the lower Siegel halfspaces. In this case, taking for S the moduli space of polarized abelian varieties of type Ψ, the map φ in (0.2) is an isomorphism. In general, the map φ is not even a local isomorphism, since the image of $d\varphi$ lies in the horizontal tangent subbundle.

One can formalize Griffiths' construction following Deligne [52]. Recall that a Hodge structure on $H_\mathbb{R}$ corresponds to a homomorphism $h : \mathbb{S} \longrightarrow \mathrm{GL}(H_\mathbb{R})$, where

$$\mathbb{S} = \mathrm{Res}_{\mathbb{C}/\mathbb{R}}(\mathbb{G}_m)$$

is the Weil restriction of the multiplicative group. Let G be a connected reductive group over $\mathbb{R}$. A homomorphism $h : \mathbb{S} \longrightarrow G$ is called *polarizable* if it factors through a maximal torus in G which is compact modulo center. Then the period domain associated to the conjugacy class of a polarizable homomorphism $h : \mathbb{S} \longrightarrow G$ is the set $\mathcal{F}(G,h)^\circ$ of conjugates under $G(\mathbb{R})$ of h. Let h_1 be the composite homomorphism

$$(\mathbb{G}_m)_\mathbb{C} \xrightarrow{(1,\mathrm{id})} \mathrm{Res}_{\mathbb{C}/\mathbb{R}}(\mathbb{G}_m)_\mathbb{C} = (\mathbb{G}_m)_\mathbb{C} \times (\mathbb{G}_m)_\mathbb{C} \longrightarrow G_\mathbb{C} .$$

Any conjugate of h_1 under $G(\mathbb{C})$ defines a filtration on the category of representations of G. The set $\mathcal{F}(G,h)$ of these filtrations can be identified with a generalized flag variety of $G_\mathbb{C}$. We obtain an open embedding

$$\mathcal{F}(G,h)^\circ \hookrightarrow \mathcal{F}(G,h) ,$$

and for suitable choices of (G,h) one obtains the examples discussed previously.

In Griffiths' presentation of the theory, the monodromy group acting on $\mathcal{F}(G,h)^\circ$ plays an important role, and the desire to form the quotient of $\mathcal{F}(G,h)^\circ$ by its action is a major reason for considering polarized Hodge structures. Here we will suppress this aspect of the theory, since there is so far no p-adic analogue of these quotients.

Background 2: p-adic Hodge structures The p-adic analogue of a Hodge structure is a *weakly admissible filtered isocrystal*, as defined by Fontaine [80]. Let L be a perfect field of characteristic $p > 0$, and let $K_0 = \mathrm{Quot}(W(L))$ be the fraction field of its ring of Witt vectors. We denote by σ the automorphism of K_0 induced by the Frobenius automorphism of L. An *isocrystal over L* is a finite-dimensional K_0-vector space V, equipped with a bijective σ-linear map $\Phi : V \longrightarrow V$. Let K be a finite field extension of K_0. A *filtered isocrystal* $(V,\Phi,\mathcal{F}^\bullet)$ *over K* is an isocrystal (V,Φ) over L equipped with a (decreasing, exhaustive and separating) $\mathbb{Z}$-filtration $\mathcal{F}^\bullet$ of the K-vector space $V \otimes_{K_0} K$. A filtered isocrystal over K is called *weakly admissible* if

$$\sum\nolimits_x x \dim \mathrm{gr}^x_{\mathcal{F}}(V' \otimes_{K_0} K) \le \mathrm{ord}\det(\Phi \mid V')$$

for any sub-isocrystal V' of V, with equality for $V' = V$.

Such a structure arises for example from an abelian variety over K with good reduction. In this case (V,Φ) is the rational Dieudonné module of its special fiber, and the filtration $\mathcal{F}^\bullet$ is given by the Hodge filtration of its generic fiber (via the comparison isomorphism with the DeRham cohomology) and has only two jumps, at $x = 0$ and at $x = 1$. Something similar is true of a p-divisible group over K with good reduction. More generally, a filtered isocrystal arises from the ith cohomology group of a smooth projective variety over K with good reduction. In this case, (V,Φ) is given by the ith crystalline cohomology group of the special fiber and again $\mathcal{F}^\bullet$ is given as the Hodge filtration of its generic fiber, which may have more than two jumps.

Note that in the previous example the extension K/K_0 is totally ramified. According to Colmez and Fontaine [44], if the finite extension K/K_0 is totally ramified, the category of weakly admissible filtered isocrystals over K is equivalent to the category of *crystalline p-adic Galois representations of* $\mathrm{Gal}(\bar{K}/K)$ under the *mysterious functor* conjectured by Grothendieck and constructed by Fontaine [80]. In the example of an abelian variety over K with good reduction, this Galois representation is given by the rational p-adic Tate module of its generic fiber, as proved by Breuil [32].

The p-adic analogue of a period domain arises by fixing the isocrystal (V, Φ) and by varying the filtration $\mathcal{F}^\bullet$. More precisely, we fix a function $g : \mathbb{Z} \longrightarrow \mathbb{Z}_{\geq 0}$ with $\sum_x g(x) = \dim V$, and consider the partial flag variety $\mathcal{F} = \mathcal{F}(V, g)$ of $\mathbb{Z}$-filtrations $\mathcal{F}^\bullet$ of type g, i.e., such that

$$\dim \mathrm{gr}^x_{\mathcal{F}}(V) = g(x)\,, \qquad \forall x \in \mathbb{Z}\,.$$

Then $\mathcal{F}$ is a smooth projective variety over $\mathbb{Q}_p$, with a transitive action of $\mathrm{GL}(V)$. The locus inside $\mathcal{F}$ corresponding to those filtrations $\mathcal{F}^\bullet$ such that $(V, \Phi, \mathcal{F}^\bullet)$ is weakly admissible is a subset $\mathcal{F}^{\mathrm{wa}}$ of $\mathcal{F} \otimes_{\mathbb{Q}_p} K_0$, which is an admissible open in the sense of rigid-analytic geometry [183]. More precisely, $\mathcal{F}^{\mathrm{wa}}$ is the complement of a p-adic family of Zariski-closed subvarieties of $\mathcal{F} \otimes_{\mathbb{Q}_p} K_0$.

More generally, and imitating Deligne's formalization of the classical period domains, one may perform this construction starting with any triple (G, b, μ). Here G is a connected reductive group over $\mathbb{Q}_p$, and b is an element of $G(K_0)$ and $\mu : (\mathbb{G}_m)_{\bar{\mathbb{Q}}_p} \longrightarrow G_{\bar{\mathbb{Q}}_p}$ is a one-parameter subgroup defined over an algebraic closure $\bar{K}_0$ of K_0. Then μ and any conjugate of μ defines a filtration on the category of representations of G and hence defines a partial flag variety $\mathcal{F}(G,\mu)$ defined over a finite extension E of $\mathbb{Q}_p$ contained in $\bar{K}_0$, the *local Shimura field* associated to (G,μ). Let $\breve{K}_0 = E.K_0$. Then the period domain $\breve{\mathcal{F}}(G, b, \mu)^{\mathrm{wa}}$ is an admissible open rigid-analytic subset of $\mathcal{F}(G,\mu) \otimes_E \breve{K}_0$. Its points parametrize weakly admissible triples (G, b, μ'), where (G, b) is fixed and μ' varies in $\mathcal{F}(G,\mu)$. According to [82], if G is quasi-split and L algebraically closed, we have $\breve{\mathcal{F}}(G, b, \mu)^{\mathrm{wa}} \neq \emptyset$ if and only if there is an inequality $\nu \leq \mu$ between the *Newton vector* ν of the isocrystal (V, Φ) and the *Hodge vector* of μ.

The best-known example of a p-adic period domain is the *Drinfeld halfspace* [71]. Let V_0 be a $\mathbb{Q}_p$-vector space of dimension n, and consider the trivial isocrystal $(V, \Phi) = (V_0 \otimes_{\mathbb{Q}_p} K_0, \mathrm{id}_{V_0} \otimes \sigma)$. Let $g : \mathbb{Z} \longrightarrow \mathbb{Z}_{\geq 0}$ be the function with

$$g(n-1) = 1, \quad g(-1) = n-1, \quad g(x) = 0 \ \text{ for } x \neq n-1, -1.$$

In this case $\mathcal{F}(V,g) = \mathbb{P}(V) \simeq \mathbb{P}^{n-1}$ is the projective space of lines in V and $\mathcal{F}(V,g)^{\mathrm{wa}}$ is the space $\Omega(V) = \Omega^n$ of all lines not contained in any $\mathbb{Q}_p$-rational hyperplane in V. Drinfeld [71] proved that Ω^n is the generic fiber of an adic formal scheme over Spf $W(L)$ which is the parameter space of certain p-divisible groups of a specific type (*special formal O_D-modules*).

In [183], this is generalized to other families of p-divisible groups, of (EL)- or of (PEL)-type. In this more general case, one obtains a formal scheme $\mathcal{M}$ over Spf $O_{\breve{K}_0}$ (no more adic in general) and a period morphism from the generic fiber of $\mathcal{M}$ to a period domain. This period morphism is an étale rigid-analytic morphism, but in contrast to the Drinfeld case, it is no longer an isomorphism. In the general case, it is conjectured that there is an open subset $\breve{\mathcal{F}}(G,b,\mu)^{\mathrm{a}}$ (analytic in the sense of Berkovich) of the analytic space associated to $\breve{\mathcal{F}}(G,b,\mu)^{\mathrm{wa}}$, the *admissible subset*, and a local system of p-adic vector spaces over $\breve{\mathcal{F}}(G,b,\mu)^{\mathrm{a}}$ such that the fiber in each point $\mathcal{F}^\bullet$ corresponds as a p-adic Galois representation with G-structure under the Fontaine functor to the filtered isocrystal with G-structure $(G,b,\mathcal{F}^\bullet)$ (this more precise version of the conjecture in [183] is due to Hartl [108]).

Let

$$J(\mathbb{Q}_p) = \{g \in G(K_0) \mid gb\sigma(g)^{-1} = b\}$$

be the automorphism group of the isocrystal with G-structure (G,b). The group $J(\mathbb{Q}_p)$ is the group of $\mathbb{Q}_p$-rational points of an algebraic group over $\mathbb{Q}_p$. It acts by naturality on $\mathcal{F}(G,\mu) \otimes_E \breve{K}_0$ preserving the period space $\breve{\mathcal{F}}(G,b,\mu)^{\mathrm{wa}}$. The action of $J(\mathbb{Q}_p)$ also preserves $\mathcal{F}(G,b,\mu)^a$ and this action is lifted to the conjectural local system mentioned above. Imposing level structures on the local system, one obtains a projective system of rigid spaces mapping by surjective étale morphisms to $\breve{\mathcal{F}}(G,b,\mu)^a$. The group $J(\mathbb{Q}_p)$ acts on each member of this projective system, and the group $G(\mathbb{Q}_p)$ acts as Hecke correspondences on the projective system as a whole. From a suitable version of ℓ-adic cohomology in the rigid context, one deduces a $\bar{\mathbb{Q}}_\ell$-vector space with a triple action of $G(\mathbb{Q}_p)$, and $J(\mathbb{Q}_p)$, and the Weil subgroup of $\mathrm{Gal}(\bar{E}/E)$. There is a conjecture by Kottwitz [185], generalizing conjectures of Carayol [37] pertaining to the Drinfeld case and the Lubin–Tate case, which describes the precise kind of Langlands correspondence between the representations of these three groups that this triple representation induces (on the *discrete Langlands parameters*; for more general parameters, cf. Harris [106]). In the Drinfeld and the Lubin–Tate cases, in which the respective period domains are Ω^n, resp. $\mathbb{P}^{n-1}$, and in which case the projective system is known to exist, these conjectures have been completely proved very recently, thanks to the results of Boyer, Dat, Faltings, Fargues, and Harris and Taylor.

Background 3: Semi-stability The basic motivation for the present monograph is Faltings' observation [73] that weak admissibility can be viewed as a semi-stability condition. Let $(V, \Phi, \mathcal{F}^\bullet)$ be a filtered isocrystal over K. For any $\mathbb{R}$-filtration $\mathcal{F}^\bullet$ of $V \otimes_{K_0} K$, let

$$\deg_{\mathcal{F}}(V) = \sum\nolimits_x x \dim \mathrm{gr}^x_{\mathcal{F}}(V \otimes_{K_0} K) .$$

Let $\mathcal{G}^\bullet$ be the slope filtration of V, i.e., the $\mathbb{Q}$-filtration

$$\mathcal{G}^x = \sum_{-\lambda \geq x} V_{-\lambda} ,$$

where $V = \bigoplus V_\lambda$ is the slope decomposition of the isocrystal (V, Φ). Its degree is given by $\deg_{\mathcal{G}}(V) = -\operatorname{ord}\det(\Phi|V)$. Then $(V, \Phi, \mathcal{F}^\bullet)$ is called *semi-stable* if

$$\frac{1}{\dim V'}(\deg_{\mathcal{F}}(V') + \deg_{\mathcal{G}}(V')) \leq \frac{1}{\dim V}(\deg_{\mathcal{F}}(V) + \deg_{\mathcal{G}}(V)) \tag{0.3}$$

for any sub-isocrystal V' of V. Hence $(V, \Phi, \mathcal{F}^\bullet)$ is weakly admissible iff it is semi-stable and if in addition the RHS of (0.3) is equal to zero.

The semi-stability condition is much more flexible than the weak admissibility condition. In particular, it lends itself to analogues in pure linear algebra and hence also to period domains in this pure linear algebra context. More precisely, let V be a finite-dimensional vector space over a field k. Let $\mathcal{F}^\bullet$ be an $\mathbb{R}$-filtration on $V \otimes_k K$, where K is a field extension of k. Then $(V, \mathcal{F}^\bullet)$ is called *semi-stable* if

$$\frac{1}{\dim V'} \deg_{\mathcal{F}'}(V') \leq \frac{1}{\dim V} \deg_{\mathcal{F}}(V)$$

for all k-subspaces V' of V. Here on the LHS, the $\mathbb{R}$-filtration $\mathcal{F}'$ on $V' \otimes_k K$ is the one induced by $\mathcal{F}^\bullet$, and the fractions on both sides are called the *slopes* of V, resp. V'.

We explicitly note that if $L = \mathbb{F}_p$, then a filtered isocrystal becomes an object of linear algebra. In particular, if Φ is the identity automorphism of V, then a filtered isocrystal over K is nothing other than an object $(V, \mathcal{F}^\bullet)$ as above (relative to $k = \mathbb{Q}_p$). It is remarkable that the first kind of object mentioned in Background 1 above can also be phrased in these terms. In fact, by Pink [177], a Hodge structure of weight n on the $\mathbb{R}$-vector space $H_{\mathbb{R}}$ is the same as a semi-stable $\mathbb{Z}$-filtration $\mathcal{F}^\bullet$ on $H_{\mathbb{C}}$ of slope $n/2$.

The filtrations on $V \otimes_k K$ of a fixed type are parametrized by the K-valued points of a partial flag variety $\mathcal{F}$ over k. If k is a finite field, there is a Zariski-open subset $\mathcal{F}^{ss}$ of $\mathcal{F}$ with K-valued points equal to the set of semi-stable pairs $(V, \mathcal{F}^\bullet)$ as above. If k is a non-archimedean local field, the set $\mathcal{F}^{ss}$ is an admissible open rigid-analytic subset of $\mathcal{F}$. These are the prototypes of the period domains in the title. In this context, the theory is stripped of its arithmetic content, and becomes geometric and more elementary. In this monograph we study the geometry of period domains in this context, and address the question of determining their cohomology. We regard the theory developed here as somewhat preliminary to the deep arithmetic questions outlined above.

The scope of this monograph In this monograph we study period domains in the context of semi-stability, in its variants of linear algebra, as well as of the isocrystal variants. Our main purpose is to bring out the analogy between period domains and the moduli spaces of vector bundles on Riemann surfaces [181]. We are especially interested in the geometry of period domains, in particular in determining their cohomology and other topological invariants. Among the topics treated we mention the following.

- The tensor product theorem of Faltings and Totaro, which states that the tensor product of two semi-stable pairs $(V, \mathcal{F}^\bullet)$ and $(V', \mathcal{F}'^\bullet)$ is again semi-stable (in the isocrystals context this was conjectured by Fontaine).
- The machinery of the Harder–Narasimhan filtration, which presents a pair $(V, \mathcal{F}^\bullet)$ as a successive extension of semi-stable pairs in a definite way.
- The relation of the concept of semi-stability of $(V, \mathcal{F}^\bullet)$ to the semi-stability concept in Geometric Invariant Theory (this is due in a special case to van der Put and Voskuil, and in general to Totaro, confirming a conjecture of Rapoport and Zink).
- The structure of the Harder–Narasimhan stratification of the partial flag variety, which reveals an interesting recursive structure of the boundary of the period domain in terms of period domains of smaller dimension.
- The ℓ-adic cohomology with compact supports of period domains (in the case of the Drinfeld halfspace Ω^n, this is due to Drinfeld for $n = 2$ and to Schneider and Stuhler, and to Dat, for arbitrary n, whereas the cohomology *complex* is due to Dat; the case of a general period domain is due to Orlik, and the determination of the Euler–Poincaré characteristic to Kottwitz and Rapoport).
- The relation between period domains over the field $\mathbb{F}_1$ with one element and thin Bruhat cells and the fibers of the moment map. Here, for the variant of the semi-stability notion over $\mathbb{F}_1$, instead of testing all k-rational subspaces of V, one tests all coordinate subspaces of V with respect to a fixed basis of V. In this variant k can be an arbitrary field.
- The generalization of the theory from $\mathrm{GL}(V)$ to arbitrary reductive groups.
- A systematic treatment of period domains including the case where L is perfect, but not necessarily algebraically closed.

The structure of this monograph The monograph consists of four parts. In the first part (Chapters I–III) we present the theory in its most elementary form. We prove the tensor product theorem in its various variants, and develop the Harder–Narasimhan machinery. We introduce period domains for GL_n over a finite field and over $\mathbb{F}_1$, i.e., in those cases that lead to Zariski-open subsets of generalized flag varieties. We study their stratifications by the Harder–Narasimhan types, resp. by their Harder–Narasimhan polygons. Also, we address the question of determining the cohomology of period domains in this context.

In the second part (Chapters V–VII) the theory in the first part is generalized to the case of an arbitrary reductive group G instead of GL_n. Again, the period domains encountered here are Zariski-open subsets of generalized flag varieties (associated to G). This part is preceded by an interlude on the Tannaka formalism in the context of algebraic groups (Chapter IV).

In the third part (Chapters VIII–X) we pass to the case of a p-adic local field as a base field. In this case we obtain period domains which are admissible rigid-analytic open subsets of generalized flag varieties. They parametrize semi-stable filtered isocrystals. The theory is analogous to that in the first two parts, but is considerably more difficult. In this sense, the first two parts of the monograph may be considered as toy models for the objects in the third part. On the other hand, the theory developed in this last part can be viewed to some degree as a warm-up for the study of the arithmetically significant covering spaces mentioned above.

In a final part (Chapter XI) we give some complements, and review some of the recent work in the area.

Content We now give a chapter-by-chapter description of the contents of the book.

Chapter I is basic for the whole book. In Section 1 we give the general concepts on which the theory rests. In Section 2 we prove the tensor product theorem of Faltings and Totaro; our proof is essentially Totaro's. In Section 3 we introduce the Harder–Narasimhan filtration; we follow Faltings in numbering its filtration steps by the slopes.

Chapter II introduces the period domains attached to a vector space and a dominant co-weight. The basic definition is given in Section 1. In Section 2 we characterize period domains through the Hilbert–Mumford inequality from Geometric Invariant Theory. Section 3 explains the stratification of the generalized flag variety according to Harder–Narasimhan types, resp. Harder–Narasimhan vectors. The subtle difference between these two stratifications – one being a refinement of the other – is a novel phenomenon that does not occur for the space of vector bundles on a Riemann surface. In Section 4 we analyze period domains "over $\mathbb{F}_1$," and connect our theory to the theory of Gelfand–Goresky–MacPherson–Serganova of thin Schubert cells and the moment map.

Chapter III addresses the problem of determining the ℓ-adic cohomology of period domains. Section 1 is devoted to an exposition of the Langlands Lemma from the theory of Eisenstein series, in which we follow closely Laumon [149] and Labesse [141]. In Section 2 we analyze the representations of $\mathrm{GL}_n(\mathbb{F}_q)$ which contain a fixed vector under the Borel subgroup. In particular, we prove by reduction to the representation theory of the symmetric group that the induced representations i_P^G and the generalized Steinberg representations v_P^G form a basis of their Grothendieck group, as P varies over the associate classes of parabolic subgroups. We give the change of basis matrix between these two bases in terms of a complex which is similar to, but in general different from, the Solomon–Tits complex. This is based on a remarkable distributivity property of the representations i_P^G, due to Cabanes. This also leads to the existence of a basis of the group algebra of the Weyl group compatible with the subspaces i_I analogous to i_P^G (however, we strongly believe that it is essentially impossible to give this basis explicitly!). In Section 3 we explain the recursion relation for the Euler–Poincaré characteristic of a period domain using stratification by Harder–Narasimhan types, and we resolve the recursion relation by using the Langlands Lemma. We also explain how, by expressing the result in terms of generalized Steinberg representations, one is led to a formula which gives the cohomology of period domains degree by degree (i.e., not merely the Euler–Poincaré characteristic), and deduce from this a precise vanishing theorem in ℓ-adic cohomology.

Chapter IV gives a brief exposition of some facts from the tannakian formalism that we will need later. We concentrate on the theory of filtrations of fiber functors, and show in particular how a filtration of the natural fiber functor on the representation category of a reductive group can be transferred naturally to any parabolic subgroup.

Chapter V transposes the theory of Chapter I, §1 to general reductive groups. There are two approaches, one *externally* through the tannakian formalism, and one *internally* through group theory – and the main point here is to show that both approaches give the same result. This is done in Sections 1 and 2. Section 3 then transfers the Harder–Narasimhan filtration to this context. We note that the external approach uses the Mumford conjecture from Geometric Invariant Theory (=Haboush's Theorem).

Chapter VI is the analogue of Chapter II for general reductive groups over finite fields. Section 1 defines period domains in this context. Section 2 relates this definition to the definition in terms of Geometric Invariant Theory. This is based on the concept of an invariant inner product on a reductive group due to Totaro, and the proof here is a simplification of Totaro's original proof. Section

3 analyzes the Harder–Narasimhan stratification, and in particular the closure relation among the strata. This is based on an analysis of the structure of the partially ordered uniquely divisible monoid of conjugacy classes of $\mathbb{Q}$-1-PS of G.

Chapter VII starts by introducing the induced representations i_P^G and generalized Steinberg representations v_P^G for general reductive groups over finite fields. It turns out that these generate in general a proper subgroup of the natural Grothendieck group, but that as in the case of GL_n, a basis of this subgroup is given by the i_P^G and also by v_P^G, as P ranges over the associate classes of parabolic subgroups. The proof of this fact is based on the observation that any finite group of Lie type has elliptic regular semi-simple elements (Lusztig informed us that this was known to him previously, but our proof here seems to be the first published one). Section 2 then is the analogue of Chapter III, §3.

Chapter VIII starts with recalling the theory of isocrystals, in particular stressing the special nature of "split semi-simple isocrystals." Then the concepts of semi-stability and of weak admissibility for isocrystals are introduced. Period domains in this new context are defined in Section 2. There are two major differences in our exposition as compared with [183]. First of all, we chose to present the theory in the context of Berkovich spaces instead of rigid-analytic varieties. This is dictated to us through our cohomology calculation in Chapter X, but also facilitates the comparison with the *admissible* set mentioned above. Second, we deal with isocrystals over arbitrary perfect fields. We also give in this section a criterion for when a period domain is non-empty (this question is vacuous in the finite field case). Section 3 introduces in this context the Harder–Narasimhan stratification; in the most general case it is an open problem to determine the set of non-empty strata. Also, the relation between the stratifications by Harder–Narasimhan type and by Harder–Narasimhan vector is difficult outside the split semi-simple case. Section 4 describes period domains for isocrystals in terms of Geometric Invariant Theory, in the split semi-simple case. Section 5 relates isocrystals with an action by a finite extension of $\mathbb{Q}_p$ to Kottwitz's σ-F-spaces, elaborating on a remark in [135].

Chapter IX is technically the most demanding. In Section 1 we introduce the concept of an isocrystal with structure in an arbitrary Tannaka category over $\mathbb{Q}_p$. The best-known example is the case when this Tannaka category is the representation category of an algebraic group over $\mathbb{Q}_p$. However, more general tannakian categories are in fact needed in order to deal with the Harder–Narasimhan formalism, even if one starts with a G-isocrystal in the usual sense. In order to deal effectively with this concept, we introduce the notion of an augmented group scheme over the Tannaka category of isocrystals over L. In Section 2 we use this formalism to transpose semi-stability and Harder–Narasimhan filtration to this context. In Section 3 we analyze the automorphism group of an isocrystal with structure in a tannakian category. In Section 4 we generalize to this context the analysis of the space of conjugacy classes of $\mathbb{Q}$-1-PS from Chapter VI, §3. In Section 5 we introduce period domains. A technically important remark is that we may change an augmented group scheme by a "weak isomorphism" without changing the period domain. In Section 6 we analyze the Harder–Narasimhan stratification in this context. The situation in general is quite complicated, but it simplifies in the split semi-simple case. The final Section 7 is devoted to the operation of restriction of scalars in this context.

Chapter X is devoted to the cohomology of period domains attached to *basic* isocrystals. In Section 1 we discuss generalized Steinberg representations in the p-adic case. The situation here is different from the finite field case, and in fact well-known and much simpler. Section 2 then treats the cohomology of period domains. Here we use basic facts on the ℓ-adic cohomology of Berkovich spaces, which we treat in an axiomatic way. By changing an augmented group scheme within its weak isomorphism class, we are reduced to a situation which is essentially identical to that over a finite field. Hence the same proof as in Chapter VII, §2 yields a result that is very similar to the one in the linear algebra context.

Chapter XI has a more informal character. In Section 1 we discuss the "fundamental complex," the main ingredient of the determination of the cohomology of period domains (and not merely their Euler–Poincaré characteristic). The geometry behind this complex also enters into Section 2, where we compare period domains over a finite field to the other class of algebraic varieties attached to finite groups of Lie type, the Deligne–Lusztig varieties. In particular, we compare properties like affineness and simple connectivity for both classes of varieties. In Section 3 we discuss some special features of the Drinfeld space. As already pointed out in [186], this period domain is quite atypical of period domains in general, but the comparison is useful to keep in mind. Section 3 concerns the conjectural local system of $\mathbb{Q}_p$-vector spaces on an open subset of a period domain, and discusses the results of Faltings [75] and Hartl [109] concerning them in the weight 1 case. In Section 4 we discuss the results of Dat [48] concerning the cohomology *complex* of the Drinfeld space. At least the *splitting theorem* holds for general period domains, but whether one can extend the main results of [48] to general period domains is an unresolved question.

Part 1

Period Domains for GL_n over Finite Fields

I. Filtered Vector Spaces

This chapter is basic for the whole monograph. Let $k \subset K$ be a field extension. We study the category of pairs consisting of a finite-dimensional vector space V over k and a filtration on the K-vector space $V \otimes_k K$. This is a quasi-abelian k-linear tensor category, whose set of objects is naturally endowed with two $\mathbb{R}$-valued functions, called *degree* and *rank*, which are additive on short exact sequences. The quotient of *degree* by *rank* is called *slope*. It is additive on tensor products and convex on short exact sequences. There is here a strong analogy with the k-linear quasi-abelian category of vector bundles over a projective smooth curve over k, endowed with the usual degree and rank functions. Keeping this analogy in mind, we introduce the notion of *semi-stable* objects, and show how any object carries a canonical filtration with semi-stable and slope-decreasing subquotients, called its *Harder–Narasimhan filtration*. Making this analogy explicit is not merely a pleasant exercise in linear algebra. It is motivated by Hodge theory, complex or p-adic, as explained in the general introduction of this monograph. Technically, the most difficult result is the tensor product theorem of Faltings and Totaro, which essentially states that the canonical filtration of a tensor product is the tensor product of the respective canonical filtrations. The importance of this theorem for the subject matter of this monograph will become apparent in Part 2.

1 Filtrations

In this section we introduce filtrations on vector spaces and their slopes. We then define semi-stability in its various variants.

Definition 1.1.1 Let K be a field and let V be a finite-dimensional vector space over K. An $\mathbb{R}$-*filtration* on V is a decreasing map

$$\mathcal{F} : \mathbb{R} \longrightarrow \{K\text{-subspaces of } V\},\ x \longmapsto \mathcal{F}^x.$$

Here, the set of subspaces of V is partially ordered by inclusion. The following conditions are imposed.

(i) There exists $x \in \mathbb{R}$ such that $\mathcal{F}^x = (0)$ and $\mathcal{F}^{-x} = V$, i.e., the filtration is exhaustive and separating.

(ii) Let $\mathcal{F}^{x-} = \bigcap_{y<x} \mathcal{F}^y$. Then $\mathcal{F}^x = \mathcal{F}^{x-}$.

Here is a picture that helps us to visualize the conventions in the case when V is one-dimensional.

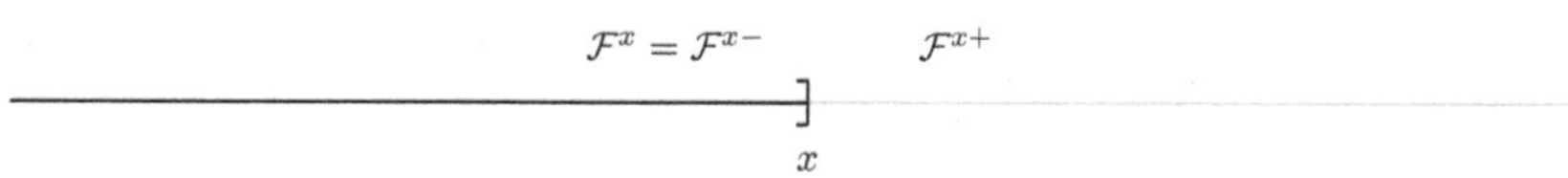

Remarks 1.1.2 (i) We also write V^x or $V^x_{\mathcal{F}}$ for $\mathcal{F}^x$. Let $\mathcal{F}^{x+} = \sum_{y>x} \mathcal{F}^y$, and $\mathrm{gr}^x_{\mathcal{F}}(V) = \mathcal{F}^x/\mathcal{F}^{x+}$. Then

$$\mathrm{gr}_{\mathcal{F}}(V) = \bigoplus_x \mathrm{gr}^x_{\mathcal{F}}(V)$$

is the *graded K-vector space* associated to $\mathcal{F}$. Obviously, since V is assumed of finite dimension, there are only a finite number of x with $\mathrm{gr}^x_{\mathcal{F}}(V) \neq 0$. These are called the *jumps of $\mathcal{F}$*. Let $\{x_1 > x_2 > \cdots > x_r\}$ be the jumps of $\mathcal{F}$. We then sometimes write an $\mathbb{R}$-filtration in the form

$$(0) \subset \mathcal{F}^{x_1} \subset \mathcal{F}^{x_2} \subset \cdots \subset \mathcal{F}^{x_r} = V.$$

Let $n = \dim V$ and $n_i = \dim \mathrm{gr}^{x_i}_{\mathcal{F}}(V)$. Let

$$(\mathbb{R}^n)_+ = \{(x_1, \ldots, x_n) \in \mathbb{R}^n \mid x_1 \geq \cdots \geq x_n\}. \tag{1.1}$$

We associate to $\mathcal{F}$ the vector $\nu(\mathcal{F}) \in (\mathbb{R}^n)_+$,

$$\nu(\mathcal{F}) = (x_1^{(n_1)}, x_2^{(n_2)}, \ldots, x_r^{(n_r)}). \tag{1.2}$$

Here the exponent (n_i) indicates that the corresponding entry is repeated n_i times. This vector is called the *type of $\mathcal{F}$*. For later use, we also introduce the notation

$$(\mathbb{Q}^n)_+ = (\mathbb{R}^n)_+ \cap \mathbb{Q}^n, \ (\mathbb{Z}^n)_+ = (\mathbb{R}^n)_+ \cap \mathbb{Z}^n, \ \mathbf{1} = (1, \ldots, 1).$$

(ii) If $I \subset \mathbb{R}$, we call $\mathcal{F}$ an *I-filtration*, if $\mathrm{gr}^x_{\mathcal{F}}(V) = 0, \forall x \in \mathbb{R} \setminus I$. Important special cases are $I = \mathbb{Q}$ and $I = \mathbb{Z}$. If $I = [x, y]$, we say that $\mathcal{F}$ is concentrated in degrees between x and y. For a $\mathbb{Z}$-filtration $\mathcal{F}$ on V, we have

$$\mathrm{gr}^x_{\mathcal{F}}(V) = \mathcal{F}^x/\mathcal{F}^{x+1}.$$

If V and W are K-vector spaces equipped with $\mathbb{R}$-filtrations, we also equip $V \oplus W$, $V \otimes_K W$ and V^* (the dual space) with $\mathbb{R}$-filtrations in the usual way,

$$\begin{aligned}(V \oplus W)^x &= V^x \oplus W^x \\ (V \otimes W)^x &= \sum_{y+z=x} V^y \otimes W^z \\ (V^*)^x &= (V^{-x+})^\perp.\end{aligned} \tag{1.3}$$

Let $U \subset V$ be a subspace. Then the induced $\mathbb{R}$-filtration on U, resp. V/U, is given as

$$\begin{aligned}U^x &= U \cap V^x \\ (V/U)^x &= (U + V^x)/U.\end{aligned} \tag{1.4}$$

By a successive application of these rules we also obtain $\mathbb{R}$-filtrations on $V^{\otimes r}$, $\mathrm{Sym}^r(V)$, $\Lambda^r V$, $r \in \mathbb{N}$.

Definition 1.1.3 Let K be a field extension of the field k. The category Fil_k^K is given by the objects $(V, \mathcal{F}_V)$, where V is a finite-dimensional k-vector space, and where $\mathcal{F} = \mathcal{F}_V$ is an $\mathbb{R}$-filtration of the K-vector space $V_K = V \otimes_k K$. A morphism in Fil_k^K is a homomorphism of k-vector spaces $f : V \longrightarrow W$ such that $f_K = f \otimes \mathrm{id}_K$ respects the filtrations, i.e.,

$$f_K(V^x) \subset W^x. \tag{1.5}$$

Sometimes we will simply write $V \in \mathrm{Fil}_k^K$ instead of $(V, \mathcal{F})$.

The category Fil_k^K is obviously an additive k-linear $\otimes$-category (see Chapter IV for an account of $\otimes$-categories). However, it is not an abelian category. Indeed, any morphism f has a kernel and a cokernel, but the canonical morphism $\mathrm{Coim}(f) \longrightarrow \mathrm{Im}(f)$ need not be an isomorphism. When it is an isomorphism, f is called a *strict* morphism in Fil_k^K. Explicitly, $\mathrm{Ker}(f)$ is the kernel of the k-linear map f with the induced filtration from V_K, and $\mathrm{Coker}(f)$ is $W/f(V)$ with the induced filtration from W_K. One checks easily that f is strict if and only if

$$\forall x \in \mathbb{R},\ f_K(V^x) = f_K(V_K) \cap W^x,$$

which means that f_K is "strictly compatible" with filtrations. We note that the class of strict epimorphisms is stable under pull-backs and that of strict monomorphisms is stable under push-forwards. Additive categories with kernels, cokernels and satisfying this condition are called *quasi-abelian* and are extensively studied in [197]. These are in particular *exact categories* (cf. [201], p. 41, for the definition of an exact category). A sequence of objects in Fil_k^K

$$0 \longrightarrow V' \xrightarrow{f'} V \xrightarrow{f''} V'' \longrightarrow 0 \tag{1.6}$$

is called a *short exact sequence* if f' is a kernel of f'' and f'' is a cokernel of f'. This is equivalent to asking that

$$0 \longrightarrow V_K'^{\,x} \longrightarrow V_K^x \longrightarrow V_K''^{\,x} \longrightarrow 0$$

is an exact sequence of K-vector spaces for any x, which in turn is equivalent to asking that

$$0 \longrightarrow \mathrm{gr}^x(V') \longrightarrow \mathrm{gr}^x(V) \longrightarrow \mathrm{gr}^x(V'') \longrightarrow 0 \tag{1.7}$$

is exact, for all x.

Definition 1.1.4 Let $V = (V, \mathcal{F}_V)$ be an object of Fil_k^K.
(i) A *subobject* of V is a k-subspace U of V with the induced filtration $\mathcal{F}_U$ on $U \otimes_k K \subset V \otimes_k K$. Note that $U \longrightarrow V$ is a strict monomorphism in the category Fil_k^K.
(ii) A *quotient object* of V is a factor space $V \longrightarrow W$, with the induced filtration $\mathcal{F}_{V/U}$ on $W = V/U$. Note that $V \longrightarrow W$ is a strict epimorphism in the category Fil_k^K.

Given two subobjects U_1 and U_2 of $(V, \mathcal{F}_V)$, the infimum of U_1 and U_2 is $U_1 \cap U_2$ with the induced filtration. Similarly, for the supremum of two quotient objects W_1 and W_2 of $(V, \mathcal{F}_V)$.

Variant 1.1.5 Instead of a single filtration, one may also consider several filtrations. Then even the case $K = k$ yields a non-trivial theory. Fix a positive integer r and consider the category Fil_k^r of k-vector spaces V equipped with $\mathbb{R}$-filtrations $\mathcal{F}_1, \dots, \mathcal{F}_r$ on V. Then we obtain as before an exact k-linear $\otimes$-category.

Variant 1.1.6 Let V be a k-vector space equipped with a decomposition $\mathbf{V} = (V = V_1 \oplus \cdots \oplus V_n)$ into lines. The set of triples $(V, \mathbf{V}, \mathcal{F})$, where $\mathcal{F}$ is an $\mathbb{R}$-filtration of V, form a category. A morphism $(V, \mathbf{V}, \mathcal{F}) \longrightarrow (V', \mathbf{V}', \mathcal{F}')$ is a morphism $(V, \mathcal{F}) \longrightarrow (V', \mathcal{F})$ in Fil_k^k, such that the matrix expressing the homomorphism $V \longrightarrow V'$ in terms of bases consisting of generators of the lines occurring in the decompositions $\mathbf{V}$ and $\mathbf{V}'$ has at most one entry in every column and in every row. This category is not additive, but has a tensor product. We will denote this category by $\mathrm{Fil}_{\mathbb{F}_1}^k$.

Of course, one may mix all three variants (consider several filtrations on $V \otimes_k K$, etc.). But to bring out better the differences between these variants, we will stick to these special cases.

Definition 1.1.7 Let $(V, \mathcal{F}) \in \mathrm{Fil}_k^K$. The *degree*, the *rank* and the *slope* of $(V, \mathcal{F})$ are defined respectively by

$$\begin{aligned}\deg(V, \mathcal{F}) &= \sum_x x \cdot \dim_K \mathrm{gr}_{\mathcal{F}}^x(V)\\ \mathrm{rk}(V, \mathcal{F}) &= \dim_k V\\ \mu(V, \mathcal{F}) &= \frac{\deg(V, \mathcal{F})}{\mathrm{rk}(V, \mathcal{F})}, \text{ if } V \neq (0).\end{aligned}$$

We also write $\deg_{\mathcal{F}}(V)$ or $\deg(V)$, resp. $\mathrm{rk}(V)$, resp. $\mu_{\mathcal{F}}(V)$ or $\mu(V)$ for these quantities.

Here is a list of elementary properties of the slope:

Lemma 1.1.8 *Let $V, W \in \mathrm{Fil}_k^K$ and $r \in \mathbb{N}$.*
(i) $\mu(V \otimes W) = \mu(V) + \mu(W)$.
(ii) $\mu(V^*) = -\mu(V)$.
(iii) $\mu(\mathrm{Sym}^r V) = r \cdot \mu(V)$, *and if $r \leq \dim V$, also* $\mu(\bigwedge^r V) = r \cdot \mu(V)$.

Proof First we note that the slope and the degree of $V \in \mathrm{Fil}_k^K$ depend only on the graded K-vector space $\mathrm{gr}^\bullet(V)$.

(i) It is well known that the functor $\mathrm{gr}^\bullet$ from Fil_k^K to the category of graded K-vector spaces is a $\otimes$-functor, i.e., we have canonical isomorphisms $\mathrm{gr}^\bullet V \otimes \mathrm{gr}^\bullet W \xrightarrow{\sim} \mathrm{gr}^\bullet(V \otimes W)$ of graded vector spaces. So we get the formula

$$\begin{aligned}\deg(V \otimes W) &= \sum_{x,y \in \mathbb{R}} (x + y) \dim(\mathrm{gr}^x(V)) \dim(\mathrm{gr}^y(W))\\ &= \sum_x (x \dim(\mathrm{gr}^x(V)) \dim(W) + \dim(\mathrm{gr}^x(V) \deg(W)))\\ &= \deg(V) \dim(W) + \dim(V) \deg(W).\end{aligned}$$

(ii) Follows immediately from the fact that $\mathrm{gr}^x(V^*) = \mathrm{gr}^{-x}(V)^*$.

(iii) Let us show that the functor $\mathrm{gr}^\bullet$ commutes with symmetric and exterior powers. By definition we have $\mathrm{Sym}^r V = \mathrm{Coim}(s_r)$ and $\bigwedge^r V = \mathrm{Coim}(a_r)$, where $s_r = \sum_{\sigma \in \mathfrak{S}_r} \sigma$, resp. $a_r = \sum_{\sigma \in \mathfrak{S}_r} \varepsilon(\sigma) \cdot \sigma$, are the symmetrization, resp. antisymmetrization, endomorphisms of $\bigotimes^r V$. We note that for any morphism f in Fil_k^K there is a canonical morphism $\mathrm{gr}^\bullet(\mathrm{Coim} f) \longrightarrow \mathrm{Coim}(\mathrm{gr}^\bullet(f))$ which is an isomorphism if and only both the underlying K-vector spaces of the target and the source have the same dimension (and iff f is strict). Now, the endomorphisms $\mathrm{gr}^\bullet(s_r)$ and $\mathrm{gr}^\bullet(a_r)$ of $\mathrm{gr}^\bullet(\bigotimes^r V)$ respectively coincide with the symmetrization and antisymmetrization maps on $\bigotimes^r \mathrm{gr}^\bullet(V)$. In particular we have $\dim(\mathrm{Coim}\, s_r) = \dim \mathrm{Coim}(\mathrm{gr}^\bullet s_r)) = \binom{r+d}{d}$ and $\dim(\mathrm{Coim}\, a_r) =$

$\dim \mathrm{Coim}(\mathrm{gr}^\bullet a_r)) = \binom{d}{r}$, where $d = \dim V$. This shows that $\mathrm{gr}^\bullet(\mathrm{Sym}^r V) \xrightarrow{\sim} \mathrm{Sym}^r(\mathrm{gr}^\bullet V)$ and $\mathrm{gr}^\bullet(\bigwedge^r V) \xrightarrow{\sim} \bigwedge^r(\mathrm{gr}^\bullet V)$ (and also that s_r and a_r are strict morphisms). Hence, denoting by $\{x_1 > x_2 > \cdots > x_l\}$ the set of jumps of V, there is an isomorphism

$$\mathrm{gr}^\bullet(\textstyle\bigwedge^r V) \xrightarrow{\sim} \displaystyle\bigoplus_{r_1+r_2+\cdots+r_l=r} \left(\bigotimes_{i=1}^{l} \textstyle\bigwedge^{r_i} \mathrm{gr}^{x_i}(V)\right)\left[-\displaystyle\sum_{i=1}^{l} r_i x_i\right],$$

which yields the formula

$$\deg(\textstyle\bigwedge^r V) = \displaystyle\sum_{r_1+r_2+\cdots+r_l=r} \left(\sum_{i=1}^{l} r_i x_i\right) \prod_{i=1}^{l} \binom{d_i}{r_i},$$

where $d_i = \dim(\mathrm{gr}^{x_i} V)$. It follows that

$$\begin{aligned} \deg(\textstyle\bigwedge^r V) &= \displaystyle\sum_{r_1+r_2+\cdots+r_l=r} \left(\sum_{i=1}^{l} d_i x_i \binom{d_i-1}{r_i-1} \prod_{j\neq i} \binom{d_j}{r_j}\right) \\ &= \sum_{i=1}^{l} d_i x_i \left(\sum_{r_1+r_2+\cdots+r_l=r} \binom{d_i-1}{r_i-1} \prod_{j\neq i} \binom{d_j}{r_j}\right) \\ &= \sum_{i=1}^{l} d_i x_i \binom{d-1}{r-1} = \deg(V) \binom{d-1}{r-1}. \end{aligned}$$

The last equality follows from the two expressions of the degree $r-1$ monomial in the polynomial $(1+X)^{d-1} = (1+X)^{d_i-1} \prod_{j\neq i}(1+X)^{d_j}$.

This shows the formula $\mu(\bigwedge^r V) = r\mu(V)$. The case of symmetric powers can be treated in a similar way and is left to the reader. □

Remark 1.1.9 As a particular case of the previous lemma we have

$$\deg(V) = \deg(\textstyle\bigwedge^{\max} V).$$

Furthermore, it is clear that when V is one-dimensional, then $\deg_{\mathcal{F}}(V)$ is the unique $x \in \mathbb{R}$ with $\mathrm{gr}^x_{\mathcal{F}}(V) \neq (0)$ (the unique jump of $\mathcal{F}$).

We also note that deg is an additive function, in the following sense.

Lemma 1.1.10 *Let* $0 \longrightarrow V' \longrightarrow V \longrightarrow V'' \longrightarrow 0$ *be a short exact sequence in* Fil_k^K. *Then*

$$\deg(V) = \deg(V') + \deg(V'').$$

Proof This follows from the property (1.7), page 6 of exact sequences in Fil_k^K. □

It is sometimes convenient to use the following formula for the degree.

Lemma 1.1.11 *Let $\{x_1 > x_2 > \cdots > x_r\}$ be a subset of $\mathbb{R}$ containing all the jumps of $(V, \mathcal{F})$. Then we have*

$$\deg(V) = \sum_{i=1}^{r-1} (x_i - x_{i+1}) \dim(V_K^{x_i}) + x_r \dim(V).$$

Proof This is an elementary computation. □

Lemma 1.1.12 *Let $f : V \longrightarrow W$ be a morphism in Fil_k^K such that f induces an isomorphism on the underlying k-vector spaces. Then $\deg(V) \leq \deg(W)$ and equality holds if and only if f is an isomorphism in Fil_k^K. More generally, for any morphism φ in Fil_k^K, we have*

$$\deg(\mathrm{Coim}(\varphi)) \leq \deg(\mathrm{Im}(\varphi))$$

with equality holding if and only if φ is strict.

Proof By assumption we have $\dim(V_K^x) \leq \dim(W_K^x)$ for all $x \in \mathbb{R}$, and moreover f is an isomorphism if and only if all these inequalities are equalities. So let $\{x_1 > x_2 > \cdots > x_r\}$ be a subset of $\mathbb{R}$ containing the jumps of V and W. The lemma follows from the formula in Lemma 1.1.11 above. □

Definition 1.1.13 Let $V \in \mathrm{Fil}_k^K$. Then V is called *semi-stable* if $\mu(W) \leq \mu(V)$, for all subobjects W $(W \neq (0))$.

Remark 1.1.14 One may also introduce the notion of *stable* objects, i.e., ask that $\mu(W) < \mu(V)$ for all subobjects $W \neq (0)$, V, resp. *polystable* objects, i.e., direct sums of stable objects of identical slopes. However, these definitions will not be used in what follows.

Proposition 1.1.15 *Let $0 \longrightarrow V' \longrightarrow V \longrightarrow V'' \longrightarrow 0$ be a short exact sequence in Fil_k^K. Then the sequence $\mu(V'), \mu(V), \mu(V'')$ is either strictly increasing or strictly decreasing or stationary. In particular we have*

$$\min\{\mu(V'), \mu(V'')\} \leq \mu(V) \leq \max\{\mu(V'), \mu(V'')\}.$$

Proof This follows by an easy calculation from the additivity of deg and of rk in short exact sequences. □

Corollary 1.1.16 *An object $V \in \mathrm{Fil}_k^K$ is semi-stable if and only if $\mu(V) \leq \mu(W)$ for all quotient objects $W \neq (0)$ of V.*

Corollary 1.1.17 *An object $V \in \mathrm{Fil}_k^K$ is semi-stable if and only if its dual V^* is semi-stable.*

Remark 1.1.18 Let $(V, \mathcal{F}) \in \mathrm{Fil}_k^K$. Let $c \in \mathbb{R}$. Consider the filtration $\mathcal{F}(c)$ on V_K given by

$$\mathcal{F}(c)^x = \mathcal{F}^{x-c}.$$

The object $V(c) := (V, \mathcal{F}(c))$ is called the *twist of V by c*. Then $\mu(V(c)) = \mu(V) + c$ and V is semi-stable if and only if $V(c)$ is semi-stable. Indeed,

$$\begin{aligned}\mu_{\mathcal{F}(c)}(V) &= \frac{1}{\dim(V)} \cdot \sum_x x \cdot \dim \mathrm{gr}^x_{\mathcal{F}(c)}(V_K) \\ &= \frac{1}{\dim(V)} \cdot \sum_x (x+c) \dim \mathrm{gr}^x_{\mathcal{F}}(V_K) \\ &= \mu_{\mathcal{F}}(V) + c. \end{aligned} \tag{1.8}$$

The same applies to any subobject of $(V, \mathcal{F})$, hence the assertion on the semi-stability. This remark allows one often to reduce to the case $\mu(V) = 0$.

Remark 1.1.19 Let K' be a field extension of K. By associating to the $\mathbb{R}$-filtration $\mathcal{F}$ on V_K the $\mathbb{R}$-filtration $\mathcal{F}' = \mathcal{F} \otimes_K K'$ on $V_{K'}$, we obtain a functor $\mathrm{Fil}_k^K \longrightarrow \mathrm{Fil}_k^{K'}$. This functor is obviously fully faithful, and preserves slopes. Furthermore the set of subobjects is unchanged under this functor. Hence $(V, \mathcal{F})$ is a semi-stable object of Fil_k^K if and only if $(V, \mathcal{F}')$ is a semi-stable object of $\mathrm{Fil}_k^{K'}$.

Proposition 1.1.20 *Let $V, W \in \mathrm{Fil}_k^K$ both be semi-stable. If $\mu(V) > \mu(W)$, then* $\mathrm{Hom}(V, W) = 0$. *If $\mu(V) = \mu(W) = \mu$, then every $f \in \mathrm{Hom}(V, W)$ is strict and* $\mathrm{Ker}\, f$, $\mathrm{Coker}\, f$, *and* $\mathrm{Im}\, f$ *are semi-stable with the same slope μ.*

Proof Let $f \in \mathrm{Hom}(V, W)$ and assume it is non-zero. Then $\mu(\mathrm{Im}\, f)$ and $\mu(\mathrm{Coim}\, f)$ are defined and we have the following inequalities

$$\mu(V) \leq \mu(\mathrm{Coim}\, f) \leq \mu(\mathrm{Im}\, f) \leq \mu(W).$$

The first and last inequalities are due to the semi-stability of V and W, and the central one follows from Lemma 1.1.12. This shows that $\mu(V) \leq \mu(W)$ and that if $\mu(V) = \mu(W) = \mu$, then $\mu(\mathrm{Coim}\, f) = \mu(\mathrm{Im}\, f) = \mu$. In this case we also deduce that $\mu(\mathrm{Ker}\, f) = \mu(\mathrm{Coker}\, f) = \mu$ from Lemma 1.1.15 and we deduce the strictness of f from Proposition 1.1.12 again. □

Corollary 1.1.21 *The full subcategory of* Fil_k^K *consisting of semi-stable objects of slope* μ *is abelian, artinian, noetherian, and stable under extensions. Its simple objects are the stable objects of slope* μ.

Proof The previous proposition shows that kernels and cokernels in the subcategory under consideration exist and coincide with kernels and cokernels in the ambient category Fil_K^k. The strictness of any morphism ensures that coimages are isomorphic to images, hence this subcategory is abelian. It is artinian and noetherian since objects are finite-dimensional over k. Now let $0 \longrightarrow V' \longrightarrow V \xrightarrow{\pi} V'' \longrightarrow 0$ be a short exact sequence in Fil_k^K with V' and V'' semi-stable of slope μ. Then $\mu(V) = \mu$ by Proposition 1.1.15. Let us show that V is semi-stable. Let W be a subobject of V. We have $\mu(W \cap V') \leq \mu$ by semi-stability of V and $\mu(W/(W \cap V')) \leq \mu(\pi(W)) \leq \mu$ by 1.1.12 and semi-stability of V''. It follows from Proposition 1.1.15 applied to the exact sequence $0 \longrightarrow V' \cap W \longrightarrow W \longrightarrow W/(V' \cap W) \longrightarrow 0$ in Fil_k^K that $\mu(W) \leq \mu$.

It is clear that a stable object of slope μ in Fil_k^K is a simple object of the subcategory under consideration. Conversely, let V be semi-stable of slope μ. Then any proper subobject W of V in Fil_k^K which has slope μ is obviously semi-stable, hence is a non-trivial proper subobject of V in the subcategory under consideration. □

Variants 1.1.22 (i) Let $\deg(V, \mathcal{F}_1, \ldots, \mathcal{F}_r)$ be an object of Fil_k^r, cf. Variant 1.1.5. We define

$$\deg(V, \mathcal{F}_1, \ldots, \mathcal{F}_r) = \sum\nolimits_{i=1}^{r} \deg(V, \mathcal{F}_i)\,,\ \mu(V, \mathcal{F}_1, \ldots, \mathcal{F}_r) = \sum\nolimits_{i=1}^{r} \mu(V, \mathcal{F}_i)\,.$$

Then all the statements above hold with trivial modifications. To check the semi-stability of an object, one has to test all k-subspaces of V. For $\mathbf{c} = (c_1, \ldots, c_r) \in \mathbb{R}^r$, let

$$V(\mathbf{c}) = V(c_1, \ldots, c_r) = (V, \mathcal{F}_1(c_1), \ldots, \mathcal{F}_r(c_r))\,.$$

Then $\mu(V(\mathbf{c})) = \mu(V) + \sum_i c_i$.

(ii) The degree and the slope of an object $(V, \mathbf{V}, \mathcal{F})$ of $\mathrm{Fil}_{\mathbb{F}_1}^k$ are defined as $\deg(V, \mathcal{F})$ and $\mu(V, \mathcal{F})$. To test the semi-stability of an object one has to test all subspaces of V, which are direct sums of lines in the fixed line decomposition $\mathbf{V}$ of V.

Examples 1.1.23 (A) The first three examples refer to the category Fil_k^K.

(i) Let $v = (x_1^{(n-1)}, x_2)$, where $x_1 > x_2$. In this case the filtering subspace of $V \otimes_k K$ has dimension $n - 1$, i.e., is a hyperplane of V_K. One easily checks that the semi-stable objects are exactly those hyperplanes which do not contain a k-rational line.

(ii) Let $\nu = (x_1, x_2^{(n-1)})$, where $x_1 > x_2$. This case is dual to the preceding one. The filtering subspace is a line in V_K. The semi-stable objects correspond to lines which are not contained in a k-rational hyperplane.
(iii) Let $n = 3$ and let $\nu = (x_1, x_2, x_3)$, where $x_1 > x_2 > x_3$. In this case one is considering full flags $(0) \subset \mathcal{F}_1 \subset \mathcal{F}_2 \subset V_K$ of V_K. The semi-stable objects are given as follows:

(a) if $x_1 - x_2 > x_2 - x_3$: $\mathcal{F}_1$ is not contained in a k-rational plane;
(b) if $x_1 - x_2 < x_2 - x_3$: $\mathcal{F}_2$ does not contain a k-rational line;
(c) if $x_1 - x_2 = x_2 - x_3$: $\mathcal{F}_1$ and $\mathcal{F}_2$ are both not k-rational.

(B) The previous examples have exact analogues in the category $\mathrm{Fil}^k_{\mathbb{F}_1}$. Case by case one obtains the following.
(i) The semi-stable objects are the hyperplanes in V which do not contain any of the lines V_i of the fixed line decomposition $\mathbf{V} = (V = V_1 \oplus \cdots \oplus V_n)$.
(ii) The semi-stable objects are the lines in V which are not contained in a hyperplane of the form $V_I = \bigoplus_{i \in I} V_i$, for some subset I of cardinality $n - 1$ in $\{1, \ldots, n\}$.
(iii) A complete flag of V is semi-stable if and only if, case by case,

(a) $\mathcal{F}_1$ is not contained in a coordinate hyperplane,
(b) $\mathcal{F}_2$ does not contain a coordinate line,
(c) $\mathcal{F}_1$ is not a coordinate line and $\mathcal{F}_2$ is not a coordinate plane.

(C) Here are two examples in the category Fil^r_k.
(i) If $r = 1$, then an object $(V, \mathcal{F})$ of Fil^r_k is semi-stable if and only if the filtration $\mathcal{F}$ has only one jump.
(ii) Let $r = 2$. Let $n = 2$, and consider an object $(V, \mathcal{F}_1, \mathcal{F}_2)$ with $\mu(\mathcal{F}_1) = \mu(\mathcal{F}_2) = 0$. (Note that by twisting we may always reduce to considering this case only, cf. Variants 1.1.22.) Then $(V, \mathcal{F}_1, \mathcal{F}_2)$ is semi-stable in exactly the following cases: either both filtrations have only one jump, or both filtrations have two jumps which occur at the same index, but the two filtering subspaces are different from each other.

Notes and References The category Fil^K_k was introduced in the context of filtered isocrystals by Faltings [73] and Totaro [218, 219] and also in the form presented here in [181, 186]. The category Fil^r_k is considered by Faltings and Wüstholz [76] in connection with a problem in diophantine approximation, see also Fujimori [84, 85]. The category $\mathrm{Fil}^k_{\mathbb{F}_1}$ is mentioned in [181]. The notation here is motivated by the fact that, as we will see later, the corresponding period spaces are the analogues "over the field with one element" of the period domains over finite fields (in the sense of Tits, i.e., where the Weyl group replaces the group of rational elements).

All statements in this section are analogues of classical facts about vector bundles on a Riemann surface [166], [200], [105]. However, the analogue of the notions of stable resp. polystable bundles has so far not played any role in the theory of filtered vector spaces, which is why we have not introduced them here. The last statement in Corollary 1.1.21 resembles the fact that the category of Fontaine's weakly admissible filtered isocrystals is abelian [80].

2 The tensor product theorem

In this section we will prove the *tensor product theorem.*

Theorem 1.2.1 *Let K be a separable field extension of k. Let $V, W \in \mathrm{Fil}_k^K$. If V and W are semi-stable, then $V \otimes W$ is semi-stable.*

Before proving this, we note the following consequence.

Corollary 1.2.2 *Let K be separable over k and let $V \in \mathrm{Fil}_k^K$. If V is semi-stable then $\mathrm{Sym}^r V$ and $\bigwedge^r V$ are semi-stable for $r \in \mathbb{N}$.*

Proof Indeed both $\mathrm{Sym}^r V$ and $\bigwedge^r V$ are defined as quotients of $\bigotimes^r V$ and have the same slope as $\bigotimes^r V$, namely $r \cdot \mu(V)$. □

To prove Theorem 1.2.1, we will need some preparations. Let, quite generally, V be a finite-dimensional vector space over a field K. Let $\mathcal{F}, \mathcal{F}'$ be two $\mathbb{R}$-filtrations on V. For $x \in \mathbb{R}$, the filtration $\mathcal{F}'$ induces a filtration on $\mathrm{gr}^x_{\mathcal{F}}(V)$ and the filtration $\mathcal{F}$ induces a filtration on $\mathrm{gr}^x_{\mathcal{F}'}(V)$. The situation is symmetric in $\mathcal{F}$ and $\mathcal{F}'$, and we have a canonical isomorphism for all $x, y \in \mathbb{R}$,

$$\mathrm{gr}^x_{\mathcal{F}}(\mathrm{gr}^y_{\mathcal{F}'}(V)) = \mathrm{gr}^y_{\mathcal{F}'}(\mathrm{gr}^x_{\mathcal{F}}(V)) .$$

Definition 1.2.3 The *scalar product* of two $\mathbb{R}$-filtrations $\mathcal{F}, \mathcal{F}'$ of V is given by

$$\langle \mathcal{F}, \mathcal{F}' \rangle = \sum_{x,y \in \mathbb{R}} xy \cdot \dim \mathrm{gr}^x_{\mathcal{F}}(\mathrm{gr}^y_{\mathcal{F}'}(V)) .$$

This terminology is justified by the following lemma. Let V be a K-vector space of dimension n, and let $\mathbf{V}$ be a decomposition into one-dimensional subspaces,

$$V = V_1 \oplus \cdots \oplus V_n .$$

We define a map

$$\begin{array}{rcl} \alpha = \alpha^{\mathbf{V}} : \mathbb{R}^n & \longrightarrow & \{\mathbb{R}\text{-filtrations on } V\} \\ \mathbf{x} = (x_1, \ldots, x_n) & \longmapsto & \mathcal{F} = \mathcal{F}^{\mathbf{V}}_{\alpha}(\mathbf{x}) \end{array}$$

by

$$\mathcal{F}^y = \bigoplus_{\substack{1\le i\le n \\ y\le x_i}} V_i \,.$$

Lemma 1.2.4 *Let* $\mathbf{x}, \mathbf{y} \in \mathbb{R}^n$. *Then* $\langle \alpha^{\mathbf{V}}(\mathbf{x}), \alpha^{\mathbf{V}}(\mathbf{y})\rangle = (\mathbf{x}, \mathbf{y})$, *where* $(\ ,\ )$ *denotes the standard scalar product on* $\mathbb{R}^n$ *.*

Proof We calculate

$$\begin{aligned}\langle \alpha^{\mathbf{V}}(\mathbf{x}), \alpha^{\mathbf{V}}(\mathbf{y})\rangle &= \sum\nolimits_{x,y} xy \cdot \dim \mathrm{gr}^x_{\alpha^{\mathbf{V}}(\mathbf{x})}(\mathrm{gr}^y_{\alpha^{\mathbf{V}}(\mathbf{y})}(V)) \\ &= \sum\nolimits_{i,j=1}^{n} x_i y_j \cdot \dim(V_i \cap V_j) \\ &= \sum\nolimits_{i} x_i y_i \,.\end{aligned}$$

□

Lemma 1.2.5 *Let* $\mathcal{F}$ *and* $\mathcal{F}'$ *be* $\mathbb{R}$*-filtrations on* V *with* $\mathcal{F}$ *a* $\mathbb{Z}$*-filtration. Then*

$$\langle \mathcal{F}, \mathcal{F}'\rangle = \sum_{x\in\mathbb{Z}} (\mu_{\mathcal{F}'}(V^x_{\mathcal{F}}) - \mu_{\mathcal{F}'}(V)) \dim(V^x_{\mathcal{F}}) + \mu_{\mathcal{F}}(V)\mu_{\mathcal{F}'}(V)\dim(V).$$

Proof Let $N_{\mathcal{F}}(V) \in \mathbb{Z}$ be the smallest number with $\mathrm{gr}^{N_{\mathcal{F}}(V)}_{\mathcal{F}}(V) \neq (0)$. Then the RHS can be written as

$$\sum_{\substack{x>N_{\mathcal{F}}(V)\\ x\in\mathbb{Z}}} \Big(\mu_{\mathcal{F}'}(V^x_{\mathcal{F}}) - \mu_{\mathcal{F}'}(V)\Big) \dim V^x_{\mathcal{F}} + \deg_{\mathcal{F}}(V)\cdot \mu_{\mathcal{F}'}(V)\,. \tag{2.1}$$

Since $\mathcal{F}$ is a $\mathbb{Z}$-filtration, we have

$$\deg_{\mathcal{F}} V = N_{\mathcal{F}}(V)\cdot \dim V + \sum_{\substack{x>N_{\mathcal{F}}(V)\\ x\in\mathbb{Z}}} \dim V^x_{\mathcal{F}}. \tag{2.2}$$

Hence the second summand in (2.1) above is equal to

$$\sum_{\substack{x>N_{\mathcal{F}}(V)\\ x\in\mathbb{Z}}} \dim V^x_{\mathcal{F}} \cdot \mu_{\mathcal{F}'}(V) + N_{\mathcal{F}}(V) \dim V \cdot \mu_{\mathcal{F}'}(V)\,.$$

Therefore (2.1) above is equal to

$$\begin{aligned}&\sum_{\substack{x>N_{\mathcal{F}}(V)\\ x\in\mathbb{Z}}} \mu_{\mathcal{F}'}(V^x_{\mathcal{F}}) \dim V^x_{\mathcal{F}} + N_{\mathcal{F}}(V)\cdot \deg_{\mathcal{F}'}(V) \\ &= \sum_{\substack{x>N_{\mathcal{F}}(V)\\ x\in\mathbb{Z}}} \Big(\sum_{y\in\mathbb{R}} y\cdot \dim \mathrm{gr}^y_{\mathcal{F}'}(V^x_{\mathcal{F}})\Big) + N_{\mathcal{F}}(V)\cdot \sum_{y\in\mathbb{R}} y\cdot \dim \mathrm{gr}^y_{\mathcal{F}'}(V) \\ &= \sum_{y\in\mathbb{R}} y \Big(\sum_{\substack{x>N_{\mathcal{F}}(V)\\ x\in\mathbb{Z}}} \dim \mathrm{gr}^y_{\mathcal{F}'}(V_{\mathcal{F}})^x + N_{\mathcal{F}}(V) \dim \mathrm{gr}^y_{\mathcal{F}'}(V)\Big)\,.\end{aligned}$$

Using again the identity (2.2), page 14 applied to $\mathrm{gr}^y_{\mathcal{F}'}(V)$ instead of V, this last expression is equal to

$$\sum_{y\in\mathbb{R}} y \cdot \deg_{\mathcal{F}}(\mathrm{gr}^y_{\mathcal{F}'}(V)) = \sum_{x,y} xy \cdot \dim \mathrm{gr}^x_{\mathcal{F}}(\mathrm{gr}^y_{\mathcal{F}'} V)$$
$$= \langle \mathcal{F}, \mathcal{F}' \rangle. \qquad \square$$

The previous lemma has a useful application to the category Fil^K_k. Let $(V, \mathcal{F}) \in \mathrm{Fil}^K_k$. Let us consider a $\mathbb{Z}$-filtration $\mathcal{F}'$ of the k-vector space V. We denote by the same symbol the filtration on V_K obtained by extension of scalars (in other words, the corresponding $\mathbb{Z}$-filtration of V_K by k-rational subspaces).

Corollary 1.2.6 *The object $(V, \mathcal{F}) \in \mathrm{Fil}^K_k$ is semi-stable if and only if the inequality $\langle \mathcal{F}, \mathcal{F}' \rangle \le \mu(V) \cdot \deg_{\mathcal{F}'}(V)$ holds for all $\mathbb{Z}$-filtrations $\mathcal{F}'$ by subobjects of $(V, \mathcal{F})$.*

Equivalently, $(V, \mathcal{F})$ is semi-stable if and only if $\langle \mathcal{F}, \mathcal{F}' \rangle \le 0$ for all $\mathbb{Z}$-filtrations $\mathcal{F}'$ by subobjects of $(V, \mathcal{F})$ with $\deg_{\mathcal{F}'}(V) = 0$.

Proof Let us prove the first assertion. Note that

$$\langle \mathcal{F}(c), \mathcal{F}' \rangle = \langle \mathcal{F}, \mathcal{F}' \rangle + c \cdot \deg_{\mathcal{F}'}(V) .$$

Hence, taking into account (1.8), page 10 and replacing V by $V(c)$, we may assume that $\mu(V) = 0$. The semi-stability of V remains unchanged under twisting, cf. Remark 1.1.18.

By the previous lemma (with the roles of $\mathcal{F}$ and $\mathcal{F}'$ interchanged), we then have for any $\mathbb{Z}$-filtration $\mathcal{F}'$ by subobjects,

$$\langle \mathcal{F}, \mathcal{F}' \rangle = \sum\nolimits_x \mu_{\mathcal{F}}(V^x_{\mathcal{F}'}) \cdot \dim V^x_{\mathcal{F}'}.$$

If $(V, \mathcal{F})$ is semi-stable, then the RHS is ≤ 0. If $(V, \mathcal{F})$ is not semi-stable, let V' be a subobject with $\mu(V') > 0$. Let $\mathcal{F}'$ be the $\mathbb{Z}$-filtration on V with

$$\mathcal{F}'^0 = V, \ \ \mathcal{F}'^1 = V', \ \ \mathcal{F}'^2 = (0) .$$

Then the RHS is strictly positive.

The second assertion follows by twisting $\mathcal{F}'$. $\square$

Now we return to the general situation. Let V be a vector space of dimension n over a field K. Let W be a subspace of V. Let $\mathbf{V} = (V = V_1 \oplus \cdots \oplus V_n)$ be a decomposition of V into one-dimensional subspaces. We are going to show the concavity of the composition of the two maps,

$$\mathbf{x} \longmapsto \alpha^{\mathbf{V}}(\mathbf{x}) \qquad \text{and} \qquad \mathcal{F} \longmapsto \deg_{\mathcal{F}}(W) .$$

Consider

$$\varphi = \varphi_W : \mathbb{R}^n \longrightarrow \mathbb{R},\ \mathbf{x} \longmapsto \deg_{\alpha^{\mathbf{v}(\mathbf{x})}}(W)\,.$$

Proposition 1.2.7 *The function φ is concave homogeneous ($\varphi(c\mathbf{x}) = c\varphi(\mathbf{x})$ for $c > 0$). More precisely, let*

$$V_I = \bigoplus_{i \in I} V_i\,,$$

and set

$$\mathcal{J}_W = \{I \subset \{1, \dots, n\} \mid\ |I| = \dim W, \text{ the projection } p_I : V \longrightarrow V_I \text{ induces an isomorphism } W \xrightarrow{\sim} V_I\}\,.$$

Then

$$\varphi_W(\mathbf{x}) = \inf_{I \in \mathcal{J}_W} \varphi_{V_I}(\mathbf{x}) = \inf_{I \in \mathcal{J}_W} \sum\nolimits_{i \in I} x_i\,.$$

Proof The morphism $p_I|W$ is a morphism of filtered vector spaces. Hence by Lemma 1.1.12 we have

$$\varphi_W(\mathbf{x}) \leq \varphi_{V_I}(\mathbf{x}).$$

Hence the LHS is smaller or equal to the RHS. On the other hand, for any $x \in \mathbb{R}$, and any fixed $\mathbf{x} = (x_1, \dots, x_n)$,

$$\mathrm{gr}^x_{\alpha^{\mathbf{v}(\mathbf{x})}}(W) \subset \mathrm{gr}^x_{\alpha^{\mathbf{v}(\mathbf{x})}}(V) = \bigoplus\nolimits_{\{i \mid x_i = x\}} V_i\,.$$

Choose for every $x \in \mathbb{R}$ a subset $I_x \subset \{i \in \{1, \dots, n\} \mid x_i = x\}$ such that $\mathrm{gr}^x_{\alpha^{\mathbf{v}(\mathbf{x})}}(W)$ projects isomorphically to $\bigoplus_{i \in I_x} V_i$. Let $I = \bigcup_x I_x$. Then $I \in \mathcal{J}_W$ and

$$\mathrm{gr}^x_{\alpha^{\mathbf{v}(\mathbf{x})}}(W) \simeq \mathrm{gr}^x_{\alpha^{\mathbf{v}(\mathbf{x})}}(V_I), \quad x \in \mathbb{R}\,.$$

Hence

$$\varphi_W(\mathbf{x}) = \varphi_{V_I}(\mathbf{x})\,. \qquad \square$$

Lemma 1.2.8 *Let $f : \mathbb{R}^n \longrightarrow \mathbb{R}$ be a concave function which takes on a positive value and such that $f(c\mathbf{x}) = cf(\mathbf{x})$, for $c > 0$. Then there exists a unique $\mathbf{x}_0 \in \mathbb{R}^n$ with $||\mathbf{x}_0|| = 1$ such that*

$$f(\mathbf{x}_0) = \max_{||\mathbf{x}||=1} f(\mathbf{x}).$$

Furthermore, for any $\mathbf{x} \in \mathbb{R}^n$,

$$f(\mathbf{x}) \leq f(\mathbf{x}_0) \cdot (\mathbf{x}_0, \mathbf{x})\,.$$

If f is of the form

$$f = f_T = \inf\{l \mid l \in T\}$$

for a finite set T of linear forms on $\mathbb{R}^n$ with rational coefficients, then there exists $c > 0$ such that $c \cdot \mathbf{x}_0 \in \mathbb{Z}^n$.

Proof For the first assertion we will prove the following stronger statement.

Let $\mathbf{y} \in S^{n-1}$ be a point with $f(\mathbf{y}) > 0$ and where f assumes a local maximum. Then $\mathbf{y} = \mathbf{x}_0$.

To see this, let $\mathbf{x} = t\mathbf{x}_0 + (1-t)\mathbf{y}$ for $0 < t < 1$. Then by concavity $f(\mathbf{x}) \geq f(\mathbf{y}) > 0$. Since $||\mathbf{x}|| < 1$, if $\mathbf{y} \neq \mathbf{x}_0$, we obtain

$$f\left(\frac{\mathbf{x}}{||\mathbf{x}||}\right) = \frac{1}{||\mathbf{x}||} \cdot f(\mathbf{x}) > f(\mathbf{y}).$$

But for small t, the point $\frac{\mathbf{x}}{||\mathbf{x}||}$ lies in an arbitrarily small neighborhood of $\mathbf{y}$. This contradicts the assumption that $\mathbf{y}$ is a local maximum of f. Hence $\mathbf{y} = \mathbf{x}_0$.

To see the second statement we may assume $||\mathbf{x}|| = 1$. Consider the point $\mathbf{x}_0 + C\mathbf{x}$, for $C > 0$. Then for any $t \in [0, 1]$ we have by the concavity of f that

$$f(t(\mathbf{x}_0 + C\mathbf{x}) + (1-t)\mathbf{x}_0) \geq tf(\mathbf{x}_0 + C\mathbf{x}) + (1-t)f(\mathbf{x}_0).$$

The argument on the LHS is equal to $\mathbf{x}_0 + tC\mathbf{x}$. Since $\mathbf{x}_0$ is the maximum of f on S^{n-1}, we have by homogeneity

$$f(\mathbf{x}_0 + tC\mathbf{x}) \leq ||\mathbf{x}_0 + tC\mathbf{x}|| f(\mathbf{x}_0).$$

The first factor on the RHS can be estimated:

$$\begin{aligned} ||\mathbf{x}_0 + tC\mathbf{x}|| &\leq (||\mathbf{x}_0||^2 + t^2C^2||\mathbf{x}||^2 + 2\langle \mathbf{x}_0, \mathbf{x}\rangle tC)^{\frac{1}{2}} \\ &= (1 + t^2C^2)^{\frac{1}{2}} \cdot \left(1 + \frac{2\langle \mathbf{x}_0, \mathbf{x}\rangle \cdot tC}{1 + t^2C^2}\right)^{\frac{1}{2}} \\ &\leq (1 + t^2C^2)^{\frac{1}{2}} \cdot \left(1 + \frac{\langle \mathbf{x}_0, \mathbf{x}\rangle \cdot tC}{1 + t^2C^2}\right). \end{aligned}$$

Putting these inequalities together, we have

$$tf(\mathbf{x}_0 + C\mathbf{x}) + (1-t)f(\mathbf{x}_0) \leq (1 + t^2C^2)^{\frac{1}{2}} \cdot \left(1 + \frac{\langle \mathbf{x}_0, \mathbf{x}\rangle}{1 + t^2C^2} \cdot tC\right) \cdot f(\mathbf{x}_0)\,.$$

Now let $t = 1/C^2$ and let $C \longrightarrow \infty$. We then obtain

$$C^{-1} \cdot f(\mathbf{x} + C^{-1}\mathbf{x}_0) + (1 - C^{-2}) \cdot f(\mathbf{x}_0) \leq (1 + C^{-2})^{\frac{1}{2}} \cdot \left(\frac{1 + \langle \mathbf{x}_0, \mathbf{x}\rangle \cdot C^{-1}}{1 + C^{-2}}\right) \cdot f(\mathbf{x}_0)\,.$$

The absolute terms on both sides are $f(\mathbf{x}_0)$. Comparing terms in C^{-1} of first order, we get

$$f(\mathbf{x}) \leq \langle \mathbf{x}_0, \mathbf{x}\rangle \cdot f(\mathbf{x}_0)\,,$$

which is the second assertion.

To see the final statement, we may replace the set of linear forms T by the subset

$$T' = \{l \in T \mid l(\mathbf{x}_0) = f(\mathbf{x}_0)\},$$

without changing $\mathbf{x}_0$. Let

$$V = \bigcap_{l,l' \in T'} \mathrm{Ker}(l - l').$$

Then V is a $\mathbb{Q}$-subspace of $\mathbb{R}^n$, and $\mathbf{x}_0 \in V$. Furthermore, $\mathbf{x}_0$ is the unique maximum on $V \cap S^{n-1}$ of the concave function

$$f_{\bar{T}'} = \inf\{l|V \mid l \in T'\},$$

on V. But the set on the RHS consists of a single element $\bar{l}$ and hence $\mathbf{x}_0$ is the unit vector in V orthogonal to $\mathrm{Ker}\,\bar{l}$. Since $\mathrm{Ker}\,\bar{l}$ is a $\mathbb{Q}$-hyperplane, the assertion follows. □

We can now prove the tensor product theorem. Let $(V, \mathcal{F})$ and $(V', \mathcal{F}')$ be two semi-stable objects of Fil_k^K. Then by Lemma 1.1.8

$$\mu(V \otimes V') = \mu(V) + \mu(V').$$

After replacing the filtrations $\mathcal{F}$ and $\mathcal{F}'$ by twists, we may therefore assume that $\mu(V) = \mu(V') = \mu(V \otimes V') = 0$. We then have to show that $\mu(W) \leq 0$ for any subobject W of $V \otimes V'$.

Fix a subobject W of $V \otimes V'$. We distinguish two cases. If for all filtrations $\mathcal{F}$ on V_K and $\mathcal{F}'$ on V'_K, we have $\deg_{\mathcal{F}\otimes\mathcal{F}'}(W) \leq 0$, then the assertion is trivially true. In the opposite case we have the following proposition.

Proposition 1.2.9 *We assume that the function* $(\mathcal{F}, \mathcal{F}') \longmapsto \deg_{\mathcal{F}\otimes\mathcal{F}'}(W)$ *on pairs of* $\mathbb{R}$*-filtrations of* V_K *resp.* V'_K*, takes positive values. Then this function takes on its maximum at a unique pair* $(\mathcal{F}_0, \mathcal{F}'_0)$ *among the pairs* $(\mathcal{F}, \mathcal{F}')$ *with*

$$\|\mathcal{F}\|^2 + \|\mathcal{F}'\|^2 = 1.$$

For any other pair $(\mathcal{F}, \mathcal{F}')$ *of filtrations of* V_K, V'_K*, there is the inequality*

$$\deg_{\mathcal{F}\otimes\mathcal{F}'}(W) \leq \deg_{\mathcal{F}_0\otimes\mathcal{F}'_0}(W) \cdot \left[\langle \mathcal{F}_0, \mathcal{F} \rangle + \langle \mathcal{F}'_0, \mathcal{F}' \rangle\right].$$

Proof Let n and n' be the dimensions of V resp. V' and let $m = \dim W$. By Proposition 1.2.7, for any line decompositions $\mathbf{V}_K = (V_K = V_1 \oplus \cdots \oplus V_n)$ and $\mathbf{V}'_K = (V'_K = V_1 \oplus \cdots \oplus V_{n'})$, there exists a set $\mathcal{J}_W$ of subsets $I \subset \{1, \ldots, n\} \times \{1, \ldots, n'\}$ of cardinality m such that

$$\deg_{\alpha^{\mathbf{V}}(\mathbf{x})\otimes\alpha^{\mathbf{V}'}(\mathbf{x}')}(W) = \inf_{I \in \mathcal{J}_W} \sum\nolimits_{(i,i') \in I} (x_i + x'_{i'}),$$

for any $\mathbf{x} = (x_1, \ldots, x_n) \in \mathbb{R}^n$ and $\mathbf{x}' = (x'_1, \ldots, x'_{n'}) \in \mathbb{R}^{n'}$. We are looking for the maximum of the LHS, as $\mathbf{V}_K, \mathbf{V}'_K$ ranges over all line decompositions of V_K and V'_K and as $\mathbf{x}, \mathbf{x}'$ satisfy $\|\mathbf{x}\|^2 + \|\mathbf{x}'\|^2 = 1$. Indeed, any filtration of V_K, resp. V'_K is of the form $\alpha^{\mathbf{V}}(\mathbf{x})$, resp. $\alpha^{\mathbf{V}'}(\mathbf{x}')$ for suitable $\mathbf{x}$, resp. $\mathbf{x}'$ and suitable $\mathbf{V}_K$, resp. $\mathbf{V}'_K$. However, there are only finitely many possibilities for $\mathcal{J}_W$, hence the existence of the maximum $(\mathcal{F}_0, \mathcal{F}'_0)$ is clear.

Let $\mathcal{F}, \mathcal{F}'$ be any two $\mathbb{R}$-filtrations on V_K resp. V'_K. Then there exist line decompositions $\mathbf{V}_K$ of V_K and $\mathbf{V}'_K$ of V'_K such that both $\mathcal{F}_0$ and $\mathcal{F}$ are in the image of $\alpha^{\mathbf{V}}$, and both $\mathcal{F}'_0$ and $\mathcal{F}'$ in the image of $\alpha^{\mathbf{V}'}$ (Bruhat lemma). From Lemma 1.2.8 it now follows that the maximum is unique and also that

$$\deg_{\mathcal{F}\otimes\mathcal{F}'}(W) \leq \deg_{\mathcal{F}_0\otimes\mathcal{F}'_0}(W) \cdot \left[\langle \mathcal{F}_0, \mathcal{F}\rangle + \langle \mathcal{F}'_0, \mathcal{F}'\rangle\right]. \qquad \square$$

End of the proof of Theorem 1.2.1. Since we are in the second alternative, we have $\deg_{\mathcal{F}_0\otimes\mathcal{F}'_0}(W) > 0$. By the previous proposition, we have

$$\deg_{\mathcal{F}\otimes\mathcal{F}'}(W) \leq \deg_{\mathcal{F}_0\otimes\mathcal{F}'_0}(W) \left[\langle \mathcal{F}_0, \mathcal{F}\rangle + \langle \mathcal{F}'_0, \mathcal{F}'\rangle\right].$$

We claim that $\mathcal{F}_0$ and $\mathcal{F}'_0$ are filtrations of V, resp. V' by subobjects, i.e., are defined over k. The principle is simple. Any k-linear endomorphism σ of K induces a k-linear endomorphism of $V_K = V \otimes_k K$ resp. V'_K. Further, if σ is a field automorphism of K then the image of a K-vector subspace of V_K is a K-vector subspace, hence σ acts on the set of $\mathbb{R}$-filtrations on V_K. Moreover we have

$$\deg_{\mathcal{F}_0\otimes\mathcal{F}'_0}(W) = \deg_{\sigma(\mathcal{F}_0)\otimes\sigma(\mathcal{F}'_0)}(\sigma(W)) = \deg_{\sigma(\mathcal{F}_0)\otimes\sigma(\mathcal{F}'_0)}(W)\,, \qquad (2.3)$$

since $\sigma(W) = W$. Since we also have $\|\sigma(\mathcal{F}_0)\|^2 + \|\sigma(\mathcal{F}'_0)\|^2 = 1$, it follows from the uniqueness property of $(\mathcal{F}_0, \mathcal{F}'_0)$, that $\sigma(\mathcal{F}_0) = \mathcal{F}_0$ and $\sigma(\mathcal{F}'_0) = \mathcal{F}'_0$. Hence by [27], V, §10.4, Prop. 7, $\mathcal{F}_0$ and $\mathcal{F}'_0$ are defined over K^σ.

Now, to prove our claim we may first assume that K is a finitely generated extension of k. Indeed $\mathcal{F}_0$ and $\mathcal{F}'_0$ are defined over a finitely generated subextension of k in K (because V and V' are finite-dimensional) and obviously descend to the optimal filtrations over this subfield, in the sense of the previous proposition. Since K is now finitely generated and is assumed separable, it is a finite separable extension of a purely transcendental extension $K_0 = k(X_1, \ldots, X_n)$. Let $\tilde{K}$ be a Galois closure of K over K_0 and let $\tilde{\mathcal{F}}_0$ and $\tilde{\mathcal{F}}'_0$ be the optimal filtrations over $\tilde{K}$. By the above principle they are invariant under $\mathrm{Gal}(K/K_0)$ hence they are defined over K_0, and obviously descend to the optimal filtrations over this field. In particular we have $\mathcal{F}_0 \otimes_K \tilde{K} = \tilde{\mathcal{F}}_0$ and $\mathcal{F}'_0 \otimes_K \tilde{K} = \tilde{\mathcal{F}}'_0$. Now assume k is infinite. Then it is the fixed subfield of K_0 under some group of k-linear field automorphisms (e.g. the group generated by the translations $X_i \mapsto X_i + \lambda$, $\lambda \in k$) hence the same descent argument shows that $\mathcal{F}_0$ and $\mathcal{F}'_0$ are defined over k. Finally, if k is finite, we choose an algebraic closure $\overline{k}$ of k, we put $\overline{K}_0 := K_0 \otimes \overline{k}$ and we denote by $\overline{\mathcal{F}}_0$, resp. $\overline{\mathcal{F}}'_0$ the optimal filtrations over $\overline{K}_0$. By Galois descent we know that they are the base change of $\mathcal{F}_0$ and $\mathcal{F}'_0$ from K_0 to $\overline{K}_0$, and by the argument above (since $\overline{k}$ is infinite) they are defined over $\overline{k}$. Then using Galois descent again we see they are defined over k.

After a homothety we may also assume by the last statement of Lemma 1.2.8 that $\mathcal{F}_0$ and $\mathcal{F}'_0$ are $\mathbb{Z}$-filtrations. Since both $(V, \mathcal{F})$ and $(V', \mathcal{F}')$ are semi-stable, we have by Corollary 1.2.6 that $\langle \mathcal{F}_0, \mathcal{F} \rangle \leq 0$ and $\langle \mathcal{F}'_0, \mathcal{F}' \rangle \leq 0$. It follows from Proposition 1.2.9 that $\deg_{\mathcal{F} \otimes \mathcal{F}'}(W) \leq 0$, a contradiction. □

Remark 1.2.10 The assumption in the statement of the tensor product theorem that K/k be a separable extension is necessary, as the following example of Pink [177], Ex. 5.16 shows.

Let k be a non-perfect field of characteristic 2 and let $z \in k^\times \setminus k^{\times,2}$. Let $K = k(\sqrt{z})$. Let V be the standard vector space of dimension 2 with basis e_1, e_2. Let $\mathcal{F}$ be the $\mathbb{Z}$-filtration of V_K given by

$$\mathcal{F}^0 = V_K, \quad \mathcal{F}^1 = \langle e_1 + \sqrt{z} e_2 \rangle, \quad \mathcal{F}^2 = (0)\,.$$

Then V is semi-stable with $\mu(V, \mathcal{F}) = 1/2$. Let $W = V \otimes V$, which we identify with $M_2(k)$ via the elementary matrices $e_{ij} = e_i \otimes e_j$. The filtration on W is given by $\mathcal{F}^0_W = W_K, \mathcal{F}^3_W = (0)$, and

$$\mathcal{F}^1_W := \left\langle \begin{pmatrix} 1 & 0 \\ \sqrt{z} & 0 \end{pmatrix}, \begin{pmatrix} 0 & 1 \\ 1 & 0 \end{pmatrix}, \begin{pmatrix} 1 & 0 \\ 0 & z \end{pmatrix} \right\rangle, \quad \mathcal{F}^2_W = \left\langle \begin{pmatrix} 1 & \sqrt{z} \\ \sqrt{z} & z \end{pmatrix} \right\rangle .$$

Consider the subspace W' of W with

$$W' = \left\langle \begin{pmatrix} 1 & 0 \\ 0 & z \end{pmatrix}, \begin{pmatrix} 0 & 1 \\ 1 & 0 \end{pmatrix} \right\rangle .$$

The induced filtration $\mathcal{F}_{W'}$ on W'_K is given by $\mathcal{F}^1_{W'} = W'_K, \mathcal{F}^3_{W'} = (0)$ and

$$\mathcal{F}^2_{W'} = \left\langle \begin{pmatrix} 1 & \sqrt{z} \\ \sqrt{z} & 1 \end{pmatrix} \right\rangle .$$

Hence $\mu(W') = 3/2 > 1 = \mu(W)$. Hence $V \otimes V$ is not semi-stable.

Variant 1.2.11 Let us indicate here the modifications needed to establish the tensor product theorem for the category $\mathrm{Fil}^k_{\mathbb{F}_1}$. The only point in the previous proof that has to be changed is the argument to show that the maximum $(\mathcal{F}_0, \mathcal{F}'_0)$ is taken on by subobjects of V resp. V' in our category. In other words, we have to show that $\mathcal{F}_0$ is equal to the direct sum $V_I = \bigoplus_{i \in I} V_i$ of some of the lines in the fixed line decomposition $\mathbf{V}$ of V, and similarly for $\mathcal{F}'_0$.

Lemma 1.2.12 *Assume that k is infinite. A subspace W of V is of the form $W = V_I$ for some $I \subset \{1, \dots, n\}$ if and only if W is $T(k)$-invariant. Here T denotes the diagonal torus in* $\mathrm{GL}(V)$ *(corresponding to the line decomposition* $\mathbf{V}$ *of V).*

Proof Let $r = \dim W$ and consider $\bigwedge^r W \subset \bigwedge^r V$. Then W is $T(k)$-invariant if and only if $\bigwedge^r W$ is an eigenvector of $T(k)$. However, the action of $T(k)$ on $\bigwedge^r V$ is diagonal: more precisely, the vectors $\{e_I \mid |I| = r\}$ are eigenvectors for pairwise distinct characters. Here $e_I = e_{i_1} \wedge \cdots \wedge e_{i_r}$ for $I = \{i_1 < \cdots < i_r\}$. Hence the basis vectors e_I are the only eigenvectors and the result follows. □

Assume now first that k is infinite. Then we have to show that $t(\mathcal{F}_0) = \mathcal{F}_0$ for any $t \in T(k)$. Here the translate of the filtration $\mathcal{F}$ by t is defined as

$$t(\mathcal{F})^x = t \cdot \mathcal{F}^x.$$

But for any two $\mathbb{R}$-filtrations $\mathcal{F}_1$ and $\mathcal{F}_2$ of V,

$$\begin{aligned}\langle t(\mathcal{F}_1), t(\mathcal{F}_2)\rangle &= \sum xy \dim \mathrm{gr}^x_{t(\mathcal{F}_1)}(\mathrm{gr}^y_{t(\mathcal{F}_2)}(V)) \\ &= \sum xy \dim \mathrm{gr}^x_{\mathcal{F}_1}(\mathrm{gr}^y_{\mathcal{F}_2}(V)) \\ &= \langle \mathcal{F}_1, \mathcal{F}_2\rangle.\end{aligned}$$

Furthermore, for $\mathbb{R}$-filtrations $\mathcal{F}$ of V, resp. $\mathcal{F}'$ of V', for any $(t, t') \in T(k) \times T'(k)$,

$$\deg_{\mathcal{F}\otimes\mathcal{F}'}(W) = \deg_{t(\mathcal{F})\otimes t'(\mathcal{F}')}((t, t')W) = \deg_{t(\mathcal{F})\otimes t'(\mathcal{F}')}(W),$$

since W is a subobject of $V \otimes V'$, which implies $(t, t')W = W$ by the previous Lemma. By the uniqueness property of $(\mathcal{F}_0, \mathcal{F}'_0)$ we deduce that $t(\mathcal{F}_0) = \mathcal{F}_0$ and $t'(\mathcal{F}'_0) = \mathcal{F}'_0$ for any $(t, t') \in T(k) \times T'(k)$. An application of Lemma 1.2.12 now shows that $\mathcal{F}_0$ and $\mathcal{F}'_0$ are filtrations by subobjects of V and V', as desired.

The case when k is finite is reduced to the case of an infinite field by replacing V and V' by $V \otimes_k k'$ and $V' \otimes_k k'$, where k' is an infinite field, cf. Corollary 1.3.10. The reader checks that the proof of Corollary 1.3.10 does not use any results in the present section.

Variant 1.2.13 Here we shall explain how the tensor product theorem can be proved for the category Fil^r_k by modifying the above proof.

Let $(V, \mathcal{F}_1, \dots, \mathcal{F}_r) \in \mathrm{Fil}^r_k$. We associate to this object the following object in Fil^k_k,

$$(\widetilde{V}, \widetilde{\mathcal{F}}) = (V^{\oplus r}, \mathcal{F}_1 \oplus \cdots \oplus \mathcal{F}_r).$$

The relation between degrees and slopes in these two categories is given by

$$\deg(\widetilde{V},\widetilde{\mathcal{F}}) = \deg(V,\mathcal{F}_1,\dots,\mathcal{F}_r),\ \mu(\widetilde{V},\widetilde{\mathcal{F}}) = \frac{1}{r}\,\mu(V,\mathcal{F}_1,\dots,\mathcal{F}_r)\,.$$

Therefore, for the semi-stability of $(V,\mathcal{F}_1,\dots,\mathcal{F}_r)$, we have to test that $\mu(\widetilde{W}) \le \mu(\widetilde{V})$ for any subobject $\widetilde{W}$ of $\widetilde{V} = V^{\oplus r}$ of the form $\widetilde{W} = W^{\oplus r}$ for some subspace W of V.

The tensor product $(V,\mathcal{F}_1,\dots,\mathcal{F}_r) \otimes (V';\mathcal{F}'_1,\dots,\mathcal{F}'_r)$ is the subobject $(\widetilde{V\otimes V'}) = (V\otimes V')^{\oplus r}$ of $\widetilde{V}\otimes\widetilde{V}'$. In other words, by writing $\widetilde{V} = V_1\oplus\cdots\oplus V_r$ and $\widetilde{V}' = V'_1\oplus\cdots\oplus V'_r$, for identical vector spaces $V_1 = \cdots = V_r = V$ resp. $V'_1 = \cdots = V'_r = V'$, the subobject $(\widetilde{V\otimes V'})$ is given by the diagonal subspace $\bigoplus_i V_i\otimes V'_i$ of $\bigoplus_{i,j} V_i\otimes V'_j$.

Now we fix a subobject $\widetilde{W}$ of the form $\widetilde{W} = W^{\oplus r}$ for some $W \subset V\otimes V'$. Let $(\mathcal{F}_0,\mathcal{F}'_0)$ be the maximizing pair of filtrations of $(\widetilde{V},\widetilde{V}')$ of $(\mathcal{F},\mathcal{F}') \longmapsto \deg_{\mathcal{F}\otimes\mathcal{F}'}(\widetilde{W})$ on the unit sphere, as before.

Claim *$\mathcal{F}_0$ is of the form $\mathcal{F}_0 = \widetilde{\overline{\mathcal{F}_0}} = \overline{\mathcal{F}_0}^{\oplus r}$, for a filtration $\overline{\mathcal{F}_0}$ of V, and similarly for $\mathcal{F}'_0$.*

We use the following Lemma.

Lemma 1.2.14 *Let k be an infinite field. A subspace of $V^{\oplus r}$ is of the form $W^{\oplus r}$ for some subspace W of V if and only if it is invariant under the group $(k^\times)^r \rtimes S_r$, acting on $V^{\oplus r}$ via*

$$(\lambda_1,\dots,\lambda_r)\times\sigma : (v_1,\dots,v_r) \longmapsto (\lambda_1 v_{\sigma(1)},\dots,\lambda_r v_{\sigma(r)})\,.$$

Proof More precisely, the invariance under $(k^\times)^r$ characterizes the subspaces of the form $W_1\oplus\cdots\oplus W_r$, and the additional invariance under S_r forces $W_1 = W_2 = \cdots = W_r$. □

Let $g \in (k^\times)^r \rtimes S_r$. Then g leaves $\widetilde{W}$ invariant. Hence

$$\deg_{g\mathcal{F}_0\otimes g\mathcal{F}'_0}(\widetilde{W}) = \deg_{g\mathcal{F}_0\otimes g\mathcal{F}'_0}(g\widetilde{W}) = \deg_{\mathcal{F}_0\otimes\mathcal{F}'_0}(\widetilde{W})\,.$$

Then the uniqueness property of $(\mathcal{F}_0,\mathcal{F}'_0)$ allows us to conclude as before that $g\mathcal{F}_0 = \mathcal{F}_0$ and $g\mathcal{F}'_0 = \mathcal{F}'_0$. Hence from the Lemma we get that $\mathcal{F}_0 = \overline{\mathcal{F}_0}^{\oplus r}$ and $\mathcal{F}'_0 = \overline{\mathcal{F}'_0}^{\oplus r}$, which proves the claim when k is infinite. We thus obtain that $\mathcal{F}_0$ and $\mathcal{F}'_0$ are filtrations by test objects for the category Fil^r_k. If $(V,\mathcal{F}_1,\dots,\mathcal{F}_r)$ and $(V',\mathcal{F}'_1,\dots,\mathcal{F}'_r)$ are semi-stable of slope zero, we therefore have, by Corollary 1.2.6,

$$\langle\widetilde{\overline{\mathcal{F}_0}},\mathcal{F}_1\oplus\cdots\oplus\mathcal{F}_r\rangle \le 0\,,\ \langle\widetilde{\overline{\mathcal{F}'_0}},\mathcal{F}'_1\oplus\cdots\oplus\mathcal{F}'_r\rangle \le 0\,.$$

It follows from Proposition 1.2.9 that $\deg_{\mathcal{F}\otimes\mathcal{F}'}(W) \le 0$, a contradiction.

The case of a finite field is reduced to the case of an infinite field as before in the case of $\mathrm{Fil}^k_{\mathbb{F}_1}$.

Notes and References The tensor product theorem has its analogue in the theory of vector bundles on a compact Riemann surface [166], where it is proved by relating stable vector bundles to irreducible unitary representations of the fundamental group. Algebraic proofs valid over any field of characteristic zero are due to Maruyama [161], and to Ramanan and Ramanathan [179]. The analogue for vector bundles on a smooth projective algebraic curve is false in characteristic $p > 0$. Theorem 1.2.1 was first proved by Faltings–Wüstholz [75] for fields of characteristic 0 in relation to a problem in diophantine approximation. Their proof is by reduction to the theorem of Narasimhan and Seshadri. Shortly after, Totaro [218] gave a new proof (in the same situation) by constructing suitable hermitian metrics on filtered vector spaces over $\mathbb{C}$. Finally, Faltings [73] resp. Totaro [219] proved the tensor product theorem for arbitrary perfect base fields around the same time. Whereas Faltings' proof is of arithmetic nature (construction of integral lattices inside vector bundles on a curve), Totaro's proof makes use of Geometric Invariant Theory and is modeled on the proof by Ramanan and Ramanathan mentioned above. Here we have followed Totaro's proof. Another proof of Theorem 1.2.1 was given by Fuijimori [84], using methods of the theory of diophantine approximations.

The tensor product theorem is closely related to Fontaine's conjecture [80] that the tensor product of weakly admissible filtered isocrystals is again weakly admissible. We will return to this topic in Part 3 of this monograph.

As observed in Remark 1.2.10, the tensor product theorem fails for non-perfect ground fields. This observation has led R. Pink to introduce in the case of $\mathbb{F}_q((t))$ a richer structure, nowadays called a *Hodge–Pink structure*, which is preserved under tensor products, [177], Thm. 5.6. We refer to papers by U. Hartl and others (see [110] and the literature quoted there) for more information on these objects.

3 The Harder–Narasimhan filtration

The Harder–Narasimhan filtration will be an $\mathbb{R}$-filtration of an object $(V, \mathcal{F}) \in \mathrm{Fil}^K_k$ by *subobjects*, i.e., a decreasing filtration $\{{}^{\alpha}V \mid \alpha \in \mathbb{R}\}$ which is exhaustive and separating, where each ${}^{\alpha}V$ is a subobject of V. It is defined by the following proposition.

Proposition 1.3.1 (a) *Let $V \in \mathrm{Fil}_k^K$. There exists an $\mathbb{R}$-filtration ${}^{\alpha}V$ of V by subobjects with the property*

$$^{\alpha}\,\mathrm{gr}(V) \neq (0) \implies {}^{\alpha}\,\mathrm{gr}(V) \text{ is semi-stable of slope } \alpha\,.$$

(b) *Any morphism $f : V \longrightarrow W$ in Fil_k^K is compatible with such filtrations,*

$$f({}^{\alpha}V) \subset {}^{\alpha}W.$$

In particular, the filtration in a) is uniquely determined.

Proof (a) We prove the existence of such a filtration by induction on the dimension of V. We first note that the set of all slopes of subobjects of V is finite, given the definition of the degree. Let us denote by $\alpha_{\max}$ its maximum. Any subobject with slope $\alpha_{\max}$ is trivially semi-stable. We claim that there is a unique *maximal* subobject ${}^{\alpha_{\max}}V$ of slope $\alpha_{\max}$. Indeed it suffices to check that, given U, W of slope $\alpha_{\max}$, their sum $U + W$ also has slope $\alpha_{\max}$. So consider the sum morphism $U \oplus W \xrightarrow{f} V$. We know that $U \oplus W$ is semi-stable with slope $\alpha_{\max}$, so $\mu(\mathrm{Coim}\, f) \geq \alpha_{\max}$. By Lemma 1.1.12 we get $\mu(U + W) = \mu(\mathrm{Im}\, f) \geq \alpha_{\max}$, hence $\mu(U + W) = \alpha_{\max}$. Now, V is semi-stable if and only if $V = {}^{\alpha_{\max}}V$ and in this case, the existence of the filtration is obvious. In general, let us put $V' := V/{}^{\alpha_{\max}}V$ and let us prove that its "maximal slope" satisfies $\alpha'_{\max} < \alpha_{\max}$. Indeed any subobject W' of V' has the form $W/{}^{\alpha_{max}}V$ for some subobject W of V. We have $\mu(W) < \mu({}^{\alpha_{\max}}V) = \alpha_{\max}$ hence by Proposition 1.1.15 we get $\mu(W') < \alpha_{\max}$. By our induction hypothesis we have a filtration $({}^{\alpha}V')_{\alpha\in\mathbb{R}}$ as in point (a) on V'. Letting ${}^{\alpha}V$ be the inverse image of ${}^{\alpha}V'$ for $\alpha < \alpha_{\max}$ and 0 for $\alpha > \alpha_{\max}$, we get the desired filtration on V.

(b) Let $({}^{\alpha}V)_{\alpha\in\mathbb{R}}$, $({}^{\alpha}W)_{\alpha\in\mathbb{R}}$ be two filtrations as in point (a). Assume that f is not compatible with these filtrations and pick an α_0 such that $f({}^{\alpha_0}V)$ is not contained in ${}^{\alpha_0}W$. Then the set $\{\beta \in \mathbb{R} \mid f({}^{\alpha_0}V) \subset {}^{\beta}W\}$ has a greatest element $\beta_0 < \alpha_0$ which is a jump of ${}^{\bullet}W$, and f induces a non-zero morphism ${}^{\alpha_0}V \longrightarrow \mathrm{gr}^{\beta_0}W$. Now the set $\{\alpha \geq \alpha_0 | {}^{\alpha}V \xrightarrow{\neq 0} \mathrm{gr}^{\beta_0}W\}$ also has a greatest element α_1 which is a jump of ${}^{\bullet}V$. Hence f induces a non-zero morphism $\mathrm{gr}^{\alpha_1}V \longrightarrow \mathrm{gr}^{\beta_0}W$, which contradicts Proposition 1.1.20. □

This filtration of V is called its *Harder–Narasimhan filtration* (HN-filtration).

Remarks 1.3.2 (i) We note that V is semi-stable if and only if the HN-filtration has only one jump.

(ii) It follows from Proposition 1.3.1 that an object V is semi-stable if and only if $\mu(W) \leq \mu(V)$ for any subobject W stable under all endomorphisms of V, i.e., it suffices to check the semi-stability condition on these subobjects.

(iii) For any $c \in \mathbb{R}$ we have ${}^{\bullet}V(c) = {}^{\bullet+c}V$.

Let $\mu_{\max}(V)$ denote the maximum slope of a subobject of V; let $\mu_{\min}(V)$ denote the minimum slope of a quotient object of V. Then

$$\mu_{\min}(V) = -\mu_{\max}(V^*) .$$

Proposition 1.3.3 *Let $\alpha_1 > \alpha_2 > \cdots > \alpha_s$ be the jumps in the* HN*-filtration of $V \in \mathrm{Fil}_k^K$. Then ${}^{\alpha_i}V/{}^{\alpha_{i-1}}V$ is the largest subobject of $V/{}^{\alpha_{i-1}}V$ with slope equal to $\mu_{\max}(V/{}^{\alpha_{i-1}}V) = \alpha_i$ (here $i = 1, \ldots, s$ and we set ${}^{\alpha_0}V = (0)$). Dually, ${}^{\alpha_i}V/{}^{\alpha_{i-1}}V$ is the largest quotient object of ${}^{\alpha_i}V$ with slope equal to $\mu_{\min}({}^{\alpha_i}V) = \alpha_i$.*

Proof It suffices to prove the first assertion. Let V' be a subobject of $V/{}^{\alpha_{i-1}}V$ with slope $\mu(V') = \mu_{\max}(V/{}^{\alpha_{i-1}}V)$. It suffices to show that $V' \subset {}^{\alpha_i}V$. But clearly, V' is semi-stable and $\mu(V') > \mu({}^{\alpha_j}V/{}^{\alpha_{j-1}}V)$ for $j = i+1, \ldots, s$. By descending induction using Proposition 1.1.20 we conclude that $V' \subset {}^{\alpha_i}V$. □

We associate to $V \in \mathrm{Fil}_k^K$ its HN-*vector* HN(V). (This vector is not to be mixed up with the type (1.2), page 4 of the filtration $\mathcal{F}$ on V_K!) If $\{\alpha_1 > \alpha_2 > \cdots > \alpha_s\}$ are the jumps in the Harder–Narasimhan filtration of V, with multiplicities $m_i = \dim {}^{\alpha_i}\mathrm{gr}(V)$, then

$$\mathrm{HN}(V) = (\alpha_1^{(m_1)}, \alpha_2^{(m_2)}, \ldots, \alpha_s^{(m_s)}) \in (\mathbb{R}^n)_+. \tag{3.1}$$

Here the exponent (m_i) indicates that the corresponding entry is repeated m_i times. It is customary to picture the HN-vector through the HN-*polygon*: this is the piecewise linear polygon connecting by line segments the points $(0,0), (\mathrm{rk}({}^{\alpha_1}V), \deg({}^{\alpha_1}V)), (\mathrm{rk}({}^{\alpha_2}V), \deg({}^{\alpha_2}V)), \ldots, (\mathrm{rk}(V), \deg(V))$ in the plane. This is a polygon with break points in $\mathbb{Z}^2$. It is convex, since the slopes of the lines form a strictly decreasing sequence. The picture on the next page illustrates the shape of an HN-polygon.

Proposition 1.3.4 *Let V' be a subobject of $V \in \mathrm{Fil}_k^K$. The point $p(V') = (\mathrm{rk}\, V', \deg V')$ lies on or below the* HN*-polygon of V.*

Proof We prove this by induction on the length of the HN-filtration of V. The induction start is when V is semi-stable, and is trivial. Now let $\alpha_1 = \mu_{\max}(V)$. Then ${}^{\alpha_1}V$ is the first step in the HN-filtration of V.

Case 1: $\mu(V') = \alpha_1$

In this case, we have $V' \subset {}^{\alpha_1}V$ by Proposition 1.3.3, and $p(V')$ lies on the first line segment of the HN-polygon of V, hence the assertion in this case.

Case 2: $\mu(V') < \alpha_1$

Let $({}^{\alpha_1}V + V')^\flat$ be the quotient object of ${}^{\alpha_1}V \oplus V'$ defined by the following exact sequence,

$$0 \longrightarrow {}^{\alpha_1}V \cap V' \longrightarrow {}^{\alpha_1}V \oplus V' \longrightarrow ({}^{\alpha_1}V \oplus V')^\flat \longrightarrow 0.$$

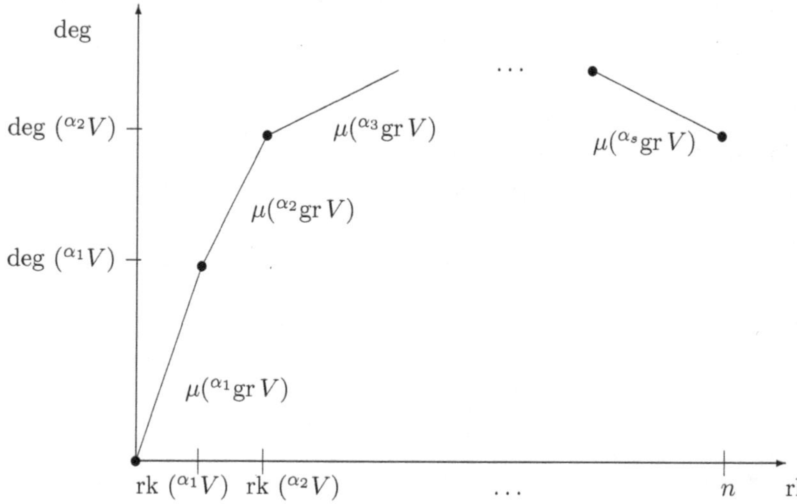

If we denote by ${}^{\alpha_1}V + V'$ the subobject of V defined by its underlying k-subspace, we have $\deg({}^{\alpha_1}V + V')^{\flat} \leq \deg({}^{\alpha_1}V + V')$. It follows that

$$\deg({}^{\alpha_1}V \cap V') + \deg({}^{\alpha_1}V + V') \geq \deg({}^{\alpha_1}V) + \deg V'.$$

Let $r = \mathrm{rk}({}^{\alpha_1}V)$, $r' = \mathrm{rk}(V')$, $s = \mathrm{rk}({}^{\alpha_1}V \cap V')$. Since $\mu({}^{\alpha_1}V \cap V') \leq \mu({}^{\alpha_1}V) = \alpha_1$, we obtain

$$\deg({}^{\alpha_1}V + V') \geq \alpha_1(r - s) + d',$$

where $d' = \deg(V')$.

Case 2a*:* $r = s$

If $V' \subset {}^{\alpha_1}V$, we are done. If ${}^{\alpha_1}V \subset V'$, then $V'/{}^{\alpha_1}V$ is a subobject of $V/{}^{\alpha_1}V$. By the induction hypothesis, $p(V'/{}^{\alpha_1}V)$ lies below the HN-polygon of $V/{}^{\alpha_1}V$. However,

$$p(V') = (\mathrm{rk}({}^{\alpha_1}V), \deg({}^{\alpha_1}V)) + p(V'/{}^{\alpha_1}V).$$

Also, the HN-polygon of V is obtained by first drawing the line segment joining the origin with the point $(\mathrm{rk}^{\alpha_1}V, \deg^{\alpha_1}V)$ and then continuing with the translated HN-polygon of $V/{}^{\alpha_1}V$. The assertion follows.

Case 2b*:* $s < r$

Then, since $\mu(V') < \alpha_1$, we have

$$\deg({}^{\alpha_1}V + V') > \frac{d'}{r'}(r - s) + d' = \frac{d'}{r'}(r + r' - s).$$

Hence

$$\mu({}^{\alpha_1}V + V') > \mu(V').$$

By induction $p(({}^{\alpha_1}V + V')/{}^{\alpha_1}V)$ lies below the HN-polygon of $V/{}^{\alpha_1}V$. As in Case 2a, we conclude that $p({}^{\alpha_1}V + V')$ lies below the HN-polygon for V. But $\mathrm{rk}(V') < \mathrm{rk}({}^{\alpha_1}V + V')$ and $\mu(V') < \mu({}^{\alpha_1}V + V')$, hence $p(V')$ lies below the line joining the origin with $p({}^{\alpha_1}V + V')$. The assertion follows. □

Corollary 1.3.5 *Let $V \in \mathrm{Fil}_k^K$. Let ${}^\bullet V$ be an arbitrary $\mathbb{R}$-filtration by subobjects in Fil_k^K. Let $p({}^\bullet V)$ be the convex polygon which interpolates the origin and the points $p({}^\alpha V), \alpha \in \mathbb{R}$. Then $p({}^\bullet V)$ lies below the* HN*-polygon of V, with the same start and the same end point.*

Proof By the previous proposition, all points $p({}^\alpha V)$ lie below the HN-polygon of V. Hence so does $p({}^\bullet V)$ which is the convex hull of these points. □

From the tensor product theorem of the previous section, it follows that the HN-filtration is compatible with the usual linear algebra constructions, in the following sense.

Proposition 1.3.6 *Let K/k be a separable field extension.*

(a) *Let $V \in \mathrm{Fil}_k^K$. Then the* HN*-filtration on V^* is the dual (in the sense of* (1.3), *page 5) of the* HN*-filtration on V. The* HN*-filtrations on $V^{\otimes r}$,* $\mathrm{Sym}^r(V)$, $\bigwedge^r V$ *are similarly induced by the* HN*-filtration on V.*

(b) *Let $V, W \in \mathrm{Fil}_k^K$. Then the* HN*-filtrations on $V \otimes W$ and $V \oplus W$ are induced by the* HN*-filtrations on V and on W.*

Proof We let $\mathrm{Fil}(\mathrm{Fil}_k^K)$, resp. $\mathrm{Grad}(\mathrm{Fil}_k^K)$, denote the category of $\mathbb{R}$-filtered, resp. $\mathbb{R}$-graded, objects in Fil_k^K. Both these categories are k-linear quasi-abelian $\otimes$-categories. An object in $\mathrm{Fil}(\mathrm{Fil}_k^K)$ is a triple $(V, \mathcal{F}_V, \Phi)$ with Φ a filtration of the k-vector space V and $\mathcal{F}_V$ a filtration of V_K, and a morphism is a k-linear map which respects both filtrations. An object in $\mathrm{Grad}(\mathrm{Fil}_k^K)$ will be written in the form $\bigoplus_{\alpha \in \mathbb{R}} (V_\alpha, \mathcal{F}_{V_\alpha})[-\alpha]$. Such an object will be called "semi-stable of slope 0" if $(V_\alpha, \mathcal{F}_{V_\alpha})$ is semi-stable of slope α, for all α (with the convention that the zero object in Fil_k^K is semi-stable of any slope). The full subcategory of $\mathrm{Grad}(\mathrm{Fil}_k^K)$ formed by such objects is additive by Corollary 1.1.21 and stable under duality by Corollary 1.1.17. By the tensor product theorem 1.2.1 and its corollary 1.2.2 it is also stable under tensor product, symmetric and exterior powers.

There is an obvious functor $\mathrm{gr}^\bullet : \mathrm{Fil}(\mathrm{Fil}_k^K) \longrightarrow \mathrm{Grad}(\mathrm{Fil}_k^K)$ which sends a triple $(V, \mathcal{F}_V, \Phi)$ to the graded object $\bigoplus_{\alpha \in \mathbb{R}} (\mathrm{gr}_\Phi^\alpha V, \mathcal{F}_V)[-\alpha]$ where $\mathcal{F}_V$ always denotes the induced filtration on a subquotient of V_K. By Proposition 1.3.1, Φ is the Harder–Narasimhan filtration of $(V, \mathcal{F}_V)$ if and only if $\mathrm{gr}^\bullet(V, \mathcal{F}_V, \Phi)$ is "semi-stable of slope 0." So the proposition follows from the next claim.

Claim *The functor* $\mathrm{gr}^\bullet$ *is an additive* $\otimes$*-functor which commutes with symmetric and exterior powers and with duality.*

The fact that the functor is additive and commutes with duality is easy. Concretely, what we have to prove to check the compatibility with tensor products is that for all α, β the canonical k-linear embedding

$$\Psi^{\alpha,\beta}_{V,W} : \mathrm{gr}^{\alpha}_{\Phi_V} V \otimes \mathrm{gr}^{\beta}_{\Phi_W} W \longrightarrow \mathrm{gr}^{\alpha+\beta}_{\Phi_V \otimes \Phi_W}(V \otimes W)$$

is *strictly compatible* with the filtrations induced by $\mathcal{F}_V \otimes \mathcal{F}_W$ on both sides. Through the canonical identifications $(V \otimes W)^* \xrightarrow{\sim} V^* \otimes W^*$ (in the category of vector spaces) the compositum $(\Psi^{-\alpha,-\beta}_{V^*,W^*})^* \circ \Psi^{\alpha,\beta}_{V,W}$ is the identity of the vector space $\mathrm{gr}^{\alpha}_{\Phi_V} V \otimes \mathrm{gr}^{\beta}_{\Phi_W} W$. Thus it suffices to prove only the *compatibility* of the maps $\Psi^{\alpha,\beta}_{V,W}$ with the filtrations induced by $\mathcal{F}_V \otimes \mathcal{F}_W$. But this follows from the commutative diagram

$$\begin{array}{ccc} \Phi^{\alpha}_{V} V \otimes \Phi^{\beta}_{W} W & \longrightarrow & (\Phi_V \otimes \Phi_W)^{\alpha+\beta}(V \otimes W) \\ \downarrow & & \downarrow \\ \mathrm{gr}^{\alpha}_{\Phi_V} V \otimes \mathrm{gr}^{\beta}_{\Phi_W} W & \longrightarrow & \mathrm{gr}^{\alpha+\beta}_{\Phi_V \otimes \Phi_W}(V \otimes W)\,. \end{array}$$

Indeed in this diagram, the left vertical arrow is a strict epimorphism in Fil^K_k, being the tensor product of strict epimorphims, while the right one obviously is filtered (and even a strict epimorphism) and the top horizontal arrow also is filtered (and even is a strict monomorphism).

The fact that $\mathrm{gr}^\bullet$ commutes with symmetric and exterior powers is proved in the same way as point (iii) of Lemma 1.1.8. The point is that the (anti)-symmetrization endomorphisms of $\bigotimes^r V$ are strict for both filtrations $\otimes^r \Phi_V$ and $\otimes^r \mathcal{F}_V$. □

Corollary 1.3.7 *Let K/k be a separable field extension. Let $V, W \in \mathrm{Fil}^K_k$. If $V \otimes W$ is semi-stable, then V and W are semi-stable.*

Proof This is a combination of Proposition 1.3.6 and Remarks 1.3.2 (i). □

Remark 1.3.8 Of course this corollary can be proved directly (without the assumption that K/k is separable) by applying Lemma 1.1.8 (i).

Corollary 1.3.9 *Consider the full subcategory of* Fil^K_k*, consisting of semi-stable objects of slope zero. This is a tannakian category with a fiber functor over k.*

Proof This combines Corollary 1.1.21 with the tensor product theorem. (For more on the concept of a tannakian category, cf. Chapter IV below.) □

Corollary 1.3.10 *The* HN-*filtration of objects in* Fil_k^K *is compatible with base change. More precisely, let k' be a field extension of k which is linearly disjoint from K. Let $K' = K \otimes_k k'$. An object $V = (V, \mathcal{F})$ of* Fil_k^K *defines an object $V \otimes_k k' = (V \otimes_k k', \mathcal{F} \otimes_K K')$ of* $\mathrm{Fil}_{k'}^{K'}$*. For the corresponding* HN-*filtrations*

$$^{\alpha}(V \otimes_k k') = {}^{\alpha}V \otimes_k k'.$$

In particular, V is semi-stable if and only if $V \otimes_k k'$ is semi-stable.

Proof The case when k' is a separable field extension follows at once from the uniqueness property of the HN-filtration. To prove the corollary in general, one easily reduces to the case when V is semi-stable and where k' is obtained by adjoining a pth root of an element of k, where p is the characteristic of k. Let ∂ be a k-derivation of k' with kernel k. Then $\nabla = \mathrm{id} \otimes_k \partial$ is a k-linear endomorphism of $V' = V \otimes_k k'$ with $\nabla(\mathcal{F}^x \otimes_K K') \subset \mathcal{F}^x \otimes_K K'$, for all x. Let $^{\alpha}V'$ be the first step in the HN-filtration of V' and consider the composed map

$$^{\alpha}V' \hookrightarrow V' \xrightarrow{\nabla} V' \longrightarrow V'/{}^{\alpha}V'.$$

From the Leibniz rule, this is a k'-linear map. Hence, by comparing slopes, this map must be trivial, which shows that $\nabla(^{\alpha}V') \subset {}^{\alpha}V'$. Hence by Jacobson descent, $^{\alpha}V'$ is defined over k, hence $\alpha = \mu(^{\alpha}V') \leq \mu(V) = \mu(V')$, due to the semi-stability of V. It follows that $^{\alpha}V' = V'$ and hence V' is semi-stable. □

Notes and References The Harder–Narasimhan filtration in the context of vector bundles appears in [105], but was also independently discovered by Quillen (cf. [95]) and by Tjurin [217]. The advantage of numbering the steps in the HN-filtration by the slopes was pointed out by Faltings [73]. Harder and Narasimhan used the HN-filtration to stratify the space of vector bundles in order to compute Betti numbers of the moduli space of semi-stable vector bundles, cf. [9]. In the next chapters we will pursue the same strategy in our context. For a similar approach in yet another context, see the paper by Reineke [188]. The proof of Corollary 1.3.10 is extracted from [73], but is in fact standard in the context of vector bundles.

The Tannaka category appearing in Corollary 1.3.9 is of the form $\mathrm{Rep}_k(G)$, where G is the automorphism group scheme over k of the natural fiber functor; cf. Chapter IV below. It seems difficult to describe the structure of the affine group scheme G. Inspired by the analogy with the theory of vector bundles on algebraic curves, and in analogy with the results of Deninger and Werner [65] on vector bundles on p-adic curves, one may at least ask for the group $\pi_0(G)$ of connected components of G, and for the maximal abelian quotient G_{ab} of G.

II. Period Domains for GL_n

In this chapter we introduce period domains in their simplest version. We start with a vector space V of dimension n over a field k, and consider the variety $\mathcal{F} = \mathcal{F}(V, \nu)$ of all $\mathbb{R}$-filtrations $\mathcal{F}$ of a given type ν. In the first section, we show that those filtrations $\mathcal{F}$ of type ν which violate the semi-stability condition along a given k-subspace V' form a Zariski-closed subset. It then follows that, if k is a finite field, the semi-stable filtrations $\mathcal{F}$ of type ν form a Zariski-open subset of $\mathcal{F}$. This open subset is the so-called *period domain*. It can be considered as a moduli space for semi-stable filtrations of type ν on V.

The analogy with vector bundles on a curve is again a useful guide. Indeed, semi-stability of vector bundles was historically introduced to define a good moduli space via Geometric Invariant Theory. The first instance of this analogy will be a characterization of the period domain in terms of the Hilbert–Mumford criterion from GIT, in Section 2. The second instance will be the description of the natural stratification of the whole flag variety $\mathcal{F}$ according to the Harder–Narasimhan type, similar to that of the moduli stack of vector bundles according to the HN type (see also the "Notes and References" of Section 3). In particular, we will clarify the structure of each HN-stratum in terms of period domains attached to vector spaces of smaller dimension. In the final section we consider the analogue over the field with one element and study the relation of the HN-stratification to the stratification by thin Schubert cells.

1 Definition and examples

Let V be a k-vector space of dimension n. There are two ways to fix the type of filtrations. The first one is to fix $\nu \in (\mathbb{R}^n)_+$. Let $\nu(1) > \cdots > \nu(r)$ be the distinct entries of ν. We also write

$$\nu = (x_1^{(n_1)}, \ldots, x_r^{(n_r)}) ,$$

where $x_i = \nu(i)$, with x_i occurring with multiplicity n_i. We denote by $\mathcal{F}(V,\nu) = \mathcal{F}(\nu)$ the functor on (Aff/k) which to $\operatorname{Spec} R$ associates the set of chains of R-submodules which are locally on $\operatorname{Spec} R$ direct summands of $V \otimes_k R$,

$$(0) = \mathcal{V}_0 \subset \mathcal{V}_1 \subset \cdots \subset \mathcal{V}_r = V \otimes_k R\,,$$

such that the rank of $\mathcal{V}_i/\mathcal{V}_{i-1}$ is equal to n_i for $i = 1, \ldots, r$. In other words, $\mathcal{F}(\nu)$ is the *variety of partial flags of type* $(n_1, n_1 + n_2, \ldots, n = n_1 + \cdots + n_r)$ of V. If $R = K$ is a field extension of k, we obtain a bijection between $\mathcal{F}(\nu)(K)$ and the set of filtrations on $V \otimes_k K$ of type ν by setting

$$\mathcal{F}^x = \mathcal{V}_i\,,$$

where $i = \max\{j \mid x_j \geq x\}$.

The other way is in some sense dual. Fix a function

$$g : \mathbb{R} \longrightarrow \mathbb{Z}_{\geq 0}$$

with $\sum_x g(x) = n$. Let $\{x_1 > x_2 > \cdots > x_r\}$ be the support of g. To g we associate the vector

$$\nu = (x_1^{(g(x_1))}, \ldots, x_r^{(g(x_r))}) \in (\mathbb{R}^n)_+\,.$$

Conversely, to $\nu = (x_1^{(n_1)}, \ldots, x_r^{(n_r)}) \in (\mathbb{R}^n)_+$, there corresponds the function g with $g(x_i) = n_i$ for $i = 1, \ldots, r$, and which vanishes for all $x \in \mathbb{R} \setminus \{x_1, \ldots, x_r\}$. In terms of g, the functor $\mathcal{F}(\nu)$ is described by

$$\mathrm{rk}(\mathcal{V}_i/\mathcal{V}_{i-1}) = g(x_i),\ i = 1, \ldots, r\,.$$

We call the function g associated to ν the *multiplicity function of* ν. Conversely, the vector ν associated to g is called the *type of* g. When we describe the partial flag variety $\mathcal{F}(\nu)$ in terms of g, we also write $\mathcal{F}(g)$ for it.

Our next aim is to show that the locus in $\mathcal{F}(\nu)$ where a given k-subspace of V violates the semi-stability condition is a Zariski-closed subset.

Proposition 2.1.1 *Let V' be a k-subspace of V. There exists a Zariski-closed subset $Y_{V'}$ of $\mathcal{F}(\nu)$ such that for any field extension K/k, and any $\mathbb{R}$-filtration $\mathcal{F}$ of type ν on $V \otimes_k K$, corresponding to the point $[\mathcal{F}] \in \mathcal{F}(\nu)(K)$, we have*

$$[\mathcal{F}] \in Y_{V'} \Longleftrightarrow \mu(V', \mathcal{F}_{V'}) > \mu(V, \mathcal{F})\,.$$

Proof By Lemma 1.1.11 we have

$$\mathrm{rk}(V') \cdot \mu(V', \mathcal{F}_{V'}) = \sum_{i=1}^{r-1} (\nu(i) - \nu(i+1)) \dim(V' \cap \mathcal{F}^i) + \nu(r) \cdot \dim(V').$$

For any $i = 1, \dots, r$, the function $[\mathcal{F}] \mapsto \dim(V' \cap \mathcal{F}^i)$ is upper semi-continuous on $\mathcal{F}(V, \nu)$ (this just means that the locus of $[\mathcal{F}] \in \mathcal{F}(V, \nu)$ where $\dim(V' \cap \mathcal{F}^i)$ is greater than a given number is Zariski-closed, which follows from elementary linear algebra).

We deduce that the function $[\mathcal{F}] \mapsto \mu(V', \mathcal{F}_{V'})$ also is upper semi-continuous, and since it affords finitely many values, the proposition follows. □

Corollary 2.1.2 *Let V be a k-vector space of dimension n and let $\nu \in (\mathbb{R}^n)_+$. Denote, for any field extension K of k, by $\mathcal{F}(\nu)(K)^{ss}$ the set of filtrations $\mathcal{F}$ of type ν on $V \otimes_k K$ such that $(V, \mathcal{F}) \in \mathrm{Fil}_k^K$ is semi-stable. Then*

$$\mathcal{F}(\nu)(K)^{ss} = \mathcal{F}(\nu)(K) \setminus \bigcup_{V'} Y_{V'} .$$

Here V' runs over all k-subspaces of V. □

Unless V has only finitely many subspaces, the RHS of the identity in Corollary 2.1.2 is not the set of K-points of an algebraic variety. (In Part 3 of this monograph we will see that if k is a p-adic field, then the RHS has a rigid-analytic interpretation).

Corollary 2.1.3 *Let k be a finite field. Then the set $\mathcal{F}(\nu)(K)^{ss}$ is the set of K-valued points of a Zariski-open subset $\mathcal{F}(\nu)^{ss}$ of $\mathcal{F}(\nu)$. In fact,*

$$\mathcal{F}(\nu)^{ss} = \mathcal{F}(\nu) \setminus \bigcup_{V'} Y_{V'} .$$

The action of the finite group $\mathrm{GL}(V)(k)$ *on $\mathcal{F}(\nu)$ preserves $\mathcal{F}(\nu)^{ss}$.* □

Definition 2.1.4 Let k be a finite field. Let V be a k-vector space of dimension n. Let $\nu \in (\mathbb{R}^n)_+$. The *period domain* associated to (V, ν) is the Zariski-open subset $\mathcal{F}(\nu)^{ss}$ of $\mathcal{F}(\nu)$.

Variant 2.1.5 Let us consider the category Fil_k^r. In this case we consider a k-vector space V of dimension n and r $\mathbb{R}$-filtrations $\mathcal{F}_1, \dots, \mathcal{F}_r$ of type $\nu_1, \dots, \nu_r \in (\mathbb{R}^n)_+$ on V. If k is finite, there is a Zariski-open subset.

$$\mathcal{F}(\nu_1, \dots, \nu_r)^{ss} \subset \mathcal{F}(\nu_1) \times \cdots \times \mathcal{F}(\nu_r)$$

such that $\mathcal{F}(\nu_1, \dots, \nu_r)^{ss}(k)$ is the set of r-tuples $(\mathcal{F}_1, \dots, \mathcal{F}_r)$ which are semi-stable.

Variant 2.1.6 Let us consider the category $\mathrm{Fil}_{\mathbb{F}_1}^k$. In this case the semi-stability of $\mathcal{F} \in \mathcal{F}(\nu)(K)$ is checked on each of the finitely many subspaces of V which are equal to the direct sum of some of the lines in the fixed decomposition $\mathbf{V} = (V = V_1 \oplus \cdots \oplus V_n)$ of V. Hence, there is a Zariski-open subset $\mathcal{F}(\nu)^{ss}$

of $\mathcal{F}(\nu)$ such that $\mathcal{F}(\nu)^{\mathrm{ss}}(K)$ consists of the semi-stable filtrations of type ν on $V \otimes_k K$ which are semi-stable with respect to the line decomposition $\mathbf{V}$ of V. Let $N(k)$ be the normalizer in $\mathrm{GL}(V)(k)$ of the decomposition $\mathbf{V}$ of V (i.e., an element of $N(k)$ preserves the decomposition as a whole). Then $N(k)$ acts on $\mathcal{F}(\nu)^{\mathrm{ss}}$. In particular, the group $W = S_n$ of permutation matrices acts on $\mathcal{F}(\nu)^{\mathrm{ss}}$.

In the remainder of this section, we give examples. We precede these examples by the following remark on normalizations. The variety $\mathcal{F}(\nu)$ and its open subvariety $\mathcal{F}(\nu)^{\mathrm{ss}}$ are unchanged by the following two operations:
(i) replace ν by $c \cdot \nu$, some $c \in \mathbb{R}_+^\times$,
(ii) replace ν by $\nu + c \cdot \mathbf{1}$, where $c \in \mathbb{R}$ and where $\mathbf{1} = (1^{(n)}) \in (\mathbb{R}^n)_+$.
In particular, we may always assume that $|\nu| = 0$. Here we set $|\nu| = \sum_{i=1}^n y_i$ for $\nu = (y_1, \ldots, y_n) \in \mathbb{R}^n$.

Example 2.1.7 Let $\nu = (n-1, (-1)^{(n-1)}) \in (\mathbb{Z}^n)_+$. In this case $\mathcal{F}(\nu) = \mathbb{P}(V)$ (the space of lines in V). The example was considered in Chapter I, Examples 1.1.23. Hence $[\mathcal{F}] \in \mathbb{P}(V)(K)^{\mathrm{ss}}$ if and only if $\mathcal{F}$ is not contained in a k-rational hyperplane. If k is finite, let

$$\Omega(V) = \mathbb{P}(V) \setminus \bigcup\nolimits_{H/k} H$$

(complement in $\mathbb{P}(V)$ of all k-rational hyperplanes). Then $\mathbb{P}(V)^{\mathrm{ss}} = \Omega(V)$. If $V = k^n$, the space $\Omega^n = \Omega(k^n)$ is the *Drinfeld halfspace* of dimension $n-1$.

Example 2.1.8 Let us consider Example 2.1.7 in variant $\mathrm{Fil}^k_{\mathbb{F}_1}$. In this case we have to remove all hyperplanes spanned by lines in the line decomposition $\mathbf{V} = (V = V_1 \oplus \cdots \oplus V_n)$ of V. Hence in this variant

$$\mathcal{F}(\nu)^{\mathrm{ss}} \simeq \Omega(\mathbf{V}) := \mathbb{P}(V) \setminus \bigcup\nolimits_{i=1}^n H_i\,,$$

where $H_i := \mathbb{P}(\bigoplus_{j \neq i} V_j)$ (complement of all coordinate hyperplanes).

Example 2.1.9 Let $n = 3$, and let $\nu = (x_1, x_2, x_3)$ with $x_1 > x_2 > x_3$. In this case $\mathcal{F}(\nu)(R)$ is the set of full flags $(0) \subsetneq \mathcal{V}_1 \subsetneq \mathcal{V}_2 \subsetneq V \otimes_k R$. The semi-stable set $\mathcal{F}(\nu)^{\mathrm{ss}}$ is described case by case by the following conditions, cf. Chapter I, Examples 1.1.23:

- if $x_1 - x_2 > x_2 - x_3$: *$\mathcal{V}_1$ is not contained in a k-rational hyperplane in V;*
- if $x_1 - x_2 < x_2 - x_3$: *$\mathcal{V}_2$ does not contain a k-rational line in V;*
- if $x_1 - x_2 = x_2 - x_1$: *$\mathcal{V}_1$ and $\mathcal{V}_2$ are both not k-rational.*

Example 2.1.10 Let us consider Example 2.1.9 in variant $\mathrm{Fil}^k_{\mathbb{F}_1}$. The conditions are case by case:

– $\mathcal{V}_1$ *is not contained in a coordinate plane in* V;
– $\mathcal{V}_2$ *does not contain a coordinate line in* V;
– $\mathcal{V}_1$ *is not a coordinate line and* $\mathcal{V}_2$ *is not a coordinate plane.*

Remark 2.1.11 In the last two examples we obtained a subdivision of $(\mathbb{R}^3)_+$ into rational polyhedral cones such that inside each relative interior the corresponding period domain remains unchanged. This is a general phenomenon. The subdivision of $(\mathbb{R}^n)_+$ into cones inside of which the open $\mathcal{F}(\nu)^{ss}$ of $\mathcal{F}(\nu)$ does not change becomes rapidly more complicated, as n increases. For $n = 4$ there are already 15 period domains inside the full flag variety. It is an open question to understand what happens to a period domain $\mathcal{F}(\nu)^{ss}$ as ν changes from one cone to a neighboring one. This question is reminiscent of the wall crossing inside the ample cone of G-line bundles in Geometric Invariant Theory, cf. [70].

Notes and References Period domains in the present linear algebra context are defined in [186]. As is explained there, not much is known about their geometry. The Drinfeld halfspace is not typical at all. In fact, as is pointed out in examples in [186], §3, most of the facts known in this special case do not generalize. The Drinfeld space is at the same time a period domain, as well as a Deligne–Lusztig variety† [58], as well as the complement of a hyperplane arrangement [167]. It is known [168] that this period domain is the unique one which also belongs to either of the other two classes of varieties. We refer to Chapter XI for more on the geometry of period domains.

2 The relation to GIT

We first recall some facts from Geometric Invariant Theory [164]. Let X be an algebraic variety over a field k. Let G be a connected reductive group over k which acts on X,

$$\sigma : G \times X \longrightarrow X\,.$$

Let $\mathcal{L}$ be a G-line bundle on X, i.e., a line bundle equipped with an action of G covering the action on X. This is given by an isomorphism of sheaves on $G \times X$

$$\phi : \sigma^*(\mathcal{L}) \longrightarrow p_2^*(\mathcal{L})$$

satisfying the usual cocycle condition, i.e., the commutativity of the following

† More precisely, Ω^n is isomorphic to a Deligne–Lusztig variety, which is by definition a subset of the variety of full flags.

diagram of homomorphisms of sheaves on $G \times G \times X$, in which $\mu : G \times G \longrightarrow G$ is the multiplication morphism:

$$\begin{array}{ccccc} (\sigma \circ (\mathrm{Id}_G \times \sigma))^* \mathcal{L} & \xrightarrow{(\mathrm{Id}_G \times \sigma)^* \phi} & (p_2 \circ (\mathrm{Id}_G \times \sigma))^* \mathcal{L} & = & (\sigma \circ p_{23})^* \mathcal{L} \\ \| & & & & \downarrow{\scriptstyle p_{23}^* \phi} \\ (\sigma \circ (\mu \times \mathrm{Id}_X))^* \mathcal{L} & \xrightarrow{(\mu \times \mathrm{Id}_X)^* \phi} & (p_2 \circ (\mu \times \mathrm{Id}_X))^* \mathcal{L} & = & (p_2 \circ p_{23})^* \mathcal{L} \,. \end{array}$$

Equivalently, the operation of G on X is lifted to the geometric line bundle associated to $\mathcal{L}$.

We denote by $\mathrm{Pic}^G(X)$ the group of isomorphism classes of G-line bundles on X (with respect to the given operation of G on X).

Every G-stable open subset U of X leads to a dual operation of G on $\Gamma(U, \mathcal{L})$ as follows:

$$\Gamma(U, \mathcal{L}) \xrightarrow{\sigma^*} \Gamma(G \times U, \sigma^* \mathcal{L}) \xrightarrow{\phi} \Gamma(G \times U, p_2^* \mathcal{L}) = \Gamma(G, \mathcal{O}_G) \otimes \Gamma(U, \mathcal{L}) \,. \quad (2.1)$$

In particular, we may speak of G-invariant sections in $\Gamma(U, \mathcal{L})$: a section s mapped under (2.1) above to $1 \otimes s$.

A geometric point x is called *semi-stable* (with respect to $\mathcal{L}$), if there exists a G-invariant section $f \in \Gamma(X, \mathcal{L}^{\otimes n})^G$ for some $n > 0$, such that the non-vanishing set X_f is an affine open neighborhood of x. There is then an open subset $X^{ss}(\mathcal{L})$ of X, such that a geometric point x of X factors through $X^{ss}(\mathcal{L})$ if and only if x is semi-stable.

We now assume that X is proper over $\mathrm{Spec}\, k$. Any 1-PS $\lambda : \mathbb{G}_m \longrightarrow G$ defined over k and any point $x \in X(k)$ define a morphism

$$\mathbb{G}_m \longrightarrow X \,,\; t \longmapsto \lambda(t) \cdot x \,. \quad (2.2)$$

Since X is proper, it follows from the valuative criterion that this morphism extends to a morphism

$$\varphi : \mathbb{A}^1 \longrightarrow X \,.$$

We write $\varphi(0) = \lim_{t \to 0} \lambda(t) x$. Then $\varphi(0) \in X(k)$ is a fixed point for the action of $\mathbb{G}_m$ via λ on X. Hence λ acts through a character $r \in \mathbb{Z}$ on the fiber $\mathcal{L}_{\varphi(0)}$. The negative of r is called the GIT-*slope* of λ in x (w.r.t. $\mathcal{L}$),

$$\mu^{\mathcal{L}}(x, \lambda) = -r \,.$$

The *Hilbert–Mumford criterion* states that *for an algebraically closed field k,*

$$x \in X^{ss}(\mathcal{L})(k) \Longleftrightarrow \mu^{\mathcal{L}}(x, \lambda) \geq 0 \,,\; \forall \text{ 1-PS } \lambda \text{ of } G \text{ defined over } k.$$

We also note the following functorial properties of the function $\mu^{\mathcal{L}}(x, \lambda)$, cf. [164], Ch. 2, §1.

• $\mu^{\mathcal{L}}(gx, \lambda) = \mu^{\mathcal{L}}(x, g^{-1}\lambda g)$, $g \in G(k)$.
• $\mu^{\mathcal{L}}(x, \lambda) = \mu^{\mathcal{L}}(\lim_{t\to 0} \lambda(t) \cdot x, \lambda)$.
• $\mu^{\mathcal{L}}(x, \lambda^r) = r \cdot \mu^{\mathcal{L}}(x, \lambda)$, for $r > 0$.
• For fixed x and λ, $\mu^{\mathcal{L}}(x, \lambda)$ defines a homomorphism from $\mathrm{Pic}^G(X)$ to $\mathbb{Z}$.
• If $f : X \longrightarrow Y$ is a G-morphism of G-varieties, then for $\mathcal{M} \in \mathrm{Pic}^G(Y)$ and setting $\mathcal{L} = f^*(\mathcal{M})$, we have $\mu^{\mathcal{L}}(x, \lambda) = \mu^{\mathcal{M}}(f(x), \lambda)$.
• If $X = X_1 \times \cdots \times X_m$ is a product of G-varieties and if $\mathcal{L}_i \in \mathrm{Pic}^G(X_i)$, then for $\mathcal{L} = \mathcal{L}_1 \boxtimes \cdots \boxtimes \mathcal{L}_m$, we have $\mu^{\mathcal{L}}(x, \lambda) = \sum_{i=1}^m \mu^{\mathcal{L}_i}(x_i, \lambda)$, for $x = (x_1, \ldots, x_m)$.

The above properties allow us to define $\mu^{\mathcal{L}}(x, \lambda) \in \mathbb{Q}$ for any $\mathcal{L} \in \mathrm{Pic}^G(X)\otimes\mathbb{Q}$ and for any $\mathbb{Q}$-1-PS λ of G. Here a $\mathbb{Q}$-1-*parameter subgroup* ($\mathbb{Q}$-1-PS) of G is a homomorphism

$$\lambda : \mathbb{D} \longrightarrow G ,$$

where $\mathbb{D} = \lim_{\leftarrow} \mathbb{G}_m$. The inverse limit is over the positive integers, by divisibility, cf. [135]. The irreducible representations of $\mathbb{D}$ are parametrized by $\mathbb{Q}$.

Now let $G \longrightarrow \mathrm{GL}(V)$ be a rational representation of G. Then G acts on $\mathbb{P}(V)$, and the tautological line bundle $\mathcal{O}(1)$ on $\mathbb{P}(V)$ is G-linearized. If X is a closed subvariety of $\mathbb{P}(V)$ invariant under G, we set $\mathcal{L} = \mathcal{O}(1)|X$ and we are in the previous situation. Fix a 1-PS λ of G defined over k. Let

$$V = \bigoplus_{i\in\mathbb{Z}} V(i) \tag{2.3}$$

be the character decomposition with respect to the action of λ. Let $\widetilde{x} \in V \setminus \{0\}$ be a representative of $x \in \mathbb{P}(V)(k)$, and let $\widetilde{x} = \sum \widetilde{x}_i$ be its eigenspace decomposition with respect to (2.3) above. Then, writing $\mu(x, \lambda)$ for $\mu^{\mathcal{L}}(x, \lambda)$, we have

$$\mu(x, \lambda) = -\min\{r \mid \widetilde{x}_r \neq 0\} .$$

In what follows we will denote by $\mathcal{F}_\lambda$ the $\mathbb{Z}$-filtration on V associated to the grading (2.3) above, i.e.,

$$\mathcal{F}_\lambda^x = \bigoplus_{i\geq x} V(i) .$$

We use similar notation for a $\mathbb{Q}$-1-PS λ. In this case $\mathcal{F}_\lambda$ is a $\mathbb{Q}$-filtration.

Let $G = \mathrm{GL}(V) \simeq \mathrm{GL}_n$. We fix $\nu \in (\mathbb{R}^n)_+$ and consider the corresponding variety $\mathcal{F}(\nu)$ of partial flags of V. Let $n_1, \ldots, n_r \in \mathbb{N}$ be the multiplicities of the entries of ν, i.e.,

$$\nu = (\nu(1)^{(n_1)}, \ldots, \nu(r)^{(n_r)}) , \text{ with } \nu(1) > \cdots > \nu(r) .$$

We therefore obtain a closed embedding into a product of Grassmannians,

$$\mathcal{F}(\nu) \hookrightarrow \mathbf{Gr}_{n_1}(V) \times \mathbf{Gr}_{n_1+n_2}(V) \times \cdots \times \mathbf{Gr}_{n_1+\cdots+n_r}(V) .$$

On each factor $X_i = \mathbf{Gr}_{n_1+\cdots+n_i}(V)$ there is a natural very ample line bundle $\mathcal{L}_i$ obtained via the Plucker embedding from the tautological line bundle $\mathcal{O}(1)$,

$$X_i \hookrightarrow \mathbb{P}\Big(\bigwedge^{n_1+\cdots+n_i} V\Big).$$

This line bundle is G-linearized. Its fiber at a point $x \in X_i$ corresponding to a subspace $\mathcal{V} \subset V$ of dimension $n_1 + \cdots + n_i$ is written heuristically as $\mathcal{L}_x = \bigwedge^{\max} \mathcal{V} = \det(\mathcal{V})$.

We will find it useful in what follows to take fractional powers of line bundles. It is obvious how to define a $\mathbb{Q}$-*line bundle* (an equivalence class of pairs $(\mathcal{L}, m)$, consisting of a line bundle $\mathcal{L}$ and a positive integer m, where $(\mathcal{L}, m) \sim (\mathcal{L}', m')$ if $\mathcal{L}^{\otimes m'} \cong \mathcal{L}'^{\otimes m}$). It is clear that the property of ampleness extends to $\mathbb{Q}$-line bundles. The isomorphism classes of $\mathbb{Q}$-line bundles are parametrized by $\mathrm{Pic}(X) \otimes \mathbb{Q}$.

Now let $\nu \in (\mathbb{Q}^n)_+$. Then we obtain an ample $\mathbb{Q}$-line bundle $\mathcal{L}(\nu)$ on $\mathcal{F}(\nu)$ as follows,

$$\mathcal{L}(\nu) = \mathcal{L}_1^{\otimes(\nu(1)-\nu(2))} \boxtimes \mathcal{L}_2^{\otimes(\nu(2)-\nu(3))} \boxtimes \cdots \boxtimes \mathcal{L}_r^{\otimes \nu(r)} \mid \mathcal{F}(\nu)\,. \tag{2.4}$$

The relation between the two semi-stability concepts, one from GIT and the other from the theory of filtered vector spaces, is based on the following two observations.

Lemma 2.2.1 *Let λ be a* 1-PS *of* $\mathrm{GL}(V)$ *defined over k. Let $[\mathcal{F}] \in \mathcal{F}(\nu)(k)$ be a point corresponding to a $\mathbb{Q}$-filtration $\mathcal{F}$ of type ν on V. Assume that $[\mathcal{F}]$ is a fixed point of λ. Then*

$$\mu^{\mathcal{L}(\nu)}([\mathcal{F}], \lambda) = -\langle \mathcal{F}, \mathcal{F}_\lambda \rangle\,.$$

Proof We choose a simultaneous splitting of the filtrations $\mathcal{F}$ and $\mathcal{F}_\lambda$. In other words, we choose a line decomposition $\mathbf{V} = (V = V_1 \oplus \cdots \oplus V_n)$ of V such that $\mathcal{F}$ is of the form $\mathcal{F} = \alpha^{\mathbf{V}}(\mathbf{x})$ for some $\mathbf{x} \in \mathbb{Q}^n$ and such that λ factors through the diagonal maximal torus T corresponding to the decomposition. Then

$$\lambda = (y_1, \ldots, y_n) \in \mathbb{Z}^n = X_*(T)\,.$$

Note that if we choose $\mathbf{x} \in (\mathbb{Q})^n_+$, as we may after permuting the basis vectors, then $\mathbf{x} = \nu = (\nu(1)^{(n_1)}, \ldots, \nu(r)^{(n_r)})$. Now

$$\langle \mathcal{F}, \mathcal{F}_\lambda \rangle = \sum_i x_i \cdot y_i = \sum_{i=1}^r \nu(i)\Big(\sum_{j=n_1+\cdots+n_{i-1}+1}^{n_1+\cdots+n_i} y_j\Big)\,. \tag{2.5}$$

The fiber of $\mathcal{L}(\nu)$ at $[\mathcal{F}]$ is isomorphic to

$$\begin{aligned} &\det(\mathcal{F}^{\nu(1)})^{\otimes(\nu(1)-\nu(2))} \otimes \cdots \otimes \det(\mathcal{F}^{\nu(r)})^{\otimes \nu(r)} \\ &= \det(\mathrm{gr}^{\nu(1)}(V))^{\otimes \nu(1)} \otimes \det(\mathrm{gr}^{\nu(2)}(V))^{\otimes \nu(2)} \otimes \cdots \otimes \det(\mathrm{gr}^{\nu(r)}(V))^{\otimes \nu(r)}\,. \end{aligned}$$

The operation of $\mathbb{G}_m$ through λ on $\det(\mathrm{gr}_{\nu(i)}(V))$ is given by $\sum_{j=n_1+\cdots+n_{i-1}+1}^{n_1+\cdots+n_i} y_j$. Hence the RHS of (2.5), page 38 is equal to the operation of λ on the fiber of $\mathcal{L}(\nu)$ at $[\mathcal{F}]$. By definition, this is the negative of $\mu^{\mathcal{L}(\nu)}([\mathcal{F}],\lambda)$. □

Lemma 2.2.2 *Let λ be a* 1-PS *of* GL(V) *defined over k. Let $[\mathcal{F}] \in \mathcal{F}(\nu)(k)$ be a point corresponding to the $\mathbb{Q}$-filtration $\mathcal{F}$ of V. Let $[\mathcal{F}_0] = \lim_{t\to 0}\lambda(t)[\mathcal{F}]$ and let $\mathcal{F}_0$ be the corresponding $\mathbb{R}$-filtration of V. Then*

$$\langle \mathcal{F}_0, \mathcal{F}_\lambda\rangle = \langle \mathcal{F}, \mathcal{F}_\lambda\rangle .$$

Proof Let $V = \bigoplus_{y\in\mathbb{Z}} V(y)$ be the eigenspace decomposition of V according to the action of $\mathbb{G}_m$ through λ. The projection $\mathcal{F}_\lambda^y V \stackrel{\pi^y}{\longrightarrow} \mathrm{gr}_{\mathcal{F}_\lambda}^y V$ has a unique $\mathbb{G}_m$-equivariant section $\iota_y : \mathrm{gr}_{\mathcal{F}_\lambda}^y V \stackrel{\sim}{\longrightarrow} V(y)$. The $\mathbb{G}_m$-family $t \mapsto t^{-y}\lambda(t)$ of endomorphisms of $\mathcal{F}_\lambda^y V$ has a limit at 0 (i.e., may be extended to a $\mathbb{A}^1$-family) which is given by $\lim_{t\to 0}(t^{-y}\lambda(t)) = \iota_y \circ \pi_y$ (i.e., the projection on $V(y)$ along $\bigoplus_{y'>y} V(y')$).

Put $[\mathcal{F}_t] := \lambda(t)[\mathcal{F}]$. The automorphism $\lambda(t)$ of V takes $\mathcal{F}^{\nu(i)}V$ into $\mathcal{F}_t^{\nu(i)}V$, hence the automorphism $t^{-y}\lambda(t)$ takes $\mathcal{F}^{\nu(i)}\mathcal{F}_\lambda^y V$ into $\mathcal{F}_t^{\nu(i)}\mathcal{F}_\lambda^y V$ for all $y \in \mathbb{Z}$. Passing to the limit, the projection $\iota_y \circ \pi_y$ thus takes $\mathcal{F}^{\nu(i)}\mathcal{F}_\lambda^y V$ into $\mathcal{F}_0^{\nu(i)}\mathcal{F}_\lambda^y V$. Since $\iota_y \circ \pi_y$ induces the identity on $\mathrm{gr}_{\mathcal{F}_\lambda}^y V$, we have in particular

$$\mathcal{F}^{\nu(i)}\,\mathrm{gr}_{\mathcal{F}_\lambda}^y V \subseteq \mathcal{F}_0^{\nu(i)}\,\mathrm{gr}_{\mathcal{F}_\lambda}^y V.$$

Furthermore, the dimension of $\mathrm{gr}_{\mathcal{F}}^{\nu(i)} V$ is given by $n_i = \sum_y \dim \mathrm{gr}_{\mathcal{F}}^{\nu(i)}\,\mathrm{gr}_{\mathcal{F}_\lambda}^y V$ hence that of $\mathcal{F}^{\nu(i)}V$ is given by $\sum_y \dim \mathcal{F}^{\nu(i)}\,\mathrm{gr}_{\mathcal{F}_\lambda}^y V$. Since $\dim(\mathcal{F}^{\nu(i)}V) = \dim(\mathcal{F}_0^{\nu(i)}V)$, this shows the equalities

$$\mathcal{F}^{\nu(i)}\,\mathrm{gr}_{\mathcal{F}_\lambda}^y V = \mathcal{F}_0^{\nu(i)}\,\mathrm{gr}_{\mathcal{F}_\lambda}^y V$$

for all y and all i. We also deduce that $\dim \mathrm{gr}_{\mathcal{F}}^{\nu(i)}\,\mathrm{gr}_{\mathcal{F}_\lambda}^y V = \dim \mathrm{gr}_{\mathcal{F}_0}^{\nu(i)}\,\mathrm{gr}_{\mathcal{F}_\lambda}^y V$ for all i and all y, hence the assertion of the Lemma follows from the definition of the scalar product of two $\mathbb{R}$-filtrations, cf. Definition 1.2.3. □

Theorem 2.2.3 *Let K be a field extension of k. Let $\mathcal{F}$ be an $\mathbb{Q}$-filtration on $V \otimes_k K$ which corresponds to a point $[\mathcal{F}] \in \mathcal{F}(\nu)(K)$. Then $(V,\mathcal{F})$ is a semi-stable object of* Fil_k^K *if and only if $[\mathcal{F}]$ satisfies the Hilbert–Mumford inequality for all* 1-PS *λ of* SL(V) *defined over k, i.e., for all such λ*

$$\mu^{\mathcal{L}(\nu)}([\mathcal{F}],\lambda) \geq 0.$$

Proof Replacing $\mathcal{F}$ by $\mathcal{F}(c)$ and ν by $\nu + c\cdot\mathbf{1}$, we may arrange that $|\nu| = 0$. In this process the semi-stability condition in Fil_k^K is not affected. Since we only test 1-PS λ factoring through $\mathrm{SL}(V)$, the same is true for the Hilbert–Mumford inequality concerning $\mu^{\mathcal{L}(\nu)}([\mathcal{F}],\lambda)$. Hence we may assume that $\mu(V,\mathcal{F}) = |\nu| = 0$.

Now $(V,\mathcal{F})$ is a semi-stable object of Fil_k^K if and only if $\langle \mathcal{F}, \mathcal{F}'_K\rangle \leq 0$ for all $\mathbb{Z}$-filtrations by subobjects $\mathcal{F}'$ of V, cf. Corollary 1.2.6 (in other words, $\mathcal{F}'$ is a $\mathbb{Z}$-filtration by k-subspaces of V). But such a filtration can be split, i.e., is of the form $\mathcal{F}' = \mathcal{F}_\lambda$ for some 1-PS λ of $\mathrm{GL}(V)$ defined over k. Conversely, any such λ defines a filtration of V by subobjects. From Lemmas 2.2.1 and 2.2.2 (in whose statements we replace k by K), we have

$$\langle \mathcal{F}, \mathcal{F}'_K\rangle = \langle \mathcal{F}, \mathcal{F}_\lambda\rangle = -\mu^{\mathcal{L}(\nu)}([\mathcal{F}],\lambda)\,. \tag{2.6}$$

Hence, if $(V,\mathcal{F})$ is a semi-stable object of Fil_k^K, we have $\mu^{\mathcal{L}(\nu)}([\mathcal{F}],\lambda) \geq 0$ for all 1-PS λ of $\mathrm{GL}(V)$ defined over k. Conversely, if this inequality holds for all 1-PS λ of $\mathrm{SL}(V)$ defined over k, it also holds for all 1-PS λ of $\mathrm{GL}(V)$ defined over k. Indeed, let us fix a common splitting for $\mathcal{F}$ and for λ. Then, after a permutation of basis vectors, $\mathcal{F} = \alpha^{\mathbf{V}}(\nu)$ and $\lambda \in \mathbb{Z}^n$. Write $\lambda = \lambda_0 + c\cdot\mathbf{1}$, where λ_0 is a $\mathbb{Q}$-1-PS of $\mathrm{SL}(V)$. Then

$$\begin{aligned}\langle \mathcal{F}, \mathcal{F}_\lambda\rangle &= \langle \mathcal{F}, \mathcal{F}_{\lambda_0}(c)\rangle = \langle \mathcal{F}, \mathcal{F}_{\lambda_0}\rangle + c\cdot(\nu,\mathbf{1})\\ &= \langle \mathcal{F}, \mathcal{F}_{\lambda_0}\rangle\,,\end{aligned} \tag{2.7}$$

since $|\nu| = 0$. Hence the converse also follows from Corollary 1.2.6. □

Variant 2.2.4 For the variant $\mathrm{Fil}_{\mathbb{F}_1}^k$, we fix a k-vector space V of dimension n and a line decomposition $\mathbf{V} = (V = V_1 \oplus \cdots \oplus V_n)$. We also fix $\nu \in (\mathbb{R}^n)_+$. Then the objects $(V,\mathbf{V},\mathcal{F})$ of $\mathrm{Fil}_{\mathbb{F}_1}^k$ of type ν are parametrized by the k-valued points of $\mathcal{F}(\nu)$. To test the semi-stability of $(V,\mathbf{V},\mathcal{F})$ in $\mathrm{Fil}_{\mathbb{F}_1}^k$, we have to consider all subspaces of V that are direct sums of lines V_i. Equivalently, we have to test filtrations of V corresponding to 1-PS of $\mathrm{GL}(V)$ factoring through the canonical maximal split torus T of $\mathrm{GL}(V)$ corresponding to the line decomposition $\mathbf{V}$ of V (the diagonal torus). The analogue of Theorem 2.2.3 in this context is therefore as follows.

Theorem 2.2.5 *Let $\mathbf{V} = (V = V_1 \oplus \cdots \oplus V_n)$ be a line decomposition of an n-dimensional k-vector space V. Let $\nu \in (\mathbb{Q}^n)_+$. Let $\mathcal{F}$ be a filtration of type ν of V and let $[\mathcal{F}] \in \mathcal{F}(\nu)(k)$ be the corresponding point. Then $(V,\mathbf{V},\mathcal{F})$ is a semi-stable object of $\mathrm{Fil}_{\mathbb{F}_1}^k$ if and only if $[\mathcal{F}]$ satisfies the Hilbert–Mumford inequality for all 1-PS λ of $\mathrm{SL}(V)$ factoring through the diagonal torus T_1 of $\mathrm{SL}(V)$, i.e., for all such λ*

$$\mu^{\mathcal{L}(\nu)}([\mathcal{F}],\lambda) \geq 0\,.$$

In particular, if k is algebraically closed, then $(V,\mathbf{V},\mathcal{F})$ is a semi-stable *object of* $\mathrm{Fil}^k_{\mathbb{F}_1}$ *if and only if the point* $[\mathcal{F}] \in \mathcal{F}(\nu)$ *is semi-stable for the action of* T_1 *on* $\mathcal{F}(\nu)$ *and for the* T_1*-linearized line bundle* $\mathcal{L}(\nu)$.

Proof The last statement follows from the Hilbert–Mumford criterion. □

Variant 2.2.6 To formulate a variant of the above theorem for Fil^r_k, fix vectors $\nu_1, \ldots, \nu_r \in (\mathbb{Q}^n)_+$. Then an r-tuple $(\mathcal{F}_1, \ldots, \mathcal{F}_r)$ of type $(\nu_1, \ldots, \nu_r)$ of the n-dimensional k-vector space V defines a point $[\mathcal{F}_\bullet] \in \prod_i \mathcal{F}(\nu_i)(k)$.

Theorem 2.2.7 *The object* $(V, \mathcal{F}_1, \ldots, \mathcal{F}_r)$ *of* Fil^r_k *is semi-stable if and only if the point* $[\mathcal{F}_\bullet]$ *of* $\prod_i \mathcal{F}(\nu_i)$ *satisfies the Hilbert–Mumford inequality for all* 1-PS λ *of* SL(V) *defined over k (w.r.t. the* SL(V)*-linearized line bundle* $\mathcal{L}(\nu_1) \boxtimes \cdots \boxtimes \mathcal{L}(\nu_r)$*).*

If k is algebraically closed, then these conditions are equivalent to the GIT-semi-stability of the point $[\mathcal{F}_\bullet]$ *for the action of* SL(V). □

Notes and References For the background on Geometric Invariant Theory, we refer to the classic [164]. The relationship (Theorem 2.2.3) between semi-stability of filtered vector spaces and that in GIT is due to Totaro [219], whose proof we have reproduced. It confirms a conjecture in [183], which was inspired by a related result of van der Put and Voskuil [220]. Totaro proved the above theorem for arbitrary reductive groups, cf. Part 2 of this monograph. The content of Theorem 2.2.7 is implicitly contained in the analysis of the Hilbert–Mumford criterion in this situation contained in [164], Ch. 4, §4. In fact, the formula in [164], Prop. 4.3. is exactly expressing the semi-stability of $(V, \mathcal{F}_1, \ldots, \mathcal{F}_r)$ in Fil^r_k.

3 The Harder–Narasimhan stratification

Let V be an n-dimensional k-vector space. Let $\nu \in (\mathbb{R}^n)_+$ and let $\mathcal{F} = \mathcal{F}(\nu)$ be the corresponding partial flag variety over Spec k, parametrizing $\mathbb{R}$-filtrations $\mathcal{F}$ of type ν of V, cf. Section 1. Let O be a complete discrete valuation ring that is a k-algebra. Set $S = \operatorname{Spec} O$, with generic point η and special point s. We denote the residue fields by $\kappa(\eta)$ and $\kappa(s)$.

Theorem 2.3.1 *Let* $f : S \longrightarrow \mathcal{F}$ *be a morphism of k-schemes. Let* $\mathcal{F}_\eta$*, resp.* $\mathcal{F}_s$*, denote the* $\mathbb{R}$*-filtration on* $V \otimes_k \kappa(\eta)$*, resp.* $V \otimes_k \kappa(s)$*, corresponding to the images* $f(\eta)$ *resp.* $f(s)$*.*

(a) *The* HN*-polygon of* $(V, \mathcal{F}_\eta)$ *lies below the* HN*-polygon of* $(V, \mathcal{F}_s)$ *and both polygons have the same endpoints. Equivalently, for the* HN*-vectors in* $(\mathbb{R}^n)_+$*, we have* $\mathrm{HN}(\mathcal{F}_\eta) \leq \mathrm{HN}(\mathcal{F}_s)$.

(b) *If* $\mathrm{HN}(\mathcal{F}_\eta) = \mathrm{HN}(\mathcal{F}_s)$*, then the* HN*-filtrations of* V *corresponding to* $\mathcal{F}_\eta$ *and to* $\mathcal{F}_s$ *coincide. Moreover, the types of the filtrations induced by* $\mathcal{F}_\eta$ *and* $\mathcal{F}_s$ *on each* ${}^\alpha\,\mathrm{gr}(V)$ *coincide.*

[Here we use the *dominance order* on $(\mathbb{R}^n)_+$, i.e.,

$$(x_1, \ldots, x_n) \le (x'_1, \ldots, x'_n) \Longleftrightarrow \sum\nolimits_{i=1}^r x_i \le \sum\nolimits_{i=1}^r x'_i$$

for all $r = 1, \ldots, n$, with equality for $r = n$.]

Proof (a) There are canonical isomorphisms of filtered vector spaces,

$$\begin{aligned} \bigwedge\nolimits^{\max}{}_{\kappa(\eta)}\big(V \otimes_k \kappa(\eta)\big) &= \big(\bigwedge\nolimits^{\max}{}_{O}\big(V \otimes_k O\big)\big) \otimes_O \kappa(\eta) \\ \bigwedge\nolimits^{\max}{}_{\kappa(s)}\big(V \otimes_k \kappa(s)\big) &= \big(\bigwedge\nolimits^{\max}{}_{O}\big(V \otimes_k O\big)\big) \otimes_O \kappa(s)\,. \end{aligned} \tag{3.1}$$

This yields that $\deg(V, \mathcal{F}_\eta) = \deg(V, \mathcal{F}_s)$, and shows that both HN-polygons have the same endpoints.

Similarly, let ${}^\bullet V(\eta)$ be the HN-filtration of V corresponding to $(V, \mathcal{F}_\eta)$. Then, for any $\alpha \in \mathbb{R}$, the previous argument shows that

$$\deg_{\mathcal{F}_\eta}({}^\alpha V) = \deg_{\mathcal{F}'_s}({}^\alpha V)\,. \tag{3.2}$$

Here ${}^\alpha V = {}^\alpha V(\eta)$ and $\mathcal{F}'_s$ denotes the filtration on ${}^\alpha V \otimes_k \kappa(s)$ which is the image via $O \longrightarrow \kappa(s)$ of the filtration on ${}^\alpha V \otimes_k O$ induced by the inclusion ${}^\alpha V \otimes_k O \subset V \otimes_k O$. Let $\mathcal{F}_s$ denote the filtration on ${}^\alpha V \otimes_k \kappa(s)$ induced from the filtration on $V \otimes \kappa(s)$. Then $({}^\alpha V, \mathcal{F}_s)$ is subobject of $(V, \mathcal{F}_s)$ and hence, by Proposition 1.3.4, the point $p(({}^\alpha V, \mathcal{F}_s)) = (\,\mathrm{rk}\,({}^\alpha V), \deg_{\mathcal{F}_s}({}^\alpha V))$ lies below the HN-polygon of $(V, \mathcal{F}_s)$. By Lemma 1.1.12, we have

$$\deg_{\mathcal{F}'_s}({}^\alpha V) \le \deg_{\mathcal{F}_s}({}^\alpha V)\,.$$

Hence the point $p(({}^\alpha V, \mathcal{F}'_s))$ also lies below the HN-polygon of $(V, \mathcal{F}_s)$. Letting α vary, we see that the HN-polygon of $(V, \mathcal{F}_\eta)$ lies below the HN-polygon of $(V, \mathcal{F}_s)$.

(b) Now assume $\mathrm{HN}(\mathcal{F}_\eta) = \mathrm{HN}(\mathcal{F}_s)$. Then we claim that the HN-filtration ${}^\bullet V(\eta)$, equipped with the filtration induced by $\mathcal{F}_s$ satisfies the conditions characterizing the HN-filtration ${}^\bullet V(s)$. By what was proved in (a), it only remains to prove the semi-stability of $({}^\alpha\mathrm{gr}(V), \mathcal{F}_s)$. Let $\alpha_1 > \alpha_2 > \cdots > \alpha_s$ be the jumps of ${}^\bullet V(\eta)$. Now $({}^{\alpha_1} V, \mathcal{F}_s)$ is semi-stable since any subobject V' of bigger slope would yield a point $p(V') = (\mathrm{rk}(V'), \deg_{\mathcal{F}_s}(V'))$ strictly above the line segment joining $(0, 0)$ with $(\mathrm{rk}({}^{\alpha_1} V), \deg_{\mathcal{F}_s}({}^{\alpha_1} V))$, which would mean that the HN-polygon of $(V, \mathcal{F}_s)$ lies strictly above the HN-polygon of $(V, \mathcal{F}_\eta)$, contrary to our assumption. Similarly one sees that $({}^{\alpha_i}\mathrm{gr}(V), \mathcal{F}_s)$ is semi-stable for $i = 2, \ldots, s$. The assertion follows.

Now let $g_{s,i}$, resp. $g_{\eta,i}$, be the multiplicity function of the filtration induced by $\mathcal{F}_s$, resp. $\mathcal{F}_\eta$, on ${}^{\alpha_i}\operatorname{gr}(V)$. We have to prove $g_{s,i} = g_{\eta,i}$ for all i. We have just seen above that $\deg_{\mathcal{F}_s}({}^{\alpha_i}\operatorname{gr}(V)) = \deg_{\mathcal{F}_\eta}({}^{\alpha_i}\operatorname{gr}(V))$ for all i. Therefore we also have $\deg_{\mathcal{F}_s}({}^{\alpha_i}V) = \deg_{\mathcal{F}_\eta}({}^{\alpha_i}V)$ for all i. Let $\{x_1 > \cdots > x_r\}$ be a set of rational numbers containing all jumps of $\mathcal{F}_s$ and $\mathcal{F}_\eta$. By Lemma 1.1.11, we get for all i

$$\sum_{l=1}^{r-1}(x_l - x_{l+1})\dim(\mathcal{F}_s^{x_l} \cap {}^{\alpha_i}V) = \sum_{l=1}^{r-1}(x_l - x_{l+1})\dim(\mathcal{F}_\eta^{x_l} \cap {}^{\alpha_i}V).$$

For all $x \in \mathbb{Q}$, the well-known degeneration rule for dimensions of intersections tells us that

$$\dim(\mathcal{F}_s^{x} \cap {}^{\alpha_i}V) \geq \dim(\mathcal{F}_\eta^{x} \cap {}^{\alpha_i}V).$$

Plugging this inequality in the previous equality yields for all x and i the following identity,

$$\dim(\mathcal{F}_s^{x} \cap {}^{\alpha_i}V) = \dim(\mathcal{F}_\eta^{x} \cap {}^{\alpha_i}V).$$

However, we have

$$\dim(\mathcal{F}_s^{x} \cap {}^{\alpha_i}V) = \sum_{y\geq x}\sum_{j\leq i} g_{s,i}(y)$$

and idem with η instead of s. From this, one deduces inductively that $g_{s,i}(x) = g_{\eta,i}(x)$ for all i and x. □

Our next aim is to enumerate all possible HN-filtrations, as $\mathcal{F}$ varies through $\mathcal{F}(\nu)$. Now the jumps of the filtration induced by $\mathcal{F}$ on each non-zero graded piece ${}^{\alpha}\operatorname{gr}(V)$ of the HN-filtration of $(V, \mathcal{F})$ are among the jumps of $\mathcal{F}$. This leads to the concept of the HN-types associated to ν. To formulate it, it is preferable to use the multiplicity function $g : \mathbb{R} \longrightarrow \mathbb{Z}_{\geq 0}$ associated to ν, cf. Section 1.

Definition 2.3.2 Let $g : \mathbb{R} \longrightarrow \mathbb{Z}_{\geq 0}$ be a multiplicity function. An HN-*type* associated to g is a tuple $(g_1, \ldots, g_s)$ of multiplicity functions which satisfies the following conditions.

(i) The integer $m_i = \sum_x g_i(x)$ is strictly positive, for each $i = 1, \ldots, s$.
(ii) $\sum_{i=1}^{s} g_i = g$.
(iii) $\frac{1}{m_1}\sum_x xg_1(x) > \frac{1}{m_2}\sum_x xg_2(x) > \cdots > \frac{1}{m_s}\sum_x xg_s(x)$.

We denote by $\Theta(g)$ or $\Theta(\nu)$ the set of HN-types associated to the multiplicity function g or the corresponding type ν. This set is finite. To an element $\theta \in \Theta(g)$ we associate its HN-*vector* $\mathrm{HN}(\theta) \in (\mathbb{R}^n)_+$ by setting

$$\mathrm{HN}((g_1, \ldots, g_s)) = \Big(\Big(\frac{1}{m_1}\sum_x xg_1(x)\Big)^{(m_1)}, \ldots, \Big(\frac{1}{m_s}\sum_x xg_s(x)\Big)^{(m_s)}\Big). \qquad (3.3)$$

Let $\mathcal{H}(\nu) = \mathcal{H}(g)$ be the set of HN-vectors thus obtained. We denote by θ_0 the distinguished element of $\Theta(g)$ consisting of the singleton g. It is the unique element with HN-vector equal to the scalar vector $\frac{|\nu|}{n} \cdot \mathbf{1}$.

Any $\mathcal{F} \in \mathcal{F}(\nu)(K)$ defines its HN-type $\theta_{\mathcal{F}}$ as follows. If $\alpha_1 > \cdots > \alpha_s$ are the jumps in the Harder–Narasimhan filtration of V, we let $\theta_{\mathcal{F}} = (g_1, \ldots, g_s)$, where g_i is the multiplicity function of the $\mathbb{R}$-filtration induced by $\mathcal{F}$ on ${}^{\alpha_i}\mathrm{gr}(V)$. Then the integer m_i is the size of the jump in α_i, i.e., the rank of ${}^{\alpha_i}\mathrm{gr}(V)$, and the sum $\sum_x x g_i(x)$ is the degree of ${}^{\alpha_i}\mathrm{gr}(V)$, and the quantities occurring in (iii) in Definition 2.3.2 are the slopes of ${}^{\alpha_i}\mathrm{gr}(V)$. Also, $\mathrm{HN}(\theta_{\mathcal{F}})$ is the Harder–Narasimhan vector of $(V, \mathcal{F})$, in the sense of Chapter I (3.1), page 25.

We now obtain disjoint decompositions into finitely many summands,

$$\mathcal{F}(\nu)(K) = \dot{\bigcup}_{\theta} \mathcal{F}(\nu)(K)_{\theta}, \quad \mathcal{F}(\nu)(K) = \dot{\bigcup}_{H} \mathcal{F}(\nu)(K)_{H} . \tag{3.4}$$

Here θ ranges over all HN-types attached to ν, and H over all HN-vectors. The first decomposition is a refinement of the second decomposition. The summand in the first decomposition corresponding to θ_0 coincides with the summand in the second decomposition corresponding to the HN-vector $\frac{|\nu|}{n} \cdot \mathbf{1}$, and both coincide with the semi-stable locus.

The next result therefore generalizes Corollary 2.1.3.

Theorem 2.3.3 *Let k be a finite field. Then for each* HN*-type θ attached to ν, there is a locally closed subset $\mathcal{F}(\nu)_{\theta}$ of $\mathcal{F}(\nu)$ with*

$$\mathcal{F}(\nu)(K)_{\theta} = \mathcal{F}(\nu)_{\theta}(K) .$$

Proof This follows in the following three steps. We let g be the multiplicity function corresponding to ν and $\theta = (g_1, \ldots, g_s)$. We adopt the notation of Definition 2.3.2.

(i) *There are only finitely many flags* $(0) \subset V_1 \subset V_2 \subset \cdots \subset V_s = V$ *of V by subspaces of dimension $m_1, m_1 + m_2, \ldots$*. This is obvious, since k is finite.

(ii) *Given a flag* $(0) \subset V_1 \subset V_2 \subset \cdots \subset V_s = V$ *as in* (i), *the set of $\mathcal{F} \in \mathcal{F}(\nu)$ which induces a filtration of type g_i on V_i/V_{i-1} is a locally closed subset of $\mathcal{F}(\nu)$*. This is easily reduced to the statement that for any $x \in \mathbb{R}$ and any subspace W of V, and any $c \in \mathbb{Z}_{\geq 0}$, the set

$$\{\mathcal{F} \in \mathcal{F}(\nu) \mid \dim \mathrm{gr}^x_{\mathcal{F}}(W) = c\} \tag{3.5}$$

is locally closed. It suffices to show that the set

$$\{\mathcal{F} \in \mathcal{F} \mid \dim \mathcal{F}^x \cap W \geq c\} \tag{3.6}$$

is closed for any $c \in \mathbb{Z}_{\geq 0}$. But this is well known and was in fact already used in the proof of Proposition 2.1.1.

(iii) Consider the locally closed subset Y of $\mathcal{F}(v)$ consisting of those $\mathcal{F}$ which induce for the given filtration $(0) \subset V_1 \subset \cdots \subset V_s = V$ as in (ii) a filtration of type g_i on V_i/V_{i-1} for $i = 1, \ldots, s$. We obtain a morphism

$$Y \longrightarrow \mathcal{F}(V_1, g_1) \times \cdots \times \mathcal{F}(V_s/V_{s-1}, g_s)\,. \tag{3.7}$$

Now the locus of $\mathcal{F}$ in $\mathcal{F}(v)$, where $V_\bullet$ is the HN-filtration of $(V, \mathcal{F})$ is the inverse image of $\mathcal{F}(g_1)^{\mathrm{ss}} \times \cdots \times \mathcal{F}(g_s)^{\mathrm{ss}}$. By Corollary 2.1.3, this is an open subset of Y. □

The locally closed subset $\mathcal{F}(v)_\theta$ is called the HN-*stratum of* $\mathcal{F}(v)$ corresponding to the HN-type θ.

We introduce a partial order on $\Theta(v)$ through the dominance order (defined right after Theorem 2.3.1) on their associated HN-vectors, cf. (3.3), page 43,

$$\theta \leq \theta' \Longleftrightarrow \mathrm{HN}(\theta) \leq \mathrm{HN}(\theta')\,. \tag{3.8}$$

Note that for all $\theta \in \Theta(v)$ we have $|\mathrm{HN}(\theta)| = |v|$. We also note that θ_0 is the unique minimal element of $\Theta(v)$.

Corollary 2.3.4 *Let k be finite. Each* HN-*stratum $\mathcal{F}(v)_\theta$ is invariant under the action of the finite group* GL$(V)(k)$*. Furthermore,*

$$\overline{\mathcal{F}(v)_\theta} \subset \bigcup\nolimits_{\theta' \geq \theta} \mathcal{F}(v)_{\theta'}\,.$$

The subset of $\mathcal{F}(v)$ with a given HN-*vector H is the scheme-theoretic disjoint sum of all* HN-*strata corresponding to the* HN-*types with* HN-*vector equal to H. In particular, there is a locally closed subset $\mathcal{F}(v)_H$ of $\mathcal{F}(v)$ such that $\mathcal{F}(v)_H(K) = \mathcal{F}(v)(K)_H$ for every field extension K of k, and*

$$\overline{\mathcal{F}(v)_H} \subset \bigcup\nolimits_{H' \geq H} \mathcal{F}(v)_{H'}\,.$$

Proof The union $X = \bigcup_{\theta' \geq \theta} \mathcal{F}(v)_{\theta'}$ is constructible. Thus we have for the topological closure $\overline{X}$ the identity $\overline{X} = \bigcup_{x \in X} \overline{\{x\}}$. To show that the union is closed, it suffices to check that the hypotheses of the valuative criterion (EGA II, Thm. 7.3.8) are satisfied. This follows from Theorem 2.3.1 (a).

On the other hand we have

$$\mathcal{F}(v)_H = \overline{\mathcal{F}(v)_H} \setminus \bigcup\nolimits_{\{\theta' \mid \mathrm{HN}(\theta') > H\}} \mathcal{F}(v)_{\theta'}\,. \tag{3.9}$$

Hence $\mathcal{F}(v)_H$ is a locally closed subset. Now Theorem 2.3.1(b) shows that $\mathcal{F}(v)_\theta$ is closed in $\mathcal{F}(v)_H$ for any $\theta \in \Theta(v)$ with $\mathrm{HN}(\theta) = H$. Hence $\mathcal{F}(v)_H$ is the disjoint sum of these closed subsets. □

Remark 2.3.5 The partial order on the set of HN-types is only a *preorder*, since $\leq + \geq$ does not imply $=$. Also, it should be pointed out that the two unions on the RHS in the inclusion relations appearing in Corollary 2.3.4 are identical.

Remarks 2.3.6 Both stratifications of $\mathcal{F}(\nu)$ (by HN-types and by HN-vectors) have bad properties.

(i) Let us consider the stratification by HN-vectors.

(a) The inclusion $\overline{\mathcal{F}(\nu)_H} \subset \bigcup_{H' \geq H} \mathcal{F}(\nu)_{H'}$ may be strict.
(b) The closure $\overline{\mathcal{F}(\nu)_H}$ may not be the union of strata.
(c) The implication

$$\overline{\mathcal{F}(\nu)_H} \cap \mathcal{F}(\nu)_{H'} \neq \emptyset \implies H \leq H'$$

cannot be reversed.

(ii) Let us consider the stratification by HN-types.

(a) The inclusion $\overline{\mathcal{F}(\nu)_\theta} \subset \bigcup_{\theta' \geq \theta} \mathcal{F}(\nu)_{\theta'}$ may be strict.
(b) The closure $\overline{\mathcal{F}(\nu)_\theta}$ may not be the union of strata.
(c) The implication

$$\overline{\mathcal{F}(\nu)_\theta} \cap \mathcal{F}(\nu)_{\theta'} \neq \emptyset \implies \theta \leq \theta'$$

cannot be reversed. Let us illustrate this by examples.

Example 2.3.7 Let $n = 4$. In this case it is easy to see that the strata by HN-vectors and by HN-types coincide. Let $\nu = (x_1 > x_2 > x_3 > x_4)$ with $|\nu| = 0$, $x_3 > 0$ and $x_1 + x_3 > 2x_2$. Let $\theta = (g_1, g_2, g_3), \theta' = (g'_1, g'_2, g'_3) \in \Theta(\nu)$ be the HN-types with

$$\mathrm{supp}(g'_1) = \{x_1, x_3\},\ \mathrm{supp}(g'_2) = \{x_2\},\ \mathrm{supp}(g'_3) = \{x_4\}$$

and

$$\mathrm{supp}(g_1) = \{x_2\},\ \mathrm{supp}(g_2) = \{x_3\},\ \mathrm{supp}(g_3) = \{x_1, x_4\}.$$

We have $\theta \leq \theta'$. A calculation shows that the corresponding strata are given by

$\mathcal{F}(\nu)_{\theta'} = \{\mathcal{F} \in \mathcal{F}(\nu) \mid \mathcal{F}^{x_1}, \mathcal{F}^{x_2}$ *are both not k-rational,* $\mathcal{F}^{x_3}$ *is k-rational and there is a 2-dim. k-rational subspace W with* $\mathcal{F}^{x_1} \subset W \subset \mathcal{F}^{x_3}\}$,

$\mathcal{F}(\nu)_\theta = \{\mathcal{F} \in \mathcal{F}(\nu) \mid \mathcal{F}^{x_1}, \mathcal{F}^{x_2}, \mathcal{F}^{x_3}$ *are all not k-rational,* $\mathcal{F}^{x_2}$ *contains a k-rational line* W_1 *and there is a 2-dim. k-rational subspace* W_2 *with* $\mathcal{F}^{x_1} \not\subset W_2$ *and* $W_1 \subset W_2 \subset \mathcal{F}^{x_3}\}$.

Let $\lambda, \mu \in \bar{k} \setminus k$ with $\lambda \neq \mu$. The filtration $\mathcal{F} \in \mathcal{F}(\nu)$ defined by

$$\mathcal{F}^{x_1} = \langle \lambda e_1 - \mu e_3 \rangle,\ \mathcal{F}^{x_2} = \langle \lambda e_1 + e_2, e_2 + \mu e_3 \rangle,\ \mathcal{F}^{x_3} = \langle e_1, e_2, e_3 \rangle$$

belongs to $\mathcal{F}(\nu)_{\theta'}$ but not to $\overline{\mathcal{F}(\nu)_\theta}$, since $\mathcal{F}^{x_2}$ does not contain a k-rational line. Thus statement (a) follows. On the other hand, the filtration $\mathcal{F}_\mu \in \mathcal{F}(\nu)$ defined by

$$\mathcal{F}_\mu^{x_1} = \langle \lambda e_2 + e_3 + \mu e_4 \rangle,\ \mathcal{F}_\mu^{x_2} = \langle e_1, \lambda e_2 + e_3 + \mu e_4 \rangle,\ \mathcal{F}_\mu^{x_3} = \langle e_1, e_2, e_3 + \mu e_4 \rangle$$

is in $\mathcal{F}(\nu)_\theta$ such that $\lim_{\mu \to 0} \mathcal{F}_\mu \in \mathcal{F}(\nu)_{\theta'}$. We conclude that $\mathcal{F}(\nu)_{\theta'} \cap \overline{\mathcal{F}(\nu)_\theta} \neq \emptyset$. Hence statement (b) follows.

Let $\theta'' = (g_1'', g_2'') \in \Theta(\nu)$ be the HN-type with $\mathrm{supp}(g_1'') = \{x_1, x_3\}$ and $\mathrm{supp}(g_2'') = \{x_2, x_4\}$. We have $\theta \leq \theta''$. The corresponding stratum is given by

$$\mathcal{F}(\nu)_{\theta''} = \{\mathcal{F} \in \mathcal{F}(\nu) \mid \mathcal{F}^{x_1}, \mathcal{F}^{x_2}, \mathcal{F}^{x_3} \textit{ are all not k-rational, there is a 2-dim. k-rational subspace } W \textit{ with } \mathcal{F}^{x_1} \subset W \subset \mathcal{F}^{x_3}\}.$$

Let $\mathcal{F} \in \overline{\mathcal{F}(\nu)_\theta}$. Then there are k-subspaces W_1, W_2 of V with $\dim W_1 = 1$, $\dim W_2 = 2$, $W_1 \subset W_2 \subset \mathcal{F}^{x_3}$ and $W_1 \subset \mathcal{F}^{x_2}$. Assume $\mathcal{F} \in \overline{\mathcal{F}(\nu)_\theta} \cap \mathcal{F}(\nu)_{\theta''}$. Then there is a k-rational subspace W of V with $\dim W = 2$ and $\mathcal{F}^{x_1} \subset W \subset \mathcal{F}^{x_3}$.

1st case: $W \neq W_2 \Rightarrow W_2 + W = \mathcal{F}^{x_3}$ is k-rational.

2nd case: $W = W_2 \Rightarrow \mathcal{F}^{x_2} = \mathcal{F}^{x_1} \oplus W_1 = W_2$ is k-rational.

Thus, in both cases we obtain a contradiction. Hence, $\mathcal{F}(\nu)_{\theta''} \cap \overline{\mathcal{F}(\nu)_\theta} = \emptyset$. Statement (c) follows.

Example 2.3.8 In general, the stratification by HN-type is strictly finer than the stratification by HN-vectors. An example is as follows, cf. [173], Example 17, p. 221. Let $n = 5$, and consider $\nu = (x_1 > x_2 > x_3 > x_4 > x_5)$ with $x_1 + x_4 = x_2 + x_3 = -\frac{1}{2}x_5$. Then $\theta_1 = (g_1, g_2)$ and $\theta_2 = (g_1', g_2')$ with

$$\mathrm{supp}(g_1) = \{x_1, x_4\}\,,\ \mathrm{supp}(g_2) = \{x_2, x_3, x_5\}\,,$$

resp.

$$\mathrm{supp}(g_1') = \{x_2, x_3\}\,,\ \mathrm{supp}(g_2') = \{x_1, x_4, x_5\}\,,$$

are two different HN-types with identical HN-vectors.

Our next purpose is to determine the structure of the locally closed subsets $\mathcal{F}(\nu)_\theta$. Let $\theta = (g_1, \ldots, g_s)$ and put as before $m_i = \sum_x g_i(x)$ for $i = 1, \ldots, s$. We fix a flag $V_\bullet$ of k-subspaces $(0) \subset V_1 \subset \cdots \subset V_s = V$, where the dimension of V_i is equal to $m_1 + \cdots + m_i$. Let $\mathcal{F}(\nu)_{\theta, V_\bullet}$ be the locally closed subset of $\mathcal{F}(\nu)$ consisting of $\mathcal{F}$ such that $V_\bullet$ is the HN-flag of $(V, \mathcal{F})$. More precisely, $\mathcal{F}$ induces on $\mathrm{gr}_i(V)$ a semi-stable filtration of type g_i, for $i = 1, \ldots, s$.

Theorem 2.3.9 *Let k be finite. The morphism*

$$\pi : \mathcal{F}(\nu)_{\theta, V_\bullet} \longrightarrow \mathcal{F}(g_1)^{\mathrm{ss}} \times \cdots \times \mathcal{F}(g_s)^{\mathrm{ss}}$$

is a vector bundle of rank $\ell(w_\theta)$. Here w_θ is the unique element of minimal length in S_n with the property that

$$w_\theta \cdot \nu = x^\theta .$$

Here

$$\nu = (x_1^{(g(x_1))}, \ldots, x_r^{(g(x_r))}) ,$$

where $x_1 > \cdots > x_r$, and

$$x^\theta = (x_1^{(g_1(x_1))}, \ldots, x_r^{(g_1(x_r))}, x_1^{(g_2(x_1))}, \ldots, x_r^{(g_2(x_r))}, \ldots, x_1^{(g_s(x_1))}, \ldots, x_r^{(g_s(x_r))}) .$$

Let us fix a basis $e_1, \ldots, e_n$ of V and consider the flag $V_\bullet^{\mathrm{std}}$ extracted from the standard flag,

$$V_i^{\mathrm{std}} = \mathrm{span}\{e_1, \ldots, e_{m_1+\cdots+m_i}\},\ i = 1, \ldots, s . \tag{3.10}$$

Let P_θ denote the standard parabolic subgroup of $G = \mathrm{GL}(V) = \mathrm{GL}_n$ which stabilizes this flag. Then we see that $\mathcal{F}(\nu)_\theta$ is the set of filtrations $\mathcal{F}$ such that the HN-filtration of $(V, \mathcal{F})$ is a flag of type $V_\bullet^{\mathrm{std}}$, i.e., of the form $hV_\bullet^{\mathrm{std}}$ for some $h \in G(k)$, which then defines a unique element in $G(k)/P_\theta(k) = (G/P_\theta)(k)$. Hence we obtain as a corollary to the above theorem the following statement.

Corollary 2.3.10 *Let k be a finite field and let $\theta = (g_1, \ldots, g_s) \in \Theta(g)$. Then $\mathcal{F}(\nu)_\theta$ is a vector bundle of rank $\ell(w_\theta)$ over the finite disjoint sum*

$$\coprod_{h \in G/P_\theta(k)} \mathcal{F}(g_1)^{\mathrm{ss}} \times \cdots \times \mathcal{F}(g_s)^{\mathrm{ss}} . \qquad \square$$

Remark 2.3.11 The morphism

$$\mathcal{F}(\nu)_\theta \longrightarrow \coprod_{h \in G/P_\theta(k)} \mathcal{F}(g_1)^{\mathrm{ss}} \times \cdots \times \mathcal{F}(g_s)^{\mathrm{ss}} \tag{3.11}$$

is equivariant for the action of $G(k)$. Here $G(k)$ acts as follows on the target space. Choose a set of representatives $\{h_i \mid i \in I\}$ for $G(k)/P_\theta(k)$. Then $h \in G(k)$ takes the point $(\mathcal{F}_1, \ldots, \mathcal{F}_s)$ in the copy of $\mathcal{F}(g_1)^{\mathrm{ss}} \times \cdots \times \mathcal{F}(g_s)^{\mathrm{ss}}$ corresponding to h_i to the point

$$\bar{p} \cdot (\mathcal{F}_1, \ldots, \mathcal{F}_s)$$

in the copy corresponding to h_j. Here j and h_j and $p \in P_\theta(k)$ are determined by the identity

$$h \cdot h_i = h_j \cdot p ,$$

and $\bar{p} \in \mathrm{GL}_{m_1}(k) \times \cdots \times \mathrm{GL}_{m_s}(k)$ is the image of p in the Levi factor group of P_θ.

Proof Let Y_θ be the locally closed subset of $\mathcal{F}(\nu)$ of filtrations $\mathcal{F}$ which induce on $V_i^{\mathrm{std}}/V_{i-1}^{\mathrm{std}}$ a filtration of type g_i for each $i = 1, \ldots, s$. It suffices to prove that the morphism

$$\pi\colon Y_\theta \longrightarrow \mathcal{F}(g_1) \times \cdots \times \mathcal{F}(g_s) \tag{3.12}$$

is a vector bundle of rank $\ell(w_\theta)$. This is true for any ground field k.

There are two issues. One is to show that the morphism (3.12) above is a vector bundle. The other issue is to determine the rank of the vector bundle.

We show that (3.12) is a vector bundle only in a simple, but representative, case. The general case works the same way but requires more notation. Namely, let us consider the case where $r = s = 2$. Hence $\mathcal{F}(\nu)$ can be identified with the Grassmannian of subspaces $\mathcal{F}$ of dimension n_1 of V. Let $W = V_{m_1}^{\mathrm{std}}$. We then have to prove that the space Y of subspaces $\mathcal{F}$ with intersection $\mathcal{F} \cap W$ of fixed dimension $g_1(x_1)$ is a vector bundle over the product $\mathcal{F}_1 \times \mathcal{F}_2$ of the Grassmannian $\mathcal{F}_1$ of subspaces of dimension $g_1(x_1)$ of W and the Grassmannian $\mathcal{F}_2$ of subspaces of dimension $n_1 - g_1(x_1) = g_2(x_1)$ of V/W. Let $\mathcal{F}_0 \in Y$, and denote its image in V/W by $\bar{\mathcal{F}}_0$. We choose a basis $\{v_1, \ldots, v_{n_1}\}$ of $\mathcal{F}_0$ by first choosing $g_1(x_1)$ vectors which form a basis of $\mathcal{F}_0 \cap W$ and then completing it to a basis of $\mathcal{F}_0$. Then all $\mathcal{F}$ in Y with $\mathcal{F} \cap W = \mathcal{F}_0 \cap W$ and with image $\bar{\mathcal{F}} = \bar{\mathcal{F}}_0$ in V/W are of the form $\mathcal{F} = \mathrm{span}\{v_1, \ldots, v_{g_1(x_1)}, v_{g_1(x_1)+1} + \varphi(v_{g_1(x_1)+1}), \ldots, v_{n_1} + \varphi(v_{n_1})\}$, for a linear map $\varphi : \bar{\mathcal{F}}_0 \longrightarrow W$, well-defined modulo $\mathcal{F}_0 \cap W$. Hence π is the vector bundle $\mathrm{Hom}(\mathcal{F}_2, W/\mathcal{F}_1)$ over $\mathcal{F}_1 \times \mathcal{F}_2$, where $\mathcal{F}_1$ resp. $\mathcal{F}_2$ are the universal bundles over $\mathcal{F}_1$ resp. $\mathcal{F}_2$.

Now let us determine the rank of the vector bundle. For this we may assume k algebraically closed. Let y be a closed point in $\mathcal{F}(g_1) \times \cdots \times \mathcal{F}(g_s)$. Then the unipotent radical $R_u(P_\theta)$ acts on the fiber $\pi^{-1}(y)$ and this action is transitive. Denote by $Q = Q_{(n_1, \ldots, n_r)}$ the standard parabolic subgroup of $G = \mathrm{GL}_n$ with block sizes $n_1 = g(x_1), \ldots, n_r = g(x_r)$. Then

$$G/Q \simeq \mathcal{F}(g)\,, \tag{3.13}$$

i.e., Q is the stabilizer of standard flag $\mathcal{F}_\bullet$,

$$(0) \subset V_{n_1}^{\mathrm{std}} \subset V_{n_1+n_2}^{\mathrm{std}} \subset \cdots \subset V\,.$$

Choose $x \in \pi^{-1}(y)$ and let $h \in G$ with $h\mathcal{F}_\bullet = x$. Then the stabilizer at x of the action of $R_u(P_\theta)$ is equal to

$$R_u(P_\theta) \cap hQh^{-1}\,. \tag{3.14}$$

Now we have

$$P(k)\backslash G(k)/Q(k) = W_P\backslash W/W_Q\,. \tag{3.15}$$

Here $W = S_n$ is the Weyl group of permutation matrices of GL_n, and $W_P = S_{m_1} \times S_{m_2} \times \cdots \times S_{m_s}$ is the Weyl group of the standard Levi component L_P of $P = P_\theta$ (with S_{m_1} permuting the first m_1 basis vectors $e_1, \ldots, e_{m_1}$, and S_{m_2} permuting the next m_2 basis vectors $e_{m_1+1}, \ldots, e_{m_2}$, etc.), and $W_Q = S_{n_1} \times \cdots \times S_{n_r}$ is the Weyl group of the standard Levi component L_Q of Q. It is obvious that the dimension of the intersection (3.14), page 49 is unchanged when h is multiplied on the left by an element of $P(k)$ and on the right by an element $Q(k)$. Hence the dimension of this intersection only depends on the double coset of h in (3.15), page 49. But from the type $(g_1, \ldots, g_s)$ of the induced filtration on $(\mathrm{gr}_1(V), \ldots, \mathrm{gr}_s(V))$ we see that the image of h in $W_P \backslash W / W_Q$ is represented by w_θ. The assertion on the rank of the vector bundle now follows from the following lemma. □

Lemma 2.3.12 *Let $P = P_{(m_1,\ldots,m_s)}$ and $Q = Q_{(n_1,\ldots,n_r)}$ be two standard parabolic subgroups of* GL_n. *Let w be a permutation matrix which is of minimal length in its double coset in*

$$W_P \backslash W / W_Q = (S_{m_1} \times \cdots \times S_{m_s}) \backslash S_n / (S_{n_1} \times \cdots \times S_{n_r}) .$$

Then

$$R_u(P)/(R_u(P) \cap wQw^{-1}) \simeq \mathbb{A}^{\ell(w)} .$$

Remark 2.3.13 We are using here that there is a unique element w of minimal length in each double coset. It is characterized by the following two properties:
(a) w^{-1} is monotone increasing on each interval

$$[1, m_1], [m_1 + 1, m_1 + m_2], \ldots, [m_1 + \cdots + m_{s-1} + 1, n] ;$$

(b) w is monotone increasing on each interval

$$[1, n_1], [n_1 + 1, n_1 + n_2], \ldots, [n_1 + \cdots + n_{r-1} + 1, n] .$$

Proof Both $R_u(P)$ and $R_u(P) \cap wQw^{-1}$ are unipotent subgroups of GL_n stable under the action of the diagonal torus. Hence both subgroups are the direct product of their root subgroups in any order [25], 14.4. Hence

$$R_u(P) = (R_u(P) \cap wQw^{-1}) \times \prod_{\alpha \in \Psi} U_\alpha . \tag{3.16}$$

Here Ψ is the set of roots occurring in $R_u(P)$, which do not occur in wQw^{-1}, and $U_\alpha \simeq \mathbb{A}^1$ is the corresponding root subgroup. Let Ψ_P resp. Ψ_Q be the set of roots occurring in $R_u(P)$ resp. $R_u(Q)$. Then

$$\Psi = \Psi_P \cap w(-\Psi_Q) . \tag{3.17}$$

Consider the set of roots,

$$\Phi_w = \{\alpha \mid \alpha > 0, w^{-1}\alpha < 0\} . \tag{3.18}$$

Then it is well-known (cf. [122], 1.7) that $|\Phi_w| = \ell(w)$. Hence it remains to prove the following assertion.

Claim $\Phi_w = \Psi$.

Let $\alpha \in \Psi$. Then $\alpha > 0$ and $\alpha = w \cdot \alpha'$, where $-\alpha' \in \Psi_Q$. Hence $w^{-1}\alpha = \alpha' < 0$, so that $\alpha \in \Phi_w$. Conversely, let $\alpha \in \Phi_w$. Now by property (b) above, w^{-1} takes a positive root occurring in the standard Levi subgroup L_P into a positive root. Hence α cannot occur in L_P, i.e., $\alpha \in \Psi_P$. By property (a), w takes a positive root occurring in the standard Levi subgroup L_Q into a positive root. Hence $\beta = -w^{-1}\alpha$ cannot occur in L_Q, i.e., $\beta \in \Psi_Q$. But then $\alpha \in w(-\Psi_Q)$, hence $\alpha \in \Psi$. □

Variant 2.3.14 Let $\mathbf{V} = (V = V_1 \oplus \cdots \oplus V_n)$ be a line decomposition. The analogue of Corollary 2.3.10 in the context of $\mathrm{Fil}^k_{\mathbb{F}_1}$ is that, for any $\theta = (g_1, \ldots, g_s) \in \Theta(v)$, the stratum $\mathcal{F}(v)_\theta$ is, in an S_n-equivariant way, a vector bundle of rank $\ell(w_\theta)$ over the disjoint union

$$\coprod_{W/W_{P_\theta}} \mathcal{F}(g_1)^{ss} \times \cdots \times \mathcal{F}(g_s)^{ss} . \qquad \square$$

Notes and References Theorem 2.3.1 is the analogue of Shatz's specialization theorem for vector bundles [200]. The HN-stratification here is analogous to the HN-stratification of the space of vector bundles on a compact Riemann surface [105, 9]. In that case also, each stratum is a vector bundle over the product of moduli spaces of semi-stable vector bundles of smaller rank. That stratification has somewhat better properties. More precisely, if the genus of the Riemann surface is equal to one, then the closure of an HN-stratum corresponding to the HN-index μ is equal to the union of all HN-strata over all HN-indices μ' with $\mu' \geq \mu$, cf. [83]. If the genus of the Riemann surface is at least 2, then the closure of a HN-stratum is generally not a union of HN-strata, but if $\mu' \geq \mu$, and μ' and μ are adjacent, then the closure of the HN-stratum for μ' meets the HN-stratum for μ, cf. [83]. The corresponding statement is not true in our context: the indices θ and θ'' in Example 2.3.7 are adjacent but $\mathcal{F}(v)_{\theta''} \cap \overline{\mathcal{F}(v)_\theta} = \emptyset$.

4 Period domains over $\mathbb{F}_1$ and thin cells

In this section we let k be an arbitrary field. Let V be a k-vector space of dimension n and let $\mathbf{V} = (V = V_1 \oplus \cdots \oplus V_n)$ be a line decomposition of

V. Put $G := \mathrm{GL}(V)$. Denote by $T \subset G$ be the diagonal torus associated to the decomposition $\mathbf{V}$. Let $\nu = (\nu_1, \ldots, \nu_n) \in (\mathbb{Z}^n)_+$. We again regard ν as an element of $X^*(T)$. Let $E(\nu)$ be the irreducible representation of G of highest weight ν, cf. [125]. Recall that it is defined as the unique irreducible G-submodule of

$$\mathrm{Sym}^{\nu_1-\nu_2}(V) \otimes \mathrm{Sym}^{\nu_2-\nu_3}(\textstyle\bigwedge^2 V) \otimes \cdots \otimes \mathrm{Sym}^{\nu_{n-1}-\nu_n}(\textstyle\bigwedge^{n-1} V) \otimes \mathrm{Sym}^{\nu_n}(\textstyle\bigwedge^n V)$$

which contains the highest weight vector

$$e_1^{\nu_1-\nu_2} \otimes (e_1 \wedge e_2)^{\nu_2-\nu_3} \otimes \cdots \otimes (e_1 \wedge \cdots \wedge e_n)^{\nu_n}.$$

Alternatively, it may be described as the unique irreducible G-submodule of the induced representation [125]

$$\mathrm{ind}_{B^-}^G(\nu) := \{f \in \Gamma(G, \mathcal{O}) \mid f(g \cdot b) = \nu(b^{-1}) \cdot f(g) \ \ \forall g \in G, b \in B^-\}.$$

Here, B^- denotes the opposite Borel subgroup of lower triangular matrices. In the following we let E be any algebraic G-representation with

$$E(\nu) \subset E \subset \mathrm{ind}_{B^-}^G(\nu).$$

(In the notation of Jantzen [125] we have $L(\nu) = E(\nu)$ and $H^0(\nu) = \mathrm{ind}_{B^-}^G(\nu)$.) We point out that if $\mathrm{char}\, k = 0$ then $E(\nu) = E = \mathrm{ind}_{B^-}^G(\nu)$.

We have the weight space decomposition of E for the action of T (cf. [29]) (we allow here for zero summands),

$$E = \bigoplus_{\beta \in X^*(T) \cap \Delta(\nu)} E_\beta \,. \tag{4.1}$$

Here

$$\Delta(\nu) = \mathrm{conv}\{w\nu \mid w \in W\} \tag{4.2}$$

denotes the convex hull in $\mathbb{R}^n = X^*(T) \otimes \mathbb{R}$ of the orbit of ν under the Weyl group $W = S_n$. Recall that $\dim E_{w\nu} = 1$, $\forall w \in W$, and that

$$E_\nu = E^{R_u(B)} = k \cdot (e_1^{\nu_1-\nu_2} \otimes (e_1 \wedge e_2)^{\nu_2-\nu_3} \otimes \cdots \otimes (e_1 \wedge \cdots \wedge e_n)^{\nu_n}) \,, \tag{4.3}$$

since this property holds for $E(\nu)$ and for $\mathrm{ind}_{B^-}^G(\nu)$, cf. [125], p. 200. Here $R_u(B)$ denotes the unipotent radical of the Borel group of upper triangular matrices in $\mathrm{GL}(V)$. The weight space E_ν defines a point $[E_\nu]$ of $\mathbb{P}(E)(k)$, and one sees easily that the stabilizer of this point under the action of G on $\mathbb{P}(E)$ is equal to the standard parabolic subgroup

$$Q = Q_{(g(x_1), g(x_2), \ldots, g(x_r))} \,. \tag{4.4}$$

Here g denotes the multiplicity function for ν and $\mathrm{supp}\, g = \{x_1 > x_2 > \cdots > x_r\}$. In other words, $\nu = (x_1^{(g(x_1))}, \ldots, x_r^{(g(x_r))})$. Hence we have an identification

$$\mathcal{F}(\nu) = G/Q \,. \tag{4.5}$$

Furthermore, the line bundle $\mathcal{L}(v)$ is equal to the homogeneous line bundle on G/Q which corresponds to the character of Q defined by the composition of homomorphisms,

$$Q_{(g(x_1),\ldots,g(x_r))} \longrightarrow \mathrm{GL}_{g(x_1)} \times \cdots \times \mathrm{GL}_{g(x_r)} \xrightarrow{(\det^{-x_1},\ldots,\det^{-x_r})}$$
$$\mathbb{G}_m \times \cdots \times \mathbb{G}_m \xrightarrow{\text{product}} \mathbb{G}_m\,. \tag{4.6}$$

This follows from the formula (2.4), page 38. On the other hand, let P be the stabilizer of the line E_v in $\mathrm{GL}(E)$. The homogeneous line bundle $O(1)$ on $\mathbb{P}(E) = \mathrm{GL}(E)/P$ is defined by the inverse of the character by which the parabolic P acts on the highest weight space, cf. [125], Part II, 2.16,

$$P \longrightarrow \mathbb{G}_m\,,\; g \longmapsto (g|E_v)^{-1}\,. \tag{4.7}$$

It follows that under the natural map $Q \to P$, this character induces just the character (4.6) above. We deduce

$$\mathcal{L}(v) = O(1)|_{\mathcal{F}(v)} \tag{4.8}$$

as homogeneous line bundles for the action of $G = \mathrm{GL}(V)$.

This identification will allow us now to introduce new stratifications of $\mathcal{F}(v)$. Let K/k be a field extension. If $x \in \mathcal{F}(v)(K)$, we let $\tilde{x} \in E \otimes_k K \setminus \{0\}$ be a representative of x and let

$$\tilde{x} = \sum\nolimits_\beta \tilde{x}_\beta \tag{4.9}$$

be the eigenspace decomposition of $\tilde{x}$ with respect to (4.1), page 52. The summands $\tilde{x}_\beta$ are the generalized *Plücker coordinates* of x. They are uniquely determined up to a common scalar. Let

$$S_x = \{\beta \in X^*(T) \mid \tilde{x}_\beta \neq 0\} \tag{4.10}$$

be the set of non-vanishing Plücker coordinates for x. This is obviously independent of the choice of the representative. Set

$$\Delta_x := \mathrm{conv}(S_x) \subset \mathbb{R}^n\,. \tag{4.11}$$

Definition 2.4.1 A *v-hypersimplex* in $\mathbb{R}^n$ is a polytope Δ with vertices in Wv such that any edge of Δ is parallel to some root $e_i - e_j \quad (i \neq j)$. Here $e_1, \ldots, e_n$ denotes the standard basis of $\mathbb{R}^n$.

In the following we denote by

$$\langle\;,\;\rangle : X_*(T) \otimes \mathbb{R} \times X^*(T) \otimes \mathbb{R} \longrightarrow \mathbb{R}$$

the non-degenerate $\mathbb{R}$-bilinear pairing which is defined on $X_*(T) \times X^*(T)$ by the identity $\chi(\lambda(t)) = t^{\langle\lambda,\chi\rangle}$, where $t \in \mathbb{G}_m(\overline{k})$.

Proposition 2.4.2 *The convex set* Δ_x *is a* ν*-hypersimplex, for any* $x \in \mathcal{F}(\nu)(K)$.

Proof Let β be a vertex of Δ_x. Then there is an affine hyperplane $H \subset \mathbb{R}^n$ with $H \cap \Delta_x = \{\beta\}$, such that Δ_x is contained in one of the two halfspaces defined by H. Let $H = H_{\lambda,r}$ be given by

$$H_{\lambda,r} = \{x \in \mathbb{R}^n \mid \langle \lambda, x\rangle = r\},$$

where $\lambda \in X_*(T)$ and $r \in \mathbb{Z}$. We may suppose that $\langle \lambda, \beta' \rangle > r$ for all $\beta' \in S_x$, $\beta' \neq \beta$. Write

$$\tilde{x} = \tilde{x}_\beta + \sum_{\beta \neq \beta'} \tilde{x}_{\beta'}.$$

Then

$$\lim_{t \to 0} \lambda(t) \cdot \tilde{x} = \tilde{x}_\beta.$$

It follows that

$$x' := \lim_{t \to 0} \lambda(t) \cdot x$$

is a fixed point in G/Q for the action of T with $\Delta_{x'} = \{\beta\}$. But these fixed points in G/Q are given by the finitely many points wQ/Q, where $w \in W/W_Q$. It follows that $\beta = w \cdot \nu$ for some $w \in W/W_Q$.

Let ϵ be an edge of Δ_x. Denote by β and γ the vertices of ϵ. Let

$$T_\epsilon := \ker(\beta - \gamma)^\circ$$

be the identity component in $\mathrm{Ker}(\beta - \gamma)$. This is a subtorus of codimension 1 in T. Let $H = H_{\mu,s} \subset \mathbb{R}^n$ be an affine hyperplane with $H \cap \Delta_x = \epsilon$. It follows that $x'' := \lim_{t \to 0} \mu(t) \cdot x$ is a fixed point for the action of T_ϵ. Without loss of generality we may suppose that $x'' = x$. Let Φ be the set of roots of G. Denote by Φ_Q the set of roots appearing in Q. Write $x \in G/Q$ as $x = g \cdot Q$, where

$$g = \prod_{\alpha \in \Phi \setminus \Phi_Q} u_\alpha.$$

Here the u_α are elements of the one-dimensional root groups $U_\alpha \subset G$. The property that x is fixed by T_ϵ is equivalent to the condition that $t \cdot g \cdot Q = g \cdot Q$ or equivalently that $t \cdot g \cdot t^{-1} \cdot Q = g \cdot Q$ for all $t \in T_\epsilon$. Since $\prod_{\alpha \in \Phi \setminus \Phi_Q} U_\alpha$ is stabilized by T we get the identity $t \cdot g \cdot t^{-1} = g$, for all $t \in T_\epsilon$. This means that $t u_\alpha t^{-1} = u_\alpha$, $\forall \alpha \in \Phi \setminus \Phi_Q$, $\forall t \in T_\epsilon$. Since T_ϵ is of codimension 1 in T there is therefore exactly one root $\alpha \in \Phi \setminus \Phi_Q$ with $u_\alpha \neq 1$. It follows that $\beta - \gamma$ is a multiple of α. □

The *list* associated to $x \in \mathcal{F}(\nu)(K)$ is the set

$$L_x = \{\beta \in W\nu \mid \tilde{x}_\beta \neq 0\}. \tag{4.12}$$

By the above proposition the ν-hypersimplex Δ_x is completely determined by L_x, as follows.

Corollary 2.4.3 *For $x, x' \in \mathcal{F}(\nu)(K)$ we have*

$$\Delta_x = \Delta_{x'} \Longleftrightarrow L_x = L_{x'} \ .$$

Proof Indeed, L_x is the set of vertices of Δ_x. □

The hypersimplices Δ_x or, equivalently, the lists are obviously invariant with respect to the action of T on $\mathcal{F}(\nu)$. The following proposition explains which T-orbits are contained in the closure of a given T-orbit.

Proposition 2.4.4 *Let $x \in \mathcal{F}(\nu)$ and let $y \in \overline{T \cdot x}$ be a point in the closure of the T-orbit of x. Then Δ_y is a face of Δ_x, and the correspondence $y \longmapsto \Delta_y$ defines a bijection between the T-orbits contained in $\overline{T \cdot x}$ and the faces of Δ_x.*

Proof Let $\lambda \in X_*(T)$ be a 1-PS and let $t \in T$ with $\lim_{s \to 0} \lambda(s) \cdot t \cdot x = y$. The existence of λ and t follows from the Bruhat decomposition. Set

$$r := \inf\{\langle \lambda, \beta \rangle \mid \beta \in S_x\}.$$

It follows that $H = H_{\lambda,r}$ is an affine hyperplane with $H \cap \Delta_x = \Delta_y$ and such that Δ_x is contained in one of the two halfspaces defined by H. Thus Δ_y is a face of Δ_x.

Conversely, let $\Delta \subset \Delta_x$ be a face. Let $H = H_{\lambda,r}$ be a hyperplane with $\Delta = H \cap \Delta_x$. Then $\Delta = \Delta_y$ for $y := \lim_{s \to 0} \lambda(s) \cdot x$. Hence every face Δ is of the form Δ_y for some $y \in \overline{T \cdot x}$. To prove the injectivity of the map $y \mapsto \Delta_y$, let $y_1, y_2 \in \overline{T \cdot x}$ be two points with $\Delta_{y_1} = \Delta_{y_2}$. Write $y_i = \lim_{s \to 0} \lambda_i(s) \cdot t_i \cdot x$, where $t_i \in T$ and $\lambda_i \in X_*(T)$ for $i = 1, 2$. Since T is commutative we may write $y_i = t_i \cdot \lim_{s \to 0} \lambda_i(s) \cdot x$. But $\Delta_{y_1} = \Delta_{y_2}$, thus the limits $\lim_{s \to 0} \lambda_i(s) \cdot x$ coincide for $i = 1, 2$. Hence y_1 and y_2 lie in the same T-orbit. □

Example 2.4.5 Let $\nu = (1^{(2)}, 0^{(2)})$. Then $\mathcal{F}(\nu) = \mathbf{Gr}_2(V)$ is the Grassmannian of 2-planes in the 4-dimensional vector space V. Then $E(\nu) = E = \mathrm{ind}_{B^-}^{G}(\nu) = \bigwedge^2 V$. For some "generic" x we have

$$\Delta_x = \Delta(\nu) = \mathrm{conv}\{e_J \mid |J| = 2\} \ . \tag{4.13}$$

Here $e_J = e_{i,j} = e_i + e_j$ for $J = \{i, j\} \subset \{1, \ldots, 4\}$. The ν-hypersimplex of a "generic" x has the following shape.

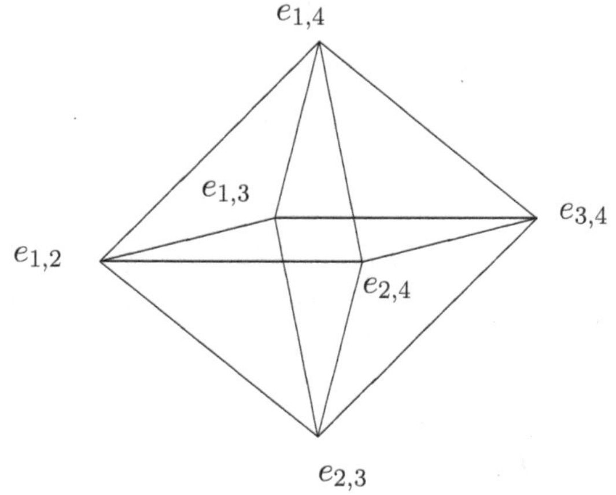

Proposition 2.4.4 says that a geometric point x' of $\mathbf{Gr}_2(V)$ with

$$\Delta_{x'} = \operatorname{conv}\{e_{1,2}, e_{1,3}, e_{2,4}, e_{3,4}\}$$

does not lie in the closure of the T-orbit of a "generic" point x. We want to reconsider this statement in this specific example. We represent a geometric point $x \in \mathbf{Gr}_2(V)$ as a 4×2-matrix with coefficients in $\overline{k}$. Consider the "generic" point x given by the matrix

$$\begin{pmatrix} 1 & 0 \\ 0 & 1 \\ x_3 & y_3 \\ x_4 & y_4 \end{pmatrix}$$

with $x_3, x_4, y_3, y_4 \neq 0$ and $x_3y_4 - x_4y_3 \neq 0$. Let

$$\lambda(t) = \operatorname{diag}(t^{r_1}, t^{r_2}, t^{r_3}, t^{r_4}).$$

Assume first that $r_1 \leq r_2 \leq r_3 \leq r_4$. Let $x' := \lim_{t\to 0} \lambda(t) \cdot x$. Then

$$x' = \lim_{t\to 0} \begin{pmatrix} 1 & 0 \\ 0 & 1 \\ t^{r_3-r_1}x_3 & t^{r_3-r_2}y_3 \\ t^{r_4-r_1}x_4 & t^{r_4-r_2}y_4 \end{pmatrix} = \begin{pmatrix} 1 & 0 \\ 0 & 1 \\ 0^{r_3-r_1}x_3 & 0^{r_3-r_2}y_3 \\ 0^{r_4-r_1}x_4 & 0^{r_4-r_2}y_4 \end{pmatrix}.$$

Here we follow the convention that $0^0 = 1$. It follows that if $e_{1,4} \notin L_{x'}$, then we also must have $e_{2,4} \notin L_{x'}$, hence $\Delta_{x'} \neq \operatorname{conv}\{e_{1,2}, e_{1,3}, e_{2,4}, e_{3,4}\}$. Something similar holds if the chain of inequalities between the r_i does not hold (let $\{r_i, r_j\}$ be two of the smallest exponents; then represent the point x by a matrix where the unit matrix is in rows i and j).

Corollary 2.4.6 $L_x = \{\beta \in W\nu \mid [E_\beta] \in \overline{T \cdot x}\}$.

Proof Let $\beta \in W\nu$ with $y = [E_\beta] \in \overline{T \cdot x}$. By semi-continuity $\{\beta\} = \Delta_y \subset \Delta_x$, hence $\beta \in L_x$. Conversely, let $\beta \in L_x$. Then

$$\tilde{x} = \tilde{x}_\beta + \sum\nolimits_{\beta' \neq \beta} \tilde{x}_{\beta'} \,. \tag{4.14}$$

We choose a Borel subgroup B' such that β is an anti-dominant weight. Then the β' occurring in (4.14) are of the form

$$\beta' = \beta + \sum\nolimits_{\alpha > 0} n_\alpha(\beta')\alpha \,, \tag{4.15}$$

where $n_\alpha(\beta') \geq 0$ for any positive simple root α with respect to B'. We choose a 1-PS λ with $\langle \lambda, \beta \rangle = 0$ and $\langle \lambda, \alpha \rangle > 0, \forall \alpha > 0$. Then $\lim_{t \to 0} \alpha(\lambda(t)) = 0, \forall \alpha > 0$. Hence

$$\lim\nolimits_{t \to 0} \lambda(t) \cdot \tilde{x} = \tilde{x}_\beta \,. \tag{4.16}$$

It follows that $[E_\beta] \in \overline{Tx}$. □

If $k = \mathbb{C}$ is the field of complex numbers there is a another description of the hypersimplices Δ_x as follows. Let $K \simeq \mathrm{U}(n)$ be a maximal compact subgroup of $G(\mathbb{C})$. We choose a K-invariant hermitian product on E. Hence all weight spaces are orthogonal to each other. The associated *moment map* is given by

$$\mu : \mathcal{F}(\nu)(\mathbb{C}) \longrightarrow \mathbb{R}^n, \; x \longmapsto \frac{1}{\|\tilde{x}\|^2} \sum\nolimits_\beta \|\tilde{x}_\beta\|^2 \cdot \beta \,. \tag{4.17}$$

Note that since the hermitian product is unique up to scalar, the moment map is independent of this choice.

Corollary 2.4.7 *For any* $x \in \mathcal{F}(\nu)(\mathbb{C})$,

$$\Delta_x = \mu(\overline{Tx}) \,.$$

Proof By the convexity theorem for the moment map [8], [100], resp. [90], we have

$$\mu(\overline{Tx}) = \mathrm{conv}(\mu((\overline{Tx})^T)) \,. \tag{4.18}$$

But the fixed points $(\overline{Tx})^T$ are given by the set $\{[E_\beta] \mid \beta \in L_x\}$ as we have seen in Corollary 2.4.6. The image of $[E_\beta]$ under μ is just β. Thus $\mu(\overline{Tx}) = \mathrm{conv}(L_x) = \Delta_x$. □

Now let us return to an arbitrary field.

The polytopes Δ_x give rise to a new stratification of $\mathcal{F}(\nu)$. A *thin Schubert cell* in $\mathcal{F}(\nu)$ is the non-empty intersection of B'-orbits, one for each Borel subgroup B' containing T [90].

Proposition 2.4.8 *The thin Schubert cells are exactly the non-empty sets* $\{x \in \mathcal{F}(\nu) \mid \Delta_x = \Delta\}$ *for a* ν*-hypersimplex* Δ *in* $\mathbb{R}^n$*. In particular, these sets are locally closed in* $\mathcal{F}(\nu)$.

Proof Let x, y be two (geometric) points lying in the same thin cell. Suppose that $\Delta_y \backslash \Delta_x \neq \emptyset$. By Corollary 2.4.3 we conclude that $L_y \backslash L_x \neq \emptyset$. Let $\alpha \in L_y \backslash L_x$. Choose a Borel subgroup B' of G containing T, such that α is dominant for B'. It follows that $\alpha \notin L_{bx}$ for all $b \in B'$. Thus x and y cannot lie in the same B'-orbit and so not in the same thin cell. This contradicts the assumption.

Suppose now that $L_x = L_y$. If x and y do not lie in the same thin cell then there is a Borel subgroup B' of G containing T and $w_1, w_2 \in W/W_Q, w_1 \neq w_2$, with $x \in B'w_1Q/Q$ and $y \in B'w_2Q/Q$. There is a 1-PS λ of T with $\langle \lambda, w_1\nu \rangle \neq \langle \lambda, w_2\nu \rangle$ and such that $\langle \lambda, \alpha \rangle > 0$ for all roots α which are B'-positive. Hence

$$\lim_{t\to 0} \lambda(t) \cdot \tilde{x} = \tilde{x}_{w_1\nu}, \text{ resp. } \lim_{t\to 0} \lambda(t) \cdot \tilde{y} = \tilde{y}_{w_2\nu}.$$

On the other hand $\lim_{t\to 0} \lambda(t) \cdot \tilde{x} = \tilde{x}_\gamma$, where $\gamma \in S_x$ satisfies $\langle \lambda, \gamma \rangle < \langle \lambda, \delta \rangle$ for all $\delta \in L_x$ with $\delta \neq \gamma$. Since $L_x = L_y$ we must have $\gamma = w_1\lambda = w_2\lambda$, a contradiction. □

Example 2.4.9 Let $\Delta(\nu) = \mathrm{conv}(W \cdot \nu)$. This ν-hypersimplex corresponds to the open stratum defined by the intersection of the big cells with respect to the Borel subgroups containing T. Any other polytope Δ_x, $x \in \mathcal{F}(\nu)$, is contained in $\Delta(\nu)$.

Remark 2.4.10 Not every ν-hypersimplex is of the form Δ_x, for some (geometric) point x of $\mathcal{F}(\nu)$ [90]. The ν-hypersimplices Δ for which the corresponding thin cell is non-empty are called *admissible*. The stratification by thin cells has the same bad properties as our stratification by HN-types, i.e., the closure of a stratum is not always a union of strata. The first counterexamples appear for $\mathcal{F}(\nu) = \mathbf{Gr}_k(V)$ with $k \geq 3$, cf. [90]. It is not known how to characterize the admissible ν-hypersimplices.

We consider the semi-stable locus $\mathcal{F}(\nu)^{\mathrm{ss}}$ of $\mathcal{F}(\nu)$ in the sense of variant $\mathrm{Fil}^k_{\mathbb{F}_1}$. By Theorem 2.2.5 we have

$$\mathcal{F}(\nu)^{\mathrm{ss}} = \mathcal{F}(T_1, \mathcal{L}(\nu))^{\mathrm{ss}}. \tag{4.19}$$

Here on the RHS of (4.19) there appears the semi-stable locus in the sense of GIT for the action of the diagonal torus T_1 of $\mathrm{SL}(V)$ and the homogeneous line bundle $\mathcal{L}(\nu)$.

Proposition 2.4.11 *A (geometric) point x of $\mathcal{F}(\nu)(\bar{k})$ lies in $\mathcal{F}(\nu)^{ss}$ if and only if $\frac{|\nu|}{n}\cdot\mathbf{1}\in\Delta_x$.*

Proof By replacing ν by a multiple and rescaling, we may assume $|\nu|=0$. We use the Hilbert–Mumford criterion applied to $\mathcal{F}(T_1,\mathcal{L})^{ss}$. Hence

$$\begin{aligned} x\in\mathcal{F}(T_1,\mathcal{L})^{ss} &\Longleftrightarrow \forall\lambda\in X_*(T_1)\ \exists\beta\in S_x : \langle\lambda,\beta\rangle\le 0\\ &\Longleftrightarrow \forall\lambda\in X_*(T_1)\ \exists \text{ non-trivial disjoint sum}\\ &\qquad \text{decomposition } S_x=S_x^+\amalg S_x^- \text{ with}\\ &\qquad \langle\lambda,\beta\rangle\ge 0, \forall\beta\in S_x^+,\ \langle\lambda,\beta\rangle\le 0, \forall\beta\in S_x^- . \end{aligned} \tag{4.20}$$

The last property shows that, if x is semi-stable, there is no hyperplane that separates 0 from S_x, hence $0\in\Delta_x$. The converse is proved in the same way. □

Corollary 2.4.12 *$\mathcal{F}(\nu)^{ss}$ is a union of thin cells, in fact the union of all thin cells corresponding to those admissible ν-hypersimplices in $\mathbb{R}^n$ which contain $\frac{|\nu|}{n}\cdot\mathbf{1}$.* □

Examples 2.4.13 (i) Let $\nu=(1,0^{(n-1)})$. Then $E(\nu)=E=\mathrm{ind}_{B^-}^G(\nu)=V$, and $W\nu=\{\nu_1,\dots,\nu_n\}$, where $\nu_i=(0^{(i-1)},1,0^{(n-i)})$. Then

$$\Omega(\mathbf{V})=\mathcal{F}(\nu)^{ss}$$

consists of those points $x\in\mathbb{P}(V)$ which are not contained in any coordinate hyperplane, cf. Examples 1.1.23. The ν-hypersimplices are all of the form

$$\Delta_I=\mathrm{conv}\{\nu_i\mid i\in I\} \tag{4.21}$$

for some non-empty subset $I\subset\{1,\dots,n\}$ (and all of them are admissible). However, only $\Delta_{\{1,\dots,n\}}$ contains $\frac{1}{n}\cdot\mathbf{1}$. Indeed, if $\frac{1}{n}\cdot\mathbf{1}\in\Delta_I$, we get

$$\frac{1}{n}\cdot\mathbf{1}=\sum\nolimits_{i\in I} r_i\nu_i\,,\ r_i\ge 0,\ \sum\nolimits_{i\in I} r_i=1\,,$$

and this implies $r_i=\frac{1}{n}\ \forall i$. Hence $\Omega(\mathbf{V})$ is a thin cell.
(ii) We reconsider Example 2.4.5. So, let $\nu=(1^{(2)},0^{(2)})$ and $\mathcal{F}(\nu)=\mathbf{Gr}_2(V)$. The period domain is given by

$\mathcal{F}(\nu)^{ss} = \{W\in\mathbf{Gr}_2(V)\mid$ *W does not contain any coordinate line and is not contained in any coordinate hyperplane* $\}$.

In this case $\mathcal{F}(\nu)^{ss}$ is the union of 10 thin cells, namely the one corresponding to $\Delta_1 = \Delta(\nu)$ and the ones corresponding to

$$\begin{aligned} \Delta_{i,j} = & \quad \mathrm{conv}(\Delta \setminus \{e_{ij}\}),\ e_{ij} \in \Delta \\ \Delta_2 = & \quad \mathrm{conv}(\{e_{12}, e_{13}, e_{24}, e_{34}\}) \\ \Delta_3 = & \quad \mathrm{conv}(\{e_{12}, e_{14}, e_{23}, e_{34}\}) \\ \Delta_4 = & \quad \mathrm{conv}(\{e_{13}, e_{14}, e_{23}, e_{24}\}). \end{aligned} \tag{4.22}$$

It turns out that not only the semi-stable locus is a union of thin cells, but that this is true for any HN-stratum.

Proposition 2.4.14 *For any* $\theta = (g_1, \ldots, g_s) \in \Theta(\nu)$, *the* HN-*stratum* $\mathcal{F}(\nu)_\theta$ *is a union of thin cells.*

Proof Let $V_\bullet^{\mathrm{std}}$ be the standard flag attached to the HN-type θ, cf. (3.10), page 48. Let $P = P_\theta \subset \mathrm{GL}(V)$ be the stabilizer of $V_\bullet^{\mathrm{std}}$ and consider the subvariety $\mathcal{F}(\nu)_{\theta, V_\bullet^{\mathrm{std}}}$ of Theorem 2.3.9 (with respect to the variant of a field with one element). Since

$$\mathcal{F}(\nu)_\theta = \coprod_{w \in W/W_P} w \cdot \mathcal{F}(\nu)_{\theta, V_\bullet^{\mathrm{std}}},$$

it suffices to show that $\mathcal{F}(\nu)_{\theta, V_\bullet^{\mathrm{std}}}$ is a union of thin cells. Consider the locally closed subvariety Y_θ in Corollary 2.3.10 consisting of all flags $\mathcal{F} \in \mathcal{F}(\nu)$, such that $\mathcal{F}|(V^i/V^{i-1})$ defines a point

$$x_i \in \mathcal{F}(g_i) = \mathcal{F}(V^i/V^{i-1}, g_i),\ i = 1, \ldots, s.$$

The variety Y_θ is a union of Schubert cells. In particular, it is a union of thin cells. Indeed, let $w = w_\theta \in W$ be the distinguished element of Theorem 2.3.9. Then Y_θ can be identified with the subvariety PwQ/Q of G/Q. Let $P = M \cdot R_u(P)$ be the Levi decomposition of P, with $M = L_P = \prod_{i=1}^s \mathrm{GL}(V^i/V^{i-1})$. Consider the P-equivariant morphism (3.12), page 49

$$\pi : Y_\theta \longrightarrow \mathcal{F}(g_1) \times \cdots \times \mathcal{F}(g_s).$$

The target of the morphism π can be identified with $M/M \cap wQw^{-1}$. Under these identifications, the map π is just the composition of the natural maps

$$PwQ/Q \to P/P \cap wQw^{-1} \to M/M \cap wQw^{-1}.$$

Let $T_i \subset \mathrm{GL}(V^i/V^{i-1})$ be the diagonal torus associated to the standard basis

$$e_{m_1+\cdots+m_{i-1}+1}, \ldots, e_{m_1+\cdots+m_i}$$

of V^i/V^{i-1}, for $i = 1, \ldots, s$. Via the inclusion $M \subset G$ we may regard T_i as a subtorus of T. Let $\nu_i \in (\mathbb{Z}^{m_i})_+$ be the type of g_i and consider the irreducible algebraic $\mathrm{GL}(V^i/V^{i-1})$-representation $E(\nu_i)$. Let

$$\mathcal{F}(g_i) \hookrightarrow \mathbb{P}(E(\nu_i)) \tag{4.23}$$

be the corresponding embedding. Let $\mathrm{ind}_{B_M^-}^M(w\nu)$ be the induced M-representation with respect to the Borel subgroup $B_M^- = B^- \cap M$ of M. By Frobenius reciprocity [125], there exists a morphism $\rho : E \to \mathrm{ind}_{B_M^-}^M(w\nu)$ of M-modules. Consider the irreducible M-representation

$$E_M(w\nu) = \otimes_{i=1}^s E(\nu_i) \tag{4.24}$$

with highest weight $w\nu$. It follows that $E_M := \rho(E)$ is a M-module with

$$E_M(w\nu) \subset E_M \subset \mathrm{ind}_{B_M^-}^M(w\nu).$$

The morphisms (4.23), and the natural morphisms $\prod_i \mathbb{P}(E(\nu_i)) \to \mathbb{P}(E_M(w\nu))$ and $\mathbb{P}(E_M(w\nu)) \hookrightarrow \mathbb{P}(E_M)$, induce an embedding

$$\mathcal{F}(g_1) \times \cdots \times \mathcal{F}(g_s) \hookrightarrow \mathbb{P}(E_M).$$

Let $E_{w\nu}$ be the 1-dimensional weight space corresponding to $w\nu$. From the above decomposition (4.24), page 61 it follows by weight reasons that $E_{w\nu} = E_{\nu_1} \otimes \cdots \otimes E_{\nu_s}$. We get a commutative diagram

$$\begin{array}{ccccc} & & G/Q & \hookrightarrow & \mathbb{P}(E) \\ & \nearrow & & & \\ Y_\theta = PwQ/Q & & & & \downarrow \\ & \searrow & & & \\ & & M/M \cap wQw^{-1} & \hookrightarrow & \mathbb{P}(E_M)\,. \end{array} \tag{4.25}$$

Here the map $\mathbb{P}(E) \to \mathbb{P}(E_M)$ is of course only rational. Let $L^i_{x_i}$ be the list of x_i, where $L^i_{x_i}$ is defined relative to $\mathcal{F}(V^i/V^{i-1}, g_i)$ and the irreducible algebraic $\mathrm{GL}(V^i/V^{i-1})$-representation $E(\nu_i)$. Let $W_i \cong S_{m_i}$ be the Weyl group of $\mathrm{GL}(V^i/V^{i-1})$. By the commutativity of (4.25) above, we get

$$L^i_{x_i} = \{\beta|T_i \mid \beta \in L_x,\ \beta|T_i \in W_i \cdot \nu_i\}.$$

Consider the ν_i-hypersimplices

$$\Delta^i_{x_i} := \mathrm{conv}(L^i_{x_i}) \subset \mathbb{R}^{n_i}.$$

By Proposition 2.4.11 we have $x_i \in \mathcal{F}(V^i/V^{i-1}, g_i)^{ss}$ if and only if $\frac{|\nu_i|}{m_i} \cdot \mathbf{1} \in \Delta^i_{x_i}$. Thus, for $x \in Y_\theta$ we have

$$x \in \mathcal{F}(\nu)_{\theta, V_\bullet^{\mathrm{std}}} \Leftrightarrow \frac{|\nu_i|}{m_i} \cdot \mathbf{1} \in \mathrm{conv}(\{\beta|T_i \mid \beta \in L_x,\ \beta|T_i \in W_i \cdot \nu_i\}),\ \forall i = 1, \ldots, s.$$

Hence $\mathcal{F}(\nu)_{\theta,\mathbf{V}_\bullet^{\rm std}}$ is the union of those thin cells contained in Y_θ which correspond to ν-hypersimplices Δ with the property that

$$\frac{|\nu_i|}{m_i}\cdot\mathbf{1}\in\operatorname{conv}(\{\beta|T_i\mid\beta\in L_\Delta,\ \beta|T_i\in W_i\cdot\nu_i\}),\ \forall i=1,\dots,s.$$

Here L_Δ denotes the set of vertices of Δ. □

Corollary 2.4.15 *For any HN-vector H, the stratum $\mathcal{F}(\nu)_H$ is a union of thin cells.* □

Examples 2.4.16 We reconsider Examples 2.4.13.

(i) Apart from $\Omega(\mathbf{V})$ there are the following strata, for $i=1,\dots,n-1$:

$\mathcal{F}(\nu)_{\theta_i}$ = $\{x\in\mathbb{P}(V)\mid$ *x is contained in a coordinate subspace of dimension i but x is not contained in any lower-dimensional coordinate subspace*$\}$.

The stratum $\mathcal{F}(\nu)_{\theta_i}$ is the union of the thin cells corresponding to the ν-hypersimplices formed by the convex hull of i vertices.

(ii) Apart from $\mathcal{F}(\nu)^{\rm ss}$ there are four other strata:

$\mathcal{F}(\nu)_{\theta_1}$ = $\{W\in\mathbf{Gr}_2(V)\mid$ *W does not contain any coordinate line but is contained in some coordinate hyperplane*$\}$.

$\mathcal{F}(\nu)_{\theta_2}$ = $\{W\in\mathbf{Gr}_2(V)\mid$ *W contains a coordinate line but is not contained in any coordinate hyperplane*$\}$.

$\mathcal{F}(\nu)_{\theta_3}$ = $\{W\in\mathbf{Gr}_2(V)\mid$ *W contains a coordinate line and is contained in some coordinate hyperplane, but W is itself not a coordinate subspace*$\}$.

$\mathcal{F}(\nu)_{\theta_4}$ = $\{W\in\mathbf{Gr}_2(V)\mid$ *W is a coordinate subspace*$\}$.

The stratum $\mathcal{F}(\nu)_{\theta_1}$ resp. $\mathcal{F}(\nu)_{\theta_2}$ is the union of the thin cells corresponding to the ν-hypersimplices $\Delta=\operatorname{conv}(\{e_I,e_J,e_K\})$ with $|I\cup J\cup K|=3$ resp. with $|I\cap J\cap K|=1$. The stratum $\mathcal{F}(\nu)_{\theta_3}$ resp. $\mathcal{F}(\nu)_{\theta_4}$ is the union of the thin cells corresponding to the edges of Δ resp. to the vertices of Δ.

Remark 2.4.17 It seems that very rarely is $\mathcal{F}(\nu)^{\rm ss}$ a thin cell, as in the Drinfeld case. Perhaps this is the only case. In other words, if $\mathcal{F}(\nu)^{\rm ss}$ is a thin cell, is then ν of the form $\nu=(x,y^{(n-1)})$ or $\nu=(x^{(n-1)},y)$?

Notes and References Thin cells and their various characterizations in the Grassmannian case were introduced by Gelfand, Goresky, MacPherson, and Serganova [89]. Their motivation was the theory of hypergeometric functions. The generalization to arbitrary projective homogeneous spaces G/P of reductive groups was treated in [90], in which the authors also consider a third characterization of thin cells using matroids, resp. projective configurations. Thin cells also appear in the work of L. Lafforgue [142] and of Hacking, Keele, and

Tevelev, see [104] and the literature cited there. We also mention the papers of Kapranov [126, 127]. The theory of thin cells was extended by Hu [120] to "nice" projective varieties over $\mathbb{C}$ admitting a torus action. The convexity property of the moment map holds more generally for symplectic manifolds, see [8], [100], and also [164]. The characterization of the semi-stable locus in terms of ν-hypersimplices is essentially due to van der Put and Voskuil [220]. They worked in the p-adic context and were interested in characterizing those cases when the center $\frac{|\nu|}{n} \cdot \mathbf{1}$ lies in the interior of Δ.

III. Cohomology of Period Domains for GL_n

In this chapter we consider the cohomology of period domains. Since these period domains are defined over a finite field, we need to use a cohomology theory that is applicable in this context. We use the ℓ-adic cohomology with compact supports. However, no detailed knowledge of this cohomology theory is needed, and we have axiomatized its use to some degree. In this way, the determination of the Euler–Poincaré characteristic becomes essentially a combinatorial task. The basic tool is the Langlands Lemma from the theory of Eisenstein series, which is the subject matter of the first section. In the second section we discuss the generalized Steinberg representations. The third section then gives formulas for the Euler–Poincaré characteristic of period domains, and discusses these results. Again, the whole strategy has an analogue in the world of moduli spaces of vector bundles, see the "Notes and References" in the end of the third section.

Throughout this chapter we will denote by T the diagonal torus in $G = GL_n$, by B the Borel group of upper triangular matrices, and by

$$\mathcal{P} = \{P \mid P \supset B\}$$

the set of *standard parabolic subgroups*.† There is a one-to-one correspondence between $\mathcal{P}$ and the set of subsets of $\{1, \ldots, n-1\}$. Here we are identifying the set $\{1, \ldots, n-1\}$ with the set Δ of simple roots $\alpha_1, \ldots, \alpha_{n-1}$, where

$$\alpha_i(t_1, \ldots, t_n) = t_i/t_{i+1}\ , \quad i = 1, \ldots n-1\ .$$

To the subset I of $\{1, \ldots, n-1\}$ there corresponds the standard parabolic P_I such that $\Delta - I$ is the set of simple roots occurring in the unipotent radical of P_I. To I there corresponds a decomposition $n_I = (n_1, \ldots, n_s)$ of n, such that

$$I = \coprod_{j=1}^{s} \{n_1 + \cdots + n_{j-1} + 1, \ldots, n_1 + \cdots + n_j - 1\}\ ,$$

† We often abbreviate this as "std psgp."

and there is also a bijection between such decompositions of n and $\mathcal{P}$. If $P = P_I$, then there is a canonical Levi decomposition $P_I = M_I N_I$, where M_I is the Levi component consisting of the diagonal blocks of size $n_I = (n_1, \dots, n_s)$, and where N_I is the unipotent radical of P_I. In this correspondence, the Borel subgroup B corresponds to $I = \emptyset$ and the non-proper parabolic G to $I = \Delta$. Also $I \subset J$ iff $P_I \subset P_J$.

1 The Langlands Lemma

For $P = P_I$, we denote by $A_I = A_{P_I}$ the maximal torus in the center of the Levi component $M_I = M_{P_I}$, and let $\mathfrak{a}_I = X_*(A_I) \otimes \mathbb{R}$. Then

$$\mathfrak{a}_\emptyset = \{x = (x_1, \dots, x_n) \in \mathbb{R}^n\}.$$

We identify the simple roots Δ (with respect to the Borel subgroup of upper triangular matrices B) with the $\mathbb{R}$-linear forms

$$x \longmapsto \alpha_i(x) = x_i - x_{i+1} , \quad i = 1, \dots, n-1$$

on $\mathfrak{a}_\emptyset$. Then for $I \subset \Delta$, we have

$$\mathfrak{a}_I = \{x \in \mathfrak{a}_\emptyset \mid \alpha_i(x) = 0 , i \in I\} .$$

If $n_I = (n_1, \dots, n_s)$ is the decomposition of n corresponding to I, the projection

$$\mathfrak{a}_I \longrightarrow \mathbb{R}^s, \qquad x \longmapsto (x_{n_1}, x_{n_1+n_2}, \dots, x_{n_1+n_2+\cdots+n_s}) ,$$

is an isomorphism.

For each $i \in \Delta$, let

$$\varpi_{\alpha_i} = \varpi_i^\Delta = \varpi_i = \frac{n-i}{n} \sum_{k=1}^{i} k\alpha_k + \frac{i}{n} \sum_{k=i+1}^{n-1} (n-k)\alpha_k$$

be the corresponding fundamental weight, so that ϖ_i is the $\mathbb{R}$-linear form on $\mathfrak{a}_\emptyset$ given by

$$\varpi_i(x) = x_1 + \cdots + x_i - \frac{i}{n}(x_1 + \cdots + x_n) \quad (\forall x \in \mathfrak{a}_\emptyset) .$$

Then for each pair $J \subset I$ of subsets of Δ, we have

$$\mathfrak{a}_I \subset \mathfrak{a}_J \subset \mathfrak{a}_\emptyset$$

and

$$\mathfrak{a}_J = \mathfrak{a}_I \oplus \mathfrak{a}_J^I ,$$

where we have set

$$\mathfrak{a}_J^I = \{x \in \mathfrak{a}_J \mid x_1 + \cdots + x_n = 0 \text{ and } \varpi_i(x) = 0, \forall i \in \Delta - I\} .$$

For any $x \in \mathfrak{a}_J$, we denote by x_I and x_J^I the canonical projections of x onto $\mathfrak{a}_I$ and $\mathfrak{a}_J^I$ respectively.

For $I \subset \Delta$ with $n_I = (n_1, \ldots, n_s)$ as the corresponding partition of n, we have

$$\mathfrak{a}_\emptyset^I = \left\{ x \in \mathfrak{a}_\emptyset \mid \sum_{k=1}^{n_j} x_{n_1+\cdots+n_{j-1}+k} = 0, \forall j = 1, \ldots, s \right\} .$$

The projections $\mathfrak{a}_\emptyset \longrightarrow \mathfrak{a}_I$, resp. $\mathfrak{a}_\emptyset \longrightarrow \mathfrak{a}_\emptyset^I$, are given explicitly by sending $x \in \mathfrak{a}_\emptyset$ to

$$x_I = \left(\frac{1}{n_1} \sum_{k=1}^{n_1} x_k, \ldots, \frac{1}{n_s} \sum_{k=1}^{n_s} x_{n_1+\cdots+n_{s-1}+k} \right) ,$$

$$x_\emptyset^I = \Bigg((x_1, \ldots, x_{n_1}) - \frac{1}{n_1} \left(\sum_{k=1}^{n_1} x_k \right) \mathbf{1}_{n_1} ; \ldots ; (x_{n_1+\cdots+n_{s-1}+1}, \ldots, x_n) - \frac{1}{n_s} \left(\sum_{k=1}^{n_s} x_{n_1+\cdots+n_{s-1}+k} \right) \mathbf{1}_{n_s} \Bigg) . \tag{1.1}$$

For each $i = n_1 + \cdots + n_{j-1} + k \in I, 1 \leq k \leq n_j - 1, 1 \leq j \leq s$, we may also consider the $\mathbb{R}$-linear form

$$\varpi_i^I = \frac{n_j - k}{n_j} \sum_{l=1}^{k} l\alpha_{n_1+\cdots+n_{j-1}+l} + \frac{k}{n_j} \sum_{l=k+1}^{n_j-1} (n_j - l)\alpha_{n_1+\cdots+n_{j-1}+l}$$

on $\mathfrak{a}_\emptyset$. If we define the coroot $\alpha_i^\vee \in \mathfrak{a}_\emptyset$ by

$$(\alpha_i^\vee)_k = \begin{cases} 1 & \text{if } k = i, \\ -1 & \text{if } k = i + 1, \\ 0 & \text{otherwise,} \end{cases}$$

for every $i \in \Delta$, then $(\alpha_i^\vee)_{i \in I}$ is a basis of the $\mathbb{R}$-vector subspace $\mathfrak{a}_\emptyset^I$ of $\mathfrak{a}_\emptyset$ and the restrictions of the linear forms $(\varpi_i^I \mid \mathfrak{a}_\emptyset^I)_{i \in I}$ form its dual basis.

Following Arthur (see [4], §2 and [143], §3), for any $J \subset I$ in Δ, we denote by

$$\tau_J^I \text{ (resp. } \widehat{\tau}_J^I) : \mathfrak{a}_\emptyset \longrightarrow \{0, 1\} \tag{1.2}$$

the characteristic function of the subset

$$\mathfrak{a}_I + \{x \in \mathfrak{a}_J^I \mid \alpha_i(x) > 0, \forall i \in I - J\} + \mathfrak{a}_\emptyset^J$$

(resp.

$$\mathfrak{a}_I + \{x \in \mathfrak{a}_J^I \mid \varpi_i^I(x) > 0, \forall i \in I - J\} + \mathfrak{a}_\emptyset^J)$$

of $\mathfrak{a}_\emptyset$. Here τ_J^I (resp. $\widehat{\tau}_J^I$) are shorthand for $\tau_{P_J}^{P_I}$ (resp. $\widehat{\tau}_{P_J}^{P_I}$). The first set is called the *acute Weyl chamber*, the second set the *obtuse Weyl chamber*. The first set is a subset of the second set, as follows from the next lemma.

Lemma 3.1.1 *Let $J \subset I$. Let $x \in \mathfrak{a}^I_J$.*

(i) *Assume for each $i \in I - J$ that $\alpha_i(x) > 0$ or $\varpi^I_i(x) > 0$. Then $\varpi^I_i(x) > 0$ for all $i \in I - J$.*

(ii) *Assume for each $i \in I - J$ that $\alpha_i(x) \leq 0$ or $\varpi^I_i(x) \leq 0$. Then $\varpi^I_i(x) \leq 0$ for all $i \in I - J$.*

Proof (i) We may reduce to the case $I = \Delta$. In fact, if we write $I = \coprod_{j=1}^s I_j$ with $I_j = \{n_1 + \cdots + n_{j-1} + 1, \ldots, n_1 + \cdots + n_j - 1\}$, then the lemma holds if and only if it holds with respect to the subsets $I_j \supset I_j \cap J,\ j = 1, \ldots, s$.

We have

$$\mathfrak{a}^\Delta_J = \{x = (x_1, \ldots, x_n) \in \mathbb{R}^n \mid \alpha_j(x) = 0\ \forall j \in J,\ x_1 + \cdots + x_n = 0\}.$$

We may suppose that $J \neq \Delta$. Let $x \in \mathfrak{a}^\Delta_J$, such that for each $i \in \Delta \setminus J$, $\alpha_i(x) > 0$ or $\varpi_i(x) > 0$. In this case we prove by induction on i that $\varpi_i(x) > 0$ for all $i \in \Delta$. Let $\Delta \setminus J = \{j_1, \ldots, j_{s-1}\}$.

Let $i = 1$. If $\varpi_{j_1}(x) > 0$, i.e., $x_1 + \cdots + x_{j_1} = j_1 x_1 > 0$ then we get $\varpi_1(x) > 0$. Otherwise $\alpha_{j_1}(x) = x_{j_1} - x_{j_1+1} > 0$. Put $j_s = n$ and $\varpi_n(x) = x_1 + \cdots + x_n = 0$. By assumption there is an index $k \leq s$ with

$$\varpi_{j_k}(x) = x_1 + \cdots + x_{j_k} \geq 0$$

and

$$\alpha_{j_l}(x) = x_{j_l} - x_{j_l+1} > 0\,, \forall l < k.$$

Then $\alpha_j(x) = x_j - x_{j+1} \geq 0$, $\forall j < j_k$. We get, by adding successively $\alpha_{j_k-1}(x)$, $2\alpha_{j_k-2}(x), \ldots, (j_k - 1)\alpha_1(x)$ to $\varpi_{j_k}(x)$, the inequalities,

$$\begin{aligned} x_1 + \cdots + 2x_{j_k-1} &\geq 0 \\ x_1 + \cdots + 3x_{j_k-2} &\geq 0 \\ &\vdots \\ x_1 + (j_k - 1)x_2 &\geq 0 \\ j_k \cdot x_1 &\geq 0. \end{aligned}$$

Since $\alpha_{j_1}(x) > 0$, we must have $j_k \cdot x_1 > 0$. Thus, we deduce that $\varpi_1(x) = x_1 > 0$.

Now we proceed with the induction step. Suppose that $\varpi_j(x) > 0$, for all $j < i$. If $i > j_{s-1}$, we also must have $\varpi_i(x) > 0$. Indeed, otherwise, since $\varpi_{i-1}(x) > 0$ we deduce $x_i < 0$ and hence also $x_k < 0$ for $k = i, \ldots, n$, which gives a contradiction to $x_1 + \cdots + x_n = 0$. So, let $i \leq j_{s-1}$. Choose $r \leq s-1$ minimal with $i \leq j_r$. If $\varpi_{j_r}(x) > 0$, we must have $\varpi_i(x) > 0$ as well, since $\varpi_{i-1}(x) > 0$ and $\alpha_k(x) = 0$ for $k = i, \ldots, j_r - 1$. Otherwise, we have $\alpha_{j_r}(x) > 0$. Again, let $k \leq s$ with $k > r$ and

$$\varpi_{j_k}(x) = x_1 + \cdots + x_{j_k} \geq 0,$$

such that

$$\alpha_{j_l}(x) = x_{j_l} - x_{j_l+1} > 0, \ \forall l \text{ with } r \leq l < k. \tag{1.3}$$

Then $\alpha_j(x) \geq 0, \ \forall j$ with $i \leq j < j_k$. In the same way as above, it follows that

$$\begin{aligned} x_1 + \cdots + x_i + \cdots + x_{j_r} + \cdots + x_{j_k-2} + 2x_{j_k-1} &\geq 0 \\ x_1 + \cdots + x_i + \cdots + x_{j_r} + \cdots + 3x_{j_k-2} &\geq 0 \\ &\vdots \\ x_1 + \cdots + x_{i-1} + x_i + (j_k - i + 2) \cdot x_{i+1} &\geq 0 \\ x_1 + \cdots + x_{i-1} + (j_k - i + 1) \cdot x_i &\geq 0 \end{aligned}$$

The term $\alpha_{j_r}(x) = x_{j_r} - x_{j_r+1}$ is among the summands for obtaining these inequalities. Since $\alpha_{j_r}(x) > 0$ by (1.3), we conclude that $x_1 + \cdots + x_{i-1} + (j_k - i+1) \cdot x_i > 0$. By the induction hypothesis we have $\varpi_{i-1}(x) = x_1 + \cdots + x_{i-1} > 0$. It follows that $(j_k - i + 1) \cdot (x_1 + \cdots + x_i) > 0$ and therefore $\varpi_i(x) > 0$.

(ii) The above proof works for $\geq$ instead of $>$ as well. The claim follows by replacing α_i and ϖ_i by their negatives $-\alpha_i$, $-\varpi_i$. □

Lemma 3.1.2 (Langlands Lemma) *For any subsets $K \subset I$ of Δ, we have*

$$\sum_{K \subset J \subset I} (-1)^{|J \setminus K|} \tau_K^J(x) \widehat{\tau}_J^I(x) = \delta_{K,I}$$

and

$$\sum_{K \subset J \subset I} (-1)^{|J \setminus K|} \widehat{\tau}_K^J(x) \tau_J^I(x) = \delta_{K,I}$$

for every $x \in \mathfrak{a}_\emptyset$. Here $\delta_{K,I} = 1$ if $K = I$ and $\delta_{K,I} = 0$ otherwise.

Proof (after [149]) Define a matrix T with coefficients $T_{K,J} = (-1)^{|K|} \tau_K^J$ if $K \subset J$ and $T_{K,J} = 0$ otherwise. Similarly, define a matrix $\widehat{T}$ with coefficients $\widehat{T}_{K,J} = (-1)^{|K|} \widehat{\tau}_K^J$ if $K \subset J$ and $\widehat{T}_{K,J} = 0$ otherwise. Then the identities in the lemma simply state that the matrices T and $\widehat{T}$ are inverses of each other. Hence it suffices to prove the first identity. For any J, such that $K \subset J \subset I$, the functions τ_K^J and $\widehat{\tau}_J^I$ are invariant under translation by $\mathfrak{a}_I$ and $\mathfrak{a}_\emptyset^K$. Therefore we may assume that $x \in \mathfrak{a}_K^I = \mathfrak{a}_J^I \oplus \mathfrak{a}_K^J$.

Let us fix $x \in \mathfrak{a}_K^I$ and let us set

$$L = K \cup \{i \in I \setminus K \mid \varpi_i^I(x) \leq 0\}$$

and

$$M = K \cup \{i \in I \setminus K \mid \alpha_i(x) > 0\} .$$

We have

$$\tau_K^J(x)\widehat{\tau}_J^I(x) \neq 0$$

for a set J with $K \subset J \subset I$ if and only if we have

$$L \subset J \subset M .$$

Moreover, if this is the case we have

$$\tau_K^J(x)\widehat{\tau}_J^I(x) = 1 .$$

Therefore we may assume that $L \subset M$ (if $L \not\subset M$ we have $I \setminus K \neq \emptyset$ and $\delta_{K,I} = 0$), and then

$$\sum_{K \subset J \subset I} (-1)^{|J \setminus K|} \tau_K^J(x)\widehat{\tau}_J^I(x) = \sum_{L \subset J \subset M} (-1)^{|J \setminus K|}$$

is equal to

$$(-1)^{|L \setminus K|}(1-1)^{|M \setminus L|} = (-1)^{|L \setminus K|}\delta_{L,M} .$$

It remains to show that $L = M$ if and only if $K = I$. If $L \neq M$ we have $K \neq I$ and we are done. Hence we must exclude the possibility that $L = M$ and that $K \neq I$. Let us show that it leads to a contradiction. Let $n_I = (n_1, \ldots, n_s)$ be the partition of n corresponding to I and let us fix $j \in \{1, \ldots, s\}$. We set

$$\begin{cases} I_j = \{1, \ldots, n_j - 1\}, \\ L_j = \{k \in I_j \mid n_1 + \cdots + n_{j-1} + k \in L\}, \\ K_j = \{k \in I_j \mid n_1 + \cdots + n_{j-1} + k \in K\}, \end{cases}$$

and

$$x_{j,k} = x_{n_1 + \cdots + n_{j-1} + k}$$

for any $k \in \{1, \ldots, n_j\}$. We have

$$\begin{cases} x_{j,1} + \cdots + x_{j,n_j} = 0, & \\ x_{j,k} = x_{j,k+1}, & \forall k \in K_j, \\ x_{j,k} > x_{j,k+1}, & \forall k \in L_j \setminus K_j, \end{cases}$$

and

$$\varpi^I_{n_1 + \cdots + n_{j-1} + k}(x) = x_{j,1} + \cdots + x_{j,k} > 0, \qquad \forall k \in I_j \setminus L_j .$$

It follows that

$$x_{j,1} + \cdots + x_{j,k} \geq 0, \qquad \forall k \in I_j .$$

Indeed, if $k \in I_j \setminus L_j$ there is nothing to prove; if

$$x_{j,1} + \cdots + x_{j,k} < 0$$

for some $k \in L_j$ and if $k \in L_j$ is minimal for this property we necessarily have $x_{j,k} < 0$; then, by an obvious induction on l, we see that

$$x_{j,l} < 0$$

and

$$x_{j,1} + \cdots + x_{j,l} < 0$$

for every $l = k, \ldots, n_j$ (if $x_{j,1} + \cdots + x_{j,l} < 0$ for some $l \in I_j$ we have $l \in L_j$ and therefore $x_{j,l+1} \leq x_{j,l}$) and we get a contradiction (take $l = n_j$). We even have

$$x_{j,1} + \cdots + x_{j,k} > 0, \qquad \forall k \in L_j \setminus K_j .$$

Indeed, if $k \in L_j \setminus K_j$ we have

$$x_{j,k} > x_{j,k+1}$$

and from the equality

$$x_{j,1} + \cdots + x_{j,k} = 0$$

we would get either

$$x_{j,1} + \cdots + x_{j,k-1} < 0 ,$$

if $x_{j,k} > 0$, or

$$x_{j,1} + \cdots + x_{j,k+1} < 0 ,$$

if $x_{j,k} \leq 0$ and this contradicts the previous assertion. But, by definition of L_j, this means that

$$L_j = K_j$$

and, by definition of $M = L$, this implies that

$$x_{j,k} \leq x_{j,k+1}, \qquad \forall k \in I_j .$$

Now only $x = 0$ satisfies at the same time

$$x_{j,1} + \cdots + x_{j,k} \geq 0, \qquad \forall k \in I_j ,$$

$$x_{j,k} \leq x_{j,k+1}, \qquad \forall k \in I_j$$

and

$$x_{j,1} + \cdots + x_{j,n_j} = 0$$

for every $j = 1, \ldots, s$. But for $x = 0$ we have $L = I$ and $M = K$ and this contradicts our assumption that $L = M$ and $K \neq I$. □

Notes and References The Langlands Lemma comes from the theory of Eisenstein series, cf. [143]. A proof is contained in Arthur's paper [4], and a simplified version in Labesse's Bourbaki talk [141]. The Langlands Lemma is also stated in [151], §1. Here we have closely followed Laumon's exposition of the proof in [149]. For generalizations of this lemma to more general convex cones than the acute and the obtuse Weyl chambers, see the appendix to [93]; see also [39] and the literature cited there. Lemma 3.1.1 is stated without proof in [141].

2 The generalized Steinberg representations

In this section we discuss some simple representations of $\mathrm{GL}_n(k)$, where k is a finite field. All representations will be in vector spaces over a fixed algebraically closed field $\mathbf{C}$ of characteristic zero. To simplify notations, we often abbreviate $G(k)$ into G for $G = \mathrm{GL}_n$ and similarly for $P \in \mathcal{P}$.

For $P \in \mathcal{P}$, we denote by i_P^G the induced representation $\mathrm{Ind}_P^G \mathbf{1}$ of the trivial representation of M_P (considered as a factor group of P). We denote by $\hat{\mathcal{G}}(i_B^G)$ the full abelian subcategory of the category of representations of G, all of whose irreducible constituents occur in i_B^G. By $\hat{G}(i_B^G)$ we denote the set of isomorphism classes of irreducible objects in $\hat{\mathcal{G}}(i_B^G)$. Then $\hat{G}(i_B^G)$ forms a basis of the Grothendieck group of $\hat{\mathcal{G}}(i_B^G)$. This Grothendieck group will be denoted by $K_0(i_B^G)$.

Before formulating the main result of this section, we recall the concept of associatedness of elements of $\mathcal{P}$. Two parabolic subgroups are called *associate*, if their Levi components are conjugate under $G(k)$. If $P = P_I$ and $P' = P_{I'}$ with associated decompositions $n_I = (n_1, \ldots, n_r)$ and $n_{I'} = (n'_1, \ldots, n'_{r'})$ of n, then P and P' are associate if and only if $r = r'$ and the entries $n_1, \ldots, n_r$ resp. $n'_1, \ldots, n'_{r'}$ are the same up to renumbering. Equivalently, let $\lambda(P_I) = \lambda(I)$ be the partition of n obtained by ordering $n_1, \ldots, n_r$ by decreasing size. Then P_I is associate to $P_{I'}$ iff $\lambda(I) = \lambda(I')$.

Theorem 3.2.1 (i) *i_P^G is equivalent to $i_{P'}^G$ if and only if P is associate to P'.*
(ii) *i_P^G contains a unique irreducible representation j_P^G which occurs with multiplicity one and such that*

$$\mathrm{Hom}_G(j_P^G, i_{P'}^G) \neq (0) \Leftrightarrow \lambda(P') \leq \lambda(P).$$

Furthermore, $\{j_{P_\lambda}^G \mid \lambda$ a partition of $n\}$ *is the set of irreducible constituents of i_B^G.*

Here, for two partitions $\lambda = (\lambda_1, \lambda_2, \ldots, \lambda_n)$ and $\lambda' = (\lambda'_1, \lambda'_2, \ldots, \lambda'_n)$ of n, we write $\lambda' \leq \lambda$ if for all $r = 1, \ldots, n$, we have

$$\sum\nolimits_{i=1}^{r} \lambda'_i \leq \sum\nolimits_{i=1}^{r} \lambda_i.$$

Proof We shall make use of the representation theory of the symmetric group. Let $W = S_n$ denote the Weyl group of G. There is a one-to-one correspondence between the elements $P = P_I \in \mathcal{P}$ and the parabolic subgroups W_I of W. Recall that a *parabolic subgroup* of W is by definition the Weyl group of the Levi component M_I of some P_I. Here $W_I = S_{n_1} \times \cdots \times S_{n_r}$ for $n_I = (n_1, n_2, \ldots, n_r)$ (with the first factor permuting the first n_1 indices, the second factor the next n_2 indices, etc.). Let $\hat{W}$ be the set of isomorphism classes of irreducible representations of W. Analogously to the definition of i_P^G, we set $i_{W'}^W := \mathrm{Ind}_{W'}^W \mathbf{1}$ for any subgroup W' of W.

Theorem 3.2.2 ([117]) *There exists a unique bijection*

$$\alpha : \hat{G}(i_B^G) \longrightarrow \hat{W}$$

characterized by the following property. An irreducible representation $\sigma \in \hat{G}(i_B^G)$ occurs in $i_{P_I}^G$ if and only if $\alpha(\sigma)$ occurs in $i_{W_I}^W$. Furthermore

$$\dim \mathrm{Hom}_G(\sigma, i_{P_I}^G) = \dim \mathrm{Hom}_W(\alpha(\sigma), i_{W_I}^W). \qquad \square$$

We now show how to deduce Theorem 3.2.1 from Theorem 3.2.2. Obviously, W_I is conjugate to $W_{I'}$ if and only if $\lambda(I) = \lambda(I')$. Since the induced representation $i_{W_I}^W$ only depends on W_I up to conjugacy, assertion (i) follows from Theorem 3.2.2. We may write $i_{W_\lambda}^W$ without ambiguity. Furthermore, it is well known [157], p.115 that, for any partition λ of n, the induced representation $i_{W_\lambda}^W$ contains an irreducible representation χ^λ with multiplicity one and with the property

$$\mathrm{Hom}_W(\chi^\lambda, i_{W_{\lambda'}}^W) \neq (0) \Leftrightarrow \lambda' \leq \lambda.$$

Let $j_{P_\lambda}^G$ be the inverse image of χ^λ under α. Then $j_{P_\lambda}^G$ has the properties required for (ii) of the theorem. Furthermore, since $\hat{W} = \{\chi^\lambda \mid \lambda \text{ partition of } n\}$, we also have

$$\hat{G}(i_B^G) = \{j_{P_\lambda}^G \mid \lambda \text{ partition of } n\}\,.$$

To prove the uniqueness assertion, by Theorem 3.2.2 we may write

$$i_{P_\lambda}^G = j_{P_\lambda}^G + \sum_{\lambda' \gneq \lambda} a_{\lambda'} \cdot j_{P_{\lambda'}}^G, \quad a_{\lambda'} \in \mathbb{Z}_{\geq 0}. \tag{2.1}$$

Hence we can also write

$$j^G_{P_\lambda} = i^G_{P_\lambda} + \sum_{\lambda' \gneqq \lambda} b_{\lambda'} \cdot i^G_{P_{\lambda'}}, \quad b_{\lambda'} \in \mathbb{Z}, \tag{2.2}$$

an identity in $K_0(i^G_B)$. If now $\tilde{j}^G_{P_\lambda}$ is another irreducible representation with the properties in (ii) for $P = P_\lambda$, then $\mathrm{Hom}(\tilde{j}^G_{P_\lambda}, i^G_{P_{\lambda'}}) = (0)$ for $\lambda' \gneqq \lambda$. Hence from (2.2) above we obtain

$$\mathrm{Hom}(\tilde{j}^G_{P_\lambda}, j^G_{P_\lambda}) = \mathrm{Hom}(\tilde{j}^G_{P_\lambda}, i^G_{P_\lambda}).$$

Since $\tilde{j}^G_{P_\lambda}$ is supposed to occur with multiplicity one in $i^G_{P_\lambda}$, we see that $\tilde{j}^G_{P_\lambda}$ is equivalent to $j^G_{P_\lambda}$. □

Corollary 3.2.3 *The elements i^G_P, as P ranges over the associate classes in $\mathcal{P}$, form a basis of $K_0(i^G_B)$.*

Proof The assertion follows from the upper triangular nature of the matrix (2.1) above expressing $i^G_{P_\lambda}$ in terms of $j^G_{P_\lambda}$. □

Next we introduce yet another basis of $K_0(i^G_B)$. Note that for $P \supset Q$ we have an inclusion of G-representations

$$i^G_P \subset i^G_Q .$$

Definition 3.2.4 The *generalized Steinberg representation* associated to $P \in \mathcal{P}$ is

$$v^G_P = i^G_P / \sum_{Q \supsetneqq P} i^G_Q.$$

Obviously, v^G_P depends up to equivalence only on the associate class of P. For $P = G$, we have $v^G_G = \mathbf{1}$.

Theorem 3.2.5 *Let $P = P_I$. Then*

$$v^G_P = (-1)^{|I|} \sum_{J \supset I} (-1)^{|J|}\, i^G_{P_J}. \tag{2.3}$$

More precisely, the following complex is an acyclic resolution of v^G_P by G-modules,

$$0 \to i^G_G \to \bigoplus_{\substack{I \subset K \subset \Delta \\ |\Delta \setminus K| = 1}} i^G_{P_K} \to \bigoplus_{\substack{I \subset K \subset \Delta \\ |\Delta \setminus K| = 2}} i^G_{P_K} \to \cdots \to \bigoplus_{\substack{I \subset K \subset \Delta \\ |K \setminus I| = 1}} i^G_{P_K} \to i^G_{P_I} \to v^G_{P_I} \to 0. \tag{2.4}$$

Here the differentials $d_{K',K} : i^G_{P_{K'}} \to i^G_{P_K}$ are defined as follows. For two subsets $K \subset K'$ of Δ, we let

$$p_{K',K} : i^G_{P_{K'}} \longrightarrow i^G_{P_K}$$

be the natural homomorphism induced by the surjection $G/P_K \to G/P_{K'}$. For arbitrary subsets $K, K' \subset \Delta$, with $|K'| - |K| = 1$ and $K' = \{k_1 < \cdots < k_r\}$, we put

$$d_{K',K} = \begin{cases} (-1)^i p_{K',K} & K' = K \cup \{k_i\} \\ 0 & K \not\subset K' \end{cases}.$$

We first note the following simple lemma.

Lemma 3.2.6 *The functor of B-invariants is an exact and faithful functor from $\hat{\mathcal{G}}(i_B^G)$ to the category of vector spaces over* $\mathbf{C}$.

Proof Exactness follows from Maschke's theorem. To prove faithfulness, it suffices to prove $V^B \neq (0)$ for any irreducible constituent V of i_B^G. By Frobenius reciprocity we have, for any G-representation V, the identity

$$V^B \cong \mathrm{Hom}_G(i_B^G, V),$$

which implies the claim, again by Maschke's theorem. □

The lemma will allow us to reduce Theorem 3.2.5 to the analogous theorem for the Weyl group. By the Bruhat decomposition $G = \bigcup_{w \in W} BwB$ we may identify the B-invariants of i_B^G with the set of functions on W. We denote this latter space by $i_{\{1\}}^W$, even though there is no natural action of W on the space of B-invariants. Similarly, for any subset K of Δ, the space of B-invariants in $i_{P_K}^G$ is identified with the space $i_{W_K}^W$ of functions on W that are invariant under W_K. Also, by the exactness of the functor of B-invariants, we deduce that

$$(v_{P_I}^G)^B = v_{W_I}^W := i_{W_I}^W / \sum_{K \supsetneq I} i_{W_K}^W.$$

The Weyl group analogue of Theorem 3.2.5 is the following.

Theorem 3.2.7 *The natural complex*

$$0 \to i_W^W \to \bigoplus_{\substack{I \subset K \subset \Delta \\ |\Delta \setminus K| = 1}} i_{W_K}^W \to \bigoplus_{\substack{I \subset K \subset \Delta \\ |\Delta \setminus K| = 2}} i_{W_K}^W \to \cdots \to \bigoplus_{\substack{I \subset K \subset \Delta \\ |K \setminus I| = 1}} i_{W_K}^W \to i_{W_I}^W \to v_{W_I}^W \to 0$$

is acyclic.

Proof We make use of the following lemma.

Lemma 3.2.8 *Let $A_1, \ldots, A_m$ be a family of subgroups of an abelian group A. Suppose that the following identities are satisfied for all subsets $I, J \subset \{1, \ldots, m\}$,*

$$\Big(\sum_{i \in I} A_i\Big) \cap \Big(\bigcap_{j \in J} A_j\Big) = \sum_{i \in I} \Big(A_i \cap \Big(\bigcap_{j \in J} A_j\Big)\Big).$$

Then the natural complex (where the differentials are given by the alternating sum of the face maps)

$$A \leftarrow \bigoplus_{i=1}^{m} A_i \leftarrow \bigoplus_{\substack{i,j=1\\ i<j}}^{m} A_i \cap A_j \leftarrow \bigoplus_{\substack{i,j,k=1\\ i<j<k}}^{m} A_i \cap A_j \cap A_k \leftarrow \cdots$$

is an acyclic resolution of the subgroup $\sum_i A_i$ of A.

Proof See [196], §2, Prop. 6. □

In order to prove Theorem 3.2.5, we want to apply this lemma to the G-submodules $i^G_{P_K}$ of $A = i^G_B$, where K ranges over the subsets of Δ with $K \supset I$ and $|\Delta \setminus K| = 1$. We have

$$i^G_{P_K} \cap i^G_{P_J} = i^G_{P_{K\cup J}}. \tag{2.5}$$

This follows from the fact that $P_{K\cup J}$ is the subgroup generated by P_K and P_J, cf. [38], Prop. 2.3.2. We also claim

$$i^G_{P_K} \cap (i^G_{P_{J_1}} + i^G_{P_{J_2}} + \cdots + i^G_{P_{J_r}}) = (i^G_{P_K} \cap i^G_{P_{J_1}}) + \cdots + (i^G_{P_K} \cap i^G_{P_{J_r}}), \tag{2.6}$$

for all subsets $K, J_1, \ldots, J_r \subset \Delta$.

Clearly the RHS of (2.6) above is contained in the LHS. Both sides of (2.6) are objects in $\hat{\mathcal{G}}(i^G_B)$. Therefore the same is true for the cokernel of the inclusion. In view of Lemma 3.2.6, it suffices to show that the above inclusion induces an identity on B-invariants. Hence it suffices to show that the following identities of subspaces of $i^W_{\{1\}}$ are satisfied

$$i^W_{W_K} \cap (i^W_{W_{J_1}} + i^W_{W_{J_2}} + \cdots + i^W_{W_{J_r}}) = (i^W_{W_K} \cap i^W_{W_{J_1}}) + \cdots + (i^W_{W_K} \cap i^W_{W_{J_r}}), \tag{2.7}$$

for all subsets $K, J_1, \ldots, J_r \subset \Delta$. This is the content of the following proposition. The condition of Lemma 3.2.8 is therefore satisfied. Theorem 3.2.5 follows. By the same token, the same argument also proves Theorem 3.2.7, since analogously to (2.5) above, we also have

$$i^W_{W_K} \cap i^W_{W_J} = i^W_{W_{K\cup J}}. \tag{2.8}$$

□

It remains to show the validity of the relations (2.7), page 76. This holds true for arbitrary finite Coxeter groups. In what follows, we denote by (W, S) a finite Coxeter system. For a subset $I \subset S$, we abbreviate $i_I = i^W_{W_I}$.

Proposition 3.2.9 *For any subsets $I, J_1, J_2, \ldots, J_r \subseteq S$, one has*

$$i_I \cap (i_{J_1} + \cdots + i_{J_r}) = (i_I \cap i_{J_1}) + \cdots + (i_I \cap i_{J_r}).$$

Proof Whenever $I \subseteq S$, we denote by e_I the sum of elements of the parabolic subgroup W_I in the group algebra $\mathbf{C}[W]$. Then we may identify i_I with the ideal $\mathbf{C}[W] \cdot e_I$. Since $i_\emptyset = \mathbf{C}[W]$, we may assume that $I, J_1, J_2, \ldots, J_r$ are all non-empty. Note also that $i_I = \{a \in \mathbf{C}[W] \mid as = a, \forall s \in I\}$, therefore $i_I \cap i_J = i_{I \cup J}$ for any $I, J \subseteq S$, cf. (2.8), page 76. Our claim now amounts to

$$i_{J_0} \cap (i_{J_1} + \cdots + i_{J_r}) = i_{J_0 \cup J_1} + \cdots + i_{J_0 \cup J_r} \tag{2.9}$$

for any non-empty subsets $J_0, J_1, J_2, \ldots, J_r \subseteq S$. We prove this by induction on r. The above shows that the case $r = 1$ is clear.

Noting that the RHS of (2.9) is clearly included in the LHS, (2.9) is equivalent to the equality of the corresponding dimensions, and therefore to

$$\dim(i_{J_0}+\cdots+i_{J_r}) = \dim(i_{J_0})+\dim(i_{J_1}+\cdots+i_{J_r})-\dim(i_{J_0 \cup J_1}+\cdots+i_{J_0 \cup J_r}). \tag{2.10}$$

Applying the induction hypothesis, (2.10) can be taken for granted to expand further the dimensions on the RHS above. Further iteration tells us that (2.10) (and therefore (2.9)) is now equivalent to

$$\dim(i_{J_0} + \cdots + i_{J_r}) = \sum_{k \geq 1} (-1)^{k+1} \sum_{0 \leq r_1 < \cdots < r_k \leq r} \dim(i_{J_{r_1} \cup \cdots \cup J_{r_k}}). \tag{2.11}$$

We will now use the following proposition, which concerns a variant of the *Coxeter complex*. Recall that the latter is associated with the coefficient system on the subsets of S with values in $\mathbf{C}$-vector spaces $(\mathbf{C}[W/W_{S'}], \varphi_{S'',S'})_{S',S'' \subseteq S}$ where the restriction map $\varphi_{S'',S'} : \mathbf{C}[W/W_{S'}] \longrightarrow \mathbf{C}[W/W_{S''}]$ is defined by $\varphi_{S'',S'}(wW_{S'}) = wW_{S''}$ whenever $w \in W$ and $S' \subseteq S'' \subseteq S$.

Proposition 3.2.10 *Let $n \geq 1$ be an integer. Denote by $\mathbf{2}^n$ the set of subsets of $\{1, \ldots, n\}$. Let $S_1, \ldots, S_n$ be non-empty subsets of S. If $I \in \mathbf{2}^n$, denote by $W(I)$ the parabolic subgroup of W associated with $\bigcup_{i \in I} S_i \subseteq S$. Let $A := (A^I, \psi_{J,I})_{I \subseteq J \in \mathbf{2}^n}$ be the coefficient system on $\mathbf{2}^n$ defined by $A^I = \mathbf{C}[W/W(I)]$ and $\psi_{JI} = \varphi_{S'',S'}$ for $S' = \bigcup_{i \in I} S_i$ and $S'' = \bigcup_{j \in J} S_j$. Then the complex associated with A*

$$C(A)^0 = \mathbf{C}[W] \longrightarrow C(A)^1 = \oplus_{i=1}^n \mathbf{C}[W/W_{S_i}] \longrightarrow \cdots$$
$$\cdots \longrightarrow C(A)^k = \oplus_{I;|I|=k} \mathbf{C}[W/W(I)] \longrightarrow \cdots \longrightarrow C(A)^n = \mathbf{C}[W/W_{S_1 \cup \cdots \cup S_n}]$$

with differential defined by

$$\partial^k(wW(I)) = \sum_{j \in \{1,\ldots,n\} \setminus I} (-1)^{|I \cap [0,j[|}\, wW(I \cup \{j\}) \quad (I \in \mathbf{2}^n \text{ of cardinality } k;\ w \in W)$$

has its homology concentrated in degree zero.

Proof Recall the length function ℓ on W with respect to S. Whenever $S' \subseteq S$, denote by $D_{S'}$ the set of representatives of $W/W_{S'}$ of minimal length (the S'-*reduced* elements). We abbreviate $D(I) = \bigcap_{i \in I} D_{S_i}$ for $I \in \mathbf{2}^n$.

We have the $\mathbf{C}$-basis $\{wW(I))_{w \in D(I)}\}$ for each A^I. This in turn yields a basis

$$\mathcal{B} = \{wW(I) \mid I \in \mathbf{2}^n,\ w \in D(I)\}$$

of $C(A) = \oplus_k C(A)^k$. Note that the restriction maps ψ_{JI} stabilize $\mathcal{B}$.

We extend the length function to $\mathcal{B}$ by $\ell(wW(I)) = \ell(w)$ whenever $w \in D(I)$.

Let us define a map

$$\tau\colon C(A) \longrightarrow C(A)$$

homogeneous of degree -1 as follows.

Whenever $w \in D_{S_1} \cup \cdots \cup D_{S_n}$, choose once and for all $i(w) \in \{1, \ldots, n\}$ such that w is $S_{i(w)}$-reduced.

We now define τ on $\mathcal{B}$ as follows. Let $I \in \mathbf{2}^n$ and $w \in D(I)$. If $I \neq \emptyset$ (therefore $i(w)$ is defined) and $i(w) \in I$, we set

$$\tau(wW(I)) = (-1)^{|I \cap [1, i(w)[|} wW(I \setminus \{i(w)\}).$$

In all other cases, i.e., when $I = \emptyset$ or $i(w) \notin I$, we set $\tau(wW(I)) = 0$.

Claim *$\tau\partial + \partial\tau$ sends any $b \in \mathcal{B}$ of degree > 0 into $b + \sum_{\ell(b') < \ell(b)} \mathbf{C} \cdot b'$.*

Write $b = wW(I)$ with $\emptyset \neq I \in \mathbf{2}^n$ and $w \in D(I)$. We abbreviate $|I \cap [1, i[| = m(I, i)$.

First case: $i(w) \in I$. Then $\tau(wW(I)) = (-1)^{m(I, i(w))} wW(I \setminus \{i(w)\})$ and

$$(\partial\tau + \tau\partial)(b) = \partial((-1)^{m(I,i(w))} wW(I \setminus \{i(w)\})) + \sum_{j \in \{1,\ldots,n\} \setminus I} (-1)^{m(I,j)} \tau(wW(I \cup \{j\}))$$

and therefore

$$(\partial\tau + \tau\partial)(b) = b + \sum_{j \in \{1,\ldots,n\} \setminus I} (-1)^{m(I,i(w)) + m(I \setminus \{i(w)\}, j)} wW(I \setminus \{i(w)\} \cup \{j\})$$
$$+ \sum_{j \in \{1,\ldots,n\} \setminus I\ ;\ i(w'_j) \in I \cup \{j\}} (-1)^{m(I,j) + m(I \cup \{j\}, i(w'_j))} w'_j . W(I \cup \{j\} \setminus i(w'_j)),$$

where w'_j denotes the unique element of $D(I \cup j) \cap wW(I \cup j)$. Note that $\ell(w'_j) \leq \ell(w)$ with equality only if $w'_j = w$, that is when w is S_j-reduced. In view of our claim, we have to care only about those j in the second sum such that $w'_j = w$ (which then implies $w \in D(I \cup j)$) and the condition is the same in the first sum. Those classes $wW(I \setminus \{i(w)\} \cup \{j\})$ then come with a coefficient $(-1)^{m(I,i(w)) + m(I \setminus \{i(w)\}, j)} + (-1)^{m(I,j) + m(I \cup \{j\}, i(w))}$ which is zero by a standard lemma (used already to check that $\partial\partial = 0$). So we get our claim in this case.

Second case: $i(w) \notin I$. Then $\tau(wW(I)) = 0$ and $b = \epsilon\tau(b')$ where $b' = wW(I \cup \{i(w)\})$ (recall that w is $S_{i(w)}$-reduced by the definition of $i(w)$) and $\epsilon = \pm 1$. Note more generally that from the definition of τ, if $\beta \in \mathcal{B}$, then $\tau(\beta)$ is either

0 or $\pm\beta'$ where $\beta' \in \mathcal{B}$ satisfies $\ell(\beta') = \ell(\beta)$. Also $\tau^2 = 0$. Returning to our b, we have

$$(\partial\tau + \tau\partial)(b) - b = \epsilon\tau((\partial\tau + \tau\partial)(b') - b') \in \epsilon\tau\left(\sum\nolimits_{\ell(b'')<\ell(b')} \mathbf{C}\cdot b''\right)$$

by the first case above. Thus our claim.

We now have an endomorphism $\rho = \partial\tau + \tau\partial$ which is homogeneous of degree zero and such that $\rho - \mathrm{Id}$ is nilpotent on any $C(A)^k$ with $k > 0$, while $\rho(C(A)^0) = 0$. Now let ρ' be the homogeneous endomorphism of degree 0 inverting ρ in each degree > 0 and zero on $C(A)^0$. Then $\tau' := \tau\rho'$ clearly satisfies $\partial\tau' + \tau'\partial = \mathrm{Id}$ in each degree > 0 since ρ commutes with both τ and ∂.

This implies our claim about the homology of $C(A)$. □

Now we can prove Proposition 3.2.9. Identifying $\mathbf{C}[W]$ with its $\mathbf{C}$-dual, it is clear that the dual of the differential $\partial^0 \colon C(A)^0 \longrightarrow C(A)^1$ is identified with the map

$$\oplus_i \mathbf{C}[W]\cdot e(S_i) \longrightarrow \mathbf{C}[W], \quad (a_i)_i \mapsto \sum\nolimits_i a_i.$$

So, applying Proposition 3.2.10 with $n = r+1$, $S_i = J_{i-1}$, we get that $\operatorname{codim}(i_{J_0} + \cdots + i_{J_r}) = \dim(H^0(A))$, while $\dim(A^i) = \sum_{0\le r_1<\cdots<r_i\le r} \dim(i_{J_{r_1}\cup\cdots\cup J_{r_i}})$. So Proposition 3.2.10 gives (2.11), page 77 by the classical expression of the Euler–Poincaré characteristic. This completes the proof. □

Proposition 3.2.9 has the following interesting consequence.

Corollary 3.2.11 *There exists a basis $\mathcal{B}$ of $i_\emptyset$ adapted to the subspaces $\{i_I \mid I \subset S\}$, i.e., each subspace i_I has a basis given by a subset of $\mathcal{B}$, as well as each sum of such subspaces.*

Proof This is a general fact on distributive complemented modular lattices, cf. [35], Prop. 3.1. □

We now return to the relation between the induced representations i_P^G and the generalized Steinberg representations v_P^G. We first note the following fact.

Corollary 3.2.12 *The representation v_B^G is the Steinberg representation. In particular, it is irreducible.*

Proof This is the case $I = \emptyset$. In this case the RHS of the identity (2.3), page 74 in Theorem 3.2.5 is the definition of the Steinberg representation, cf. [38], p.187. □

Remark 3.2.13 In general, the generalized Steinberg representations v_P^G are not irreducible (in contrast to the p-adic case, cf. Part 3 of this monograph).

For instance for $n = 4$, and denoting the elements $P_I \in \mathcal{P}$ by their associated decompositions n_I of $n = 4$, we have

$$i^G_{P_{(2,2)}} = j^G_{P_{(2,2)}} + j^G_{P_{(3,1)}} + j^G_{P_{(4)}}$$

$$v^G_{P_{(2,2)}} = j^G_{P_{(2,2)}} + j^G_{P_{(3,1)}} \ :$$

this follows from the corresponding decomposition of $\mathrm{Ind}^W_{W_\lambda} \mathbf{1}$, cf. [157], table on p. 239.

Corollary 3.2.14 *For any $P \in \mathcal{P}$,*

$$i^G_P = \sum_{Q \supset P} v^G_Q \, .$$

In particular, the generalized Steinberg representations v^G_P, as P ranges over the associate classes in $\mathcal{P}$, form a basis of $K_0(i^G_B)$.

Proof We write i_I resp. v_I instead of $i^G_{P_I}$ resp. $v^G_{P_I}$. From Theorem 3.2.5 we obtain for $P = P_I$,

$$\begin{aligned} \sum_{Q \supset P} v^G_Q &= \sum_{J \supset I} (-1)^{|J|} \cdot \sum_{K \supset J} (-1)^{|K|} \cdot i_K \\ &= \sum_K i_K \cdot \sum_{I \subset J \subset K} (-1)^{|K \setminus J|} = i_I = i^G_P. \end{aligned}$$

□

Notes and References The complex of Theorem 3.2.7 in the case $I = \emptyset$ coincides with the homology complex of the *Coxeter complex*, and the assertion is in this case a classical result, cf. [46]. Similarly, there are a number of different proofs of Theorem 3.2.5 in the case $I = \emptyset$, cf. [45, 152, 88, 202, 138].

For general I, Theorems 3.2.5 and 3.2.7 are due to Lehrer [152]. His proof is different. He shows that the complex (2.4), page 74 is the cohomology cochain complex of a subcomplex of the combinatorial Tits building attached to (G, B). He then starts with the base simplex and pastes successively adjacent maximal simplices. The resulting complex has the same cohomology as the preceeding one, as long as the attached simplex has a proper face of the same distance. The maximal simplices which do not have this property are exactly the ones which contribute to the cohomology, cf. also Lemma 3.3.19 below.

The proof given here is an adaptation of the proof in the p-adic case, cf. [149], p. 194, and [24], p. 202. It is ultimately simplicial in nature, since the proof of Schneider and Stuhler [196] of Lemma 3.2.8 uses simplicial methods. The proof of the crucial distributivity property in Proposition 3.2.9 and Proposition 3.2.10 was communicated to us by M. Cabanes. Note that the assertion of Proposition 3.2.10 in the case when $S = \{s_1, \ldots, s_n\}$ and $S_i = \{s_i\}, i = 1, \ldots, n,$

coincides with the Coxeter complex. It would be interesting to have a topological interpretation of the acyclicity statement of Proposition 3.2.10, similar to the topological interpretation of the acyclicity of the Coxeter complex. It should be pointed out that, even though Corollary 3.2.11 gives the existence of an adapted basis of $\mathbf{C}[W]$, no explicit such basis is known to us. The situation is reminiscent of the analogous problem of constructing an adapted basis for the cohomology of Springer fibers, or of the affine rings of the schemes of "nilpotent diagonal matrices", see [35].

We mention that in [59], Deligne and Lusztig define a complex which looks superficially similar to the complex (2.4), page 74. For a representation V of $G(k)$, they construct a complex

$$0 \longrightarrow V_{\emptyset} \longrightarrow \bigoplus_{|I|=1} V_{(I)} \longrightarrow \bigoplus_{|I|=2} V_{(I)} \longrightarrow \cdots \longrightarrow V_{(S)} \longrightarrow 0,$$

where $V_{(I)} = \oplus_{P \in \mathcal{P}_I} V^{U_P}$ is the direct sum of the invariants under the unipotent radicals of the parabolic subgroups of type I. Let i_0 be the smallest integer such that $V_{(I)} \neq 0$ for some I with $|I| = i_0$. Then they show by elementary topological arguments that when V is irreducible, this complex has a single homology group in degree i_0. For $i_0 = 0$ this again leads to the Coxeter complex, but for $i_0 > 0$ these complexes are not related to (2.4), page 74.

For an exposition of the Steinberg representation in terms of the combinatorial building of G for an arbitrary finite group with BN-pair, cf. [46], §66, and the literature cited there. The Steinberg representation can also be realized as the dual object of the trivial representation under the duality operator in the representation theory of finite groups of Lie type [68]. In [38] the character of the Steinberg representation is determined.

Generalized Steinberg representations were introduced by Solomon [202] and also studied by Surowski [213] and Stanley [207], for arbitrary reductive groups over a finite field. In [152] Lehrer uses the resolution (2.4), page 74 to determine the value of the character of v_P^G on regular semi-simple elements.

In contrast to the finite field case, the generalized Steinberg representations behave differently for a p-adic field. In the p-adic case, all generalized Steinberg representations are irreducible smooth representations and are pairwise non-isomorphic [24]. The complex analogous to (2.4), page 74 is again acyclic [24], [149], [172].

3 The Euler–Poincaré characteristic

In this section we determine the Euler–Poincaré characteristic of period domains over a finite field. We begin by collecting some facts about the Euler–Poincaré characteristic of varieties. In fact, we will axiomatize these facts to some degree to show that our approach is also applicable in other contexts.

Let k be a finite field of characteristic p with q elements. Let $\overline{k}$ be an algebraic closure of k. We denote by Γ_k the Galois group $\mathrm{Gal}(\overline{k}/k)$. We fix a prime number $\ell \neq p$. Let X be an algebraic variety over k. We consider the ℓ-adic cohomology groups with compact supports $H^i_c(X \otimes_k \overline{k}, \overline{\mathbb{Q}}_\ell)$. For brevity, we will denote them by $H^i_c(X)$. Recall [53] that $H^i_c(X)$ is a finite-dimensional $\overline{\mathbb{Q}}_\ell$-vector space with a continuous action of the Galois group Γ_k, for each i. Also $H^i_c(X) = 0$ for $i > 2\dim X$ or $i < 0$. We may therefore introduce the *Euler–Poincaré characteristic* $\chi_c(X) = \sum_i (-1)^i H^i_c(X)$ as an element of the Grothendieck group $K_0(\Gamma_k\text{-mod})$ of Γ_k-modules in finite-dimensional $\mathbf{C}$-vector spaces, where $\mathbf{C} = \overline{\mathbb{Q}}_\ell$.

The cohomology groups $H^i_c(X)$ are functorially associated to X. Hence, if a finite group G acts on X, we may also view $\chi_c(X)$ as an element of the Grothendieck group of $G\times\Gamma_k$-modules. We now list the properties of the Euler–Poincaré characteristic that we are going to use.

(i) If $X = \dot{\bigcup} X_i$ is a disjoint partition into locally closed subsets, then

$$\chi_c(X) = \sum\nolimits_i \chi_c(X_i)\,.$$

This follows by induction from the long exact cohomology sequence relating the cohomology with compact supports of a Zariski-open subset U, of the total space X and of the closed complement $Y = X - U$.

(ii) If $\mathcal{V}$ is a vector bundle of rank n over X, then

$$\chi_c(\mathcal{V}) = \chi_c(X) \otimes t^n\,.$$

Here $t = t_k$ denotes the *Tate twist*, i.e., the one-dimensional representation of Γ_k such that the geometric Frobenius (the inverse of the Frobenius substitution) acts by multiplication by the cardinality q of k. By t^n we denote the n-fold tensor power of t (the geometric Frobenius acts by q^n).
This follows from the fact that $H^{i+2n}_c(\mathcal{V}) = H^i_c(X) \otimes t^n$, for all i.

(iii) If $X = X_1 \times X_2$, then

$$\chi_c(X) = \chi_c(X_1) \cdot \chi_c(X_2)\,.$$

This follows from the Künneth formula

$$H^n_c(X) = \bigoplus_{i+j=n} H^i_c(X_1) \otimes H^j_c(X_2)\,.$$

Finally, we shall use the calculation of the Euler–Poincaré characteristic of a generalized flag variety.

(iv) Let V be a vector space of dimension n over k and let $\nu \in (\mathbb{R}^n)_+$. Let $\mathcal{F}(\nu)$ be the corresponding partial flag variety over k, cf. Chapter II, §1. Let

$W = S_n$ and let W_ν be the stabilizer of ν in W. Then

$$\chi_c(\mathcal{F}(\nu)) = \sum_{w \in W/W_\nu} t^{\ell(w)} . \tag{3.1}$$

Here $\ell(w)$ denotes the minimal length of a representative in W of the coset wW_ν.

This follows from the decomposition of $\mathcal{F}(\nu)$ into Bruhat cells, which shows that there is no cohomology in odd degrees, and

$$H_c^{2i}(\mathcal{F}(\nu)) = \bigoplus_{\substack{w \in W/W_\nu \\ \ell(w)=i}} t^i . \tag{3.2}$$

The action of $G(k)$ on each summand of (3.2) is trivial.

Remark 3.3.1 The Euler–Poincaré characteristic is a map from the "set" of k-varieties into the Grothendieck ring of Γ_k-modules. Another map from the set of k-varieties into a commutative ring which shares the properties (i)–(iv) is given by the *virtual Poincaré polynomial* $P_X(t^{\frac{1}{2}}) \in \mathbb{Z}[t^{\frac{1}{2}}]$. Recall [86], p. 92, that $P_X(t^{\frac{1}{2}})$ is given as

$$P_X(t^{\frac{1}{2}}) = \sum\nolimits_m (-1)^m \chi_c^m(X) \cdot t^{\frac{m}{2}} .$$

Here

$$\chi_c^m(X) = \sum (-1)^i \dim \operatorname{gr}_W^m(H_c^i(X)) ,$$

where W denotes the weight filtration on $H_c^i(X)$. Naturally, our method also applies to this variant.

Yet another such map is given by forgetting the Γ_k-action, and just retaining the dimension of the cohomology groups: this is then the ordinary Euler–Poincaré characteristic for the cohomology with compact supports. Note [148] that the Euler–Poincaré characteristic for the cohomology with compact supports is identical to the Euler–Poincaré characteristic for ℓ-adic cohomology without supports. Our result also gives, of course, the ordinary Euler–Poincaré characteristic of period domains.

Remark 3.3.2 One use of the Euler–Poincaré characteristic in the Grothendieck group of Γ_k-modules is the Grothendieck–Lefschetz fixed point formula.

Let X be a scheme over $k = \mathbb{F}_q$. Then there is the Frobenius $\sigma_{X/k} : X \longrightarrow X$. If $X = \operatorname{Spec} A$, then $\sigma_{X/k}$ is induced by the k-algebra endomorphism of A which sends an element a of A to a^q. The Frobenius is the identity on the set of points of X, and the q-power map on the structure sheaf. If X is of the form $X_0 \otimes_k \bar{k}$, for a k-scheme X_0 of finite type, then

$$\sigma_{X/k} = (\mathrm{Id}_{X_0} \otimes \sigma_{\bar{k}/k}) \circ \mathrm{Fr}_X ,$$

where the *Frobenius endomorphism* $\mathrm{Fr}_X : X \longrightarrow X$ is a morphism of $\bar{k}$-schemes. If $X_0 = \mathbb{A}^n = \mathrm{Spec}\, k[T_1, \dots, T_n]$, then $\mathrm{Fr}_X^*(T_i) = T_i^q$, for $i = 1, \dots, n$. On the ℓ-adic cohomology with compact support $H_c^i(X_0 \otimes_k \bar{k}, \bar{\mathbb{Q}}_\ell)$ the Frobenius automorphism $\sigma_{\bar{k}/k}$ acts by functoriality through the second factor (this is the action that was mentioned in the beginning of this section). On the other hand, the Frobenius endomorphism Fr_X, being a proper morphism (in fact, finite), induces an endomorphism Fr_X^* on $H_c^i(X_0 \otimes_k \bar{k}, \bar{\mathbb{Q}}_\ell)$. Since $\sigma_{X/k}^*$ is the identity on $H_c^i(X_0 \otimes_k \bar{k}, \bar{\mathbb{Q}}_\ell)$, we deduce that

$$\mathrm{Fr}_X^* = (\sigma_{\bar{k}/k}^*)^{-1} = \varphi \,,$$

where φ denotes the geometric Frobenius action, i.e., the action of the inverse of $\sigma_{\bar{k}/k}$. Now the fixed points of Fr_X^m on $X(\bar{k})$ can be identified with $X_0(\mathbb{F}_{q^m})$. Hence the *Grothendieck–Lefschetz fixed point formula* for the Frobenius endomorphism [53] gives

$$|X_0(\mathbb{F}_{q^m})| = \sum_i (-1)^i \mathrm{trace}(\varphi^m, H_c^i(X)) = \mathrm{trace}(\varphi^m, \chi_c(X)) \,. \tag{3.3}$$

Let us evaluate this formula for $\mathcal{F}(\nu)$ and compare it to the formula (3.2), page 83 for the cohomology of $\mathcal{F}(\nu)$. Now the number of points of $\mathcal{F}(\nu)$ over $\mathbb{F}_{q^m}$ can be calculated by elementary means. Indeed, if $\nu = (x_1^{(n_1)}, \dots, x_r^{(n_r)})$, then $\mathcal{F}(\nu) = G/Q_\nu$ where $Q_\nu = Q_{(n_1,\dots,n_r)}$ is the standard parabolic subgroup with blocks of size $n_1, \dots, n_r$, and we obtain

$$|\mathcal{F}(\nu)(\mathbb{F}_{q^m})| = \frac{|G(\mathbb{F}_{q^m})|}{|Q_\nu(\mathbb{F}_{q^m})|} = \frac{P_n(q^m)}{P_{n_1}(q^m) \cdot \,\cdots\, \cdot P_{n_r}(q^m)} \,. \tag{3.4}$$

Here

$$P_n(T) = (T-1)(T^2-1)\cdots(T^n-1) \,. \tag{3.5}$$

Comparison with (3.2), page 83 gives the following identity in the Grothendieck ring of Γ_k-modules,

$$\sum_{w \in W/W_\nu} t^{\ell(w)} = \frac{P_n(t)}{P_{n_1}(t) \cdot \,\cdots\, \cdot P_{n_r}(t)} \,. \tag{3.6}$$

Of course, this may also be deduced as a formal consequence of the corresponding identity of polynomials in the indeterminate t, which is well-known [210], p. 130, and may be proved by induction on n. Note that, conversely, the identity of polynomials follows from the identity (3.6), page 84 in the Grothendieck group of Γ_k-modules. Indeed, if $P(T) \in \mathbb{Z}[T]$ with $P(t) = 0$, then $\mathrm{trace}(\varphi^m, P(t)) = 0$ for $m = 1, 2, \dots$. Hence $P(q^m) = 0$ for all $m > 0$, which implies $P(T) = 0$.

We now turn to the cohomology of period domains. Let V be a k-vector space of dimension n. We fix $\nu \in (\mathbb{R}^n)_+$. Let $\mathcal{F}(\nu)$ be the corresponding flag variety over k. Let $Q_\nu \in \mathcal{P}$ be the standard-parabolic subgroup defining this flag variety, i.e., $\mathcal{F}(\nu) = G/Q_\nu$. Let W_ν be the stabilizer of ν in W. For any $P \in \mathcal{P}$, we denote by W_P the Weyl group of the Levi component M_P of P. Then $W_\nu = W_{Q_\nu}$.

Recall the HN-stratification of $\mathcal{F}(\nu)$ from Chapter II (3.4), page 44. We obtain an implicit recursion relation for the Euler–Poincaré characteristic of $\mathcal{F}(\nu)^{\mathrm{ss}}$, as follows. First of all, since

$$\mathcal{F}(\nu) = \dot{\bigcup}_\theta \mathcal{F}(\nu)_\theta ,$$

we have by property (i) of the Euler–Poincaré characteristic above,

$$\chi_c(\mathcal{F}(\nu)) = \sum\nolimits_\theta \chi_c(\mathcal{F}(\nu)_\theta) . \tag{3.7}$$

On the other hand, by Corollary 2.3.10, for each HN-type $\theta = (g_1, \dots, g_s)$, the corresponding HN-stratum $\mathcal{F}(\nu)_\theta$ is a $G(k)$-equivariant vector bundle of rank $\ell(w_\theta)$ over the disjoint sum of copies of the product $\mathcal{F}(g_1)^{\mathrm{ss}} \times \cdots \times \mathcal{F}(g_s)^{\mathrm{ss}}$ enumerated by the index set $G(k)/P_\theta(k)$, with the obvious action of $G(k)$, cf. Remark 2.3.11. We deduce from the properties (ii) and (iii) of the Euler–Poincaré characteristic that

$$\chi_c(\mathcal{F}(\nu)_\theta) = \mathrm{Ind}_{P_\theta(k)}^{G(k)}(\chi_c(\mathcal{F}(g_1)^{\mathrm{ss}}) \times \chi_c(\mathcal{F}(g_2)^{\mathrm{ss}}) \times \cdots \times \chi_c(\mathcal{F}(g_s)^{\mathrm{ss}})) \cdot t^{\ell(w_\theta)} . \tag{3.8}$$

Here $\chi_c(\mathcal{F}(g_1)^{\mathrm{ss}}) \times \chi_c(\mathcal{F}(g_2)^{\mathrm{ss}}) \times \cdots \times \chi_c(\mathcal{F}(g_s)^{\mathrm{ss}})$ is considered as an element of the Grothendieck group of representations of $\Gamma_k \times M_{P_\theta}(k)$, and we have used the obvious abbreviation for the induction from $P_\theta(k)$ to $G(k)$ (where the Levi $M_{P_\theta}(k)$ is considered as a factor group of $P_\theta(k)$).

Henceforth, we will denote by $[\mathcal{P}]$ the set of associate classes in $\mathcal{P}$. For $P \in \mathcal{P}$, we denote by $[P]$ the image of P in $[\mathcal{P}]$.

Lemma 3.3.3 *In the Grothendieck group of representations of* $\Gamma_k \times G(k)$,

$$\chi_c(\mathcal{F}(\nu)^{\mathrm{ss}}) = \sum\nolimits_{[P] \in [\mathcal{P}]} N_{[P]}^G(\nu) \cdot i_P^G ,$$

with uniquely defined elements $N_{[P]}^G(\nu) \in \mathbb{Z}[t]$.

Proof We use induction on n. The start of the induction is clear. For the induction step we use the identity (3.7) above, in which $\chi_c(\mathcal{F}(\nu)^{\mathrm{ss}})$ occurs as the summand corresponding to the unique minimal element θ_0 among the indices. We need to prove that all other terms are linear combinations of representations i_P^G with coefficients in $\mathbb{Z}[t]$. This is clear for the LHS of (3.7) above by property (iv) of the Euler–Poincaré characteristic above. On the other hand, for the summands on the RHS for $\theta \neq \theta_0$ this follows by induction from the identity (3.8), page 85 and the transitivity of the induction procedure.

The uniqueness of the coefficients $N^G_{[P]}(\nu)$ follows from Corollary 3.2.3. □

The following theorem gives an explicit expression for the coefficients $N^G_{[P]}(\nu)$.

Theorem 3.3.4 *For $[P] \in [\mathcal{P}]$, the coefficient $N^G_{[P]}(\nu)$ is given by*

$$N^G_{[P]}(\nu) = (-1)^{a^G_P} \cdot \sum_{P \in [P]} \sum_{w \in W/W_\nu} t^{\ell(w)} \cdot \hat{\tau}^G_P(w\nu).$$

Equivalently,

$$\chi_c(\mathcal{F}(\nu)^{\rm ss}) = \sum_{P \in \mathcal{P}} (-1)^{a^G_P} \cdot \sum_{w \in W/W_\nu} t^{\ell(w)} \cdot \hat{\tau}^G_P(w\nu) \cdot i^G_P \, .$$

Here we denote by $\ell(w)$ the minimal length of a representative in W of $w \in W/W_\nu$. Also the value of a^G_P is the same for all $P \in [P]$ and is equal to $a^G_P = \dim \mathfrak{a}^G_P = \dim \mathfrak{a}^\Delta_I = |\Delta| - |I|$ if $P = P_I$, in the notation of Section 1.

Proof We proceed by induction on n. The start of the induction is trivial.

Denote by $n^G_P(\nu)$ the coefficient

$$n^G_P(\nu) = (-1)^{a^G_P} \cdot \sum_{w \in W/W_\nu} t^{\ell(w)} \cdot \hat{\tau}^G_P(w\nu) \tag{3.9}$$

of i^G_P in the above expression of $\chi_c(\mathcal{F}(\nu)^{\rm ss})$, so that we have to prove that

$$\chi_c(\mathcal{F}(\nu)^{\rm ss}) = \sum_{P \in \mathcal{P}} n^G_P(\nu) \cdot i^G_P \, . \tag{3.10}$$

We need to make the summands on the RHS of (3.7), page 85 explicit. However, an HN-type $\theta = (g_1, \ldots, g_s)$ corresponds uniquely to the conjugate

$$w_\theta \nu = (x_1^{(g_1(x_1))}, \ldots, x_r^{(g_1(x_r))}; x_1^{(g_2(x_1))}, \ldots, x_r^{(g_2(x_r))}; \ldots; x_1^{(g_s(x_1))}, \ldots, x_r^{(g_s(x_r))})$$

of ν. Also, w_θ is of minimal length in its double coset in $W_{P_\theta} \backslash W/W_\nu$. Furthermore, by the explicit expression for the image of $w_\theta \nu$ in $\mathfrak{a}_I$ given by (1.1), page 67, the positivity property of θ (property (iii) in Definition 2.3.2) translates into $\alpha_{m_1+\cdots+m_j}(w_\theta \nu) > 0$ for $j = 1, \ldots, s-1$, i.e., property (iii) is equivalent to $\tau^G_{P_\theta}(w_\theta \nu) = 1$. Hence we obtain a bijection between the set $\Theta(\nu)$ of HN-types associated to ν and the set of pairs (Q, w) where $Q \in \mathcal{P}$ and $w \in W_Q \backslash W/W_\nu$, such that $\tau^G_Q(w\nu) = 1$. Taking into account the expression (3.8), page 85 for the Euler–Poincaré characteristic of the HN-strata and applying the induction hypothesis, the recursion relation (3.7), page 85 therefore takes the following form,

$$\chi_c(\mathcal{F}(\nu)) = \chi_c(\mathcal{F}(\nu)^{\mathrm{ss}}) + \sum_{Q \neq G} \sum_{w \in W_Q \backslash W / W_\nu} \tau_Q^G(w\nu) \cdot t^{\ell(w)} \cdot \mathrm{Ind}_{Q(k)}^{G(k)} \Big(\sum_{P \subset Q} n_P^Q(w\nu) \cdot i_P^Q \Big) . \tag{3.11}$$

Here for $w \in W_Q \backslash W / W_\nu$ we denote by $\ell(w)$ the minimal length of a representative in W.

In order to prove the theorem, all we need to show is that the following identity (3.12) is satisfied, for all $P \in \mathcal{P}$,

$$\sum_{Q \supset P} \sum_{w \in W_Q \backslash W / W_\nu} \tau_Q^G(w\nu) \cdot t^{\ell(w)} \cdot n_P^Q(w\nu) = \begin{cases} \sum_{w \in W/W_\nu} t^{\ell(w)} & P = G \\ 0 & P \neq G \end{cases} . \tag{3.12}$$

Indeed, then we would obtain

$$\begin{aligned} &\sum_{P \in \mathcal{P}} n_P^G(\nu) \cdot i_P^G + \sum_{Q \neq G} \sum_{w \in W_Q \backslash W / W_\nu} \tau_Q^G(w\nu) \cdot t^{\ell(w)} \cdot \mathrm{Ind}_{Q(k)}^{G(k)} \Big(\sum_{P \subset Q} n_P^Q(w\nu) \cdot i_P^Q \Big) \\ &\qquad = \sum_{Q} \sum_{w \in W_Q \backslash W / W_\nu} \tau_Q^G(w\nu) \cdot t^{\ell(w)} \cdot \mathrm{Ind}_{Q(k)}^{G(k)} \Big(\sum_{P \subset Q} n_P^Q(w\nu) \cdot i_P^Q \Big) \\ &\qquad = \sum_{P} i_P^G \cdot \Big(\sum_{Q \supset P} \sum_{w \in W_Q \backslash W / W_\nu} \tau_Q^G(w\nu) \cdot t^{\ell(w)} \cdot n_P^Q(w\nu) \Big) \\ &\qquad = \Big(\sum_{w \in W/W_\nu} t^{\ell(w)} \Big) \cdot i_G^G \\ &\qquad = \chi_c(\mathcal{F}(\nu)) , \end{aligned}$$

where in the last line we used property (iv) of the Euler–Poincaré characteristic. Hence looking back at (3.11), we deduce the desired identity (3.10), page 86.

To prove the identity (3.12) above, we calculate for $P \in \mathcal{P}$, by inserting the values for $n_P^Q(w\nu)$,

$$\begin{aligned} &\sum_{Q \supset P} \sum_{w \in W_Q \backslash W / W_\nu} \tau_Q^G(w\nu) \cdot t^{\ell(w)} \cdot (-1)^{a_P^Q} \cdot \sum_{w_1 \in W_Q/(W_Q)_{w\nu}} t^{\ell(w_1)} \cdot \hat{\tau}_P^Q(w_1 w\nu) \\ &\qquad = \sum_{\tilde{w} \in W/W_\nu} t^{\ell(\tilde{w})} \cdot \sum_{Q \supset P} (-1)^{a_P^Q} \cdot \tau_Q^G(\tilde{w}\nu) \cdot \hat{\tau}_P^Q(\tilde{w}\nu) \\ &\qquad = \begin{cases} \sum_{w \in W/W_\nu} t^{\ell(w)} & P = G \\ 0 & P \neq G . \end{cases} \end{aligned}$$

Here in the first identity we used the fact ([38], Prop. 2.7.5) that the element $\tilde{w}$ in W of shortest length in its class in W/W_ν is the product of the shortest element w in its class in $W_Q \backslash W / W_\nu$ and an element w_1 in W_Q which is of minimal length in its class in $W_Q/(W_Q)_{w\nu}$, in a unique way, and that $\ell(\tilde{w}) =$

$\ell(w)+\ell(w_1)$. The last identity follows from the Langlands Lemma 3.1.2, where we note that $a_{P_J}^{P_I} = |I \setminus J|$. □

The above theorem expresses the Euler–Poincaré characteristic of $\mathcal{F}(\nu)$ in terms of induced representations i_P^G. One may also express the Euler–Poincaré characteristic in terms of generalized Steinberg representations v_P^G, cf. Section 2. By the upper triangular nature of the base change matrix between these two bases of $K_0(i_B^G)$, we deduce from Corollary 3.2.14 that

$$\chi_c(\mathcal{F}(\nu)^{\mathrm{ss}}) = \sum\nolimits_{[P]\in[\mathcal{P}]} M_{[P]}^G(\nu)\cdot v_P^G\,, \tag{3.13}$$

with uniquely defined elements $M_{[P]}^G(\nu) \in \mathbb{Z}[t]$. It turns out that the coefficients $M_{[P]}^G(\nu)$ have a somewhat more convenient expression than the coefficients $N_{[P]}^G(\nu)$.

For $w \in W/W_\nu$, set

$$\Delta(w\nu) := \{\alpha \in \Delta \mid \langle w\nu, \varpi_\alpha\rangle > 0\}\,.$$

Denote by $P(w\nu)$ the unique standard parabolic subgroup P with $\Delta_P = \Delta(w\nu)$, i.e., $\Delta(w\nu)$ is the set of simple roots occurring in the unipotent radical of $P(w\nu)$.

Theorem 3.3.5 *In the Grothendieck group of $\Gamma_k \times G(k)$-modules, the following expression holds:*

$$\chi_c(\mathcal{F}(\nu)^{\mathrm{ss}}) = \sum_{w\in W/W_\nu} (-1)^{a_{P(w\nu)}^G}\, t^{\ell(w)}\cdot v_{P(w\nu)}^G.$$

In other words, for any $[P] \in [\mathcal{P}]$,

$$M_{[P]}^G(\nu) = (-1)^{a_P^G} \sum_{\{w\in W/W_\nu \mid P(w\nu)\in[P]\}} t^{\ell(w)}\,.$$

Proof By Theorem 3.3.4 and using Corollary 3.2.14, we may write

$$\chi_c(\mathcal{F}(\nu)) = \sum_{Q\in\mathcal{P}} m_Q^G\cdot v_Q^G\,,$$

where, using the coefficients (3.9), page 86, we may write by Corollary 3.2.14

$$\begin{aligned} m_Q^G(\nu) &= \sum\nolimits_{P\subset Q} n_P^G \\ &= \sum\nolimits_{P\subset Q} (-1)^{a_P^G} \sum\nolimits_{w\in W/W_\nu} t^{\ell(w)}\cdot\hat{\tau}_P^G(w\nu) \\ &= \sum\nolimits_{w\in W/W_\nu} t^{\ell(w)} \sum\nolimits_{P\subset Q} (-1)^{a_P^G}\cdot\hat{\tau}_P^G(w\nu)\,. \end{aligned} \tag{3.14}$$

Now $\hat{\tau}_P^G(w\nu) = 1$ iff $\langle w\nu, \varpi_\alpha\rangle > 0\ \forall\alpha \in \Delta_P$, i.e., iff $\Delta_P \subset \Delta(w\nu)$. Furthermore,

$a_P^G = |\Delta_P|$. Hence the inner sum in (3.14), page 88 is equal to

$$\sum_{\Delta_Q\subset\Delta_P\subset\Delta(w\nu)} (-1)^{|\Delta_P|} = (-1)^{|\Delta_Q|} \sum_{\Delta_Q\subset\Delta_P\subset\Delta(w\nu)} (-1)^{(|\Delta_P|-|\Delta_Q|)}$$
$$= (-1)^{|\Delta_Q|}\cdot\delta_{Q,P(w\nu)}\,.$$

□

Variant 3.3.6 Consider the category $\mathrm{Fil}^k_{\mathbb{F}_1}$. In this context we have a natural action of the Weyl group $W = S_n$ on the ℓ-adic cohomology with compact support $H_c^*(\mathcal{F}(\nu)^{ss})$. For $w \in W/W_\nu$, let $W(w\nu) := W_{\Delta\setminus\Delta(w\nu)}$. Then $W(w\nu) = W_{P(w\nu)}$. Put $a^W_{W(w\nu)} = a^G_{P(w\nu)}$. The analogue of the above theorems is as follows. In the Grothendieck group of $\Gamma_k\times S_n$-representations we have

$$\chi_c(\mathcal{F}(\nu)^{ss}) = \sum_{w\in W/W_\nu} (-1)^{a^W_{W(w\nu)}}\cdot t^{\ell(w)}\cdot v^W_{W(w\nu)}.$$

In terms of the induced representations $i^W_{W_\lambda}$, where λ ranges over the partitions of n, we have

$$\chi_c(\mathcal{F}(\nu)^{ss}) = \sum_\lambda N^W_\lambda(\nu)\cdot i^W_{W_\lambda}\,,$$

where $N^W_\lambda(\nu) := N^G_{[P_\lambda]}(\nu)$.

Examples 3.3.7 (i) Let $n = 3$ and let $\nu = (x_1, x_2, x_3)$ such that $x_1 > x_2 > x_3$. First we consider the case where $x_2 - x_3 > x_1 - x_2$. Let $s_1 = (1,2)$ and $s_2 = (2,3)$ be the simple reflections generating W. In this case the sets $\Delta(w\nu)$ are given by the following table.

w	1	s_1	s_2	s_1s_2	s_2s_1	$s_1s_2s_1$
$\Delta(w\nu)$	Δ	Δ	$\{\alpha_1\}$	$\emptyset$	$\{\alpha_1\}$	$\emptyset$

There are four standard parabolic subgroups, which we enumerate according to the decomposition n_I of n: the Borel subgroup $B = P_{(1,1,1)}$, the two maximal proper parabolics $P_{(2,1)}$ and $P_{(1,2)}$ and G. The coefficients m_P^G of (3.14), page 88 are given as follows:

P	B	$P_{(2,1)}$	$P_{(1,2)}$	G
m_P^G	$1+t$	0	$-(t+t^2)$	t^2+t^3

The case where $x_1 - x_2 > x_2 - x_3$ is dual to the case discussed above, i.e., the parabolics $P_{(1,2)}$ and $P_{(2,1)}$ have to be interchanged in the above result.

(ii) Let again $n = 3$ and $\nu = (x_1, x_2, x_3)$ with $x_1 > x_2 > x_3$, but now consider the case $x_2 - x_3 = x_1 - x_2$. The sets $\Delta(w\nu)$ are given by the following table.

w	1	s_1	s_2	s_1s_2	s_2s_1	$s_1s_2s_1$
$\Delta(w\nu)$	Δ	$\{\alpha_2\}$	$\{\alpha_1\}$	$\emptyset$	$\emptyset$	$\emptyset$

The contributions m_P^G are given as follows:

P	B	$P_{(2,1)}$	$P_{(1,2)}$	G
m_P^G	1	$-t$	$-t$	$2t^2+t^3$

(iii) Let $\mathcal{F}(\nu)^{ss} = \Omega^n$ be Drinfeld's upper halfspace of dimension $n-1$, cf. Example 2.1.7. In other words, we take $\nu = (n-1, (-1)^{(n-1)})$. In this case the elements in W of shortest length in their class in W/W_ν are $w_0 = 1, w_1, \ldots, w_{n-1}$, with

$$w_i = s_i \cdot s_{i-1} \cdot \cdots \cdots s_1 \,.$$

It follows that

$$\Delta(w_i\nu) = \{\alpha_{i+1}, \ldots, \alpha_{n-1}\} \; (= \emptyset \text{ if } i = n-1)\,.$$

In this case the parabolics associated to different elements in W/W_ν are different: to w_0 corresponds the Borel subgroup, to w_i for $i = 1, \ldots, n-2$ the parabolic $P_{(i+1,1^{(n-i-1)})}$, and to w_{n-1} the parabolic G. Denoting these parabolics by $P_0, \ldots, P_{n-1}$, we obtain

$$M_{[P_i]}^G = (-1)^{n-1-i} \cdot t^i\,. \tag{3.15}$$

Now Ω^n is the complement of an arrangement of hyperplanes in $\mathbb{A}_k^{n-1}$, see [167]. We may therefore apply the following proposition.

Proposition 3.3.8 *Let $U \subset \mathbb{A}^n$ be the complement of an arrangement of hyperplanes. Then*

$$\mathrm{H}_c^{2n-i}(U) = \begin{cases} m_i \cdot t^{2n-2i} \;, \; m_i \in \mathbb{Z}_{\geq 0}, \text{ for } i \in [0,n] \\ (0) \;\; \text{for } i \notin [0,n]\,. \end{cases}$$

Proof Since U is affine of dimension n, the range of vanishing cohomology follows from a general feature of the ℓ-adic cohomology with compact supports of smooth affine varieties [7]. The rest follows from [167], p. 194. □

In particular, the cohomology group $H_c^{2(n-1)-i}(\Omega^n)$ is a pure Galois module of weight $2(n-1)-2i$ for $i = 0, \ldots, 2(n-1)$, and hence there can be no cancellation when passing from the cohomology to the Euler–Poincaré characteristic. Since the summand $M_{[P_i]}^G \cdot v_{P_i}^G$ of the expression for the Euler–Poincaré characteristic is pure of weight $2i$, it has to contribute in degree $n-1+i$. We have therefore proved the following.

Proposition 3.3.9 *For $i = 0, \ldots, n-1$,*

$$H_c^{2(n-1)-i}(\Omega^n) = v_{P_{(n-i,1^{(i)})}}^G \otimes t^{n-1-i}\,.$$

In particular, the G-representation on $H_c^{2(n-1)-i}(\Omega^n)$ *is irreducible in this range. Furthermore,* $H_c^j(\Omega^n) = (0)$ *for* $j = 0, \ldots, n-2$.

In particular, the lowest degree cohomology occurs in degree $n-1$ *and is given by the Steinberg representation.*

Proof The only assertion that remains to be proved is the irreducibility of $v^G_{P_{(n-i,1^{(i)})}}$, or, equivalently by Theorem 3.2.2, the irreducibility of $v^W_{W_{(n-i,1^{(i)})}}$ for $0 \leq i \leq n-1$. However, $v^W_{W_{(n-i,1^{(i)})}}$ contains the irreducible representation χ^λ with $\lambda = (n-i, 1^{(i)})$, which is simply the ith exterior power of the standard representation of dimension $n-1$ of W, cf. [87], Exercise 4.6. Therefore it suffices to show that $\dim v^W_{W_{(n-i,1^{(i)})}} = \binom{n-1}{i}$. By Lemma 3.3.19 below, we need to show that $|Y_{\{1,\ldots,n-i-1\}}| = \binom{n-1}{i}$. Here

$$\begin{aligned} Y_{\{1,\ldots,n-i-1\}} = \{ & w \in W \mid w\alpha_j > 0 \text{ for } j = 1, \ldots, n-i-1; \\ & w\alpha_j < 0 \text{ for } j = n-i, \ldots, n-1\}. \end{aligned}$$

This follows easily from the fact that for any permutation $w \in Y_{\{1,\ldots,n-i-1\}}$ we have $w(n-i) = n$ and that $w(1), \ldots, w(n-i-1)$ are in increasing order, and $w(n-i+1), \ldots, w(n)$ are in decreasing order. □

It turns out that the expression of the Euler–Poincaré characteristic given by Theorem 3.3.5 is well-suited to determining the individual cohomology groups $H_c^i(\mathcal{F}(v)^{ss})$. Namely, the summand corresponding to $w \in W/W_v$ contributes in a single degree $2\ell(w) + |\Delta(wv)|$. More precisely, the cohomology of $\mathcal{F}(v)^{ss}$ is given by the following theorem, cf. [169].

Theorem 3.3.10

$$H_c^*(\mathcal{F}(v)^{ss}) = \bigoplus_{w \in W/W_v} t^{\ell(w)} \cdot v^G_{P(wv)}[-2\ell(w) - |\Delta(wv)|].$$

Here the symbol $[-m], m \in \mathbb{N}$, indicates that the corresponding module is shifted into degree m of the graded cohomology ring.

We will not prove this theorem in this monograph, but refer to the original source [169]. Some of the geometry behind this proof is discussed in Chapter XI, §1. Here we will discuss the nature of this formula and point out interesting consequences.

Remark 3.3.11 Consider, for each $w \in W/W_v$, the complex

$$C_w^\bullet : 0 \to i_G^G \to \bigoplus_{\substack{\Delta\setminus\Delta(wv)\subset K\subset\Delta \\ |\Delta\setminus K|=1}} i^G_{P_K} \to \bigoplus_{\substack{\Delta\setminus\Delta(wv)\subset K\subset\Delta \\ |\Delta\setminus K|=2}} i^G_{P_K} \to \cdots \to \bigoplus_{\substack{\Delta\setminus\Delta(wv)\subset K\subset\Delta \\ |K\setminus(\Delta\setminus\Delta(wv))|=1}} i^G_{P_K} \to i^G_{P(wv)}. \tag{3.16}$$

Here, the representation i_G^G is in degree 0. Note that $P_{\Delta\setminus\Delta(wv)} = P(wv)$. By Theorem 3.2.5 the cohomology of this complex is concentrated in degree $|\Delta(wv)|$ and yields the generalized Steinberg representation $v_{P(wv)}^G$. Thus, if we consider the twisted and shifted complex $t^{l(w)} \cdot C_w^\bullet[-2\ell(w)]$ then the cohomology of this complex gives exactly the w-contribution in the cohomology of $H_c^*(\mathcal{F}(\nu)^{ss})$.

Let us consider the object $R\Gamma_c(\mathcal{F}(\nu)^{ss}) = R\Gamma_c(\mathcal{F}(\nu)^{ss} \otimes_k \bar{k}, \bar{\mathbb{Q}}_\ell)$ in the derived category of $G(k)\times\Gamma_k$-representations. It is tempting to expect a decomposition

$$R\Gamma_c(\mathcal{F}(\nu)^{ss}) = \bigoplus_{w\in W/W_\nu} t^{l(w)} \cdot C_w^\bullet[-2\ell(w)]. \tag{3.17}$$

If we knew that the Frobenius acts semi-simply, then since $G(k)$ is a finite group, it would follow that the complex $R\Gamma_c(\mathcal{F}(\nu)^{ss})$ splits in the derived category of $G(k) \times \Gamma_k$-modules, so that the identity (3.17) above would indeed be indeed true.

In the case of p-adic period domains the situation is much more complicated, because of the inertia subgroup in the absolute Galois group. In the case of the p-adic Drinfeld halfspace, one can recover the missing monodromy operator on the complex $R\Gamma_c(\Omega^n)$ induced by the inertia group (which is expected by the Langlands correspondence), cf. [48]. We will come back to this in Part 4 of this monograph.

We may deduce from Theorem 3.3.10 the following remarkable vanishing property, which generalizes the case of the Drinfeld upper halfspace, cf. Proposition 3.3.9 above.

Corollary 3.3.12 *In low degrees,*

$$H_c^i(\mathcal{F}(\nu)^{ss}) = 0 \textit{ for } i \textit{ with } 0 \le i \le n-2$$

and

$$H_c^{n-1}(\mathcal{F}(\nu)^{ss}) = v_B^G.$$

Furthermore, the top cohomology is

$$H_c^{2\ell(w_0)}(\mathcal{F}(\nu)^{ss}) = t^{\ell(w_0)} \cdot i_G^G,$$

where w_0 is of maximal length in W/W_ν, i.e., $\dim \mathcal{F}(\nu) = \ell(w_0)$.

To prove this corollary we need some combinatorial lemmas in Weyl groups. Let W^ν be the set of representatives of shortest length in their cosets W/W_ν.

Lemma 3.3.13 *Let $w \in W^\nu$. Let $s = s_\alpha$ be a simple reflection associated to a simple root $\alpha \in \Delta$, such that $sw \in W^\nu$. Suppose that $\ell(sw) = \ell(w) + 1$. Then*

$$\Delta(swv) \setminus \{\alpha\} = \Delta(wv) \setminus \{\alpha\}$$

and

$$\Delta(swv) \subset \Delta(wv).$$

In particular, the sets $\Delta(swv)$ and $\Delta(wv)$ differ at most by one element.

Proof Recall that $\Delta(wv) = \{\alpha \in \Delta \mid \langle wv, \varpi_\alpha\rangle\ > 0\}$. For $\beta \in \Delta$, we have

$$\langle swv, \varpi_\beta\rangle \ = \ \langle wv, s\varpi_\beta\rangle$$

and

$$s\varpi_\beta = \begin{cases} \varpi_\beta & ; \ \beta \neq \alpha \\ \varpi_\alpha - \alpha & ; \ \beta = \alpha \end{cases}$$

([28], VI, §1.10). The first statement follows. The assumption $\ell(sw) = \ell(w)+1$ is equivalent to the positivity $w^{-1}(\alpha) > 0$ ([38], Prop. 2.2.8). Since v is contained in the closure of the acute Weyl chamber, we deduce that

$$\langle wv, \alpha\rangle \ = \ \langle v, w^{-1}\alpha\rangle \ \geq 0.$$

Thus, for $\alpha \in \Delta(swv)$, we get

$$0 < \langle swv, \varpi_\alpha\rangle \ = \ \langle wv, \varpi_\alpha - \alpha\rangle \ = \ \langle wv, \varpi_\alpha\rangle - \langle wv, \alpha\rangle.$$

It follows that

$$\langle wv, \varpi_\alpha\rangle > \langle wv, \alpha\rangle \geq 0,$$

hence $\alpha \in \Delta(wv)$. □

Lemma 3.3.14 *Let $w \in W^v$. Then*

$$|\Delta \setminus \Delta(wv)| \leq \ell(w).$$

Proof Let $w = s_1 \cdots s_r$, $r = \ell(w)$, be a reduced decomposition of w into simple reflections s_i. Since

$$W^v = \{w \in W \mid \ell(ws) = \ell(w) + 1 \ \text{ for all simple reflections } s \in W_v\}$$

([38], Prop. 2.3.3 and 2.2.8), we conclude that all expressions

$$s_r, \ s_{r-1}s_r, \ldots, s_2 \cdots s_r$$

are contained in W^v, as well. The statement follows by applying inductively Lemma 3.3.13. □

Proof (*of Corollary 3.3.12*) Applying the previous lemma, we get for all $w \in W^v$ the inequality

$$|\Delta(wv)| + 2\ell(w) \geq (|\Delta| - \ell(w)) + 2\ell(w) = |\Delta| + \ell(w) \geq |\Delta|.$$

Equality holds in this chain if and only if $w = 1$. Since $|\Delta| = n - 1$, the first two assertions follow. For the last statement, we note that $\Delta(w_0\nu) = \emptyset$, since $w_0\nu = (x_r^{(n_r)}, \ldots, x_1^{(n_1)})$. It can also be deduced from the connectedness of $\mathcal{F}(\nu)^{ss}$. □

We stress the opposite behavior of the quantities $\ell(w)$ and $\Delta(w\nu)$ with respect to the *Bruhat order* $\leq$ on W/W_ν. If $w' \leq w$ we have $\ell(w') \leq \ell(w)$, but $\Delta(w\nu) \subset \Delta(w'\nu)$. The latter inclusion can be seen as follows. Let $\geq$ be the semi-order on $X^*(T)$ defined by

$$\chi_1 \geq \chi_2 \text{ if } \chi_1 - \chi_2 = \sum_{\alpha\in\Delta} n_\alpha \cdot \alpha \text{ with } n_\alpha \in \mathbb{N}_0.$$

If $w' \leq w$ (or equivalently $w'^{-1} \leq w^{-1}$) then $w'^{-1}\varpi_\alpha \geq w^{-1}\varpi_\alpha$ for all $\alpha \in \Delta$, cf. [79], §3. Thus $w'^{-1}\varpi_\alpha - w^{-1}\varpi_\alpha = \sum_{\alpha\in\Delta} n_\alpha \cdot \alpha$ with $n_\alpha \geq 0$. But ν is contained in the closure of the acute Weyl chamber which means that $\langle \nu, \alpha\rangle \geq 0$ for all $\alpha \in \Delta$. We deduce that if $\langle w\nu, \varpi_\alpha\rangle = \langle \nu, w^{-1}\varpi_\alpha\rangle \geq 0$ then $\langle w'\nu, \varpi_\alpha\rangle = \langle \nu, w'^{-1}\varpi_\alpha\rangle \geq 0$.

The following example shows that for $w' \leq w$, the contribution corresponding to w' can occur in larger degree than that of w.

Example 3.3.15 Let $n = 5$ and let $\nu = (x_1 > x_2 > x_3 > x_4 > x_5)$ with $\sum_{i=1}^5 x_i = 0$ and $x_4 > 0$. Then $W^\nu = W = S_5$. Consider the cyclic permutations $w, w' \in W$,

$$w' = (2, 3, 4) \text{ resp. } w = (2, 3, 4, 5).$$

We have $w' < w$ and $\ell(w') = 2$ resp. $\ell(w) = 3$. Further, we compute

$$\Delta(w'\nu) = \{\alpha_1, \alpha_2, \alpha_3, \alpha_4\} \text{ resp. } \Delta(w\nu) = \{\alpha_1\}.$$

Thus,

$$2\ell(w') + |\Delta(w'\nu)| = 8\,, \quad \text{but } 2\ell(w) + |\Delta(w\nu)| = 7.$$

Example 3.3.16 In contrast to Ω^n, the cohomology of arbitrary period domains is not pure in general. Consider Example 3.3.7 (i). Theorem 3.3.10 gives

$$H_c^0(\mathcal{F}(\nu)^{ss}) = 0 \qquad H_c^4(\mathcal{F}(\nu)^{ss}) = t \cdot v_B^G \oplus t^2 \cdot i_G^G$$

$$H_c^1(\mathcal{F}(\nu)^{ss}) = 0 \qquad H_c^5(\mathcal{F}(\nu)^{ss}) = t^2 \cdot v_{P_{(1,2)}}^G$$

$$H_c^2(\mathcal{F}(\nu)^{ss}) = v_B^G \qquad H_c^6(\mathcal{F}(\nu)^{ss}) = t^3 \cdot i_G^G$$

$$H_c^3(\mathcal{F}(\nu)^{ss}) = t \cdot v_{P_{(1,2)}}^G \qquad H_c^i(\mathcal{F}(\nu)^{ss}) = 0 \;\; \forall i > 6\,.$$

Variant 3.3.17 Consider the category $\mathrm{Fil}^k_{\mathbb{F}_1}$. In this context we get

$$H^*_c(\mathcal{F}(\nu)^{\mathrm{ss}}) = \bigoplus_{w \in W/W_\nu} t^{\ell(w)} \cdot v^W_{W(w\nu)}[-2\ell(w) - |\Delta(w\nu)|].$$

Remark 3.3.18 The problem of computing the cohomology of $\mathcal{F}(\nu)^{\mathrm{ss}}$ in the context of $\mathrm{Fil}^{\mathbb{F}_1}_k$ makes sense for arbitrary fields k, even when a Frobenius automorphism no longer exists. If $k = \mathbb{C}$ one may work in the context of the theory of mixed Hodge structures [60]. We replace ℓ-adic cohomology with compact supports by singular cohomology with compact supports, with rational coefficients. In this situation the Tate twist is replaced by the 1-dimensional Hodge structure $\mathbb{Q}(1)$ with underlying vector space $(2\pi i)\mathbb{Q}$ and Hodge type $(-1, -1)$. The nth Tate twist t^n is replaced by $\mathbb{Q}(n) := \mathbb{Q}(1)^{\otimes n}$. With these trivial modifications, the results above carry over *mutatis mutandis*.

We finally use the Grothendieck–Lefschetz fixed point formula (3.3), page 84 to determine the number of points of $\mathcal{F}(\nu)^{\mathrm{ss}}$ over any finite extension $\mathbb{F}_{q^m}$ of $k = \mathbb{F}_q$. We may alternatively use either of the formulas for the Euler–Poincaré characteristic obtained in Theorem 3.3.5 and in Theorem 3.3.4.

We first use the formula for the Euler–Poincaré characteristic given by Theorem 3.3.5. Then

$$|\mathcal{F}(\nu)^{\mathrm{ss}}(\mathbb{F}_{q^m})| = \sum_{w \in W/W_\nu} (-1)^{a^G_{P(w\nu)}} \cdot \mathrm{trace}(\varphi^m, t^{\ell(w)} \cdot v^G_{P(w\nu)}).$$

Now φ^m acts on $t^{\ell(w)}$ by multiplication with $q^{m\ell(w)}$. On the generalized Steinberg representations $v^G_{P(w\nu)}$, $w \in W/W_\nu$, the action of φ is via the identity. Thus we get

$$\mathrm{trace}(\varphi^m, t^{\ell(w)} \cdot v^G_{P(w\nu)}) = q^{m\ell(w)} \cdot \dim v^G_{P(w\nu)}. \tag{3.18}$$

The dimension of $v^G_{P(w\nu)}$ may be computed by Theorem 3.2.5, as follows.

Lemma 3.3.19 *Let $I \subset \Delta$. Set $Y_I = \{w \in W \mid w\alpha > 0\ \forall \alpha \in I,\ w\alpha < 0\ \forall \alpha \in \Delta \setminus I\}$. Then*

(i) $\dim v^G_{P_I} = \sum_{w \in Y_I} q^{\ell(w)}$.

(ii) $\dim v^W_{W_I} = |Y_I|$.

Proof Consider the short exact sequence cut out of the complex (2.4), page 74,

$$\bigoplus_{\substack{I \subset K \subset \Delta \\ |K \setminus I| = 1}} i^G_{P_K} \to i^G_{P_I} \to v^G_{P_I} \to 0.$$

The underlying vector space of $i_{P_I}^G$ may be identified with the space of $\overline{\mathbb{Q}}_\ell$-valued functions on $(G/P_I)(k)$. For $w \in W$, let

$$N_w := \prod_{\substack{\alpha>0 \\ w\alpha<0}} U_\alpha$$

be the unipotent subgroup of B generated by the positive root subgroups made negative by w. Let $W^I \subset W$ be the set of representatives of shortest length in their cosets W/W_I. Consider the *Bruhat decomposition* of $(G/P_I)(k)$,

$$(G/P_I)(k) = \bigcup_{w\in W^I} N_w(k)wP_I(k)/P_I(k). \tag{3.19}$$

A basis of $i_{P_I}^G$ is given by the characteristic functions $\chi_{n,w}$ for $w \in W^I, n \in N_w(k)$ on the singletons $\{nwP_I(k)/P_I(k)\}$. Analogously, the characteristic functions $\chi_{n,w}$ for $w \in W^K, n \in N_w(k)$, form a basis of $i_{P_K}^G$. But Y_I coincides with the set $W^I\setminus(\bigcup_{K\supsetneq I} W^K)$, cf. [38], Prop. 2.3.3. On the other hand, these basis vectors are mapped into one another (up to sign) under the natural map, $\bigoplus_{\substack{I\subset K\subset\Delta \\ |K\setminus I|=1}} i_{P_K}^G \to i_{P_I}^G$. It follows that the functions $\chi_{n,w}$ with $w \in Y_I$ are precisely the basis vectors of $i_{P_I}^G$ which are not in the image of this map. Since $|N_w(k)| = q^{\dim N_w} = q^{\ell(w)}$, the claim (i) follows.

The claim (ii) follows in the same way, replacing the complex (2.4), page 74 by its W-analogue appearing in Theorem 3.2.7. □

Remark 3.3.20 If $I = \emptyset$, i.e., $P_I = B$ then $Y_I = \{w_0\}$, where w_0 is the longest Weyl group element in W. Thus $\dim v_B^G = q^{\ell(w_0)} = |R_u(B)(k)|$ which coincides with the formula in [38], Cor. 6.4.3., cf. also [46], Prop. 67.12.

We set $Y_w := Y_{\Delta\setminus\Delta(wv)}$ for $w \in W/W_v$.

Proposition 3.3.21 *We have*

$$|\mathcal{F}(v)^{\mathrm{ss}}(\mathbb{F}_{q^m})| = \sum_{w\in W/W_v} (-1)^{a^G_{P(wv)}} q^{m\ell(w)} \cdot \sum_{y\in Y_w} q^{\ell(y)}.$$

Proof By Lemma 3.3.19 we have $\dim v_{P_I}^G = \sum_{y\in Y_I} q^{\ell(y)}$. From (3.18), page 95 we see that

$$\mathrm{trace}(\varphi^m, t^{\ell(w)} \cdot v^G_{P(wv)}) = q^{m\ell(w)} \cdot \sum_{y\in Y_w} q^{\ell(y)}.$$

Thus the claim follows from Theorem 3.3.5. □

Variant 3.3.22 In the context of $\mathrm{Fil}^k_{\mathbb{F}_1}$, by Lemma 3.3.19 (ii) and Variant 3.3.6,

$$|\mathcal{F}(v)^{\mathrm{ss}}(\mathbb{F}_{q^m})| = \sum_{w\in W/W_v} (-1)^{a^{\mathrm{W}}_{W(wv)}} q^{m\ell(w)} \cdot |Y_w|.$$ □

Now we count the rational points of $\mathcal{F}(\nu)^{ss}$, starting with the formula for the Euler–Poincaré characteristic given by Theorem 3.3.4. We have $\dim i_P^G = |G(k)/P(k)|$ and by the Bruhat decomposition $|G(k)/P(k)| = \sum_{w\in W/W_P} q^{\ell(w)}$. It follows that

$$\begin{aligned}|\mathcal{F}(\nu)^{ss}(\mathbb{F}_{q^m})| &= \sum_{P\in\mathcal{P}}(-1)^{a_P^G}\cdot\sum_{w'\in W/W_P} q^{\ell(w')}\cdot\sum_{w\in W/W_\nu} q^{m\ell(w)}\cdot\hat{\tau}_P^G(w\nu).\\ &= \sum_{P\in\mathcal{P}}(-1)^{a_P^G}\cdot\sum_{w'\in W/W_P} q^{\ell(w')}\cdot\sum_{\substack{w\in W/W_\nu\\ \Delta_P\subset\Delta(w\nu)}} q^{m\ell(w)}\\ &= \sum_{P\in\mathcal{P}}(-1)^{a_P^G}\cdot\sum_{w'\in W/W_P}\sum_{\substack{w\in W/W_\nu\\ \Delta_P\subset\Delta(w\nu)}} q^{m\ell(w)+\ell(w')}.\end{aligned}\tag{3.20}$$

One may also use the formula (3.4), page 84 for the number of points of $(G/P)(k)$. For an s-tuple of positive integers $(m_1,\dots,m_s)$ with $\sum_1^s m_i = n$, the cardinality of $|G(k)/P_{(m_1,\dots,m_s)}(k)|$ is given by

$$\mathrm{Pol}_{(m_1,\dots,m_s)}(q) := \frac{P_n(q)}{P_{m_1}(q)\cdot\;\cdots\;\cdot P_{m_s}(q)},$$

where

$$P_n(X) = (X-1)(X^2-1)\cdots(X^n-1).$$

Denote by Dec_n the set of such decompositions of n (for varying s). Then

$$|\mathcal{F}(\nu)^{ss}(\mathbb{F}_{q^m})| = \sum_{(m_1,\dots,m_s)\in\mathrm{Dec}_n}(-1)^{s-1}\cdot\mathrm{Pol}_{(m_1,\dots,m_s)}(q)\cdot\sum_{\substack{w\in W/W_\nu\\ \Delta_{P_{(m_1,\dots,m_s)}}\subset\Delta(w\nu)}} q^{m\ell(w)}.\tag{3.21}$$

Here we used that $(-1)^{s-1} = (-1)^{a^G_{P_{(m_1,\dots,m_s)}}}$.

Examples 3.3.23 Let us reconsider Examples 3.3.7.

(i) Let $\nu = (x_1,x_2,x_3)$ with $x_1 > x_2 > x_3$ and $x_2 - x_3 > x_1 - x_2$. The sets Y_w are given by the following table.

w	1	s_1	s_2	s_1s_2	s_2s_1	$s_1s_2s_1$
Y_w	$\{s_1s_2s_1\}$	$\{s_1s_2s_1\}$	$\{s_2s_1, s_1\}$	$\{1\}$	$\{s_2s_1, s_1\}$	$\{1\}$
$\Delta(w\nu)$	Δ	Δ	$\{\alpha_1\}$	$\emptyset$	$\{\alpha_1\}$	$\emptyset$

We get by using Proposition 3.3.21

$$\begin{aligned}&|\mathcal{F}(\nu)^{ss}(\mathbb{F}_{q^m})|\\ &= q^3 + q^{m+3} - q^m(q^2+q) + q^{2m} - q^{2m}(q^2+q) + q^{3m}\\ &= q^{3m} - q^{2m+2} - q^{2m+1} + q^{2m} - q^{m+2} - q^{m+1} + q^{m+3} + q^3.\end{aligned}\tag{3.22}$$

To apply formulas (3.20), page 97 and (3.21), page 97, we consider the following tables.

P	B	$P_{(2,1)}$	$P_{(1,2)}$	G
$(m_1,\dots,m_s)$	$(1,1,1)$	$(2,1)$	$(1,2)$	(3)
W/W_P	W	$\{1,(2,3),(1,2,3)\}$	$\{1,(1,2),(1,3,2)\}$	$\{1\}$
Δ_P	Δ	$\{\alpha_2\}$	$\{\alpha_1\}$	$\emptyset$
$\mathrm{Pol}_{(m_1,\dots,m_s)}(q)$	q^3+2q^2+2q+1	q^2+q+1	q^2+q+1	1

(3.23)

w	1	s_1	s_2	s_1s_2	s_2s_1	$s_1s_2s_1$
$\Delta(wv)$	Δ	Δ	$\{\alpha_1\}$	$\emptyset$	$\{\alpha_1\}$	$\emptyset$

The contributions of the parabolic subgroups (resp. of the elements in Dec_3) to (3.20), page 97 (resp. (3.21), page 97) are given as follows.

B	$(1,1,1)$	$(1+2q+2q^2+q^3)\cdot(1+q^m)$
$P_{(2,1)}$	$(2,1)$	$-(q^2+q+1)\cdot(1+q^m)$
$P_{(1,2)}$	$(1,2)$	$-(q^2+q+1)\cdot(1+2q^m+q^{2m})$
G	(3)	$1+2q^m+2q^{2m}+q^{3m}$.

Thus we get by applying (3.20), page 97 resp. (3.21), page 97,

$$\begin{aligned}|\mathcal{F}(v)^{\mathrm{ss}}(\mathbb{F}_{q^m})| &= (1+2q+2q^2+q^3)\cdot(1+q^m)-(q^2+q+1)\cdot(1+q^m)\\ &\quad -(q^2+q+1)\cdot(1+2q^m+q^{2m})+(1+2q^m+2q^{2m}+q^{3m}).\end{aligned}$$

The total sum coincides with (3.22), page 97.

(ii) Let $v=(x_1,x_2,x_3)$ with $x_1>x_2>x_3$ and $x_2-x_3=x_1-x_2$. In this case, the sets Y_w are given by the following table.

w	1	s_1	s_2	s_1s_2	s_2s_1	$s_1s_2s_1$
Y_w	$\{s_1s_2s_1\}$	$\{s_1s_2,s_2\}$	$\{s_2s_1,s_1\}$	$\{1\}$	$\{1\}$	$\{1\}$
$\Delta(wv)$	Δ	$\{\alpha_2\}$	$\{\alpha_1\}$	$\emptyset$	$\emptyset$	$\emptyset$.

We get by using Proposition 3.3.21

$$\begin{aligned}|\mathcal{F}(v)^{\mathrm{ss}}(\mathbb{F}_{q^m})| &= q^3-2q^m(q+q^2)+2q^{2m}+q^{3m}\\ &= q^{3m}+2q^{2m}-2q^{m+2}-2q^{m+1}+q^3. \qquad (3.24)\end{aligned}$$

To apply formulas (3.20), page 97 and (3.21), page 97, we have to consider table (3.23) again. The contributions of the parabolic subgroups (resp. of the elements in Dec_3) in (3.20), page 97 (resp. (3.21), page 97) are as follows.

B	$(1,1,1)$	$1+2q+2q^2+q^3$
$P_{(2,1)}$	$(2,1)$	$-(q^2+q+1)\cdot(1+q^m)$
$P_{(1,2)}$	$(1,2)$	$-(q^2+q+1)\cdot(1+q^m)$
G	(4)	$1+2q^m+2q^{2m}+q^{3m}$

Thus we get by applying (3.20), page 97 resp. (3.21), page 97

$$\begin{aligned} |\mathcal{F}(\nu)^{ss}(\mathbb{F}_{q^m})| &= 1 + 2q + 2q^2 + q^3 - (q^2 + q + 1) \cdot (1 + q^m) \\ &\quad -(q^2 + q + 1) \cdot (1 + q^m) + 1 + 2q^m + 2q^{2m} + q^{3m}. \end{aligned}$$

This formula coincides with (3.24), page 98.

(iii) Let us consider the Drinfeld case $\nu = (n-1, (-1)^{n-1})$. First we use Proposition 3.3.21. For this we note that $W/W_\nu = \{w_0, \ldots, w_{n-1}\}$, cf. Examples 3.3.7, (ii). Also note that for $i = 0, \ldots, n-1$, we have $\Delta(w_i\nu) = \{\alpha_{i+1}, \ldots, \alpha_{n-1}\}$ so that

$$\begin{aligned} Y_i = Y_{w_i} = \{y \in S_n \mid y(i+1) = n,\ y|[1,i] \text{ monotone increasing}, \\ y|[i+1,n] \text{ monotone decreasing}\}\,. \end{aligned} \tag{3.25}$$

The shortest element in Y_i is the element y_0 which maps

$$(1, \ldots, n) \mapsto (1, \ldots, i, n, n-1, \ldots, i+2, i+1)\,,$$

and its length is $\ell(y_0) = \binom{n-i}{2}$. An element $y \in Y_i$ is determined uniquely by the tuple of non-negative integers $(r_1, \ldots, r_i)$ given by the amount that the first, the second, etc. entry of y_0 is moved to the right. In other words, $y(j) = j + r_j$ for $j = 1, \ldots, i$. Here the entries $(r_1, \ldots, r_i)$ have to satisfy

$$0 \le r_1 \le r_2 \le \cdots \le r_i < n - i\,, \tag{3.26}$$

and, conversely, to every such tuple $\mathbf{r} = (r_1, \ldots, r_i)$ corresponds an element $y \in Y_i$ of length $\ell(y) = \ell(y_0) + |\mathbf{r}| = \binom{n-i}{2} + \sum_{j=1}^{i} r_j$. Let us denote by R_i the set of such tuples. Since $a^G_{P(w_i\nu)} = n - 1 - i$, we obtain from Proposition 3.3.21 the following expression:

$$|\Omega^n(\mathbb{F}_{q^m})| = \sum_{i=0}^{n-1} (-1)^{n-1-i} q^{mi} \sum_{\mathbf{r} \in R_i} q^{\binom{n-i}{2} + |\mathbf{r}|}\,. \tag{3.27}$$

Here for $i = 0$, the inner sum consists of one summand equal to $q^{\binom{n}{2}}$.

Somewhat dually, we may represent an element $y \in Y_i$ by an $(n-1-i)$-tuple of non-negative integers $(r_n, r_{n-1}, \ldots, r_{i+2})$ with

$$0 \le r_{i+2} \le r_{i+3} \le \cdots \le r_n < i + 1\,,$$

with corresponding element $y \in Y_i$ characterized by $y(j) = n + i + 1 - j - r_j$ for $j = i+2, \ldots, n$. In other words, we may also identify Y_i with R_{n-1-i}. In the same way as before we obtain the formula

$$|\Omega^n(\mathbb{F}_{q^m})| = \sum_{i=0}^{n-1} (-1)^{n-1-i} q^{mi} \sum_{\mathbf{r} \in R_{n-1-i}} q^{\binom{n-i}{2} + |\mathbf{r}|}\,. \tag{3.28}$$

Now let us use formula (3.21), page 97. For $(m_1, \ldots, m_s) \in \mathrm{Dec}_n$, we have

$$\Delta_{P_{(m_1,\ldots,m_s)}} \subset \Delta(w_i\nu) \iff m_1 \geq i+1\,.$$

Hence we get, applying (3.21), page 97,

$$|\Omega^n(\mathbb{F}_{q^m})| = \sum_{(m_1,\ldots,m_s)\in\mathrm{Dec}_n} (-1)^{s-1}\mathrm{Pol}_{(m_1,\ldots,m_s)}(q) \sum\nolimits_{i=0}^{m_1-1} q^{mi}\,. \tag{3.29}$$

Deligne and Lusztig count the points of $\Omega^n(\mathbb{F}_{q^m})$ in an elementary way, but also implicitly making use of the Harder–Narasimhan stratification. Their result is

$$|\Omega^n(\mathbb{F}_{q^m})| = (q^m - q)(q^m - q^2)\cdots(q^m - q^{n-1})\,, \tag{3.30}$$

cf. [58], Prop. 2.3. Expanding in terms of powers of q^m, we obtain the expression (3.28), page 99 (which is why we also gave the dual expression of (3.27), page 99).

Remark 3.3.24 A third basis of $K_0(i_B^G)$ is given by the irreducibles $j_{P_\lambda}^G$, where λ ranges over the partitions of n, cf. Theorem 3.2.1. Now

$$i_{P_\lambda}^G = j_{P_\lambda}^G + \sum_{\lambda'>\lambda} K_{\lambda',\lambda} \cdot j_{P_{\lambda'}}^G\,,$$

where $K_{\lambda',\lambda}$ is a *Kostka number*, as follows from Theorem 3.2.2 and the corresponding expression of the induced representation in terms of irreducibles for the symmetric group, cf. [157], p. 115. Plugging this into the expression of $\chi_c(\mathcal{F}(\nu))$ in terms of induced representations given by Theorem 3.3.4 yields an expression of the Euler–Poincaré characteristic in terms of irreducibles. However, this expression seems quite unwieldy.

Notes and References The observation that the Langlands Lemma can be applied to solve the recursion relation for the Euler–Poincaré characteristic is due to Kottwitz, and is mentioned in [185], Thm. 5.3. A similar recursion relation that can also be solved by applying the Langlands Lemma arises in the cohomology of the moduli space of G-bundles [151], cf. also [115]. Using this method, del Baño Rollin was able to determine the Hodge numbers of the moduli space of vector bundles on a compact Riemann surface [51]. The Betti numbers of this moduli space were determined earlier by Zagier [223]. Proposition 3.3.9 is the analogue over a finite field of the theorem of Schneider and Stuhler [196], which holds over local fields. The proof given here is different from the original proof. Theorem 3.3.10 was conjectured by the third author, and was proved by the second author in his thesis [169]. The second author subsequently generalized this result to the case of an arbitrary reductive group over a finite field, cf. [170]. This result will be discussed in Part 2 of this monograph.

In [153] Lehrer considers the cohomology of the open set of regular semi-simple elements in a split semi-simple algebraic group, and determines the number of its elements over a finite field. His method is completely different, although the problems look somewhat similar. For another example of point counting see the paper by Kisin and Lehrer [131].

Assertion (ii) of Lemma 3.3.19 is in Lehrer [152], for an arbitrary reductive group over a finite field.

Part 2

Period Domains for Reductive Groups over Finite Fields

IV. Interlude on the Tannakian Formalism

This chapter is preliminary for the next chapters in which we generalize to arbitrary reductive groups the theory developed in the first three chapters for GL_n. The purpose of this chapter is to recall those facts about tannakian categories that are used later on. There are essentially no proofs in this chapter, for which we refer to the original sources, the texts by Saavedra [193], by Deligne and Milne [62], by Deligne [55, 56], as well as [191].

1 Tensor categories

We recall briefly the concept of a tensor category ($\otimes$-category). For details we refer to [62, 193].

A *tensor category* is a category C, together with a bi-functor (called the *tensor product*)

$$\otimes : C \times C \longrightarrow C, \quad (X, Y) \longmapsto X \otimes Y \ ,$$

and functorial isomorphisms

$$\Psi_{X,Y} : X \otimes Y \xrightarrow{\sim} Y \otimes X \qquad X, Y \in \mathrm{ob}\, C$$

$$\Phi_{X,Y,Z} : X \otimes (Y \otimes Z) \xrightarrow{\sim} (X \otimes Y) \otimes Z \qquad X, Y, Z \in \mathrm{ob}\, C \ ,$$

which are called the *commutativity constraint* and the *associativity constraint*. These data are submitted to various conditions, of which we mention only the following samples.

(i) $\Psi_{Y,X} \circ \Psi_{X,Y} = \mathrm{id}_{X \otimes Y} \ , \qquad X, Y \in \mathrm{ob}\, C \ ,$

(ii) the commutativity of

$$\begin{array}{ccccc}
 & (X \otimes Y) \otimes Z & \xrightarrow{\Psi_{X \otimes Y,Z}} & Z \otimes (X \otimes Y) & \\
{\scriptstyle \Phi_{X,Y,Z}} \nearrow & & & & \searrow {\scriptstyle \Phi_{Z,X,Y}} \\
X \otimes (Y \otimes Z) & & & & (Z \otimes X) \otimes Y\,. \\
{\scriptstyle \mathrm{id}_X \otimes \Psi_{Y,Z}} \searrow & & & & \nearrow {\scriptstyle \Psi_{X,Z} \otimes \mathrm{id}_Y} \\
 & X \otimes (Z \otimes Y) & \xrightarrow{\Psi_{X,Z,Y}} & (X \otimes Z) \otimes Y &
\end{array}$$

Examples 4.1.1 (i) Let R be a commutative ring. Then the category $C = \mathrm{Mod}_R$ of R-modules *of finite type* with the usual tensor product (and the usual commutativity and associativity constraints) is a $\otimes$-category.
(ii) Let G be an affine group scheme over a field k. Let $\mathrm{Rep}_k(G)$ be the category of *finite-dimensional* representations of G over k. This is a $\otimes$-category. If (V, ρ) and (V', ρ') are objects in $\mathrm{Rep}_k(G)$ then $V \otimes V'$ is supplied with the diagonal action of G.
(iii) The category $\mathrm{Fil}^k_{\mathbb{F}_1}$ from Variant 1.1.6 is a $\otimes$-category.

Let C be a $\otimes$-category. A *unit object* of C is a pair $(\mathbf{1}, e)$, consisting of an object $\mathbf{1}$ of C and an isomorphism $e : \mathbf{1} \longrightarrow \mathbf{1} \otimes \mathbf{1}$ such that the functor $X \longmapsto \mathbf{1} \otimes X$ is an equivalence of categories from C to itself.

Remarks 4.1.2 (i) A unit object $(\mathbf{1}, e)$ defines a functorial isomorphism $e_X : X \xrightarrow{\sim} \mathbf{1} \otimes X$.
(ii) Let $(\mathbf{1}', e')$ be another unit object of C. Then there exists a unique isomorphism $a : \mathbf{1} \longrightarrow \mathbf{1}'$ making the following diagram commutative:

$$\begin{array}{ccc}
\mathbf{1} & \xrightarrow{e} & \mathbf{1} \otimes \mathbf{1} \\
{\scriptstyle a} \big\downarrow & & \big\downarrow {\scriptstyle a \otimes a} \\
\mathbf{1}' & \xrightarrow{e'} & \mathbf{1}' \otimes \mathbf{1}'\,.
\end{array}$$

We next recall the notion of an *internal Hom object* of two objects of a $\otimes$-category C. Let $X, Y \in \mathrm{ob}\, C$ and consider the functor

$$T \longmapsto \mathrm{Hom}(T \otimes X, Y)\,, \ T \in \mathrm{ob}\, C\,.$$

If this functor is representable, the representing object is denoted by $\underline{\mathrm{Hom}}(X, Y)$, i.e.,

$$\mathrm{Hom}(T \otimes X, Y) = \mathrm{Hom}(T, \underline{\mathrm{Hom}}(X, Y))\,.$$

The universal homomorphism corresponding to the identity homomorphism

$\mathrm{id}_{\underline{\mathrm{Hom}}(X,Y)}$ is called the *evaluation morphism* and is denoted by $\mathrm{ev}_{X,Y}$,

$$\mathrm{ev}_{X,Y} : \underline{\mathrm{Hom}}(X, Y) \otimes X \longrightarrow Y .$$

The morphism $f : T \longrightarrow \underline{\mathrm{Hom}}(X, Y)$ corresponding by the universal property to a given morphism $g : T \otimes X \longrightarrow Y$ is characterized by the commutativity of the following diagram:

$$\begin{array}{ccc} T \otimes X & \xrightarrow{\quad g \quad} & Y . \\ {\scriptstyle f \otimes \mathrm{id}} \big\downarrow & \nearrow_{\mathrm{ev}_{X,Y}} & \\ \underline{\mathrm{Hom}}(X, Y) \otimes X & & \end{array}$$

Let $(\mathbf{1}, e)$ be a unit object. The *dual* of an object $X \in \mathrm{ob}\, C$ is defined by

$$\check{X} = \underline{\mathrm{Hom}}(X, \mathbf{1}) ,$$

provided this last internal Hom is representable. In this case we abbreviate $\mathrm{ev}_{X,\mathbf{1}}$ into ev_X,

$$\mathrm{ev}_X : \check{X} \otimes X \longrightarrow \mathbf{1} .$$

Assume further that $\check{\check{X}}$ exists. Then we have bijections

$$\mathrm{Hom}(X, \check{\check{X}}) \xrightarrow{\sim} \mathrm{Hom}(X \otimes \check{X}, \mathbf{1}) \xrightarrow{\sim} \mathrm{Hom}(\check{X} \otimes X, \mathbf{1}) , \tag{1.1}$$

where the last arrow is induced by the commutativity constraint $\Psi_{X,\check{X}}$. The object X is called *reflexive* if the morphism $X \longrightarrow \check{\check{X}}$ which under (1.1) above corresponds to ev_X is an isomorphism.

Example 4.1.3 Let $C = \mathrm{Mod}_R$. Then internal Homs exist for any R-modules X, Y, and

$$\underline{\mathrm{Hom}}(X, Y) = \mathrm{Hom}_R(X, Y), \quad \check{X} = \mathrm{Hom}_R(X, R) .$$

An object $X \in \mathrm{Mod}_R$ is reflexive if and only if X is a projective R-module of finite type.

Let X_1, X_2, Y_1, Y_2 be objects of a $\otimes$-category C, and assume the existence of $\underline{\mathrm{Hom}}(X_i, Y_i)$ for $i = 1, 2$ and of $\underline{\mathrm{Hom}}(X_1 \otimes X_2, Y_1 \otimes Y_2)$. By the universal property of the evaluation morphism we get a morphism

$$\underline{\mathrm{Hom}}(X_1, Y_1) \otimes \underline{\mathrm{Hom}}(X_2, Y_2) \longrightarrow \underline{\mathrm{Hom}}(X_1 \otimes X_2, Y_1 \otimes Y_2) , \tag{1.2}$$

which is not necessarily an isomorphism.

This leads us to the concept of a rigid $\otimes$-category.

Definition 4.1.4 A *rigid tensor category* is a $\otimes$-category C with a unit object, satisfying the following conditions.

(i) Internal Homs $\underline{\mathrm{Hom}}(X, Y)$ exist for all $X, Y \in \mathrm{ob}\, C$.
(ii) All objects are reflexive.
(iii) All morphisms (1.2), page 107 are isomorphisms.

In [55] it is shown that the mere existence of dual objects, with the natural morphisms $\mathrm{ev}_X : \check{X} \otimes X \longrightarrow \mathbf{1}$ and $\delta : \mathbf{1} \longrightarrow \check{X} \otimes X$ satisfying some natural identities, implies all properties required above.

Examples 4.1.5 (i) The category $\mathrm{Modprtf}_R$, the full subcategory of Mod_R with objects the projective R-modules of finite type, is a rigid $\otimes$-category.
(ii) The $\otimes$-category $\mathrm{Rep}_k(G)$, where G is an affine group scheme over k (cf. 4.1.1, (ii)) is a rigid $\otimes$-category. The dual object of $(V, \rho) \in \mathrm{Rep}_k(G)$ is the contragredient representation.

Let C be a tensor category with unit object, in which every object X has a dual $\check{X}$. To any $f \in \mathrm{Hom}(X, Y)$ there corresponds a ${}^t f \in \mathrm{Hom}(\check{Y}, \check{X})$ characterized by the commutativity of the following diagram:

$$\begin{array}{ccc} \check{Y} \otimes X & \xrightarrow{\mathrm{id} \otimes f} & \check{Y} \otimes Y \\ {\scriptstyle {}^t f \otimes \mathrm{id}} \downarrow & & \downarrow {\scriptstyle \mathrm{ev}_{\check{Y}}} \\ \check{X} \otimes X & \xrightarrow[\mathrm{ev}_{\check{X}}]{} & \mathbf{1}\,. \end{array}$$

Hence we obtain a contravariant functor $\check{}: C \longrightarrow C$, which sends an object of C into its dual.

Proposition 4.1.6 *Let C be a rigid $\otimes$-category. Then the functor $\check{}$ is a contravariant equivalence of categories.* □

We next come to the characteristic functors between $\otimes$-categories.

Definition 4.1.7 Let C and C' be $\otimes$-categories with unit objects. A *tensor functor* from C to C' is a pair (F, c) consisting of a functor $F : C \longrightarrow C'$ and an isomorphism of functors

$$c: \quad \otimes \circ (F, F) \xrightarrow{\sim} F \circ \otimes\,.$$

We require $F(\mathbf{1}, e) = (F(\mathbf{1}), F(e))$ to be a unit object of C' and (F, c) to be compatible with the commutativity and the associativity constraints of C resp. C'.

We refer to [193], Ch. I, 4.2.1 and 4.2.2, resp. [62], 1.8 for the details on these last conditions. For instance, the compatibility with the commutativity constraints is the commutativity of the following diagram:

$$\begin{array}{ccc} F(X) \otimes F(Y) & \xrightarrow{\Psi_{F(X),F(Y)}} & F(Y) \otimes F(X) \\ {\scriptstyle c}\downarrow & & \downarrow{\scriptstyle c} \\ F(X \otimes Y) & \xrightarrow{F(\Psi_{X,Y})} & F(Y \otimes X) . \end{array}$$

Proposition 4.1.8 *Let $F : C \longrightarrow C'$ be a tensor functor between rigid $\otimes$-categories. Then there are canonical isomorphisms for all $X, Y \in C$,*

$$F(\underline{\mathrm{Hom}}(X, Y)) \xrightarrow{\sim} \underline{\mathrm{Hom}}(F(X), F(Y))$$
$$F(\check{X}) \xrightarrow{\sim} F(X)\check{} \, .$$

Proof See [193], Ch. I, 5.2.2. □

We next introduce morphisms between tensor functors.

Definition 4.1.9 Let (F, c) and (G, d) be tensor functors from the $\otimes$-category C into the $\otimes$-category C'. A *morphism of tensor functors* from (F, c) to (G, d) is a functor morphism $\lambda : F \longrightarrow G$ such that the following diagrams commute.

(i)

$$\begin{array}{ccc} F(X) \otimes F(Y) & \xrightarrow{c_{X,Y}} & F(X \otimes Y) \\ {\scriptstyle \lambda_X \otimes \lambda_Y}\downarrow & & \downarrow{\scriptstyle \lambda_{X \otimes Y}} \\ G(X) \otimes G(Y) & \xrightarrow{d_{X,Y}} & G(X \otimes Y) . \end{array}$$

(ii)

$$\begin{array}{ccc} F(\mathbf{1}) & & \\ & \searrow{\scriptstyle \sim} & \\ {\scriptstyle \lambda_1}\downarrow & & \mathbf{1}' . \\ & \nearrow{\scriptstyle \sim} & \\ G(\mathbf{1}) & & \end{array}$$

Here the oblique isomorphisms are the unique isomorphisms between unit objects mentioned in Remarks 4.1.2, (ii).

We denote by $\mathrm{Hom}^{\otimes}((F, c), (G, d))$, or simply by $\mathrm{Hom}^{\otimes}(F, G)$, the set of tensor morphisms from (F, c) to (G, d). We similarly have the notations $\mathrm{End}^{\otimes}(F)$, $\mathrm{Isom}^{\otimes}(F, G)$ and $\mathrm{Aut}^{\otimes}(F)$.

Proposition 4.1.10 *Let (F, c) and (G, d) be two tensor functors between rigid $\otimes$-categories. Then every tensor morphism from (F, c) to (G, d) is an isomorphism, i.e.,*

$$\mathrm{Hom}^{\otimes}(F, G) = \mathrm{Isom}^{\otimes}(F, G).$$

Proof (See [193], Ch. I 5.2.3, or [145], 1.1.9.) The inverse $\mu : G \longrightarrow F$ to $\lambda : F \longrightarrow G$ is defined by the commutative diagrams for varying X:

$$\begin{array}{ccc} F(\check{X}) & \xrightarrow{\lambda_{\check{X}}} & G(\check{X}) \\ \simeq \downarrow & & \downarrow \simeq \\ F(X)\check{} & \overset{{}^t\mu_X}{\dashrightarrow} & G(X)\check{}\,. \end{array}$$

□

We next consider abelian $\otimes$-categories.

Definition 4.1.11 Let R be a commutative ring. An *abelian* $\otimes$-category (resp. a *R-linear abelian* $\otimes$-category) is a $\otimes$-category which is abelian (resp. is abelian and R-linear) such that the bi-functor $\otimes$ is bilinear (resp. is R-bilinear).

Let C be an abelian $\otimes$-category, with unit object $\mathbf{1}$. Let $R = \mathrm{End}(\mathbf{1})$. Then R "acts" on every object $X \in \mathrm{ob}\, C$. Indeed, using the identification $X \xrightarrow{\sim} \mathbf{1} \otimes X$, we have a ring homomorphism

$$\mathrm{End}(\mathbf{1}) \longrightarrow \mathrm{End}(X)\,. \tag{1.3}$$

Now the image of (1.3) lies in the center of $\mathrm{End}(X)$. Hence R is commutative and the $\otimes$-category C is an abelian R-linear category.

Proposition 4.1.12 *Let C be a rigid abelian $\otimes$-category. Then the bi-functor $\otimes$ commutes with $\varinjlim$ and $\varprojlim$ in both variables. In particular, $\otimes$ is exact in both variables.*

Proof (see [145] 1.1.4) The functor $X \longmapsto X \otimes Y$ has a right adjoint $Z \longmapsto \underline{\mathrm{Hom}}(Y, Z)$, i.e.,

$$\mathrm{Hom}(X \otimes Y, Z) = \mathrm{Hom}(X, \underline{\mathrm{Hom}}(Y, Z))\,.$$

Hence the functor $X \longmapsto X \otimes Y$ commutes with $\varinjlim$. Similarly, $X \longmapsto X \otimes Y$ has a left adjoint $Z \longmapsto \underline{\mathrm{Hom}}(Y, Z)$, i.e.,

$$\mathrm{Hom}(Z, X \otimes Y) = \mathrm{Hom}(\underline{\mathrm{Hom}}(Y, Z), X)\,.$$

Hence the functor $X \longmapsto X \otimes Y$ commutes with $\varprojlim$. The assertions concerning the second factor are obtained by symmetry. □

For the structure of $R = \mathrm{End}(\mathbf{1})$ we mention the following fact.

Proposition 4.1.13 *Let C be a rigid abelian $\otimes$-category, with unit object* $\mathbf{1}$. *For any subobject U of* $\mathbf{1}$ *we have*

$$\mathbf{1} = U \oplus U^{\perp} ,$$

where $U^{\perp} = \ker(\mathbf{1} \longrightarrow \check{U})$. *This sets up a one-to-one correspondence between subobjects of* $\mathbf{1}$ *and idempotents in* $\mathrm{End}(\mathbf{1})$. *In particular, if* $\mathrm{End}(\mathbf{1})$ *is a field, then* $\mathbf{1}$ *is simple.*

Proof See [62], Prop. 1.17, and Rem. 1.18. □

Proposition 4.1.14 *Let C and C' be rigid abelian $\otimes$-categories such that* $\mathrm{End}(\mathbf{1})$ *is a field, and with $\mathbf{1}' \neq 0$. Every exact $\otimes$-functor from C to C' is faithful.*

Proof In C, we have

$$X \neq 0 \Longleftrightarrow \check{X} \otimes X \longrightarrow \mathbf{1} \qquad \text{is an epimorphism.}$$

This property is preserved by an exact $\otimes$-functor. □

Definition 4.1.15 (a) Let k be a field and let C be a rigid k-linear abelian $\otimes$-category. A *fiber functor* on C with values in a commutative k-algebra R is a k-linear exact faithful tensor functor

$$\omega : C \longrightarrow \mathrm{Mod}_R .$$

(b) Let ω and ω' be two fiber functors on C with values in R. A *morphism of fiber functors* from ω to ω' is simply a morphism of tensor functors from ω to ω'.

This concept now leads us to the notion of a tannakian category.

Definition 4.1.16 (a) A *Tannaka category* over a field k is a k-linear rigid abelian $\otimes$-category with $\mathrm{End}(\mathbf{1}) = k$, such that there exists a fiber functor on C with values in some k-algebra R.
(b) A Tannaka category is called *neutral* if it admits a fiber functor with values in k.

Example 4.1.17 Let G be an affine group scheme over k. The $\otimes$-category $\mathrm{Rep}_k(G)$ is a neutral Tannaka category with natural fiber functor

$$\omega = \omega^G : \mathrm{Rep}_k(G) \longrightarrow \mathrm{Vec}_k , \quad (V, \rho) \longmapsto V .$$

However, we will see below that there may be other fiber functors on $\mathrm{Rep}_k(G)$, non-isomorphic to ω^G.

Let (C, ω) be a neutral Tannaka category with a fiber functor over k. Then $\mathrm{Aut}^{\otimes}(\omega)$ is a group attached to (C, ω) with

$$\begin{aligned}\mathrm{Aut}^{\otimes}(\omega) = &\{(\lambda_X \in \mathrm{Aut}(\omega(X))_{X\in C} \mid \lambda_1 = \mathrm{id}_k, \lambda_{X\otimes Y} = \lambda_X \otimes \lambda_Y, \\ &\lambda_Y \circ \omega(\alpha) = \omega(\alpha) \circ \lambda_X, \forall \alpha \in \mathrm{Hom}(X, Y)\}\,.\end{aligned}$$

This definition can be made functorial. Define a group-valued functor $\underline{\mathrm{Aut}}^{\otimes}(\omega)$ on the category of k-algebras as follows:

$$\begin{aligned}\underline{\mathrm{Aut}}^{\otimes}(\omega)(R) = &\{\lambda_X \in \mathrm{Aut}(\omega(X) \otimes R)_{X\in C} \mid \lambda_1 = \mathrm{id}_R, \\ &\lambda_{X\otimes Y} = \lambda_X \otimes_R \lambda_Y, \\ &\lambda_Y \circ (\omega(\alpha) \otimes R) = (\omega(\alpha) \otimes R) \circ \lambda_X,\ \forall \alpha \in \mathrm{Hom}(X, Y)\}\,.\end{aligned} \tag{1.4}$$

The following theorem will be crucial for the use of the tannakian formalism in this part of the monograph.

Theorem 4.1.18 *Let G be an affine group scheme over the field k. The obvious map of group-valued functors*

$$G \longrightarrow \underline{\mathrm{Aut}}^{\otimes}(\omega^G)$$

is an isomorphism.

Proof See [62], Prop. 2.8. □

A homomorphism $f : G \longrightarrow G'$ of affine group schemes over k defines a tensor functor

$$\omega^f : \mathrm{Rep}_k(G') \longrightarrow \mathrm{Rep}_k(G), \quad (V', \rho') \longmapsto (V', \rho' \circ f)\,.$$

Corollary 4.1.19 *Let G and G' be affine group schemes over k and let $F : \mathrm{Rep}_k(G') \longrightarrow \mathrm{Rep}_k(G)$ be a tensor functor with $\omega^G \circ F = \omega^{G'}$. Then F is of the form $F = \omega^f$ for a unique homomorphism $f : G \longrightarrow G'$.* □

We next come to the main theorem on neutral Tannaka categories.

Theorem 4.1.20 *Let C be a neutral Tannaka category over k and let ω be a fiber functor on C defined over k.*

(i) *The functor $\underline{\mathrm{Aut}}^{\otimes}(\omega)$ on the category of k-algebras is representable by an affine group scheme G over k.*

(ii) *There exists a unique tensor equivalence $F : C \longrightarrow \mathrm{Rep}_k(G)$ such that $\omega^G \circ F = \omega$.*

Proof See [62], Thm. 2.11. □

We thus see that a neutral Tannaka category is equivalent to $\mathrm{Rep}_k(G)$ for an affine group scheme; the fiber functors allow us to keep track of the various natural equivalences. It is natural in this context to ask how properties of G are reflected in $\mathrm{Rep}_k(G)$; and similarly for homomorphisms $G \longrightarrow G'$. We mention the two following results in this direction.

Proposition 4.1.21 *Let G be an affine group scheme over k.*

(i) *G is a finite group scheme if and only if there exists $X \in \mathrm{Rep}_k(G)$ such that every object of $\mathrm{Rep}_k(G)$ is isomorphic to a subquotient of X^n for some $n \geq 0$.*

(ii) *G is algebraic (i.e., of finite type over k) if and only if there exists $X \in \mathrm{Rep}_k(G)$ such that every object of $\mathrm{Rep}_k(G)$ is isomorphic to a subquotient of a finite direct sum $\bigoplus_\alpha X^{\otimes m_\alpha} \otimes \check{X}^{\otimes n_\alpha}$ for some $m_\alpha, n_\alpha \geq 0$. Furthermore, if Y is a faithful representation of G, then $X = Y \oplus \check{Y}$ has the stated property.*

(iii) *Assume* char $k = 0$. *Then G is a pro-reductive group scheme if and only if $\mathrm{Rep}_k(G)$ is a semi-simple abelian category.*

Proof For (i) and (ii), see [62], Prop. 2.20, and its proof. For (iii), see [62], Prop. 2.23. □

Proposition 4.1.22 *Let $f : G \longrightarrow G'$ be a homomorphism of affine group schemes over k, with associated tensor functor $\omega^f : \mathrm{Rep}_k(G') \longrightarrow \mathrm{Rep}_k(G)$.*

(i) *f is faithfully flat if and only if ω^f is fully faithful and every subobject of $\omega^f(X')$ for $X' \in \mathrm{Rep}_k(G')$ is isomorphic to an object of the form $\omega^f(Y')$ for a subobject Y' of X'.*

(ii) *f is a closed immersion if and only if every object of $\mathrm{Rep}_k(G)$ is isomorphic to a subquotient of an object of the form $\omega^f(X')$ for $X' \in \mathrm{Rep}_k(G')$.*

Proof See [62], Prop. 2.21. □

We now investigate the relationship between different fiber functors ω and ω' on a Tannaka category C. Recall that $\underline{\mathrm{Aut}}^{\otimes}(\omega)$ was defined in two steps. Similarly, one defines in a first step the set $\mathrm{Isom}^{\otimes}(\omega, \omega')$ of isomorphisms of fiber functors, which is a principal homogeneous $\mathrm{Aut}^{\otimes}(\omega)$-set (possibly empty), and then, in a second step, a functor $\underline{\mathrm{Isom}}^{\otimes}(\omega, \omega')$ on the category of k-algebras.

Theorem 4.1.23 *Let ω, ω' be two fiber functors on C, defined over k, and let $G := \underline{\mathrm{Aut}}^{\otimes}(\omega)$.*

(i) *The functor $\underline{\mathrm{Isom}}^{\otimes}(\omega, \omega')$ is representable by a G-torsor.*

(ii) *The map* $\omega' \mapsto \underline{\mathrm{Isom}}^{\otimes}(\omega, \omega')$ *is part of an equivalence of categories from the groupoid of tensor functors on C over k and the groupoid of G-torsors.*

In particular the isomorphism classes of fiber functors on $\mathrm{Rep}_k(G)$ over k are parametrized by the cohomology set $H^1_{\mathrm{fpqc}}(\mathrm{Spec}\, k, G)$. If G is a smooth group scheme, this cohomology set may be replaced by Galois cohomology $H^1(k, G)$, cf. [99], §6.

We end this section with some remarks on non-neutral Tannaka categories. Let C be an Tannaka category over k. Then it is isomorphic to the representation category of an affine gerb [62]. We do not want to discuss these objects. But we will need the following.

For any k-algebra R, there is a *base change of C from k to R*, cf. [136], 3.3. This is an R-linear tensor category C_R with the same objects as C, but where the morphisms are given by

$$\mathrm{Hom}_{C_R}(X, Y) = \mathrm{Hom}_C(X, Y) \otimes_k R. \tag{1.5}$$

It is immediate that if $R = k'$ is a field extension of k, then $C_{k'}$ is a Tannaka category over k'. If $\omega : C \longrightarrow \mathrm{Vec}_{k'}$ is a fiber functor over k', then ω factors in a unique way as $\omega = \omega' \circ \beta$, where $\beta : C \longrightarrow C_{k'}$ is the obvious functor, and where $\omega' : C_{k'} \longrightarrow \mathrm{Vec}_{k'}$ is a fiber functor of $C_{k'}$. Hence its automorphism group $\underline{\mathrm{Aut}}^{\otimes}(\omega')$ is an affine group scheme over k'. If C is neutral, and ω is a fiber functor over k and $\omega' = \omega \otimes_k k'$, then there is a natural isomorphism of group schemes over k',

$$\underline{\mathrm{Aut}}^{\otimes}(\omega) \otimes_k k' \simeq \underline{\mathrm{Aut}}^{\otimes}(\omega'). \tag{1.6}$$

Finally we consider the following variant of automorphism groups defined by fiber functors. Let $\omega : \mathcal{T} \longrightarrow C$ be an exact $\otimes$-functor of Tannaka categories over k. We consider for any k-algebra R, its induced $\otimes$-functor $\omega^R : \mathcal{T} \longrightarrow C_R$ given by the composition of ω with the natural morphism $C \longrightarrow C_R$. Let $\mathrm{Aut}^{\otimes}(\omega^R)$ be its group of $\otimes$-automorphisms.

Proposition 4.1.24 (i) *The functor* $R \mapsto \mathrm{Aut}^{\otimes}(\omega^R)$ *is representable by an affine group scheme* $\underline{\mathrm{Aut}}^{\otimes}(\omega)$ *over k.*

(ii) *Let* $\mathcal{T} = \mathrm{Rep}_k(G')$ *and* $C = \mathrm{Rep}_k(G)$, *and let* $\omega = \omega^f : \mathcal{T} \longrightarrow C$ *for the homomorphism* $f : G \longrightarrow G'$. *Then* $\underline{\mathrm{Aut}}^{\otimes}(\omega) = \mathrm{Cent}_f(G')$.

Here $\mathrm{Cent}_f(G')$ denotes the centralizer of f in G'.

Proof See [136], App. A. □

Notes and References The tannakian formalism, as presented here, originated with Grothendieck and has had an enormous influence on subsequent developments. The formalism is used in the theory of mixed Hodge structures, the theory of F-crystals and the theory of motives, and many other contexts. The theory of tannakian categories is treated in Saavedra [193] and also by Deligne and Milne in [62]. It was completed by Deligne [55] by correcting a mistake in the original definition of a Tannaka category (Deligne observed the necessity of adding the condition $\mathrm{End}(\mathbf{1}) = k$ to its definition). Other references, which also include the theory of non-neutral tannakian categories, are [56, 191, 145]. For a survey of the theory, see [33].

2 Gradings and filtrations

Let Ab be the category of abelian groups. We associate to any abelian group M the group functor $D(M)$ on the category of k-algebras defined by

$$D(M)(R) = \mathrm{Hom}_{\mathrm{Ab}}(M, R^\times)\ .$$

It is representable by an affine group scheme, cf. [5, 64]. Namely, $D(M) = \mathrm{Spec}\, A$ with

$$A = k^{(M)} \qquad \text{(direct sum)}\ ,$$

with multiplication given in terms of the natural basis $(e_m)_{m\in M}$ of $k^{(M)}$ by

$$A \otimes_k A \longrightarrow A\ , \qquad e_m \otimes e_{m'} \longmapsto e_{m+m'}\ .$$

The group structure on $\mathrm{Spec}\, A$ is defined by a co-multiplication map $A \longrightarrow A \otimes_k A$ and an antipod (or coinverse) $A \longrightarrow A$, making A into a *Hopf algebra* [25]. Here the co-multiplication is given by

$$A \longrightarrow A \otimes_k A\ , \qquad e_m \longmapsto e_m \otimes e_m$$

and the antipod maps e_m to e_{-m}. The isomorphism of functors $D(M) \xrightarrow{\sim} \mathrm{Spec}\, A$ is defined by

$$\begin{aligned} D(M)(R) = \mathrm{Hom}_{\mathrm{Ab}}(M, R^\times) &\longrightarrow \mathrm{Hom}(A, R) \\ \varphi &\longmapsto f_\varphi(e_m) = \varphi(m)\ . \end{aligned}$$

An affine group scheme isomorphic to some $D(M)$ is called *diagonalizable*. The functor $M \longmapsto D(M)$ defines an anti-equivalence of categories between the category Ab and the category of diagonalizable affine group schemes over k.

Examples 4.2.1 (i) $D(\mathbb{Z}) = \mathbb{G}_m$.
(ii) $D(\mathbb{Z}/n\mathbb{Z}) = \mu_n$.
(iii) $\mathbb{D} = D(\mathbb{Q}) = D(\lim_{\rightarrow_n} \frac{1}{n}\mathbb{Z}) = \lim_{\leftarrow_n} \mathbb{G}_m$, cf. [135], 4.1.

(iv) We avoid the diagonalizable group $D(\mathbb{R})$ because it is difficult to visualize this group scheme.
(v) $D(M)$ is a torus if and only if M is a free abelian group of finite type.

Let M be an abelian group. We consider the category $M\text{-Grad}_k$ of *M-graded* finite-dimensional k-vector spaces; that is, finite-dimensional vector spaces V equipped with a direct sum decomposition

$$V = \bigoplus_{m \in M} V_m \, .$$

Morphisms in this category are k-linear maps of the underlying vector spaces which respect the M-gradings, i.e., which are of degree 0.

Proposition 4.2.2 *The category $M\text{-Grad}_k$ is a neutral Tannaka category over k, and the obvious functor $\omega_0 : M\text{-Grad}_k \longrightarrow \text{Vec}_k$ is a fiber functor over k. There is a natural equivalence of Tannaka categories between $M\text{-Grad}_k$ and $\text{Rep}_k(D(M))$, given by*

$$\begin{array}{rcl} \text{Rep}_k(D(M)) & \longrightarrow & M\text{-Grad}_k \\ (V, \rho) & \longmapsto & V = \bigoplus_{m \in M} V_m \, , \end{array}$$

where

$$V_m = \{ v \in V \mid \rho(g)v = \chi_m(g)v, \; g \in D(M) \} \, .$$

Here $\chi_m : D(M) \longrightarrow \mathbb{G}_m$ is the character equal to $\chi_m = D(\mathbb{Z} \stackrel{m}{\longrightarrow} M)$. □

In particular, the proposition gives a natural identification $\underline{\text{Aut}}^{\otimes}(\omega) = D(M)$. Hence we obtain from Corollary 4.1.19 the following useful fact.

Corollary 4.2.3 *Let G be an affine group scheme over k. There is a one-to-one correspondence between homomorphisms $D(M) \longrightarrow G$ and factorizations of tensor functors,*

$$\begin{array}{ccc} \text{Rep}_k(G) & \xrightarrow{\quad \omega^G \quad} & \text{Vec}_k \, . \\ & \searrow \; \; \nearrow_{\omega_0} & \\ & M\text{-Grad}_k & \end{array}$$

(Here ω_0 is the forgetful functor.) □

Definition 4.2.4 An *M-grading* of the natural fiber functor ω^G is a tensor functor $F : \text{Rep}_k(G) \longrightarrow M\text{-Grad}_k$ with $\omega^G = \omega_0 \circ F$.

We note that such a functor F is automatically exact and faithful.

Remarks 4.2.5 (i) Let F be an M-grading of ω^G, and for $(V,\rho) \in \mathrm{Rep}_k(G)$, let $F((V,\rho)) = \bigoplus_{m\in M} V_m$. This direct sum decomposition is not in general respected by the G-operation, i.e., this is not a direct sum of G-representations.

(ii) An M-grading of ω^G is equivalent to a grading $V = \bigoplus_{m\in M} V_m$ for each $(V,\rho) \in \mathrm{Rep}_k(G)$ with the following properties.

(a) It is functorial, i.e. for $\alpha \in \mathrm{Hom}_G((V,\rho),(V',\rho'))$ one has $\alpha(V_m) \subset V'_m$, $\forall m \in M$.

(b) It is compatible with the tensor product, i.e.

$$(V' \otimes V'')_m = \bigoplus_{m'+m''=m} V'_{m'} \otimes V''_{m''} .$$

(c) $\mathbf{1}_0 = k$.

(iii) Let $G = \mathrm{GL}(V)$. Then an M-grading of ω^G is equivalent to an M-grading of V. Indeed, an M-grading of ω^G defines by the previous remark an M-grading of V' for each $(V',\rho') \in \mathrm{Rep}_k(G)$. In particular, this applies to the natural representation (V,ρ) of G. Conversely, an M-grading $V = \bigoplus V_m$ defines a homomorphism

$$D(M) \longrightarrow G\ ,\ g \longmapsto \sum_{m\in M} \chi_m(g) \cdot \mathrm{id}_{V_m} \quad .$$

By Corollary 4.2.3, this homomorphism corresponds to an M-grading of ω^G. One checks easily that these constructions are inverse to one another.

(iv) Similarly, let G be the automorphism group of a finite-dimensional k-vector space V equipped with a non-degenerate alternating or symmetric bilinear form. Then an M-grading of ω^G is equivalent to an M-grading $V = \bigoplus_{m\in M} V_m$ such that $V_m \perp V_{m'}$ for $m \neq -m'$.

(v) Let K/k be a field extension. An *M-grading of ω^G over K* is a factorization of tensor functors,

$$\begin{array}{ccc}
\mathrm{Rep}_k(G) & \xrightarrow{\ \omega^G\otimes_k K\ } & \mathrm{Vec}_K \ . \\
 & \searrow\ (\text{dashed}) \qquad \nearrow_{\omega_0} & \\
 & M\text{-}\mathrm{Grad}_K &
\end{array}$$

Equivalently, it is the specification of an M-grading of $V \otimes_k K$ for every $(V,\rho) \in \mathrm{Rep}_k(G)$ satisfying the obvious analogues of conditions (a)–(c) in (ii) above.

We next turn to filtrations. Item (ii) of the Remarks 4.2.5 above motivates the following definition, cf. [193], Ch. IV. 2.†

† We only consider filtrations that Saavedra calls *exact*.

Definition 4.2.6 Let G be an affine group scheme over k. A *$\mathbb{Q}$-filtration $\mathcal{F}$* of ω^G is the specification of a $\mathbb{Q}$-filtration $\mathcal{F}V = \mathcal{F}^\bullet V$ on the underlying vector space of every object $(V,\rho) \in \mathrm{Rep}_k(G)$ (in the sense of Definition 1.1.1). This specification is subject to the following requirements.

(i) It is functorial, i.e., for each $\alpha : (V',\rho') \longrightarrow (V'',\rho'')$, one has

$$\alpha(\mathcal{F}^x V') \subset \mathcal{F}^x V'' , \quad \forall x \in \mathbb{Q} .$$

(ii) It is compatible with tensor products,

$$\mathcal{F}^x(V' \otimes V'') = \sum_{x'+x''=x} \mathcal{F}^{x'} V' \otimes \mathcal{F}^{x''} V'' .$$

(iii)

$$\mathcal{F}^x \mathbf{1} = \begin{cases} k & \text{for} \quad x \leq 0 \\ (0) & \text{for} \quad x > 0 \end{cases}$$

(iv) The functor

$$\mathrm{Rep}_k(G) \longrightarrow \mathbb{Q}\text{-}\mathrm{Grad}_k , \quad (V,\rho) \longmapsto \mathrm{gr}^\bullet_{\mathcal{F}}(V)$$

is exact.

Let $\mathbb{Q}\text{-}\mathrm{Fil}_k^k$ be the category of Definition 1.1.3 with $K = k$, where, however, we insist on $\mathbb{Q}$-filtrations instead of $\mathbb{R}$-filtrations. There are two functors:

$$\begin{array}{ccc} & & \mathrm{Vec}_k \\ & \overset{\omega_0}{\nearrow} & \\ \mathbb{Q}\text{-}\mathrm{Fil}_k^k & & \\ & \underset{\mathrm{gr}}{\searrow} & \\ & & \mathbb{Q}\text{-}\mathrm{Grad}_k \end{array} .$$

Here ω_0 is the forgetful functor $(V, \mathcal{F}^\bullet) \longmapsto V$, and gr is the associated graded, $(V, \mathcal{F}^\bullet) \longmapsto \mathrm{gr}^\bullet_{\mathcal{F}}(V)$. As in the case of gradings of ω^G one sees that a $\mathbb{Q}$-filtration of ω^G is equivalent to a tensor functor factorization

$$\begin{array}{ccc} \mathrm{Rep}_k(G) & \xrightarrow{\quad \omega^G \quad} & \mathrm{Vec}_k \\ & {\scriptstyle F}\searrow \quad \nearrow{\scriptstyle \omega_0} & \\ & \mathbb{Q}\text{-}\mathrm{Fil}_k^k & \end{array}$$

such that the functor $\mathrm{gr} \circ F : \mathrm{Rep}_k(G) \longrightarrow \mathbb{Q}\text{-}\mathrm{Grad}_k$ is exact.

Remark 4.2.7 In Definition 4.2.6, we may equivalently remove axiom (iv) and replace axiom (i) by :

(i′) it is *strictly* functorial, i.e., for each $\alpha : (V',\rho') \longrightarrow (V'',\rho'')$, one has $\alpha(\mathcal{F}^x V') = \alpha(V') \cap \mathcal{F}^x V''$, $\forall x \in \mathbb{Q}$.

Remark 4.2.8 A $\mathbb{Q}$-grading of ω^G defines a $\mathbb{Q}$-filtration of ω^G. In one guise, this association is given by composition with the functor

$$\mathrm{Fil} : \mathbb{Q}\text{-Grad}_k \longrightarrow \mathbb{Q}\text{-Fil}_k^k, \quad V = \bigoplus_{x\in\mathbb{Q}} V_x \longmapsto \mathcal{F}^x V = \sum_{y\geq x} V_y \,.$$

In the other guise, it is given by the diagram of functors:

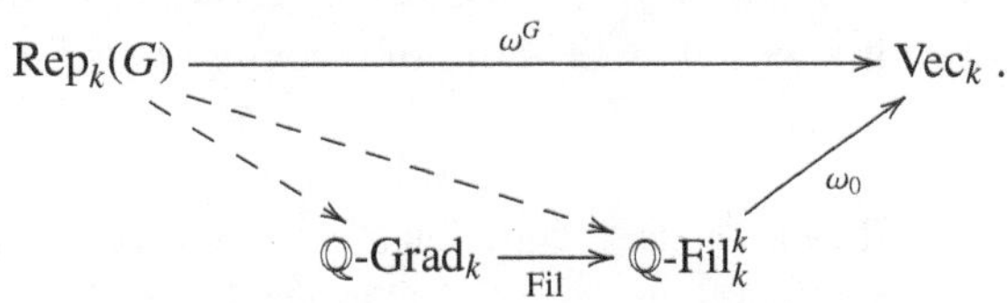

Variants 4.2.9 (i) One similarly defines $\mathbb{Z}$-filtrations of ω^G, or $\mathbb{R}$-filtrations. We will avoid the latter, since the diagonalizable group $D(\mathbb{R})$ is difficult to visualize, cf. Examples 4.2.1, (iv), cf. Theorem 4.2.13 below.

(ii) Let K be a field extension of k. A $\mathbb{Q}$-*filtration* $\mathcal{F}$ of ω^G *over* K is the specification of a $\mathbb{Q}$-filtration $\mathcal{F}V = \mathcal{F}^\bullet V$ of the K-vector space $V \otimes_k K$, for each object $(V,\rho) \in \mathrm{Rep}_k(G)$, satisfying the obvious analogues of conditions (i)–(iv) in Definition 4.2.6 above. Note that the functor in (iv) becomes

$$\mathrm{Rep}_k(G) \longrightarrow \mathbb{Q}\text{-Grad}_K, \, (V,\rho) \longmapsto \mathrm{gr}^\bullet_{\mathcal{F}}(V) \,.$$

Equivalently, a $\mathbb{Q}$-filtration of ω^G over K is a tensor functor factorization

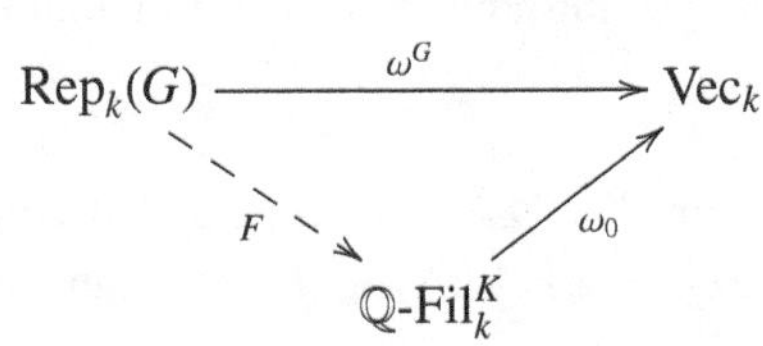

such that the tensor functor $\mathrm{gr} \circ F : \mathrm{Rep}_k(G) \longrightarrow \mathbb{Q}\text{-Grad}_K$ is exact. As in Remark 4.2.8, a $\mathbb{Q}$-grading of ω^G over K (cf. Remark 4.2.5, (v)) defines a $\mathbb{Q}$-filtration of ω^G over K.

(iii) More generally, let C be a Tannaka category over k, and ω a fiber functor on C defined over an extension K of k. A $\mathbb{Q}$-filtration of ω is a factorization $\omega = \omega_0 \circ F$ with F a tensor functor $C \longrightarrow \mathbb{Q}\text{-Fil}_k^K$ such that $\mathrm{gr} \circ F$ is exact.

Remarks 4.2.10 (i) We note the following functoriality property. Let $f : G \longrightarrow G'$ be a homomorphism of affine group schemes over k, with corresponding tensor functor $\omega^f : \mathrm{Rep}_k(G') \longrightarrow \mathrm{Rep}_k(G)$. Then a $\mathbb{Q}$-filtration $\mathcal{F}$ of ω^G induces in the obvious way a $\mathbb{Q}$-filtration $\mathcal{F}' = f_*(\mathcal{F})$ of $\omega^{G'}$, compatibly with the passage to the associated gradings, via Corollary 4.2.3,

$$\begin{array}{ccc} \mathbb{D} & \xrightarrow{\ \mathrm{gr}\circ F_{\mathcal{F}'}\ } & G' . \\ & {\scriptstyle \mathrm{gr}\circ F_{\mathcal{F}}} \searrow \quad \nearrow {\scriptstyle f} & \\ & G & \end{array}$$

(ii) The motivation for imposing the exactness condition (iv) in Definition 4.2.6 is the desire to describe filtrations of ω^G in terms of gradings of ω^G, cf. Theorem 4.2.13 below.

Remark 4.2.11 Let $G = \mathrm{GL}(V)$. Then a $\mathbb{Q}$-filtration of ω^G defines a $\mathbb{Q}$-filtration $\mathcal{F}^\bullet V$ of V through the natural representation of G. Conversely, let $\mathcal{F}^\bullet V$ be a $\mathbb{Q}$-filtration of V. A *splitting of* $\mathcal{F}^\bullet V$ is a direct sum decomposition $V = \bigoplus_{x\in\mathbb{Q}} V_x$ with associated filtration $V^x = \sum_{y\geq x} V_y$ equal to $\mathcal{F}^\bullet V$. Such a splitting always exists. By Remark 4.2.5 (ii) it defines a $\mathbb{Q}$-grading of ω^G. The associated $\mathbb{Q}$-filtration $\mathcal{F}$ of ω^G (see Remark 4.2.8) is independent of the choice of the splitting of $\mathcal{F}^\bullet V$. We therefore see that a $\mathbb{Q}$-filtration of ω^G is equivalent to a $\mathbb{Q}$-filtration of V. Similarly, a $\mathbb{Q}$-filtration of ω^G over an extension field K of k is equivalent to a $\mathbb{Q}$-filtration of the K-vector space $V \otimes_k K$.

We leave it to the reader to formulate a similar statement concerning symplectic and orthogonal groups, cf. Remark 4.2.5 (iv), as far as $\mathbb{Q}$-gradings are concerned.

The existence of a splitting of a filtration is also true in a more general context.

Definition 4.2.12 A $\mathbb{Q}$-filtration $\mathcal{F}$ of ω^G is *splittable* if it comes from a $\mathbb{Q}$-grading of ω^G, in the sense of Remark 4.2.8. A choice of any such $\mathbb{Q}$-grading of ω^G is called a *splitting* of $\mathcal{F}$.

We note that by Corollary 4.2.3, a splitting of the $\mathbb{Q}$-filtration $\mathcal{F}$ of ω^G is given by a $\mathbb{Q}$-1-PS $\lambda : \mathbb{D} \longrightarrow G$. If $\mathcal{F}$ is a $\mathbb{Z}$-filtration, then a splitting is given by a 1-PS $\lambda : \mathbb{G}_m \longrightarrow G$. We make the analogous definitions for a $\mathbb{Q}$-filtration of ω^G over an extension field K of k. A splitting then corresponds to a $\mathbb{Q}$-1-PS that is defined over K, i.e., $\lambda : \mathbb{D}_K \longrightarrow G_K$.

Let $\mathcal{F}$ be a $\mathbb{Q}$-filtration of ω^G. We introduce the following group-valued subfunctors of G. Let $P = P_{\mathcal{F}} = \underline{\mathrm{Aut}}^{\otimes}_{\mathcal{F}}(\omega^G)$, with

$$\underline{\mathrm{Aut}}^{\otimes}_{\mathcal{F}}(\omega^G)(R) = \{\lambda \in G(R) \mid \lambda(\mathcal{F}^x V \otimes R) \subset \mathcal{F}^x V \otimes R\,, \quad \forall (V,\rho) \in \mathrm{Rep}_k(G)\,,\ \forall x \in \mathbb{Q}\}\,, \tag{2.1}$$

for any k-algebra R. Similarly, let $U = U_{\mathcal{F}} = \underline{\mathrm{Aut}}^{\otimes !}_{\mathcal{F}}(\omega^G)$, with

$$\underline{\mathrm{Aut}}^{\otimes !}_{\mathcal{F}}(\omega^G)(R) = \{\lambda \in P(R) \mid (\lambda - \mathrm{id})(\mathcal{F}^x V \otimes R) \subset \mathcal{F}^{x+} V \otimes R\,, \quad \forall (V,\rho) \in \mathrm{Rep}_k(G)\,,\ \forall x \in \mathbb{Q}\}\,. \tag{2.2}$$

We may now formulate the main result on filtrations of ω^G.

Theorem 4.2.13 *Let G be an algebraic group over k.*

(a) *If either G is reductive or k has characteristic* 0, *then every $\mathbb{Q}$-filtration $\mathcal{F}$ of ω^G is splittable.*

Assume further that G is reductive.

(b) *The group-functor $P_{\mathcal{F}}$ is representable by a parabolic subgroup of G, and $U_{\mathcal{F}}$ by the unipotent radical of $P_{\mathcal{F}}$. For the Lie algebras*

$$\mathrm{Lie}\, P_{\mathcal{F}} = \mathcal{F}^0 \mathrm{Lie}\, G\,,\ \mathrm{Lie}\, U_{\mathcal{F}} = \mathcal{F}^{0+} \mathrm{Lie}\, G\,.$$

Here Lie G is equipped with a $\mathbb{Q}$-filtration through the adjoint representation of G.

(c) *Let $\lambda : \mathbb{D} \longrightarrow G$ be a splitting of the $\mathbb{Q}$-filtration $\mathcal{F}$ of ω^G. The centralizer $Z(\lambda)$ of λ is a Levi subgroup of $P_{\mathcal{F}}$, i.e., $Z(\lambda) \xrightarrow{\sim} P_{\mathcal{F}}/U_{\mathcal{F}}$ and the induced homomorphism $\overline{\lambda} : \mathbb{D} \longrightarrow P_{\mathcal{F}}/U_{\mathcal{F}}$ is central.*

(d) *Let λ and λ' be two* $\mathbb{Q}$-1-PS *with associated $\mathbb{Q}$-filtrations $\mathcal{F}$ resp. $\mathcal{F}'$ of ω^G. Then the following statements are equivalent:*

- $\mathcal{F} = \mathcal{F}'$
- $P_{\mathcal{F}} = P_{\mathcal{F}'}$ *and* $\overline{\lambda} = \overline{\lambda}' : \mathbb{D} \longrightarrow P_{\mathcal{F}}/U_{\mathcal{F}}$
- $\exists\, g \in P_{\mathcal{F}}(k)$ *with* $\lambda' = \mathrm{Int}(g) \circ \lambda$
- $\exists!\, u \in U_{\mathcal{F}}(k)$ *with* $\lambda' = \mathrm{Int}(u) \circ \lambda$

Here Int(g) denotes the inner automorphism $x \longmapsto gxg^{-1}$ of G.

Proof See [193], Ch. IV, Prop. 2.2.5, cf. also [164], Ch. 2, §2. □

We leave it to the reader to formulate the analogous statements for splittings of $\mathbb{Z}$-filtrations of ω^G, and also for $\mathbb{Q}$-filtrations of ω^G over an extension field K of k.

Remark 4.2.14 If k has characteristic 0, then statement (a) still holds true when G is not reductive, cf. [193], Ch. IV 2.2. But not the other statements.

Example 4.2.15 Let $G = T$ be a torus. Then a $\mathbb{Q}$-filtration $\mathcal{F}$ of ω^T splits in a unique way, i.e., is given by a $\mathbb{Q}$-1-PS $\lambda : \mathbb{D} \longrightarrow T$. In other words, the $\mathbb{Q}$-filtrations $\mathcal{F}$ of ω^T are in one-to-one correspondence with $X_*(T)_{\mathbb{Q}}^{\Gamma_k} = X_*(T)_k \otimes \mathbb{Q}$. Similarly, the $\mathbb{Q}$-filtrations of ω^T over an extension field K of k correspond bijectively to the elements of $X_*(T)_{\mathbb{Q}}^{\Gamma_K}$. Here Γ_K denotes the absolute Galois group of K.

Definition 4.2.16 Let λ and λ' be $\mathbb{Q}$-1-PS of G. Then λ and λ' are said to be *par-equivalent*, if the four equivalent conditions in Theorem 4.2.13 (d) are satisfied.

We make a similar definition if λ and λ' are defined over some extension field K of k.

Let $\lambda : \mathbb{D} \to G$ be a splitting of a given $\mathbb{Q}$-filtration $\mathcal{F}$ of ω^G. In this case we also write P_λ instead of $P_{\mathcal{F}}$, resp. U_λ instead of $U_{\mathcal{F}}$. By Theorem 4.2.13 (d), these subgroups only depend on the par-equivalence class of λ. For any field extension K of k, their K-valued points are given by (cf. [164], Def. 2.3/Prop. 2.6)

$$P_\lambda(K) = \{g \in G(K) \mid \lim_{t\to 0} \mathrm{Int}(\lambda(t)) \circ g \text{ exists}\}, \tag{2.3}$$

resp.

$$U_\lambda(K) = \{g \in G(K) \mid \lim_{t\to 0} \mathrm{Int}(\lambda(t)) \circ g = 1\}.$$

To end this section, we record a result on filtrations of fiber functors which will be useful in the next chapters.

Proposition 4.2.17 *Let G be a reductive group over k and P a psgp defined over k. Let $\mathcal{F}$ be a $\mathbb{Q}$-filtration on ω^G over an extension field K of k. Then there exists a unique splittable $\mathbb{Q}$-filtration $\mathcal{F}_P$ on ω^P over K which induces $\mathcal{F}$ on ω^G.*

Proof We first prove the existence statement. Let Q be the K-parabolic of G_K associated to the filtration $\mathcal{F}$. Then $Q \cap P_K$ contains a maximal K-split torus T of G, cf. [26], 4.18. Let $\lambda : \mathbb{D}_K \longrightarrow G_K$ be a $\mathbb{Q}$-1-PS whose associated $\mathbb{Q}$-filtration is equal to $\mathcal{F}$. Since all maximal K-split tori of Q are conjugate under $Q(K)$ [26], 8.2, we may choose λ factoring through T. But then λ factors through P_K, and hence defines a $\mathbb{Q}$-filtration of ω^P over K which induces by construction the given filtration $\mathcal{F}$ on ω^G.

Now we prove the uniqueness statement. Let $\lambda, \lambda' : \mathbb{D}_K \longrightarrow P_K$ be two $\mathbb{Q}$-1-PS inducing the same $\mathbb{Q}$-filtration $\mathcal{F}$ on ω^G over K. Then λ and λ' factor through $P_K \cap Q$. Since all maximal K-split tori of $P_K \cap Q$ are conjugate, we may conjugate λ and λ' by elements of $P(K) \cap Q(K)$ such that both $\mathbb{Q}$-1-PS factor through T. By the definition of Q, this does not affect the associated filtrations on ω^P over K. Now we are going to prove $\lambda = \lambda'$. Now there is a unique Levi subgroup M of Q containing T, namely the centralizer of $T \cap \mathrm{rad}(Q)$, cf. [26], 4.15 (b). Since both centralizers of λ and λ' are Levi subgroups of Q and contain T, it follows that $Z_G(\lambda) = Z_G(\lambda') = M$. Furthermore, λ and λ' are conjugate under some element $g \in Q(K)$. Hence such an element g normalizes M. But $N_Q(M) = M$ (after multiplying an element in $N_Q(M)(K)$ by an element in $M(K)$, we may assume that the element also normalizes T; then the assertion follows from the Bruhat decomposition, [26], 5.15). Hence $g \in M(K)$, and since λ and λ' are centralized by M, we deduce $\lambda = \lambda'$. □

Notes and References The theory of filtrations of fiber functors is expounded in Saavedra [193], Ch. IV. Par-equivalence between 1-PS is almost the same as the equivalence relation leading to the *rational flag complex* of G, cf. [164], or to the *rational vector building* of Rousseau [192]: in that equivalence relation, λ is equivalent to λ' if λ^n is par-equivalent to $\lambda'^{n'}$, for natural numbers n and n'.

V. Filtrations on $\mathrm{Rep}_k(G)$

In this chapter we transpose the theory of the first part of this monograph from GL_n to arbitrary reductive groups. There are two approaches to this task. The *external approach* is to use the tannakian formalism of the previous chapter to reduce everything to the case of GL_n. The *internal approach* is to express everything in terms of the group under consideration, and the essential content of the present chapter is to show that both approaches are compatible. Again, such considerations also arise in the realm of vector bundles, in order to define and study G-bundles. Besides this analogy, the main motivation for transposing the theory to more general reductive groups comes from the expected arithmetical applications of period domains in the p-adic case. For example, the case of symplectic groups comes in naturally when dealing with p-adic uniformization of Siegel varieties.

Since we will use the tensor product theorem of Chapter I, §2, we will assume throughout this chapter that the ground field k is perfect. We fix an algebraic closure $\bar{k}$ of k and denote by Γ_k its absolute Galois group. The symbols $X^*(H)$, resp $X_*(H)$, will denote the group of all $\bar{k}$-characters, resp. the set of all $\bar{k}$-cocharacters, of a k-group scheme H.

We will use the following notational conventions regarding reductive groups G over k. We denote the adjoint group of G by G_{ad}, and the derived group by G_{der}. We denote by Z_G the center of G, and by Z_G^o its connected component. Then Z_G^o is also the maximal torus contained in Z_G. We let A_G be the maximal split torus contained in Z_G. Hence

$$X_*(A_G) = X_*(Z_G^o)^{\Gamma_k} .$$

We recall that the abelianization $G_{\mathrm{ab}} := G/[G,G]$ of G is a torus, the maximal quotient torus of G. We denote by

$$A'_G = \mathrm{Hom}(X_k^*(G_{ab}), \mathbb{G}_m)$$

the maximal *split* torus quotient of G. Hence

$$X^*(A'_G) = X^*(G_{\mathrm{ab}})^{\Gamma_k} = X^*(G)^{\Gamma_k}\ .$$

Most importantly, the compositions

$$Z_G^o \longrightarrow G_{\mathrm{ab}} \quad , \quad A_G \longrightarrow A'_G$$

are both isogenies, i.e., surjective with finite kernels.

Finally we will denote by $\mathrm{Fil}_K(\omega^G)$ the set of $\mathbb{Q}$-filtrations of ω^G over an extension field K. If $\rho : G \longrightarrow G'$ is a morphism of reductive groups, we have a push-forward map $\rho_* : \mathrm{Fil}_K(\omega^G) \longrightarrow \mathrm{Fil}_K(\omega^{G'})$, cf. Remarks 4.2.10 (i).

1 Slopes

Let G be a reductive group over the field k and let K be an extension field of k. We first give an internal definition of slopes and degrees.

Definition 5.1.1 Let $\mathcal{F} \in \mathrm{Fil}_K(\omega^G)$.
(i) The *degree* $\deg(\mathcal{F}) \in X_*(A'_G)_{\mathbb{Q}}$ of $\mathcal{F}$ is the unique splitting $\mathbb{D}_K \longrightarrow A'_G$ of the push-forward of $\mathcal{F}$ to A'_G.
(ii) The *slope* $\mu(\mathcal{F}) \in X_*(A_G)_{\mathbb{Q}}$ is the preimage of $\deg(\mathcal{F})$ via the canonical isomorphism $X_*(A_G)_{\mathbb{Q}} \xrightarrow{\sim} X_*(A'_G)_{\mathbb{Q}}$.

Remark 5.1.2 When $G = \mathrm{GL}(V)$ for some k-vector space V, the determinant morphism allows us to identify $A'_G \xrightarrow{\sim} \mathbb{G}_m$ and $X_*(A'_G)_{\mathbb{Q}} \xrightarrow{\sim} \mathbb{Q}$. Via this identification, and using Remark 4.2.11, the degree in the new sense above matches with the degree in the sense of Part 1, i.e., $\deg(\mathcal{F}) = \deg(V, \mathcal{F}^\bullet V)$, as is easily checked by choosing a splitting of $\mathcal{F}$. Furthermore, the action of scalars allows one to identify $\mathbb{G}_m \xrightarrow{\sim} A_G$ and $\mathbb{Q} \xrightarrow{\sim} X_*(A_G)_{\mathbb{Q}}$ and the isomorphism $X_*(A_G)_{\mathbb{Q}} \xrightarrow{\sim} X_*(A'_G)_{\mathbb{Q}}$ is then identified with the multiplication by $\dim(V)$. Hence the slope in the new sense matches with the slope in the old sense, i.e., $\mu(\mathcal{F}) = \mu(V, \mathcal{F}^\bullet V)$. So the above definition is a natural generalization of the $\mathrm{GL}(V)$-case.

Now, we look at functorial properties of degrees and slopes. Let $\rho : G \longrightarrow G'$ be a homomorphism of reductive groups over k and let $\mathcal{F} \in \mathrm{Fil}_K(\omega^G)$. Then ρ induces also a morphism $\rho_* : X_*(A'_G) \longrightarrow X_*(A'_{G'})$ and from the definitions we have

$$\deg(\rho_*\mathcal{F}) = \rho_* \deg(\mathcal{F}). \tag{1.1}$$

The morphism ρ is called *homogeneous* if $\rho(A_G) \subset A_{G'}$. In this case, we also have an induced map $\rho_* : X_*(A_G) \longrightarrow X_*(A_{G'})$ and from the definitions we get

$$\mu(\rho_*\mathcal{F}) = \rho_*\mu(\mathcal{F}).$$

In the particular case where ρ is a rational representation $\rho : G \longrightarrow \mathrm{GL}(V)$, and going through the identifications in Remark 5.1.2, we get the formula

$$\deg(V, \mathcal{F}^{\bullet}V) = \langle \deg(\mathcal{F}), \textstyle\bigwedge^{\max} V \rangle. \tag{1.2}$$

Here $\mathcal{F}^{\bullet}V = \rho_*\mathcal{F}$ is the filtration on V induced by $\mathcal{F}$, and the one-dimensional representation $\bigwedge^{\max} V$ is considered as a character of A'_G (since it necessarily factors through A'_G). If furthermore, ρ is homogeneous, then (V, ρ) has a central character given by

$$\zeta_V := \rho_{|A_G} : A_G \longrightarrow \mathbb{G}_m = A_{\mathrm{GL}(V)},$$

and we get the formula

$$\mu(V, \mathcal{F}^{\bullet}V) = \langle \mu(\mathcal{F}), \zeta_V \rangle. \tag{1.3}$$

Either (1.2) or (1.3) above may serve as *external* definitions of degrees and slopes.

As a particular case we note the following fact.

Remark 5.1.3 If G has anisotropic center, then for any rational representation $(V, \rho) \in \mathrm{Rep}_k(G)$, we have $\mu(V, \mathcal{F}^{\bullet}V) = 0$.

In the case where $G = T$ is a torus, the filtration $\mathcal{F}$ has a unique splitting $\lambda : \mathbb{D}_K \longrightarrow T$. If we fix an embedding of the algebraic closure of k inside K into $\bar{k}$, we may view λ as an element of $X_*(T)$. The following result makes explicit the relation between λ and $\mu(\mathcal{F})$. Recall that the action of Γ_k on $X_*(T)$ factors through a finite quotient, hence the stabilizer Γ_λ of λ in Γ_k has finite index.

Proposition 5.1.4 *With the foregoing notation we have*

$$\mu(\mathcal{F}) = [\Gamma_k : \Gamma_\lambda]^{-1} \cdot \sum\nolimits_{\tau \in \Gamma_k/\Gamma_\lambda} {}^{\tau}\lambda\,.$$

Proof Let Γ be a finite quotient of Γ_k such that the action on $X_*(T)$ factors through Γ. Consider the averaging endomorphism A_Γ of $X_*(T)_{\mathbb{Q}}$ given by $x \mapsto \frac{1}{|\Gamma|}\sum_{\tau \in \Gamma} {}^{\tau}x$. Then obviously the RHS of the above expression is equal to $A_\Gamma(\lambda)$.

Now we have $X_*(A'_T)_{\mathbb{Q}} = X_*(T)_{\mathbb{Q},\Gamma_k} = X_*(T)_{\mathbb{Q},\Gamma}$ (coinvariants), so $\deg(\mathcal{F})$ is just the image of λ in the Γ-coinvariants of $X_*(T)_{\mathbb{Q}}$. But the averaging map A_Γ descends to an isomorphism $X_*(T)_{\mathbb{Q},\Gamma} \xrightarrow{\sim} X_*(T)_{\mathbb{Q}}^{\Gamma}$ which is inverse to the induced isomorphism $X_*(T)_{\mathbb{Q}}^{\Gamma} \xrightarrow{\sim} X_*(T)_{\mathbb{Q},\Gamma}$. □

Notes and References The slope homomorphism introduced here is similar to the one in the theory of G-bundles on a curve [9, 151]. In the latter case these homomorphisms are considered as $\mathbb{Q}$-1-PS of the maximal split torus quotient A'_G. This notion is compatible with our terminology via the isomorphism $X_*(A_G)_{\mathbb{Q}} \xrightarrow{\sim} X_*(A'_G)_{\mathbb{Q}}$. On the other hand, the slope homomorphism is analogous to the weight homomorphism $w : \mathbb{G}_{m,\mathbb{R}} \longrightarrow G_{\mathbb{R}}$ in the theory of Hodge structures [193]. It is also analogous to the homomorphism $\nu_b : \mathbb{D} \longrightarrow G$ that Kottwitz associates to an isocrystal with G-structure (G, b), cf. [135]. When $G = T$ is a torus, Kottwitz gives an explicit formula for ν_b, which is very similar to the formula in Proposition 5.1.4, cf. [135], 2.5 and 2.8.

2 Semi-stability

Having now slopes at our disposal, we next come to the concept of semi-stability. Our definition is half-internal and half-external.

Definition 5.2.1 Let $\mathcal{F}$ be a $\mathbb{Q}$-filtration of ω^G over some extension field K/k. Then $\mathcal{F}$ is called *semi-stable* if (Lie G, $\mathcal{F}^\bullet$Lie G) is a semi-stable object of Fil_k^K.

Here Lie $G \otimes_k K$ acquires a filtration $\mathcal{F}^\bullet$Lie $G = \rho_*(\mathcal{F})$ through the adjoint representation ρ of G on Lie G.

Example 5.2.2 Let $G = \mathrm{GL}(V)$ for some k-vector space V. In this case $\mathrm{Lie}(G) = V \otimes V^*$ is equipped with the filtration $\mathcal{F}^\bullet V \otimes \mathcal{F}^\bullet V^*$. By Theorem 1.2.1 and Corollary 1.3.7, we know that $V \otimes V^*$ is semi-stable if and only if V is semi-stable. Hence the definition above is compatible with the definition of Chapter I, in the sense that $\mathcal{F}$ is semi-stable if and only if $(V, \mathcal{F}^\bullet V)$ is a semi-stable object of Fil_k^K.

Example 5.2.3 Let $G = T$ be a torus. Then any $\mathbb{Q}$-filtration of ω^T is semi-stable since T acts trivially on Lie T.

Remark 5.2.4 The adjoint representation factors through G_{ad}. By Remark 5.1.3 and functoriality (1.1), page 126 we have $\mu(\mathrm{Lie}\ G, \mathcal{F}^\bullet\mathrm{Lie}\ G) = 0$.

Proposition 5.2.5 *Let $\mathcal{F}$ be a $\mathbb{Q}$-filtration of ω^G over K and let $\mathcal{F}_{\mathrm{ad}}$ be the induced $\mathbb{Q}$-filtration of $\omega^{G_{\mathrm{ad}}}$. Then $\mathcal{F}$ is semi-stable if and only if $\mathcal{F}_{\mathrm{ad}}$ is semi-stable.*

Proof There is an exact sequence in $\mathrm{Rep}_k(G)$,

$$0 \longrightarrow \mathrm{Lie}\ Z_G \longrightarrow \mathrm{Lie}\ G \longrightarrow \mathrm{Lie}\ G_{\mathrm{ad}} \longrightarrow 0\,.$$

Now Lie Z_G is isomorphic to a direct sum $\mathbf{1}^n$ of copies of the trivial representation, hence (Lie $Z_G, \mathcal{F}^\bullet$Lie Z_G) is isomorphic to a direct sum of copies of neutral objects $\mathbf{1}$ as filtered vector space.

Let $W \subset$ Lie G be a k-subspace. We obtain an exact sequence

$$0 \longrightarrow (\text{Lie } Z_G) \cap W \longrightarrow W \longrightarrow W_{\text{ad}} \longrightarrow 0\,,$$

where W_{ad} denotes the image of W in Lie G_{ad}. From the above we have

$$\deg(W) = \deg(W_{\text{ad}})\,.$$

Hence $\mu(W) \leq 0 \iff \mu(W_{\text{ad}}) \leq 0$. By Remark 5.1.3 this proves the assertion, since every subspace of Lie G_{ad} is of the form W_{ad} for some subspace W of Lie G. □

We next study functoriality properties of semi-stability.

Proposition 5.2.6 *Let $\rho : G \longrightarrow G'$ be a morphism of reductive groups over k and let $\mathcal{F} \in \text{Fil}_K(\omega^G)$.*

(i) *If ρ is injective, then $\rho_*\mathcal{F}$ semi-stable implies $\mathcal{F}$ semi-stable.*

(ii) *If ρ is homogeneous, then $\mathcal{F}$ semi-stable implies $\rho_*\mathcal{F}$ semi-stable.*

Proof (i) In this case ρ induces an embedding of Lie G into Lie G' and the filtration $\mathcal{F}^\bullet$Lie G is induced by the filtration $(\rho_*\mathcal{F})^\bullet$Lie G'. Since Lie G' is semi-stable and both Lie G and Lie G' have the same slope 0, we conclude that Lie G is semi-stable.

(ii) By assumption we get a morphism $G/A_G \longrightarrow G'/A_{G'}$. Now by the same proof as in Proposition 5.2.5, we see that $\mathcal{F}$ is semi-stable iff its induced filtration on ω^{G/A_G} is semi-stable. Thus we may assume that G and G' have anisotropic center. Hence for any rational representation (V, ρ) of G, we have the identity $\mu(V, \mathcal{F}^\bullet V) = 0$, cf. Remark 5.1.3. In order to prove that (Lie $G', \mathcal{F}^\bullet$Lie G') is semi-stable, we choose a faithful representation (W, χ) of G_{ab}. We consider the G-representation $U =$ Lie $G \oplus W$ given by the adjoint representation on Lie G and by the representation on W. Then $(U, \mathcal{F}^\bullet U)$ is semi-stable since both summands are semi-stable of the same slope 0, cf. Corollary 1.1.21. Further, it is up to a finite kernel a faithful representation. Again, by dividing out this kernel, we may assume that U is indeed a faithful G-representation. By Proposition 4.1.21, the representation Lie G' of G is a quotient of some subrepresentation V of a representation of the following type,

$$\widetilde{V} = \bigoplus_\alpha U^{\otimes m_\alpha} \otimes \check{U}^{\otimes n_\alpha}\,.$$

By the tensor product theorem (Theorem 1.2.1), we know that $(\widetilde{V}, \mathcal{F}^\bullet \widetilde{V})$ is semi-stable. Since $(V, \mathcal{F}^\bullet V)$ is a subobject of the same slope 0 (cf. Remark 5.1.3), it is also semi-stable. Again, since $(\mathrm{Lie}\, G', \mathcal{F}^\bullet \mathrm{Lie}\, G')$ is a quotient object of $(V, \mathcal{F}^\bullet V)$ of the same slope 0, it is semi-stable. □

The above proposition allows us to compare the definition of semi-stability above to a possible external definition of semi-stability.

Corollary 5.2.7 *Let $\mathcal{F}$ be a $\mathbb{Q}$-filtration on ω^G over an extension field K of k.*
(i) *Let $(V, \rho) \in \mathrm{Rep}_k(G)$ be homogeneous. If $\mathcal{F}$ is semi-stable, then $(V, \mathcal{F}^\bullet V) = (V, \rho_*(\mathcal{F}))$ is a semi-stable object of Fil_k^K*
(ii) *Conversely, let $(V, \rho) \in \mathrm{Rep}_k(G)$ be a faithful representation. If $(V, \mathcal{F}^\bullet V)$ is a semi-stable object of Fil_k^K, then $\mathcal{F}$ is semi-stable.* □

Remark 5.2.8 Statement (i) is obviously false if we drop the assumption that (V, ρ) is homogeneous (even for faithful representations). In particular if $f : G \to G'$ is a homomorphism and $\mathcal{F}$ is a semi-stable filtration of ω^G, then the induced filtration $\mathcal{F}' = f_*(\mathcal{F})$ on $\omega^{G'}$ need not be semi-stable. This obvious disadvantage of our definition of semi-stability is outweighed by other advantages, e.g., the validity of Theorem 5.3.1 and Corollary 5.3.3 in the next section.

We next want to give an internal characterization of semi-stability. Let P be a k-psgp of G, with unipotent radical U_P and Levi quotient M_P. Recall from Proposition 4.2.17 that any filtration $\mathcal{F} \in \mathrm{Fil}_K(\omega^G)$ comes from a unique filtration $\mathcal{F}_P$ of ω^P over K. The degree $\deg(\mathcal{F}_P) \in X_*(A'_P)_{\mathbb{Q}}$ is defined in the same way as for a reductive group. Here A'_P is the maximal k-split torus quotient of P, or equivalently of its Levi quotient $M = M_P$. In the following statement we write $\mathfrak{u}_P$ for the Lie algebra $\mathrm{Lie}\, U_P$, which is a P-stable subspace of $\mathrm{Lie}\, G$. As usual, the determinant $\bigwedge^{\max} \mathfrak{u}_P$ factors through A'_P.

Proposition 5.2.9 *Let $\mathcal{F} \in \mathrm{Fil}_K(\omega^G)$. Then $\mathcal{F}$ is semi-stable if and only if for any maximal proper k-psgp P of G we have*

$$\left\langle \deg(\mathcal{F}_P), \textstyle\bigwedge^{\max} \mathfrak{u}_P \right\rangle \le 0.$$

Proof Let $\mathcal{F}$ be semi-stable and let P be any k-psgp of G. The subspace $\mathfrak{u}_P$ of $\mathrm{Lie}\, G$ is P-stable. By the condition imposed on $\mathcal{F}_P$, the filtrations $\mathcal{F}_P^\bullet \mathfrak{u}_P$ and $\mathcal{F}^\bullet \mathrm{Lie}\, G_{|\mathfrak{u}_P}$ coincide. By the semi-stability of $\mathrm{Lie}\, G$ and the fact that its slope is 0, we therefore have

$$\deg\left(\mathfrak{u}_P, \mathcal{F}_P^\bullet \mathfrak{u}_P\right) \le 0\,.$$

Hence we get

$$\langle \deg(\mathcal{F}_P), \bigwedge^{\max} \mathfrak{u}_P \rangle = \deg\left(\bigwedge^{\max} \mathfrak{u}_P, \mathcal{F}^\bullet_{M_P} \bigwedge^{\max} \mathfrak{u}_P\right) = \deg\left(\mathfrak{u}_P, \mathcal{F}^\bullet_P \mathfrak{u}_P\right) \leq 0,$$

as desired. To prove the converse, we use the Harder–Narasimhan filtration introduced in the next section. Therefore we postpone this proof until right below Proposition 5.3.5. □

It is useful to have a more explicit criterion, using standard parabolic subgroups and roots. Our notation will be modeled on that used in Chapter III, §1. We fix a minimal parabolic subgroup P_0 *defined over* k and a Levi subgroup M_0 of P_0. A *standard parabolic subgroup* (*std psgp* for short) is a parabolic P with $P \supset P_0$. There is a unique Levi subgroup M of P containing M_0. We also write A_P resp. A'_P for A_M resp. A'_M. Let $\mathfrak{a}_P = X_*(A_P)_\mathbb{Q}$. For std psgps $P \subset Q$ we have a series of homomorphisms

$$A_Q \subset A_P \longrightarrow A'_P \twoheadrightarrow A'_Q \,.$$

This induces a direct sum decomposition

$$\mathfrak{a}_P = \mathfrak{a}_Q \oplus \mathfrak{a}_P^Q \,, \tag{2.1}$$

where $\mathfrak{a}_P^Q = \ker(X_*(A'_P)_\mathbb{Q} \longrightarrow X_*(A'_Q)_\mathbb{Q})$. Let

$$\Phi_P = \{\text{characters of } A_P \text{ which occur in Lie } G\} \,.$$

Hence $\Phi_P \subset (\mathfrak{a}_P^G)^* \subset \mathfrak{a}_P^*$.

We also introduce

$$\Phi_P^+ = \{\text{characters of } A_P \text{ which occur in Lie } U_P\} \,. \tag{2.2}$$

If $P = P_0$, then Φ_{P_0} is a root system with $\Phi_{P_0}^+$ as a set of positive roots. Let Δ_{P_0} be the set of simple roots in $\Phi_{P_0}^+$.

If $P \neq P_0$, then Φ_P is not a root system. We define Δ_P as the set of non-trivial restrictions of elements of Δ_{P_0} to A_P ($\subset A_{P_0}$). Then Δ_P is a basis of the $\mathbb{Q}$-vector space $(\mathfrak{a}_P^G)^* \subset X^*(A_P)_\mathbb{Q}$. If P is a maximal std psgp, then Δ_P consists of a single element.

Let $\mathcal{F}_P$ be the filtration associated to $\mathcal{F}$ via Proposition 4.2.17. By push-forwarding it to the Levi quotient M_P of P followed by a lift to the std Levi subgroup M of P, we get a filtration $\mathcal{F}_M \in \mathrm{Fil}_K(\omega^M)$.

Corollary 5.2.10 *Let $\mathcal{F} \in \mathrm{Fil}_K(\omega^G)$. Then $\mathcal{F}$ is semi-stable if and only if for any maximal std psgp P, the following inequality holds,*

$$\langle \mu(\mathcal{F}_M), \alpha \rangle \leq 0 \ \textit{for the unique element}\ \alpha \in \Delta_P \,.$$

Proof Indeed this follows from Proposition 5.2.9, since the restriction of the character $\bigwedge^{\max} \mathfrak{u}_P$ to A_M is a positive integral multiple of α in $X_*(A_M)$, and since we have

$$\langle \mu(\mathcal{F}_M), (\textstyle\bigwedge^{\max} \mathfrak{u}_P)_{|A_M} \rangle = \langle \deg(\mathcal{F}_P), \textstyle\bigwedge^{\max} \mathfrak{u}_P \rangle. \qquad \square$$

Notes and References Corollary 5.2.10 is analogous to a similar statement in the theory of G-bundles on Riemann surfaces, cf. [180, 151]. In fact, the paper [180] of Ramanathan generalizes the work of Narsasimhan and Seshadri [166] to the case of G-bundles. He observes that it suffices to check the semi-stability of these objects with respect to maximal parabolic subgroups. In [9], §10, Atiyah and Bott consider the notion of semi-stability of G-bundles in our sense. More precisely, they prove that a G-bundle ξ is semi-stable if and only if the corresponding adjoint vector bundle $ad(\xi)$ is semi-stable. They also examine the functoriality of semi-stability under homomorphism of reductive groups. They prove that the canonical parabolic reduction of G-bundles (which corresponds to the notion of the HN-filtration) is compatible with respect to group homomorphisms. We will return to this in the next section.

3 The Harder–Narasimhan filtration

The Harder–Narasimhan filtration for filtered vector spaces (Chapter I, §3) has an analogue in our context. It is given by the following theorem.

Theorem 5.3.1 *Let $\mathcal{F}$ be a $\mathbb{Q}$-filtration of ω^G over K. There exists a unique $\mathbb{Q}$-filtration ${}^\bullet\omega^G_{\mathcal{F}}$ of ω^G over k, such that for every $(V,\rho) \in \mathrm{Rep}_k(G)$, the induced filtration $\rho_*({}^\bullet\omega^G_{\mathcal{F}})$ is the Harder–Narasimhan filtration ${}^\bullet V$ of V.*

Proof For every $(V,\rho) \in \mathrm{Rep}_k(G)$, the k-vector space V is equipped with its Harder–Narasimhan filtration ${}^\bullet V$, cf. Chapter I, §2. This association is functorial and compatible with tensor products, cf. Proposition 1.3.6. This proves properties (i) and (ii) of Definition 4.2.6, and property (iii) is trivial. It remains to prove the exactness property of the Harder–Narasimhan filtration. Let

$$0 \longrightarrow V' \stackrel{\iota}{\longrightarrow} V \stackrel{\pi}{\longrightarrow} V'' \longrightarrow 0 \tag{3.1}$$

be a short exact sequence in $\mathrm{Rep}_k(G)$. We need to show that the induced sequence

$$0 \longrightarrow {}^\bullet V' \stackrel{{}^\bullet\iota}{\longrightarrow} {}^\bullet V \stackrel{{}^\bullet\pi}{\longrightarrow} {}^\bullet V'' \longrightarrow 0 \tag{3.2}$$

is a short exact sequence in Fil^k_k. We note that when k has characteristic 0, the exact sequence (3.1) above is *split* since G is reductive, hence (3.2) splits as well (by compatibility of HN-filtrations with direct sums, see Proposition 1.3.6(b)) and we are done. In general, however, we need to invoke a non-trivial result of Haboush [103], formerly conjectured by Mumford, which is fundamental in Geometric Invariant Theory over a field of positive characteristic.

Indeed, to prove the exactness of the sequence (3.2) above, we have to show that ${}^\bullet\iota$ is a strict monomorphism and ${}^\bullet\pi$ is a strict epimorphism. By compatibility of the HN-filtration with duality, we may only deal with the monomorphism ${}^\bullet\iota$.

By definition, ${}^\bullet\iota$ is strict if and only if its graded associate $\mathrm{gr}^\bullet({}^\bullet\iota)$ is injective. Hence by compatibility of $\mathrm{gr}^\bullet$ with exterior powers, we see that ${}^\bullet\iota$ is strict if and only if $\bigwedge^{d'}({}^\bullet\iota)$ is strict, where $d' = \dim V'$. But by Proposition 1.3.6, we have $\bigwedge^{d'}({}^\bullet\iota) = {}^\bullet(\bigwedge^{d'}\iota)$, hence we are reduced to the case $\dim V' = 1$.

Now, by compatibility of $\mathrm{gr}^\bullet$ with symmetric powers, we see that ${}^\bullet\iota$ is strict if and only if $\mathrm{Sym}^r({}^\bullet\iota)$ is strict for all, or equivalently for one, positive integer r. On the other hand, Proposition 1.3.6 again tells us that $\mathrm{Sym}^r({}^\bullet\iota) = {}^\bullet(\mathrm{Sym}^r\iota)$.

At this point we invoke Haboush's result [103], Thm. 5.2. It ensures the existence of a positive integer r such that the monomorphism $\mathrm{Sym}^r\,\iota$ is split in $\mathrm{Rep}_k(G)$. As already explained, this implies that ${}^\bullet(\mathrm{Sym}^r\,\iota)$ is strict. □

The filtration ${}^\bullet\omega^G_{\mathcal{F}}$ is called the *Harder–Narasimhan filtration* of ω^G corresponding to $\mathcal{F}$. The psgp $P_{\mathrm{HN}} = P_{{}^\bullet\omega^G_{\mathcal{F}}}$ of G (defined over k) associated to ${}^\bullet\omega^G_{\mathcal{F}}$ is called the HN-*parabolic* associated to $\mathcal{F}$.

Remark 5.3.2 Let $\rho : G \longrightarrow G'$ be a morphism of reductive groups and let $\mathcal{F} \in \mathrm{Fil}_K(\omega^G)$. By its very definition, the HN-filtrations enjoy the following functoriality property:

$$ {}^\bullet\omega^{G'}_{\rho_*\mathcal{F}} = \rho_*({}^\bullet\omega^G_{\mathcal{F}}). $$

In particular, if ρ is faithful (i.e., injective), then the HN-psgps P_{HN} and P'_{HN} of $\mathcal{F}$ and $\rho_*\mathcal{F}$ satisfy $P_{\mathrm{HN}} = G \cap P'_{HN}$.

Corollary 5.3.3 *Let $\mathcal{F} \in \mathrm{Fil}_K(\omega^G)$. Then the following properties are equivalent.*

(i) *$\mathcal{F}$ is semi-stable.*
(ii) *The* HN-*filtration ${}^\bullet\omega^G_{\mathcal{F}}$ is split by $\mu(\mathcal{F}) : \mathbb{D} \longrightarrow A_G \subset G$.*
(iii) *The* HN-*psgp P_{HN} is equal to G.*

Proof (i) implies (ii). Indeed, let $\mathcal{F}$ be semi-stable and let λ be a splitting of the HN-filtration of $\mathcal{F}$. By the previous remark applied to $G' = G_{\mathrm{ad}}$ and since the HN-filtration of $(\mathrm{Lie}\,G, \mathcal{F}^\bullet\mathrm{Lie}\,G)$ is trivial, λ factors through A_G, and therefore it is uniquely defined. To compute it, we apply again Remark 5.3.2 with $G' = A'_G$.

(ii) implies (iii) obviously.

(iii) implies (i). Indeed, assume that the HN-psgp of $\mathcal{F}$ equals G. Equivalently, the HN-filtration has a unique splitting and it factors through A_G. Therefore the HN-filtration of $(\mathrm{Lie}\, G, \mathcal{F}^\bullet \mathrm{Lie}\, G)$ is trivial. □

Example 5.3.4 Let $(V, (\ , \ ))$ be a symplectic or quadratic vector space over k, i.e., $(\ , \ )$ is a non-degenerate alternating or symmetric form on V. Let $G = \mathrm{Aut}(V, (\ , \ ))$ be the corresponding symplectic or orthogonal group.
(a) A filtration $\mathcal{F}$ on ω^G defined over K corresponds in a one-one way to a selfdual filtration $\mathcal{F}^\bullet V$ of $V \otimes_k K$, i.e.,

$$(\mathcal{F}^x)^\perp = \mathcal{F}^{-x} ,$$

cf. Definition 4.2.6.
(b) The filtration $\mathcal{F}$ is semi-stable if and only if

$$\mu(V') \leq \mu(V) ,$$

for all totally isotropic subspaces V' of V. Indeed, this follows from Corollary 5.3.3 and the fact that the psgp P_{HN} is the stabilizer of a self-dual flag of V.

The definition of the HN-filtration of ω^G associated to a $\mathbb{Q}$-filtration $\mathcal{F}$ of ω^G over K is purely external. One can also give an internal characterization.

Proposition 5.3.5 *Let $\mathcal{F} \in \mathrm{Fil}_K(\omega^G)$. Let ${}^\bullet\omega$ be a $\mathbb{Q}$-filtration of ω^G with associated psgp $P = P_{{}^\bullet\omega}$, Levi quotient $M = P_{{}^\bullet\omega}/U_{{}^\bullet\omega}$, and central cocharacter $\overline{\lambda} \in X_*(A_M)_{\mathbb{Q}}$ (see Theorem 4.2.13 (c)). Let $\mathcal{F}_M \in \mathrm{Fil}_K(\omega^M)$ be the push-forward to M of the filtration $\mathcal{F}_P$ on ω^P over K given by Proposition 4.2.17. Then ${}^\bullet\omega = {}^\bullet\omega^G_{\mathcal{F}}$ if and only if $\mathcal{F}_M$ is semi-stable of slope $\mu(\mathcal{F}_M) = \overline{\lambda}$.*

Proof We first check the statement in the $\mathrm{GL}(V)$-case. In this case, ${}^\bullet\omega$ is given by a filtration ${}^\bullet V$ of the k-vector space V. Then $M = \prod_\alpha \mathrm{GL}({}^\alpha\mathrm{gr}(V))$ and $\overline{\lambda} = \prod_\alpha \overline{\lambda}_\alpha$ with $\overline{\lambda}_\alpha$ corresponding to α via the identification $\mathbb{Q} \xrightarrow{\sim} X_*(A_{\mathrm{GL}({}^\alpha\mathrm{gr}(V))})_{\mathbb{Q}}$. On the other hand, the filtration $\mathcal{F}_M$ is given by the collection $(\mathcal{F}^\bullet({}^\alpha\mathrm{gr}(V)))_\alpha$ of filtrations on each ${}^\alpha\,\mathrm{gr}(V)_K$ which are induced by $\mathcal{F}^\bullet V$. Its slope $\mu(\mathcal{F}_M)$ decomposes as a product $\prod_\alpha \mu(\mathcal{F}_M)_\alpha$ with $\mu(\mathcal{F}_M)_\alpha$ corresponding to $\mu({}^\alpha\mathrm{gr}(V), \mathcal{F}^\bullet({}^\alpha\mathrm{gr}(V)))$ via the same identification as above. Moreover, $\mathcal{F}_M$ is semi-stable if and only if each $({}^\alpha\mathrm{gr}(V), \mathcal{F}^\bullet({}^\alpha\mathrm{gr}(V)))$ is a semi-stable object in Fil_k^K (same argument as in Example 5.2.2). Therefore the assertion of the Proposition formally follows from the definition of the HN-filtration of $(V, \mathcal{F}^\bullet V)$ in Proposition 1.3.1.

Now we shall reduce the general case to the GL(V)-case. We start with quite general considerations. Let $\rho : G \longrightarrow G'$ be a morphism of reductive groups and let $\mathcal{F}'$ and ${}^\bullet\omega'$ be the push-forward of $\mathcal{F}$ and ${}^\bullet\omega$ to G'. We define $P' := P_{{}^\bullet\omega'}$, M' and $\overline{\lambda}'$ as in the proposition, but regarding G' and ${}^\bullet\omega'$. Then by the characterization of P and P' in Theorem 4.2.13 (b), we have a commutative diagram

$$\begin{array}{ccc} G & \xrightarrow{\ \rho\ } & G' \\ \uparrow & & \uparrow \\ P & \longrightarrow & P' \\ \downarrow & & \downarrow \\ M & \xrightarrow[\rho_M]{} & M' . \end{array}$$

If ρ is injective, then so is ρ_M, because of the equality $U_{{}^\bullet\omega} = G \cap U_{{}^\bullet\omega'}$. Moreover it follows from the definition that $\mathcal{F}'_{M'} = \rho_{M,*}(\mathcal{F}_M)$.

Now we show that ${}^\bullet\omega^G_{\mathcal{F}}$ has the properties stated in the proposition. Indeed choose a faithful representation $\rho : G \longrightarrow G' = \mathrm{GL}(V)$. Then $\rho_*({}^\bullet\omega^G_{\mathcal{F}})$ is the HN-filtration of $\rho_*(\mathcal{F})$ hence by the GL(V)-case treated above, $\rho_{M,*}(\mathcal{F}_M)$ is semi-stable of slope $\rho_*\overline{\lambda}$ in $X_*(A_{M'})_{\mathbb{Q}}$. Since $\rho_M : M \longrightarrow M'$ is injective, the semi-stability of $\mathcal{F}_M$ follows from Proposition 5.2.6 (i). To compute the slope $\mu(\mathcal{F}_M)$ we note that, in this situation where both $\mathcal{F}_M$ and $\mathcal{F}'_{M'}$ are semi-stable, the $\mathbb{Q}$-1-PS $\rho_M \circ \mu(\mathcal{F}_M) : \mathbb{D} \longrightarrow M'$ factors through $A_{M'}$ and coincides with $\mu(\mathcal{F}'_{M'})$. Indeed this follows from Corollary 5.3.3 and Remark 5.3.2. Then, using the injectivity of ρ again, we conclude that $\mu(\mathcal{F}_M) = \overline{\lambda}$.

Finally, let ${}^\bullet\omega$ be a filtration as in the statement of the proposition and let $\rho : G \longrightarrow G' = \mathrm{GL}(V)$ be any rational representation of G. By Corollary 5.3.3, $\mu(\mathcal{F}_M)$ is the unique splitting of the HN-filtration of $\mathcal{F}_M$ and therefore the $\mathbb{Q}$-1-PS $\rho_M \circ \mu(\mathcal{F}_M)$ of M' is a splitting of the HN-filtration of $\mathcal{F}'_{M'}$. But by construction, this $\mathbb{Q}$-1-PS $\rho_M \circ \mu(\mathcal{F}_M)$ coincides with $\rho_M \circ \overline{\lambda} = \overline{\lambda}'$, hence it factors through $A_{M'}$. Therefore, again by Corollary 5.3.3, $\mathcal{F}'_{M'}$ is semi-stable, and moreover, because of $\deg(\mathcal{F}'_{M'}) = \rho_* \deg(\mathcal{F}_M)$, we have $\mu(\mathcal{F}'_{M'}) = \rho \circ \mu(\mathcal{F}_M) = \overline{\lambda}'$ in $X_*(A_{M'})_{\mathbb{Q}}$. By the GL(V)-case treated above we conclude that ${}^\bullet\omega'$ is the HN-filtration of $\mathcal{F}'$. Since this is true for any representation (V, ρ) of G, we deduce that ${}^\bullet\omega$ is the HN-filtration of $\mathcal{F}$. □

Now we are able to finish the proof of Proposition 5.2.9. So let $\mathcal{F} \in \mathrm{Fil}_K(\omega^G)$ and let us assume that $\mathcal{F}$ is not semi-stable. Then by Corollary 5.3.3 the HN-psgp P of $\mathcal{F}$ is proper. Let Q be a maximal proper psgp containing P. The filtration $\mathcal{F}_Q$ associated to $\mathcal{F}$ via Proposition 4.2.17 is also functorially induced by $\mathcal{F}_P$ via the morphism $\iota : P \longrightarrow Q$. Whence we get the equality $\deg(\mathcal{F}_Q) = \iota_* \deg(\mathcal{F}_P)$ in $X_*(A'_Q)_{\mathbb{Q}}$. We thus have

$$\left\langle \deg(\mathcal{F}_Q), \bigwedge^{\max} \mathfrak{u}_Q \right\rangle_{A'_Q} = \left\langle \deg(\mathcal{F}_P), (\bigwedge^{\max} \mathfrak{u}_Q)_{|A'_P} \right\rangle_{A'_P}$$
$$= \left\langle \mu(\mathcal{F}_{M_P}), (\bigwedge^{\max} \mathfrak{u}_Q)_{|A_{M_P}} \right\rangle_{A_{M_P}}.$$

Now, let λ be a splitting of the HN-filtration of $\mathcal{F}$ and let $\overline{\lambda}$ be its image in $X_*(A_{M_P})_{\mathbb{Q}}$. By Proposition 5.3.5 above, we have $\overline{\lambda} = \mu(\mathcal{F}_{M_P})$. However, by Theorem 4.2.13 (b), $\mathfrak{u}_P$ is the sum of the eigensubspaces of positive weight for the action of $\mathbb{D}$ on $\mathrm{Lie}(G)$ via λ. Therefore, since $\mathfrak{u}_Q \subset \mathfrak{u}_P$, we have $\langle \overline{\lambda}, (\bigwedge^{\max} \mathfrak{u}_Q)_{|A_{M_P}} \rangle > 0$, whence $\langle \mu(\mathcal{F}_{M_Q}), \bigwedge^{\max} \mathfrak{u}_Q \rangle > 0$ as desired.

Notes and References The Harder–Narasimhan filtration approach for arbitrary reductive groups appeared first in the context of G-bundles on a curve [9], cf. also [151], Lemma 3.1. Atiyah and Bott consider for a given G-bundle ξ the vector bundle $\mathrm{ad}(\xi)$ associated with the adjoint representation. Then the HN-parabolic subgroup P and the HN-type are induced by the HN-filtration of $\mathrm{ad}(\xi)$. In this way they get a P-bundle ξ_P which they call the canonical parabolic reduction of ξ. Further, they prove that the canonical reduction is compatible with respect to group homomorphisms.

Haboush's theorem [103], often still referred to as the Mumford conjecture, states that for any semi-simple algebraic group G over a field k, and for any rational representation $(V, \rho) \in \mathrm{Rep}_k(G)$, the following property is satisfied. Let $v \in V, v \neq 0$, be fixed by the action of G. Then there is a G-invariant polynomial F on V such that $F(v) \neq 0$. When k has characteristic 0 the polynomial F can be taken to be linear by the complete reducibility of the representation.

VI. Period Domains for Reductive Groups

In this chapter we introduce period domains for arbitrary reductive groups over finite fields k. We consider a conjugacy class $\mathcal{N}$ of $\mathbb{Q}$-1-PS of $G_{\bar{k}}$ which gives rise to a flag variety $\mathcal{F}(G,\mathcal{N})$. These varieties are defined over a finite extension of k, the reflex field of $(G,\mathcal{N})$. We shall show in the first section that the semi-stable filtrations inside $\mathcal{F}(G,\mathcal{N})$ form a Zariski-open subset $\mathcal{F}(G,\mathcal{N})^{ss}$. In the second section we treat the connection to Geometric Invariant Theory. In the third section we discuss, in analogy to the GL_n-case, the stratifications of $\mathcal{F}(G,\mathcal{N})$ by HN-*types* resp. HN-*vectors*. Again, the strata form locally closed subsets which, for the first kind of strata, are vector bundles over period domains of smaller dimension.

1 Definitions

Let G be a reductive group over the perfect field k. We fix an algebraic closure $\bar{k}$ of k and denote by $\Gamma = \Gamma_k$ the corresponding absolute Galois group of k. This group acts discretely, i.e., with open stabilizers, on the set $X_*(G)_{\mathbb{Q}}$ of all $\mathbb{Q}$-1-PS of G defined over $\bar{k}$. Concretely the action is given by

$$^{\sigma}\nu(t) = \sigma(\nu(\sigma^{-1}(t)))\ .$$

We will denote by $X_*(G)_{\mathbb{Q}}/G$ the set of $G(\bar{k})$-conjugacy classes in $X_*(G)_{\mathbb{Q}}$. It is again a discrete Γ_k-set.

Let T be a maximal torus in G, defined over k, and with absolute Weyl group $W := N_G(T)(\bar{k})/Z_G(T)(\bar{k})$. Then the inclusion of T into G induces a Γ_k-equivariant isomorphism

$$X_*(T)_{\mathbb{Q}}/W \xrightarrow{\sim} X_*(G)_{\mathbb{Q}}/G.$$

The relevance of this set to our subject comes from the following diagram:

$$X_*(G)_{\mathbb{Q}} \xrightarrow{\lambda\mapsto\mathcal{F}_\lambda} \mathrm{Fil}_{\bar{k}}(\omega^G) \xrightarrow{\mathcal{F}\mapsto N(\mathcal{F})} X_*(G)_{\mathbb{Q}}/G \xrightarrow{\deg} X_*(A_G')_{\mathbb{Q}}. \tag{1.1}$$

Here the first map takes a $\mathbb{Q}$-1-PS λ to its associated filtration $\mathcal{F}_\lambda$, hence it is the quotient map for the par-equivalence relation on $X_*(G)_\mathbb{Q}$, by Theorem 4.2.13. The second map takes a filtration $\mathcal{F}$ to its $G(\bar{k})$-conjugacy class, which is by definition the $G(\bar{k})$-conjugacy class of any splitting of $\mathcal{F}$. The third map, denoted by deg, is the push-forward induced by the projection $G \longrightarrow A'_G$. It is compatible with the degree of a filtration as defined in Definition 5.1.1, since we have $\deg(\mathcal{F}) = \deg(\mathcal{N}(\mathcal{F}))$ for any $\mathcal{F} \in \mathrm{Fil}_{\bar{k}}(\omega^G)$.

Example 6.1.1 Let $G = \mathrm{GL}(V)$ for an n-dimensional k-vector space V. Choose a k-basis of V, whence a maximal torus T and an isomorphism $X_*(T)_\mathbb{Q} \xrightarrow{\sim} \mathbb{Q}^n$. Then W is the symmetric group S_n on n letters and acts in the obvious way on $\mathbb{Q}^n$. Therefore $(\mathbb{Q}^n)_+$ is naturally a fundamental domain for this action. In other words, we have a canonical identification $X_*(G)_\mathbb{Q}/G \xrightarrow{\sim} (\mathbb{Q}^n)_+$. Moreover the map $\mathcal{F} \longrightarrow \mathcal{N}(\mathcal{F})$ takes a filtration $\mathcal{F}$ to its *type* in the sense of Chapter I. Identifying further $X_*(A'_G)_\mathbb{Q}$ with $\mathbb{Q}$ via the determinant, the third map in (1.1), page 137 is then given by $(x_1, \dots, x_n) \mapsto \sum_i x_i$.

We see from this example that elements in $X_*(G)_\mathbb{Q}/G$, i.e., conjugacy classes of $\mathbb{Q}$-1-PS of $G_{\bar{k}}$, may be considered as *types* of filtrations for the group G.

Definition 6.1.2 A *PD-pair* is a pair $(G, \mathcal{N})$ consisting of a reductive group G over k and a conjugacy class of $\mathbb{Q}$-1-PS $\mathcal{N} \in X_*(G)_\mathbb{Q}/G$. The *Shimura field* or *reflex field* of the PD-pair $(G, \mathcal{N})$ is the finite extension $E(G, \mathcal{N}) = \bar{k}^{\Gamma_\mathcal{N}}$ of k inside $\bar{k}$. Here $\Gamma_\mathcal{N}$ is the stabilizer of $\mathcal{N}$ in Γ_k.

Concretely, let T be a maximal torus of G defined over k and let $\nu \in X_*(T)_\mathbb{Q}$ lie in $\mathcal{N}$. Then the reflex field E is characterized by its Galois group,

$$\Gamma_E = \{\sigma \in \Gamma \mid {}^\sigma\nu = w\nu, \text{ some } w \in W\} .$$

Examples 6.1.3 (i) If G is a split group, then the action of Γ_k on $X_*(G)_\mathbb{Q}/G$ is trivial. Indeed, in this case we may choose T to be a split torus. It follows that $E(G, \mathcal{N}) = k$ for any conjugacy class $\mathcal{N}$.

(ii) Let $G = T$ be a torus. In this case $\mathcal{N}$ consists of a single element $\nu \in X_*(T)_\mathbb{Q}$. Then $\Gamma_E = \mathrm{Stab}_\Gamma(\nu)$.

(iii) Let k'/k be quadratic extension. Let $V' = k'^n$, with the hermitian form given by

$$\left(\sum x_i e_i, \sum y_i e_i\right) = \sum x_i \bar{y}_i .$$

Let $G = \mathrm{U}(V', (\ ,\ ))$ be the corresponding unitary group over k. As maximal torus we take

$$T = \{\mathrm{diag}(x_1, \dots, x_n) \mid x_1\bar{x}_1 = \cdots = x_n\bar{x}_n = 1\} .$$

Then

$$X_*(T) = \mathbb{Z}^n ,$$

with Γ-action factoring through $\mathrm{Gal}(k'/k) = \mathbb{Z}/2$ with

$$(x_1, \ldots, x_n) \longmapsto (-x_1, \ldots, -x_n) .$$

Furthermore, we may identify W with S_n acting by permuting the entries in $\mathbb{Z}^n$, hence we have an identification,

$$X_*(T)_{\mathbb{Q}}/W = \overline{C} := \{(x_1, \ldots, x_n) \in \mathbb{Q}^n \mid x_1 \geq x_2 \geq \cdots \geq x_n\} .$$

The action of Γ on $\overline{C}$ is given by

$$(x_1, \ldots, x_n) \longmapsto (-x_n, \ldots, -x_1) .$$

It follows that $E = k'$, unless $[\nu] \in \overline{C}$ is of the form $\nu = (x_1, \ldots, x_n)$ with $x_i + x_{n+1-i} = 0$, for $i = 1, \ldots, n$. In this last case we have $E = k$.

Theorem 6.1.4 *Let $(G, \mathcal{N})$ be a* PD*-pair, with associated reflex field $E = E(G, \mathcal{N})$. There exists a smooth projective variety $\mathcal{F}(G, \mathcal{N})$ over E with*

$$\begin{aligned} \mathcal{F}(G, \mathcal{N})(\overline{k}) &= \{\nu \in \mathcal{N} \text{ modulo par-equivalence}\} \\ &= \{\mathcal{F} \in \mathrm{Fil}_{\overline{k}}(\omega^G) \text{ of type } \mathcal{N}\} . \end{aligned}$$

The variety is homogeneous under G_E, and is a generalized flag variety for G_E.

Proof Assume first that $\mathcal{N}$ contains a fixed point ν_0 under Γ_E. This assumption is obviously satisfied when G is split, cf. Examples 6.1.3 (i). It is also satisfied when G is quasi-split, cf. Lemma 6.1.5 below. Let Q_{ν_0} be the psgp of G_E defined by the par-equivalence class of ν_0. Then

$$\mathcal{F}(G, \mathcal{N}) = G_E/Q_{\nu_0}$$

obviously has all the required properties.

In general, there is a projective scheme $\mathcal{P}$ over k parametrizing all parabolic subgroups, cf. [6], Exp. 26, §3. The union of those geometric connected components of $\mathcal{P}$ which correspond to $\mathcal{N}$ is defined over E. The corresponding open and closed subscheme $\mathcal{F}(G, \mathcal{N})$ of $\mathcal{P}_E$ is the required variety. However, it may happen that $\mathcal{F}(G, \mathcal{N})(E) = \emptyset$. □

Lemma 6.1.5 (Kottwitz Lemma) *Let G be quasi-split. Then there exists a* $\mathbb{Q}$-1-PS $\nu \in \mathcal{N}$ *which is defined over $E = E(G, \mathcal{N})$.*

Proof See [134], Lemma 1.1.3. □

Remark 6.1.6 Let K be any field extension of E. Observe that any choice ι of a k-linear field embedding of the algebraic closure of k in K into $\overline{k}$ induces a map $\mathrm{Fil}_K(\omega^G) \xrightarrow{\mathcal{N}_\iota} X_*(G)_{\mathbb{Q}}/G$ (factorising through $\mathrm{Hom}(\mathbb{D}_K, G_K)/G(K)$). So $\mathcal{N}_\iota(\mathcal{F})$ is somehow the "type of $\mathcal{F}$ along ι." Moreover, for any two E-linear embedding ι, ι' as above, we have $\mathcal{N}_\iota(\mathcal{F}) = \mathcal{N} \Leftrightarrow \mathcal{N}_{\iota'}(\mathcal{F}) = \mathcal{N}$. This gives sense to the sentence "$\mathcal{F}$ has type $\mathcal{N}$" for $\mathcal{F} \in \mathrm{Fil}_K(\omega^G)$. With this definition we have

$$\mathcal{F}(G,\mathcal{N})(K) = \{\mathcal{F} \in \mathrm{Fil}_K(\omega^G) \text{ of type } \mathcal{N}\}.$$

We note that for any $\mathcal{F} \in \mathcal{F}(G,\mathcal{N})(K)$, we have $\deg(\mathcal{F}) = \deg(\mathcal{N})$, see diagram (1.1), page 137. We may now also define

$$\mathcal{F}(G,\mathcal{N})(K)^{\mathrm{ss}} := \{\mathcal{F} \in \mathcal{F}(G,\mathcal{N})(K) \mid \mathcal{F} \text{ is semi-stable}\}.$$

Example 6.1.7 Let $G = \mathrm{GL}(V)$ for some n-dimensional k-vector space V. As already explained, $E = k$ and $\mathcal{N}$ corresponds to some element $\nu \in (\mathbb{Q}^n)_+$. Then what is now denoted by $\mathcal{F}(\mathrm{GL}(V),\mathcal{N})$ is nothing but what was denoted by $\mathcal{F}(V,\nu)$ in Chapter II. Similarly, the notation $\mathcal{F}(\mathrm{GL}(V),\mathcal{N})(K)^{\mathrm{ss}}$ corresponds to the former notation $\mathcal{F}(V,\nu)(K)^{\mathrm{ss}}$.

We now look at functorial properties of these objects. We fix a PD-pair $(G,\mathcal{N})$ and set $E = E(G,\mathcal{N})$. Let $\rho : G \longrightarrow G'$ be a morphism of reductive groups over k. Then we have a push-forward map

$$\rho_* : X_*(G)_{\mathbb{Q}}/G \longrightarrow X_*(G')_{\mathbb{Q}}/G'.$$

Since it is Γ_k-equivariant, we have $E(G',\rho_*\mathcal{N}) \subset E$. Then the functorial map $\mathrm{Fil}_{\overline{k}}(\omega^G) \longrightarrow \mathrm{Fil}_{\overline{k}}(\omega^{G'})$ underlies a morphism of projective algebraic varieties over Spec E,

$$\mathcal{F}(G,\mathcal{N}) \longrightarrow \mathcal{F}(G',\rho_*(\mathcal{N}))_E\ . \tag{1.2}$$

Proposition 6.1.8 *Keep the foregoing notation and assume that ρ is injective and homogeneous. Then the morphism* (1.2) *above is a closed immersion and for every field extension K of E, we have*

$$\mathcal{F}(G,\mathcal{N})(K)^{ss} = \mathcal{F}(G,\mathcal{N})(K) \cap \mathcal{F}(G',\rho_*(\mathcal{N}))(K)^{ss}.$$

Proof The closed immersion property is a standard fact. The other statement follows from Proposition 5.2.6. □

Proposition 6.1.9 *Let G_{ad} be the adjoint group of G, and let $\mathcal{N}_{\mathrm{ad}}$ be the induced conjugacy class of* $\mathbb{Q}$-1-PS *of G_{ad}. Then for every extension K of $E(G,\mathcal{N})$*

$$\mathcal{F}(G,\mathcal{N})(K)^{ss} \xrightarrow{\sim} \mathcal{F}(G_{\mathrm{ad}},\mathcal{N}_{\mathrm{ad}})(K)^{ss}\ .$$

Proof The map $\mathcal{F}(G, \mathcal{N})(K) \longrightarrow \mathcal{F}(G_{\text{ad}}, \mathcal{N}_{\text{ad}})(K)$ is bijective. The assertion follows from Proposition 5.2.5. □

Corollary 6.1.10 *Let k be a finite field. Then there exists a Zariski-open subset $\mathcal{F}(G, \mathcal{N})^{ss}$ of $\mathcal{F}(G, \mathcal{N})$ such that $\mathcal{F}(G, \mathcal{N})^{ss}(K) = \mathcal{F}(G, \mathcal{N})(K)^{\text{ss}}$ for every extension K of $E = E(G, \mathcal{N})$. The action of the finite group $G(k)$ on $\mathcal{F}(G, \mathcal{N})$ preserves $\mathcal{F}(G, \mathcal{N})^{ss}$.*

Proof By Proposition 5.2.5 we may assume that G is of adjoint type. By Proposition 6.1.8 above applied to a faithful representation $\rho : G \longrightarrow G' = \text{GL}(V)$, we are reduced to the case of GL_n, which is treated in Chapter II, §1. □

Definition 6.1.11 Let $(G, \mathcal{N})$ be a PD-pair over the finite field k. The *period domain* associated to $(G, \mathcal{N})$ is the Zariski-open subset $\mathcal{F}(G, \mathcal{N})^{ss}$ of $\mathcal{F}(G, \mathcal{N})$.

Variants 6.1.12 (i) Let $\mathcal{N} = (\mathcal{N}_1, \ldots, \mathcal{N}_r)$ be a tuple of r conjugacy classes of $\mathbb{Q}$-1-PS of G. Let $E = E(G, \mathcal{N}) = E(G, \mathcal{N}_1) \cdot \cdots \cdot E(G, \mathcal{N}_r)$ be the composite of the individual reflex fields. Then we may consider the product of flag varieties

$$\mathcal{F}(G, \mathcal{N}) = \prod_{i=1}^{r} \mathcal{F}(G, \mathcal{N}_i),$$

which is defined over $E(G, \mathcal{N})$. For every faithful homogeneous representation $(V, \rho) \in \text{Rep}_k(G)$, let $\rho_*(\mathcal{N}) = (\rho_*(\mathcal{N}_1), \ldots, \rho_*(\mathcal{N}_r))$. The resulting morphism

$$\mathcal{F}(G, \mathcal{N}) \longrightarrow \mathcal{F}(\text{GL}(V), \rho_*(\mathcal{N}))_E$$

is a closed immersion and for every field extension K of E, the intersection

$$\mathcal{F}(G, \mathcal{N})(K)^{ss} = \mathcal{F}(G, \mathcal{N})(K) \cap \mathcal{F}(\text{GL}(V), \rho_*(\mathcal{N}))(K)^{ss}$$

is the set of semi-stable filtrations $\mathcal{F}$ in $\mathcal{F}(G, \mathcal{N})(K)$. The analogous statements of Proposition 6.1.9 and Corollary 6.1.10 hold.

(ii) Now we describe the variant of the category $\text{Fil}^k_{\mathbb{F}_1}$ in the case of arbitrary reductive groups. Fix a maximal split torus $S \subset G$. Let $\mathcal{F}$ be a $\mathbb{Q}$-filtration of ω^G over an extension K/k. Then $\mathcal{F}$ is semi-stable if $(\text{Lie}\, G, \mathcal{F}^\bullet \text{Lie}\, G)$ is semi-stable with respect to all S-stable subspaces of $\text{Lie}\, G$. Here we consider the adjoint action of S on $\text{Lie}\, G$. If $G = \text{GL}_n$ then this definition is, by the tensor product theorem of Variant 1.2.11, equivalent to our original definition in 1.1.22.

Let $\mathcal{N}$ be a conjugacy class of $\mathbb{Q}$-1-PS of $G_{\bar{k}}$ with reflex field $E = E(G, \mathcal{N})$. Denote by $\mathcal{F}(G, S, \mathcal{N})(\bar{k})^{ss}$ the set of semi-stable filtrations in the above sense. In order to see that this set is induced by a subvariety of $\mathcal{F}(G, \mathcal{N})$ one proceeds as follows. First of all, one can consider in the above setting instead of a maximal split torus any split torus. In particular, this applies to $G = \mathrm{GL}_n$. In the latter case one shows similarly to the case of variant $\mathrm{Fil}^k_{\mathbb{F}_1}$ that for a given split torus $S' \subset \mathrm{GL}_n$ the set $\mathcal{F}(\mathrm{GL}_n, S', \mathcal{N})(\bar{k})^{ss}$ is induced by a subvariety $\mathcal{F}(\mathrm{GL}_n, S', \mathcal{N})^{ss}$. Then it is checked by the same proof that Proposition 5.2.7 holds, i.e., for every faithful homogeneous representation $(V, \rho) \in \mathrm{Rep}_k(G)$, the morphism $\mathcal{F}(G, \mathcal{N}) \longrightarrow \mathcal{F}(\mathrm{GL}(V), \rho_*(\mathcal{N}))_E$ is a closed immersion and for every field extension K of E the intersection

$$\mathcal{F}(G, S, \mathcal{N})(K)^{ss} = \mathcal{F}(G, \mathcal{N})(K) \cap \mathcal{F}(\mathrm{GL}(V), \rho(S), \rho_*(\mathcal{N}))(K)^{ss}$$

is the set of semi-stable filtrations $\mathcal{F}$ in $\mathcal{F}(G, \mathcal{N})(K)$. The analogous statements to Proposition 6.1.9 and Corollary 6.1.10 hold.

Examples 6.1.13 (i) Let $G = \mathrm{GSp}_4$ be the group of symplectic similitudes of the 4-dimensional symplectic vector space $V = k^4$, with symplectic form

$$\left\langle \sum x_i e_i, \sum y_i e_i \right\rangle = x_1 y_4 + x_2 y_3 - x_3 y_2 - x_4 y_1 .$$

As maximal torus T we take the diagonal matrices in G. Then

$$X_*(T) = \{(x_1, x_2, x_3, x_4) \in \mathbb{Z}^4 \mid x_1 + x_4 = x_2 + x_3\} ,$$

and

$$X_*(T)_{\mathbb{Q}}/W \cong \bar{C}_{\mathbb{Q}} := \{(x_1, \ldots, x_4) \in \mathbb{Q}^4 \mid x_1 + x_4 = x_2 + x_3, x_1 \geq x_2 \geq x_3 \geq x_4\} .$$

Let $\mathcal{N}$ be regular, i.e., correspond to $\nu = (x_1, \ldots, x_4) \in \bar{C}_{\mathbb{Q}}$ with $x_1 > x_2 > x_3 > x_4$. Then $\mathcal{F}(G, \mathcal{N})$ is the variety of full *symplectic* flags

$$0 \underset{\neq}{\subset} \mathcal{F}_1 \underset{\neq}{\subset} \mathcal{F}_2 \underset{\neq}{\subset} \mathcal{F}_3 \underset{\neq}{\subset} \bar{k}^4 ,$$

i.e., $\mathcal{F}_1^{\perp} = \mathcal{F}_3, \mathcal{F}_2^{\perp} = \mathcal{F}_2$. Then $\mathcal{F}(G, \mathcal{N})^{ss}$ is described as follows:

$\mathcal{F}_2$ contains no k-rational line, and $\mathcal{F}_1$ is not contained in a totally isotropic k-rational plane.

(ii) Let $G = \mathrm{GSp}_6$ be the group of symplectic similitudes of the 6-dimensional symplectic vector space $V = k^6$, with symplectic form

$$\left\langle \sum x_i e_i, \sum y_i e_i \right\rangle = x_1 y_6 + x_2 y_5 + x_3 y_4 - x_4 y_3 - x_5 y_2 - x_6 y_1 .$$

Again we take for T the diagonal torus in G. Then

$$X_*(T) = \{(x_1, x_2, x_3, x_4, x_5, x_6) \in \mathbb{Z}^6 \mid x_1 + x_6 = x_2 + x_5 = x_3 + x_4\} ,$$

and

$$X_*(T)_{\mathbb{Q}}/W \cong \bar{C}_{\mathbb{Q}} := \\ \{(x_1, \ldots, x_6) \in \mathbb{Q}^6 \mid x_1 + x_6 = x_2 + x_5 = x_3 + x_4, x_1 \geq \cdots \geq x_6\} .$$

Let $\mathcal{N}$ be regular, i.e., correspond to $v = (x_1, \ldots, x_6) \in \bar{C}_{\mathbb{Q}}$ with $x_1 > \cdots > x_6$. Then $\mathcal{F}(G, \mathcal{N})$ is the variety of full symplectic flags

$$0 \underset{\neq}{\subset} \mathcal{F}_1 \underset{\neq}{\subset} \mathcal{F}_2 \underset{\neq}{\subset} \mathcal{F}_3 \underset{\neq}{\subset} \mathcal{F}_4 \underset{\neq}{\subset} \mathcal{F}_5 \underset{\neq}{\subset} \bar{k}^6 ,$$

i.e., $\mathcal{F}_1^\perp = \mathcal{F}_5, \mathcal{F}_2^\perp = \mathcal{F}_4$ and $\mathcal{F}_3^\perp = \mathcal{F}_3$. Then $\mathcal{F}(G, \mathcal{N})^{ss}$ is described as follows:

$\mathcal{F}_3$ contains no k-rational line, and $\mathcal{F}_4$ does not contain a totally isotropic k-rational plane E with $\mathcal{F}_2 \cap E \neq (0)$, and $\mathcal{F}_5$ does not contain a totally isotropic k-rational plane E with $\mathcal{F}_1 \subset E \subset \mathcal{F}_5$, and $\mathcal{F}_4$ contains no 3-dimensional totally isotropic k-rational subspace, and $\mathcal{F}_2$ is not contained in a 3-dimensional totally isotropic k-rational space E, and $\mathcal{F}_1$ is not contained in a 3-dimensional totally isotropic k-rational space E with $\dim E \cap \mathcal{F}_3 \geq 2$, and $\mathcal{F}_5$ does not contain a 3-dimensional totally isotropic k-rational space E with $\dim E \cap \mathcal{F}_3 \geq 2$.

Let $\alpha_1 = \epsilon_1 - \epsilon_2$, $\alpha_2 = \epsilon_2 - \epsilon_3$, $\alpha_3 = 2\epsilon_3$ denote the set of std simple roots. If $\alpha_1(v) > \alpha_2(v) + \alpha_3(v)$, i.e., $x_1 > x_2 + x_3$, resp. if $\alpha_1(v) < \alpha_2(v) + \alpha_3(v)$, i.e., $x_1 < x_2 + x_3$, we have to impose the following further condition if $x_1 \neq x_2 + x_3$.

$x_1 > x_2 + x_3$: *There is no 3-dimensional totally isotropic k-rational space E with $\mathcal{F}_1 \subset E \subset \mathcal{F}_5$.*

$x_1 < x_2 + x_3$: *There is no 3-dimensional totally isotropic k-rational space E with $\dim \mathcal{F}_3 \cap E \geq 2$.*

(iii) Let $G = \mathrm{GSp}_{2n}$ be the group of symplectic similitudes of the symplectic form on $V = k^{2n}$ given by

$$\left\langle \sum x_i e_i, \sum y_i e_i \right\rangle = \sum_{i=1}^{n} x_i \cdot y_{2n+1-i} - \sum_{i=n+1}^{2n} x_i \cdot y_{2n+1-i} .$$

Let T be the diagonal maximal torus, so that

$$X_*(T) = \{(x_1, \ldots, x_{2n}) \in \mathbb{Z}^{2n} \mid x_i + x_{2n+1-i} = a, \forall i\}.$$

Let $\mathcal{N}$ be the conjugacy class of the element $(1^{(n)}, 0^{(n)}) \in X_*(T)$. Then $\mathcal{F}(G, \mathcal{N})$ is the variety of lagrangian subspaces $\mathcal{V} \subset V$. Then $\mathcal{F}(G, \mathcal{N})^{ss}$ consists of the lagrangian subspaces $\mathcal{V}$ such that for all totally isotropic k-subspaces $W \in V$ the following inequality holds,

$$\dim \mathcal{V} \cap (W \otimes_k K) \leq 1/2 \cdot \dim W .$$

To check this, one uses Example 5.3.4.

Notes and References The reflex field was first introduced in the theory of Shimura varieties [61] (there it is the field of definition of a pro-variety and is a finite extension of $\mathbb{Q}$).

2 The relation to GIT

In this section we will transpose the results of Chapter II, §2 from GL_n to an arbitrary reductive group. In the case of GL_n, the fixed element $\nu \in (\mathbb{Q}^n)_+$ could be interpreted as defining a conjugacy class $\mathcal{N}$ of $\mathbb{Q}$-1-PS, as well as defining a homogeneous $\mathbb{Q}$-line bundle on $\mathcal{F}(\nu)$. In the general case, such a double use of ν can also be made, but one has to spell out on what it depends. Let $\bar{k}$ be an algebraic closure of k and let $\Gamma = \Gamma_k$.

Definition 6.2.1 An *invariant inner product* on the reductive group G over k is a positive-definite bilinear form on $X_*(T)_{\mathbb{Q}}$ for any maximal torus T of G defined over $\bar{k}$. The following conditions are required.

(i) For $g \in G(\bar{k})$, the inner automorphism $\mathrm{Int}(g)$ induces an isometry

$$\mathrm{Int}(g) : X_*(T)_{\mathbb{Q}} \longrightarrow X_*({}^gT)\,,\ {}^gT = g \cdot T \cdot g^{-1}\,.$$

(ii) Any $\sigma \in \Gamma$ induces an isometry

$$\sigma : X_*(T)_{\mathbb{Q}} \longrightarrow X_*({}^{\sigma}T)_{\mathbb{Q}}\,.$$

We can make this more explicit using the *L-action* of Γ, cf. [22]. Fix a maximal torus T_0 defined over $\bar{k}$ and a Borel subgroup B_0 containing T_0. For $\sigma \in \Gamma$, there exists $g \in G(\bar{k})$ with

$$g\,{}^{\sigma}T_0 g^{-1} = T_0\,,\ g\,{}^{\sigma}B_0 g^{-1} = B_0\,.$$

Therefore $\sigma_* := \mathrm{Int}(g) \circ \sigma$ is an automorphism of $X_*(T_0)_{\mathbb{Q}}$. This automorphism is independent of the choice of g, since any other choice is of the form $g' = tg$, for $t \in T_0(\bar{k})$. Also, the automorphism σ_* is independent of B_0, and

$$(\sigma\tau)_* = \sigma_*\tau_*\,,\quad \sigma, \tau \in \Gamma\,. \tag{2.1}$$

In this way we obtain an action, the L-action, of the semi-direct product $\Gamma \ltimes W$ on $X_*(T_0)$. Here W denotes the Weyl group of T_0.

Note that if G is quasi-split over k, we may choose T_0 and B_0 defined over k, in which case the L-action of Γ coincides with the usual action of Γ on $X_*(T_0)_{\mathbb{Q}}$, stemming from the fact that T_0 is defined over k. In this case Γ acts on W by its usual Galois action, and it is this action that is used in forming the semi-direct product $\Gamma \ltimes W$.

An invariant inner product on G obviously defines a positive-definite bilinear form on $X_*(T_0)_{\mathbb{Q}}$ which is invariant under $\Gamma \ltimes W$. The converse is given by the following Lemma.

Lemma 6.2.2 *Let* $(\ ,\ )$ *be a positive-definite* $\Gamma \ltimes W$*-invariant bilinear form on* $X_*(T_0)_{\mathbb{Q}}$*. There exists a unique invariant inner product on G which induces* $(\ ,\ )$.

Proof This is an immediate consequence of the conjugacy of maximal tori in $G_{\overline{k}}$. □

We note that there always exists a positive-definite $\Gamma \ltimes W$-invariant bilinear form since Γ acts through the finite quotient Γ/Γ_E where E is the splitting field of T_0. Indeed, let (,) be any positive-definite bilinear form on $X_*(T_0)_{\mathbb{Q}}$. Then

$$(v,w)' = \frac{1}{\#((\Gamma/\Gamma_E) \ltimes W)} \sum_{g \in (\Gamma/\Gamma_E) \ltimes W} (gv, gw)$$

defines a positive-definite $\Gamma \ltimes W$-invariant bilinear form on $X_*(T_0)_{\mathbb{Q}}$.

Examples 6.2.3 (i) Let G be semi-simple. There is a natural choice of an invariant inner product (,) on $X_*(T_0)_{\mathbb{Q}}$, given by the *Killing form*,

$$(v, v') = \sum_{\chi \in \Phi} \langle v, \chi \rangle \cdot \langle v', \chi \rangle .$$

Here Φ denotes the set of roots of T_0.
(ii) Let $G = T$ be a torus. Then the L-action of Γ on $X_*(T)$ coincides with the usual action. Any Γ-invariant positive-definite bilinear form on $X_*(T)_{\mathbb{Q}}$ is an invariant inner product on T; there is no natural choice for it.
(iii) Any reductive group G is the almost direct product of a central torus and G_{der}. Therefore the next lemma gives a complete overview of the invariant inner products on G.

Lemma 6.2.4 (i) *An invariant inner product on G is the orthogonal direct sum of invariant inner products on Z_G^o and on G_{der}.*

(ii) *Let G be semi-simple and let $G_1, \dots, G_n$ be the set of k-simple normal subgroups of G, so that $G_1 \times \cdots \times G_n \longrightarrow G$ is a central isogeny. Then an invariant inner product on G is the orthogonal direct sum of inner products on each factor $G_i, i = 1, \dots, n$.*

(iii) *An invariant inner product on a k-simple semi-simple group is a positive rational multiple of the Killing form.*

Proof Let first $k = \overline{k}$ be algebraically closed. Then, if G is simple, the Weyl group acts by a non-trivial irreducible representation on $X_*(T_0)_{\mathbb{Q}}$, cf. [28], Ch. V, §4.8. In particular, there is only one invariant inner product, up to scalar. Therefore, for an arbitrary reductive group, the W-representation on $X_*(T_0)_{\mathbb{Q}}$ decomposes into a direct sum of irreducible representations, of which the direct sum of copies of the trivial representation corresponds to $X_*(Z_G^o)_{\mathbb{Q}}$, and the non-trivial irreducible representations correspond to the simple factor of G_{der}. The lemma follows in the case of an algebraically closed field.

We deduce (i) and (ii) over an arbitrary field k. For (iii) we note that if G is k-simple, then Γ acts transitively on the $\bar{k}$-simple factors of $G_{\bar{k}}$. Hence the result follows from the previous case by property (ii) in the Definition 6.2.1 of an invariant inner product. □

We now start with a reductive group G and fix an invariant inner product on G. Let $\mathcal{N}$ be a conjugacy class of $\mathbb{Q}$-1-PS of $G_{\bar{k}}$, and denote by $E = E(G, \mathcal{N})$ the corresponding reflex field. Let $\mathcal{F} = \mathcal{F}(G, \mathcal{N})$ be the corresponding partial flag variety over E. Our aim is to define a homogeneous ample $\mathbb{Q}$-line bundle $\mathcal{L}$ on $\mathcal{F}$.

Let $x \in \mathcal{F}(\bar{k})$ and let $\nu \in \mathcal{N}$ be a $\mathbb{Q}$-1-PS giving rise to x. Let $Q = Q_x$ be the corresponding psgp and $M = M_x$ its Levi factor and D_M its maximal torus factor group, and finally A'_Q its maximal split torus factor group. Then ν induces a central $\mathbb{Q}$-1-PS $\bar{\nu}$ of M. Now any maximal torus $\overline{T}$ in M is the isomorphic image of a unique torus T of G contained in Q, hence the invariant inner product on G induces an invariant inner product on M. Hence $\bar{\nu}$ can also be considered as an element $\bar{\nu}^*$ of $X^*(\overline{T})_{\mathbb{Q}}$. Since $\bar{\nu}$ is central, $\bar{\nu}^*$ can be considered as an element of $X^*(D_M)_{\mathbb{Q}}$, since

$$X^*(\overline{T})_{\mathbb{Q}} = X^*(\overline{T} \cap M_{\mathrm{der}})_{\mathbb{Q}} \oplus X^*(D_M)_{\mathbb{Q}}$$

is an orthogonal sum decomposition, by Lemma 6.2.4 (i).

Now $\mathcal{F} \otimes_k \bar{k} = G_{\bar{k}}/Q$ and the character $-\bar{\nu}^*$ of Q defines a homogeneous $\mathbb{Q}$-line bundle $\mathcal{L}$ on $\mathcal{F} \otimes_k \bar{k}$, via

$$\mathcal{L} = G \times^Q \mathbb{G}_{a,-\nu^*} \ . \tag{2.2}$$

The minus sign here appears in order to have an ample line bundle (rather than an anti-ample line bundle), as the next lemma shows.

Lemma 6.2.5 (i) *The line bundle $\mathcal{L}$ is independent of the choice of the base point, and is ample.*

(ii) *$\mathcal{L}$ is defined over E, i.e., it comes by base change from a (unique) line bundle on $\mathcal{F}(G, \mathcal{N})$ over E.*

Proof (i) Let $Q' = {}^gQ$ be a conjugate psgp. Then $\mathrm{Int}(g)$ induces an isomorphism between M and $M' = Q'/R_u(Q')$, and hence also between D_M and $D_{M'}$. This last isomorphism is independent of the choice of g. This shows the required independence. For the ampleness, we may assume by the faithfully flatness criterion, EGA IV, Cor. 2.7.2, that k is algebraically closed. Then we apply [125], Rem. 1, p. 232. This remark says that a homogeneous line bundle on G/Q induced by a character χ of Q is ample† if and only if $\langle \chi, \alpha^\vee \rangle < 0$ for all $\alpha \in \Delta_Q$. Here $\alpha^\vee$ is the co-root of $\alpha \in \Delta$.

(ii) By Hilbert's Theorem 90, it suffices to show that the isomorphism class of $\mathcal{L}$ is defined over E. But for any $\sigma \in \Gamma_E$, the two $\mathbb{Q}$-1-PS ${}^\sigma\nu$ and ν are conjugate. Hence under the canonical isomorphism between D_M and $D_{{}^\sigma M}$, the two elements $-\overline{\nu}^*$ and ${}^\sigma(-\overline{\nu}^*)$ are identical. Hence ${}^\sigma\mathcal{L}$ is isomorphic to $\mathcal{L}$ (even as homogeneous line bundles). □

Remark 6.2.6 Let $\rho : G \hookrightarrow G'$ be a closed embedding, and let $\mathcal{N}' = \rho_*(\mathcal{N})$. Assume that the invariant inner product on G is the restriction of the invariant inner product on G'. Then the homogeneous $\mathbb{Q}$-line bundle $\mathcal{L}$ on $\mathcal{F}(G, \mathcal{N})$ is isomorphic to the restriction of the homogeneous $\mathbb{Q}$-line bundle $\mathcal{L}'$ under the closed embedding

$$\mathcal{F}(G, \mathcal{N}) \hookrightarrow \mathcal{F}(G', \mathcal{N}') \otimes_{E'} E \, .$$

Example 6.2.7 Let $G = \mathrm{GL}_n$ and let $\mathcal{N}$ be given by the element $\nu \in \mathbb{Q}^n = X_*(T)_\mathbb{Q}$, with

$$\nu = (\nu(1)^{(n_1)}, \nu(2)^{(n_2)}, \ldots, \nu(r)^{(n_r)}) \, ,$$

where $\nu(1) > \nu(2) > \cdots > \nu(r)$. Here T denotes the diagonal torus in G. The psgp corresponding to the base point $[\mathcal{F}_\nu] \in \mathcal{F}(\nu)$ is the std psgp $Q = Q_{(n_1,\ldots,n_r)}$ with Levi factor $M = \mathrm{GL}_{n_1} \times \cdots \times \mathrm{GL}_{n_r}$ and with $D_M = A'_Q \cong \mathbb{G}_m^r$, via $(\det_1, \ldots, \det_r)$. We have a natural identification

$$X^*(D_M)_\mathbb{Q} = \mathbb{Q}^r \, .$$

As invariant inner product on G we take the standard one on $X_*(T)_\mathbb{Q} = \mathbb{Q}^n$. Then one sees immediately that

$$\nu^* = (\nu(1), \nu(2), \ldots, \nu(r)) \in \mathbb{Q}^r = X^*(D_M)_\mathbb{Q} \, .$$

It follows that the homogeneous $\mathbb{Q}$-line bundle $\mathcal{L}$ coincides with the homogeneous $\mathbb{Q}$-line bundle corresponding to the character $(\det_1^{-\nu(1)} \cdot \cdots \cdot \det_r^{-\nu(r)})$ of $M = \mathrm{GL}_{n_1} \times \cdots \times \mathrm{GL}_{n_r}$. Hence $\mathcal{L}$ coincides with the line bundle $\mathcal{L}(\nu)$ considered in Chapter II, §2, as follows from (2.4) therein, since the 1-dimensional T-representations on the fiber are the same.

Theorem 6.2.8 *Let $(G, \mathcal{N})$ be a PD-pair, and fix an invariant inner product on G. Let $\mathcal{L}$ denote the corresponding ample homogeneous $\mathbb{Q}$-line bundle on $\mathcal{F} = \mathcal{F}(G, \mathcal{N})$ over $E = E(G, \mathcal{N})$. Let K be a field extension of E, and let*

† Note that the std psgps in [125] are lower triangular which implies that there > is used instead of <.

$x \in \mathcal{F}(K)$, with corresponding $\mathbb{Q}$-filtration $\mathcal{F}_x$ of ω^G over K. Then $\mathcal{F}_x$ is semi-stable if and only if for all 1-PS *λ of G_{der} defined over k the Hilbert–Mumford inequality holds, i.e.,*

$$\mu^{\mathcal{L}}(x, \lambda) \geq 0 .$$

Proof By Proposition 5.2.5 we may assume that G is semi-simple. Let $(V, \rho) \in \mathrm{Rep}_k(G)$ be a faithful representation,

$$\rho : G \hookrightarrow \mathrm{GL}(V) .$$

We assume first that the invariant inner product on G is induced by the scalar product on $\mathrm{GL}(V)$, as used in Chapter II, §2. In this case, the homogeneous ample $\mathbb{Q}$-line bundle $\mathcal{L}$ is isomorphic to the pullback of the homogeneous ample $\mathbb{Q}$-line bundle $\widetilde{\mathcal{L}}$ on $\widetilde{\mathcal{F}} = \mathcal{F}(\mathrm{GL}(V), \rho_*(\mathcal{N}))$ under the closed embedding

$$\rho : \mathcal{F}(G, \mathcal{N}) \hookrightarrow \mathcal{F}(\mathrm{GL}(V), \rho_*(\mathcal{N}))_E ,$$

cf. Remark 6.2.6.

Let $\mathcal{F}_x \in \mathcal{F}(K)$ be semi-stable. Then, since G is semi-simple, the object $(V, \mathcal{F}_x^\bullet V)$ of Fil_k^K is semi-stable, cf. Corollary 5.2.7. In other words, $\rho(x) \in \widetilde{\mathcal{F}}(K)^{ss}$. It follows from Theorem 2.2.3 that $\mu^{\widetilde{\mathcal{L}}}(\rho(x), \widetilde{\lambda}) \geq 0$ for all 1-PS $\widetilde{\lambda}$ of $\mathrm{SL}(V)$, defined over k. If λ is a 1-PS of G, then $\widetilde{\lambda} = \rho \circ \lambda$ factors through $\mathrm{SL}(V)$ since G is semi-simple. We obtain

$$\mu^{\mathcal{L}}(x, \lambda) = \mu^{\widetilde{\mathcal{L}}}(\rho(x), \widetilde{\lambda}) \geq 0 ,$$

as required.

To prove the converse, assume that $\mathcal{F}_x$ is not semi-stable. We need to find a $\mathbb{Q}$-1-PS λ of G over k with $\mu^{\mathcal{L}}(x, \lambda) < 0$. We take λ to be a splitting of the HN-filtration $\mathcal{F}_{\mathrm{HN}}$ associated to $\mathcal{F}_x$. Then

$$\begin{aligned}
\mu^{\mathcal{L}}(x, \lambda) &= \mu^{\widetilde{\mathcal{L}}}(\rho(x), \widetilde{\lambda}) = -\langle \mathcal{F}_x, \mathcal{F}_{\mathrm{HN}} V \rangle \\
&= -\sum_{x,y} xy \dim \mathrm{gr}^x_{\mathcal{F}_x} \mathrm{gr}^y_{\mathcal{F}_{\mathrm{HN}}}(V) \\
&= -\sum_{y} y \cdot \sum_{x} x \dim \mathrm{gr}^x_{\mathcal{F}_x} \mathrm{gr}^y_{\mathcal{F}_{\mathrm{HN}}}(V) \\
&= -\sum_{y} y \deg_{\mathcal{F}_x}(\mathrm{gr}^y_{\mathcal{F}_{\mathrm{HN}}}(V)) \\
&= -\sum_{y} y \cdot \mu(\mathrm{gr}^y_{\mathcal{F}_{\mathrm{HN}}}(V)) \cdot \dim \mathrm{gr}^y_{\mathcal{F}_{\mathrm{HN}}}(V) \\
&= -\sum_{y} y^2 \cdot \dim \mathrm{gr}^y_{\mathcal{F}_{\mathrm{HN}}}(V) . \qquad (2.3)
\end{aligned}$$

Since $\mathcal{F}_x$ is not semi-stable, the HN-filtration $\mathcal{F}_{\mathrm{HN}}$ is non-trivial, cf. Corollary 5.3.3. Hence the RHS of (2.3) above is strictly negative.

This proves the theorem for this particular choice of invariant inner product. Now let us consider an arbitrary invariant inner product. We may assume G to be adjoint. Let $G_1, \ldots, G_n$ be the k-simple normal factors of G. Then $G = G_1 \times \cdots \times G_n$ and accordingly

$$\mathcal{F}(G, \mathcal{N}) = \prod\nolimits_i \mathcal{F}(G_i, \mathcal{N}_i)\,,$$

where $\mathcal{N}_i = (\mathrm{pr}_i)_*(\mathcal{N})$. Also, by Lemma 6.2.4, the invariant inner product on G is the product of invariant inner products on the factors G_i and hence accordingly, $\mathcal{L}$ is an exterior tensor product,

$$\mathcal{L} = \mathcal{L}_1 \boxtimes \mathcal{L}_2 \boxtimes \cdots \boxtimes \mathcal{L}_n\,.$$

This easily reduces the assertion to the case when G is k-simple. But then, by Lemma 6.2.4 (iii) any two invariant inner products differ by a scalar in $\mathbb{Q}_+^\times$. Therefore the general assertion follows from the case of the special invariant inner product considered above. □

Notes and References The notion of an invariant inner product is due to Totaro [219], §7. Theorem 6.2.8 is just Thm. 3 in Totaro [219], and proves a conjecture of Rapoport and Zink [183], p. 46. This conjecture was inspired by a result of van der Put and Voskuil [220], Prop. 2.6. The proof given here of the implication that the Hilbert–Mumford inequality implies semi-stability is a simplification of the argument in [219].

3 The Harder–Narasimhan stratification

We fix a PD-pair $(G, \mathcal{N})$. Let $P \subset G$ be a k-parabolic subgroup and let $\mathcal{N}_P \in X_*(P)_{\mathbb{Q}}/P$ be a conjugacy class of $\mathbb{Q}$-1-PS of $P_{\bar{k}}$. As in diagram (1.1), page 137 we define its degree $\deg(\mathcal{N}_P)$ as the push-forward of $\mathcal{N}_P$ to $X_*(A'_P)_{\mathbb{Q}}$.

Definition 6.3.1 An HN-*pair* $(P, \mathcal{N}_P)$ is a pair consisting of a k-psgp P of G and a conjugacy class $\mathcal{N}_P \in X_*(P)_{\mathbb{Q}}/P$ which satisfy the following conditions.

(i) Under the inclusion $P \subset G$ the conjugacy class $\mathcal{N}_P$ is mapped to the conjugacy class $\mathcal{N}$.

(ii) For any (or equivalently, one) lifting $\lambda : \mathbb{D} \longrightarrow P$ of $\deg(\mathcal{N}_P)$ to a $\mathbb{Q}$-1-PS of P, we have the equality $P = P_\lambda$. (Here P_λ is the k-psgp of G associated to λ.)

If $g \in G(k)$, then conjugation by g induces an isomorphism $A'_P \xrightarrow{\sim} A'_{gP}$ which takes $\deg(\mathcal{N}_P)$ to $\deg(\mathcal{N}_{gP})$.

Definition 6.3.2 An HN-*type* associated to $(G, \mathcal{N})$ is a $G(k)$-conjugacy class of HN-pairs $(P, \mathcal{N}_P)$.

Furthermore, the HN-*vector* of an HN-type θ is the $G(k)$-conjugacy class $\mathrm{HN}(\theta) \in X_*(G)_{\mathbb{Q}}^{\Gamma_k}/G(k)$ of any $\mathbb{Q}$-1-PS λ as in Definition 6.3.1 (ii) for any HN-pair $(P, \mathcal{N}_P)$ in θ.

We denote by $\Theta(G, \mathcal{N})$ the set of HN-types associated to $(G, \mathcal{N})$ and by θ_0 or $\theta_{G,\mathcal{N}}$ the distinguished element $(G, \mathcal{N}) \in \Theta(G, \mathcal{N})$. Correspondingly, we denote by $\mathcal{H}(G, \mathcal{N})$ the set of HN-vectors. We will see below that these sets are finite. However, we record before some functorial properties.

First, any morphism $\rho : G \longrightarrow G'$ of reductive groups over k induces a map

$$\rho_* : \Theta(G, \mathcal{N}) \longrightarrow \Theta(G', \rho_*(\mathcal{N})).$$

More precisely, if $(P, \mathcal{N}_P)$ is an HN-pair for $(G, \mathcal{N})$ and λ is a $\mathbb{Q}$-1-PS as in condition (ii) of Definition 6.3.1, then the pair $(P_{\rho_*(\lambda)}, \rho_*(\mathcal{N}_P))$ is an HN-pair for $(G', \rho_*(\mathcal{N}))$. Moreover, this map ρ_* is compatible with taking HN-vectors in an obvious way.

Next, the Galois group Γ_E of the reflex field $E = E(G, \mathcal{N})$ acts on $\Theta(G, \mathcal{N})$. Indeed, any $\sigma \in \Gamma_E$ leaves $\mathcal{N}$ stable. Hence for any HN-pair $(P, \mathcal{N}_P)$, the pair $(P, {}^{\sigma}\mathcal{N}_P)$ satisfies condition (i) in Definition 6.3.1. Furthermore, since the push-forward map $X_*(P)_{\mathbb{Q}}/P \mapsto X_*(A'_P)_{\mathbb{Q}}$ is Γ_k-equivariant, we have $\deg({}^{\sigma}\mathcal{N}_P) = \deg(\mathcal{N}_P)$, hence condition (ii) of Definition 6.3.1 also is satisfied. We denote the action of Γ_E on $\Theta(G, \mathcal{N})$ by $\theta \longmapsto {}^{\sigma}\theta$.

Now we give a more concrete description of these objects. We fix a minimal psgp P_0 over k and a Levi subgroup M_0 of P_0. Let $\mathcal{P}$ be the set of std psgps $P \supset P_0$. We use the same notation as in Corollary 5.2.10. In particular, for $P \in \mathcal{P}$ with std Levi component M, we write $A_P = A_M$. In this case we may define the "standard slope" $\mu(\mathcal{N}_P) \in X_*(A_P)_{\mathbb{Q}}$ of a conjugacy class $\mathcal{N}_P$ to be the preimage of $\deg(\mathcal{N}_P)$ by the isomorphism $X_*(A_P)_{\mathbb{Q}} \xrightarrow{\sim} X_*(A'_P)_{\mathbb{Q}}$. Then condition (ii) in Definition 6.3.1 may be reformulated in the following way:

$$\langle \mu(\mathcal{N}_P), \alpha \rangle > 0\,, \quad \forall \alpha \in \Delta_P\,. \tag{3.1}$$

To make this even more explicit, let us also fix a maximal torus T in M_0 *defined over* k, with absolute Weyl group W. Let $\mathcal{N}$ correspond to the element $\nu = \nu_G \in X_*(T)_{\mathbb{Q}}/W$. Any HN-type $\theta \in \Theta(G, \mathcal{N})$ has a unique HN-pair representative $(P, \mathcal{N}_P)$ with $P \in \mathcal{P}$. Then $\mathcal{N}_P$ corresponds to a unique element $\nu_P \in X_*(T)_{\mathbb{Q}}/W_M$, where W_M denotes the absolute Weyl group of T in M. In this way we obtain an identification of $\Theta(G, \mathcal{N})$ with the set of pairs (P, ν_P) with $P \in \mathcal{P}$ and $\nu_P \in X_*(T)_{\mathbb{Q}}/W_{M_P}$, satisfying the following two conditions.

(i) $\nu_P \equiv \nu \bmod W$.

(ii) Let $\mu(\nu_P) \in X_*(A_P)_\mathbb{Q}$ be the image of ν_P under $X_*(T)_\mathbb{Q} \longrightarrow X_*(A'_P)_\mathbb{Q} \xrightarrow{\sim} X_*(A_P)_\mathbb{Q}$. (Note that this image is independent of the representative modulo W_{M_P}.) Then

$$\langle \mu(\nu_P), \alpha \rangle > 0 \,, \quad \forall \alpha \in \Delta_P \,.$$

Note that in this more concrete description of $\Theta(G, \mathcal{N})$, the action of Γ_E is given by $\theta = (P, \nu_P) \mapsto {}^\sigma\theta = (P, {}^\sigma\nu_P)$, where ${}^\sigma\nu_P$ is the class modulo W_M of the conjugate under σ of a representative of ν_P in $X_*(T)_\mathbb{Q}$.

There is also a more explicit version of the HN-vector HN(θ) of a HN-type θ. Indeed, let

$$\bar{C}_\mathbb{Q} = \{x \in X_*(A_{P_0})_\mathbb{Q} \mid \langle x, \alpha \rangle \geq 0 \,, \quad \forall \alpha \in \Delta_{P_0}\} \tag{3.2}$$

be the *k-rational positive closed Weyl chamber*. It is a fundamental domain for the action of the *relative Weyl group* $W_0 = N(A_{P_0})(k)/Z(A_{P_0})(k)$ on $X_*(A_{P_0})_\mathbb{Q}$, cf. [28], Ch. VI, §10. The canonical embedding $X_*(A_{P_0}) \hookrightarrow X_*(G)^{\Gamma_k}$ induces a bijection

$$\bar{C}_\mathbb{Q} \xrightarrow{\sim} X_*(G)^{\Gamma_k}_\mathbb{Q}/G(k). \tag{3.3}$$

Hence we may consider HN(θ) as an element of $\bar{C}_\mathbb{Q}$. In this setting, if θ has as its standard representative HN-pair $(P, \mathcal{N}_P)$, then HN(θ) is the image of $\mu(\mathcal{N}_P)$ under the inclusion $X_*(A_P)_\mathbb{Q} \hookrightarrow X_*(A_{P_0})_\mathbb{Q}$. Indeed, by condition (3.1), page 150, this image lies in $\bar{C}_\mathbb{Q}$.

Example 6.3.3 Let $G = \mathrm{GL}_n$ and let $P_0 = B$ be the Borel subgroup of upper triangular matrices. Let $M_0 = T = A_{P_0}$ be the group of diagonal matrices. Via the identification $X_*(T)_\mathbb{Q}/W \cong (\mathbb{Q}^n)_+$, the conjugacy class $\mathcal{N}$ is given by a vector

$$\nu = \nu_G = (\nu(1)^{(n_1)}, \ldots, \nu(r)^{(n_r)}) \in \mathbb{Q}^n$$

with $\nu(1) > \nu(2) > \cdots > \nu(r)$ and $n_i \geq 1$, $\forall i$. Let $\theta = (P, \nu_P)$ be an HN-type, with $P \in \mathcal{P}$ and $\nu_P \in X_*(T)_\mathbb{Q}/W_{M_P}$. Let $(m_1, \ldots, m_s)$ be the decomposition of n such that $P = P_{(m_1,\ldots,m_s)}$. Thus under the natural identification $W \cong S_n$ we get $W_{M_P} \cong S_{m_1} \times \cdots \times S_{m_s}$. Then $X_*(T)_\mathbb{Q}/W_{M_P}$ has a natural set of representatives in $\mathbb{Q}^n$ given by the vectors $(x_1, \ldots, x_n) \in \mathbb{Q}^n$ with

$$\begin{aligned}
&x_1 \geq x_2 \geq \cdots \geq x_{m_1}, \\
&x_{m_1+1} \geq x_{m_1+2} \geq \cdots \geq x_{m_1+m_2}, \\
&\vdots \\
&x_{m_1+\cdots m_{s-1}+1} \geq x_{m_1+\cdots m_{s-1}+2} \geq \cdots \geq x_n.
\end{aligned}$$

If we choose for ν_P such a representative, then condition (i) means that ν_P is a permutation of ν, whereas condition (3.1), page 150 translates into the *strict* inequalities

$$\frac{x_1 + \cdots + x_{m_1}}{m_1} > \frac{x_{m_1+1} + \cdots + x_{m_1+m_2}}{m_2} > \cdots > \frac{x_{m_1+\cdots m_{s-1}+1} + \cdots + x_n}{m_s}.$$

We note that the appearance of denominators follows from the identification $X_*(A_P)_\mathbb{Q} \simeq X_*(A'_P)_\mathbb{Q}$. Thus we see that for $G = \mathrm{GL}_n$ the notion of an HN-type defined here is identical to the notion of a HN-type defined in Definition 2.3.2.

In this situation the HN-vector associated to $\theta = (P, \mathcal{N}_P)$ is given via the identification

$$\bar{C}_\mathbb{Q} = \{(x_1, \ldots, x_n) \in \mathbb{Q}^n \mid x_1 \geq x_2 \geq \cdots \geq x_n\}$$

by

$$\Big((\frac{x_1 + \cdots + x_{m_1}}{m_1})^{(m_1)}, (\frac{x_{m_1+1} + \cdots + x_{m_1+m_2}}{m_2})^{(m_2)}, \ldots,$$
$$(\frac{x_{m_1+\cdots m_{s-1}+1} + \cdots + x_n}{m_s})^{(m_s)}\Big).$$

Let K be a field extension of E contained in $\bar{k}$. Any $\mathcal{F} \in \mathcal{F}(G, \mathcal{N})(K)$ defines its HN-type $\theta_\mathcal{F} \in \Theta(G, \mathcal{N})$ as follows. Let ${}^\bullet\omega^G_\mathcal{F}$ be the HN-filtration of ω^G associated to $\mathcal{F}$, cf. Theorem 5.3.1, and let $P = P_{\mathrm{HN}}$ be the HN-psgp attached to ${}^\bullet\omega^G_\mathcal{F}$. Let $\mathcal{F}_P$ be the unique $\mathbb{Q}$-filtration of ω^P over K inducing $\mathcal{F}$ as in Proposition 4.2.17, and let $\mathcal{N}_P$ be the conjugacy class of $\mathcal{F}_P$. Then by Proposition 5.3.5, the pair $(P, \mathcal{N}_P)$ is an HN-pair and we denote by $\theta_\mathcal{F}$ its class in $\Theta(G, \mathcal{N})$. We obtain in this way disjoint decompositions into finitely many summands,

$$\mathcal{F}(G, \mathcal{N})(K) = \bigcup_\theta \mathcal{F}(G, \mathcal{N})(K)_\theta\,, \quad \mathcal{F}(G, \mathcal{N})(K) = \bigcup_H \mathcal{F}(G, \mathcal{N})(K)_H\,.$$

Here H runs through the HN-vectors of HN-types. By Corollary 5.3.3, we have

$$\mathcal{F}(G, \mathcal{N})(K)_{\theta_0} = \mathcal{F}(G, \mathcal{N})(K)^{ss} = \mathcal{F}(G, \mathcal{N})(K)_{H_0},$$

where $H_0 = \mathrm{HN}(\theta_0) = \mu(\mathcal{N})$. The next result therefore generalizes Corollary 6.1.10.

Let $\theta \in \Theta(G, \mathcal{N})$ be an HN-type and choose an HN-pair $(P, \mathcal{N}_P)$ as representative. Let $M = M_P$ be the Levi quotient and denote by $\mathcal{N}_M$ the induced conjugacy class of $\mathbb{Q}$-1-PS of M. We define the reflex field $E(P, \mathcal{N}_P)$ of $(P, \mathcal{N}_P)$ to be the reflex field $E(M, \mathcal{N}_M)$ of the PD-pair $(M, \mathcal{N}_M)$. This is a finite extension of k. Finally, we set

$$E_\theta := E(P, \mathcal{N}_P)$$

which is independent of the chosen HN-pair $(P, \mathcal{N}_P) \in \theta$. It is the *reflex field of the HN-type* θ .

Lemma 6.3.4 *For any* HN-*type* $\theta \in \Theta(G, \mathcal{N})$*, the reflex field* E_θ *of* θ *is an extension of* $E = E(G, \mathcal{N})$.

Proof We have to prove that the absolute Galois group Γ_{E_θ} of E_θ is contained in Γ_E. More concretely, if $\sigma \in \Gamma_{E_\theta}$ then we have to find for any 1-PS $\lambda \in \mathcal{N}$, an element $g \in G(\bar{k})$ with ${}^\sigma\lambda = g\lambda g^{-1}$. If $\lambda \in \mathcal{N}_P$ then this property is satisfied by assumption. In general, there is some $g \in G(\bar{k})$ such that $\lambda' = g\lambda g^{-1} \in \mathcal{N}_P$. But then there is some $p \in P(\bar{k})$ such that ${}^\sigma\lambda' = p\lambda' p^{-1}$. We compute ${}^\sigma\lambda = {}^\sigma(g^{-1}\lambda' g) = \sigma(g)^{-1} p\lambda p^{-1}\sigma(g) = g'\lambda g'^{-1}$ with $g' = \sigma(g)^{-1}pg$. The claim follows. □

Theorem 6.3.5 *Assume that k is a finite field. For each* HN-*type* $\theta \in \Theta(G, \mathcal{N})$*, there is a locally closed subvariety* $\mathcal{F}(G, \mathcal{N})_\theta$ *of* $\mathcal{F}(G, \mathcal{N}) \otimes_E E_\theta$ *with*

$$\mathcal{F}(G, \mathcal{N})_\theta(\bar{k}) = \mathcal{F}(G, \mathcal{N})(\bar{k})_\theta\ .$$

Moreover, any $\sigma \in \Gamma_k$ *induces an isomorphism* ${}^\sigma\mathcal{F}(G, \mathcal{N})_\theta \xrightarrow{\sim} \mathcal{F}(G, \mathcal{N})_{\sigma\theta}$.

We will prove this theorem along the same lines as the corresponding result for GL_n, namely Theorem 2.3.3. For this, we need a generalization of item (iii) in the proof of that theorem to the case of an arbitrary reductive group.

For the present discussion, k may be again an arbitrary perfect field. For a PD-pair $(G, \mathcal{N})$ over k with reflex field E, we denote by

$$\overline{\mathcal{F}}(G, \mathcal{N}) = \mathcal{F}(G, \mathcal{N}) \otimes_E \bar{k}$$

the base change of $\mathcal{F}(G, \mathcal{N})$ from E to $\bar{k}$. Let P be a k-parabolic subgroup of G and let $\mathcal{N}_P$ be a conjugacy class of $\mathbb{Q}$-1-PS of $P_{\bar{k}}$ which is mapped to $\mathcal{N}$ under the inclusion $P \subset G$. Then there is a unique locally closed subvariety $\overline{\mathcal{F}}(P, \mathcal{N}_P)$ of $\overline{\mathcal{F}}(G, \mathcal{N})$ (a generalized Bruhat cell) such that

$$\overline{\mathcal{F}}(P, \mathcal{N}_P)(\bar{k}) = \{\mathcal{F} \in \overline{\mathcal{F}}(G, \mathcal{N})(\bar{k}) \mid \mathcal{F}_P \text{ has type } \mathcal{N}_P\}.$$

Here $\mathcal{F}_P$ denotes the filtration on ω^P over $\bar{k}$ associated to $\mathcal{F}$ by Proposition 4.2.17. Further, let $\mathcal{N}_{M_P}$ be the push-forward of $\mathcal{N}_P$ to the Levi quotient M_P of P. Then the map $\mathcal{F}_P \mapsto \mathcal{F}_{M_P}$ induces a morphism of $\bar{k}$ varieties

$$\overline{\mathcal{F}}(P, \mathcal{N}_P) \longrightarrow \overline{\mathcal{F}}(M_P, \mathcal{N}_{M_P}).$$

Moreover, $\overline{\mathcal{F}}(P, \mathcal{N}_P)$ together with the morphisms to $\overline{\mathcal{F}}(G, \mathcal{N})$ and $\overline{\mathcal{F}}(M_P, \mathcal{N}_{M_P})$ are defined over the reflex field $E(P, \mathcal{N}_P)$ of $(P, \mathcal{N}_P)$.

To make this more explicit, fix a Borel pair (T, B) with T a maximal torus defined over k and such that $T \subset B \subset P \subset G$. Let $\nu \in X_*(T)_\mathbb{Q}$ be the unique representative of $\mathcal{N}$ lying in the Weyl chamber defined by B. Further, let M be the unique Levi component of P containing T, and let $\nu_P \in X_*(T)_\mathbb{Q}$ be the unique representative of $\mathcal{N}_P$ lying in the Weyl chamber defined by $B \cap M$. Then, under the identification of $\overline{\mathcal{F}}(G, \mathcal{N})$ with $G_{\bar{k}}/Q_\nu$, the subvariety $\overline{\mathcal{F}}(P, \mathcal{N}_P)$ can be identified with the $P_{\bar{k}}$-homogeneous locally closed subvariety $P_{\bar{k}} w Q_\nu / Q_\nu$, for any element $w \in W$ such that $w\nu = \nu_P$. If we denote by $W_P, W_{Q_\nu} \subset W$ the subgroups defined by the Levi factors of P resp. Q_ν, then the double coset $w \in W_P \backslash W / W_{Q_\nu}$ is uniquely determined by this property. Moreover, denoting by $\pi : P \longrightarrow M_P$ the projection and by $Q^M_{\nu_P}$ the psgp of M_P associated with $\pi \circ \nu_P$, the morphism $\overline{\mathcal{F}}(P, \mathcal{N}_P) \longrightarrow \overline{\mathcal{F}}(M_P, \mathcal{N}_{M_P})$ can be identified with the projection induced by π,

$$P_{\bar{k}}/(P_{\bar{k}} \cap {}^w Q_{\nu_G}) = P_{\bar{k}}/(P_{\bar{k}} \cap Q_{\nu_P}) \longrightarrow M_{P,\bar{k}}/Q^M_{\nu_P}.$$

Lemma 6.3.6 *The morphism $\overline{\mathcal{F}}(P, \mathcal{N}_P) \to \overline{\mathcal{F}}(M_P, \mathcal{N}_{M_P})$ is an affine bundle of rank $\ell(w)$.*

Proof (sketch) To prove the bundle property, the first observation is that the morphism $P_{\bar{k}} \cap Q_{\nu_P} \longrightarrow Q^M_{\nu_P}$ induced by the projection $\pi : P \longrightarrow M_P$ is surjective. Then, since its kernel is $U_{P,\bar{k}} \cap Q_{\nu_P}$, we get a cartesian diagram

$$\begin{array}{ccc} P_{\bar{k}}/(U_{P,\bar{k}} \cap Q_{\nu_P}) & \longrightarrow & \overline{\mathcal{F}}(P, \mathcal{N}_P) \\ \downarrow & & \downarrow \\ M_{P,\bar{k}} & \longrightarrow & \overline{\mathcal{F}}(M_P, \mathcal{N}_{M_P}) . \end{array}$$

The bottom map is smooth and surjective, hence in particular it is faithfully flat. Now let $Q_{-\nu_P}$ be the T-opposite psgp to Q_{ν_P}. The second ingredient comes from the theory of the "big cell" which implies that the multiplication in G induces an isomorphism of varieties $(U_{P,\bar{k}} \cap U_{Q_{-\nu_P}}) \times (U_{P,\bar{k}} \cap Q_{\nu_P}) \xrightarrow{\sim} U_{P,\bar{k}}$. Let us abbreviate $U'_{P,\bar{k}} := U_{P,\bar{k}} \cap U_{Q_{-\nu_P}}$. We note that the above diagram is $U'_{P,\bar{k}}$-equivariant if we let $U'_{P,\bar{k}}$ act on $P_{\bar{k}}$ by left multiplication and trivially on $M_{P,\bar{k}}$. Since the left vertical map is a $U'_{P,\bar{k}}$-torsor, it follows by descent that the right vertical one also is a $U'_{P,\bar{k}}$-torsor. But since U'_P is unipotent, this torsor has to be trivial. (In fact we already have a section $M_{P,\bar{k}}/Q^M_{\nu_P} \xrightarrow{\sim} M_{\bar{k}}/(M_{\bar{k}} \cap Q_{\nu_P}) \hookrightarrow P_{\bar{k}}/(P_{\bar{k}} \cap Q_{\nu_P})$.) Now U'_P is smooth and unipotent, hence is isomorphic to an affine space, as a variety, cf. [64], §2, 3.9, §4, 4.1. The formula for the rank follows as in the GL_n-case from the following lemma. □

Lemma 6.3.7 *Let $P, Q \subset G$ be two std psgps. Let $w \in W$ be of minimal length in its double coset in $W_P \backslash W / W_Q$. Then $U_P / U_P \cap wQw^{-1} \simeq \mathbb{A}^{\ell(w)}$.*

Proof The proof is the same as in the GL_n-case, cf. Lemma 2.3.12. □

Proof of Theorem 6.3.5 Let $(P, \mathcal{N}_P)$ be any HN-pair in the HN-type θ. We use the same notation as above. Inside $\mathcal{F}(M_P, \mathcal{N}_{M_P})$ we have the period domain of semi-stable points $\mathcal{F}(M_P, \mathcal{N}_{M_P})^{ss}$ which is an open subvariety defined over $E(P, \mathcal{N}_P) = E(M_P, \mathcal{N}_{M_P})$ by Theorem 6.1.10. Put

$$\mathcal{F}(P, \mathcal{N}_P)(\bar{k})_\theta := \mathcal{F}(G, \mathcal{N}_P)(\bar{k})_\theta \cap \mathcal{F}(P, \mathcal{N}_P)(\bar{k}).$$

Then by Proposition 5.3.5, we see that $\mathcal{F}(P, \mathcal{N}_P)(\bar{k})_\theta$ is just the preimage of $\mathcal{F}(M_P, \mathcal{N}_{M_P})^{ss}(\bar{k})$ under the projection map $\mathcal{F}(P, \mathcal{N}_P) \longrightarrow \mathcal{F}(M_P, \mathcal{N}_{M_P})$. Thus $\mathcal{F}(P, \mathcal{N}_P)(\bar{k})_\theta$ is induced by an open subvariety $\mathcal{F}(P, \mathcal{N}_P)_\theta$ of $\mathcal{F}(P, \mathcal{N}_P)$. But then, since $\mathcal{F}(G, \mathcal{N})(\bar{k})_\theta = \bigcup_{g \in G/P(k)} g \cdot \mathcal{F}(P, \mathcal{N}_P)(\bar{k})_\theta$, we see that $\mathcal{F}(G, \mathcal{N})(\bar{k})_\theta$ is induced by the locally closed subvariety of $\mathcal{F}(G, \mathcal{N}) \otimes_E E_\lambda$ obtained by the corresponding scheme-theoretic disjoint union. □

Corollary 6.3.8 *Let k be finite. Let $\theta \in \Theta(G, \mathcal{N})$ and let $(P, \mathcal{N}_P) \in \theta$. Then $\mathcal{F}(G, \mathcal{N})_\theta$ is a $G(k)$-equivariant affine bundle over the finite disjoint sum*

$$\coprod_{G(k)/P(k)} \mathcal{F}(M_P, \mathcal{N}_{M_P})^{ss}.$$

(The action of $G(k)$ on the target space is described in Remark 2.3.11.) *Its rank is the length $\ell(w)$ of the element w defined as in Lemma* 6.3.6 *(and which depends on some Borel pair such that $T \subset B \subset P$).*

Now we look for an analogue of Corollary 2.3.4 in this more general setting. In that corollary we used a partial ordering on the set of HN-vectors which was induced by the dominance order on the set $(\mathbb{Q}^n)_+$. Similarly we need here a partial ordering on the set of HN-vectors. However, here we may consider HN-vectors in two ways: either as elements of $X_*(G)^{\Gamma_k}/G(k)$, which is better suited for functoriality, or as elements of a closed Weyl chamber $\overline{C}_\mathbb{Q}$, which is more concrete, but apparently depends on some choices. To simplify our discussion we will identify $X_*(G)^{\Gamma_k}/G(k)$ with $\overline{C}_\mathbb{Q}$ via the isomorphism (3.3), page 151. Then there are two candidates for the ordering in question. First, there is an "external" ordering defined on $X_*(G)^{\Gamma_k}/G(k)$ by

$$H \leq H' \Leftrightarrow \big(\rho_* H \leq \rho_* H', \ \forall (V, \rho) \in \mathrm{Rep}_k(G)\big). \tag{3.4}$$

Here the latter $\leq$ is the dominance order with respect to $\mathrm{GL}(V)$, recalled right after Theorem 2.3.1. On the other hand, there is the "internal" ordering, the *dominance order* defined on the chamber $\overline{C}_\mathbb{Q}$ singled out by the choice of the minimal psgp P_0, cf. [9, 133], by

$$x \leq x' \Leftrightarrow \Big(x' - x = \sum_{\alpha \in \Delta_{P_0}} n_\alpha \cdot \alpha^\vee, \quad n_\alpha \in \mathbb{Q}_{\geq 0}\Big). \tag{3.5}$$

Here $\alpha^\vee$ denotes the co-root corresponding to α. The following proposition says that these two candidates coincide.

Proposition 6.3.9 *Let $x, x' \in \bar{C}_{\mathbb{Q}}$. The following conditions are equivalent.*

(i) *x lies in the convex hull of the finite set $W_0 x'$.*

(ii) *$x \leq x'$.*

Hence the relation $\leq$ extends to arbitrary $G(k)$-conjugacy classes of $\mathbb{Q}$-1-PS of G.

(iii) *For any irreducible $(V, \rho) \in \mathrm{Rep}_k(G)$, we have*

$$\rho(x) \leq \rho(x'),$$

where $\rho(x)$ and $\rho(x')$ are considered as $\mathbb{Q}$-1-PS of $\mathrm{GL}(V)$.

Proof For the proof, we need the following set

$$\bar{C}_{\mathbb{Q}}^\vee = \{y \in X^*(A_{P_0})_{\mathbb{Q}} \mid \langle y, \alpha^\vee \rangle \geq 0, \quad \forall \alpha \in \Delta_{P_0}\}.$$

Again, this is a fundamental domain for the action of W_0 on $X^*(A_{P_0})_{\mathbb{Q}}$.

(i) $\Rightarrow$ (ii): We write $x = \sum_w a_w \cdot wx'$ with $a_w \geq 0$ and $\sum_w a_w = 1$. Since $x, x' \in \bar{C}_{\mathbb{Q}}$, we may assume that $a_w \in \mathbb{Q}$ for all w. Since $x' \in \bar{C}_{\mathbb{Q}}$, by [28], Ch. VI, Prop. 18, we know that for $w \in W_0$

$$x' - wx' = \sum\nolimits_{\alpha \in \Delta_{P_0}} n_\alpha^w \cdot \alpha^\vee \quad , \quad n_\alpha^w \in \mathbb{Q}_{\geq 0}\,.$$

Thus we get

$$x' - x = (\sum\nolimits_w a_w)x' - \sum\nolimits_w a_w \cdot wx = \sum\nolimits_{\alpha \in \Delta_{P_0}} \sum\nolimits_w a_w \cdot n_\alpha^w \cdot \alpha^\vee.$$

(ii) $\Rightarrow$ (i): Assume that x is not contained in the convex hull of $W_0 x'$. Then there is an affine hyperplane separating $W_0 x'$ from x. Thus there is a vector $\beta \in X^*(A_{P_0})_{\mathbb{Q}}$ with $\langle wx', \beta \rangle < \langle x, \beta \rangle, \forall w \in W_0$. Let $w \in W_0$ such that $w^{-1}\beta$ is contained in the Weyl chamber $\bar{C}_{\mathbb{Q}}^\vee$. Since $x \in \bar{C}_{\mathbb{Q}}$, we deduce as in (i) that $x - w^{-1}x = \sum_{\alpha \in \Delta_{P_0}} m_\alpha \cdot \alpha^\vee$ with $m_\alpha \geq 0$. Thus we get $x' - w^{-1}x = (x' - x) + (x - w^{-1}x) = \sum_{\alpha \in \Delta_{P_0}} (n_\alpha + m_\alpha) \cdot \alpha^\vee$. But then

$$\langle wx', \beta \rangle - \langle x, \beta \rangle = \langle x', w^{-1}\beta \rangle - \langle w^{-1}x, w^{-1}\beta \rangle \geq 0\,,$$

a contradiction.

For the proof of (ii) $\Leftrightarrow$ (iii), we need the following lemma. For a representation $(V, \rho) \in \mathrm{Rep}_k(G)$ and $x \in \bar{C}_{\mathbb{Q}}$, we consider the induced grading $V = \oplus_{r \in \mathbb{Q}} V(x)_r$. We set

$$\max(\rho_*(x)) = \max\{r \mid V(x)_r \neq 0\}.$$

By [215], Thm. 7.2 we know that for each dominant weight $\chi \in X^*(T)/W$, there is an irreducible k-rational G-representation (V_χ, ρ_χ) with the following property. Over the algebraic closure $\bar{k}$ the representation V_χ splits into a direct sum of irreducible representation, such that all highest weights are contained in $\Gamma_*(\chi)$. Here Γ_* denotes the L-action of Γ on $X^*(T)$, cf. (2.1), page 144. All these highest weights give rise by restriction to A_{P_0} to the same element in $\bar{C}^\vee_{\mathbb{Q}}$. Moreover, two such representations (V_χ, ρ_χ) and $(V_{\chi'}, \rho'_\chi)$ are k-isomorphic if and only if χ and χ' are conjugate under the L-action of Γ.

Lemma 6.3.10 *Let $x, x' \in \bar{C}_{\mathbb{Q}}$. Then $x \leq x'$ if and only if* $\max(\rho_*(x)) \leq \max(\rho_*(x'))$ *for any* $(V, \rho) \in \mathrm{Rep}_k(G)$.

Proof We start by observing that if $(V, \rho) = (V_\chi, \rho_\chi)$ is an irreducible representation corresponding to the integral dominant weight vector $\chi \in X^*(T)/W$, then $\max(\rho_*(x)) = \langle x, \chi \rangle$. In fact, let Δ be the set of absolutely simple roots of G with respect to $T \subset P_0$. For any other weight χ' appearing in V there is some $\sigma \in \Gamma$ such that $\sigma_*(\chi) - \chi' = \sum_{\beta \in \Delta} n_\beta \beta$ with $n_\beta \geq 0$. This implies $\langle x, \sigma_*(\chi) - \chi' \rangle \geq 0$ since $x \in \bar{C}_{\mathbb{Q}}$ and the relative simple roots are given by Δ via the restriction of T to A_{P_0}. Thus $\langle x, \chi \rangle = \langle x, \sigma_*(\chi) \rangle \geq \langle x, \chi' \rangle$.

Let $x \leq x'$, i.e., $x' - x = \sum_{\alpha \in \Delta_{P_0}} n_\alpha \cdot \alpha^\vee$, $n_\alpha \in \mathbb{Q}_{\geq 0}$. In order to show the assertion, we may assume that $(V, \rho) = (V_\chi, \rho_\chi)$ is irreducible for some dominant weight χ. Further we may assume that χ is even Γ-invariant since $x' - x$ is Γ-invariant. But then $\langle x' - x, \chi \rangle \geq 0$, i.e., $\max(\rho_*(x)) \leq \max(\rho_*(x'))$.

Conversely, let us assume that $\max(\rho_*(x)) \leq \max(\rho_*(x'))$ for every irreducible $(V, \rho) = (V_\chi, \rho_\chi)$, where $\chi \in \bar{C}^\vee_{\mathbb{Q}} \subset X^*(T)/W$. Since those Γ-invariant χ generate $\bar{C}^\vee_{\mathbb{Q}}$, we conclude by duality that $x' - x$ has the desired shape. □

(ii) ⇒ (iii): This follows from the above lemma by restricting representations of $\mathrm{GL}(V)$ to G.

(iii) ⇒ (ii): We apply again the above lemma. The claim follows from the fact that if $G = \mathrm{GL}_n$ and $\rho = \mathrm{id}$ and $x = (x_1 \geq x_2 \geq \cdots \geq x_n) \in \bar{C}_{\mathbb{Q}}$, then x_1 is nothing else but $\max(\rho_*(x))$. □

Remark 6.3.11 It follows from the proof of Proposition 6.3.9, more precisely from the proof of Lemma 6.3.10, that there is a finite set I of rational representations of G such that for any $H, H' \in X_*(G)^{\Gamma_k}/G(k)$, we have

$$H \leq H' \Leftrightarrow \left(\rho_* H \leq \rho_* H', \ \forall (V, \rho) \in I\right).$$

In fact, one may choose any I consisting of irreducible representations such that their highest weights span the closed Weyl chamber. On the other hand, it is in general not possible to find some singleton $I = \{(V, \rho)\}$ even if (V, ρ) is a faithful representation.

Now, we transfer the above partial order to $\Theta(G, \mathcal{N})$ via the associated HN-vectors,

$$\theta \leq \theta' \iff \mathrm{HN}(\theta) \leq \mathrm{HN}(\theta') . \tag{3.6}$$

We note that all $\theta \in \Theta(G, \mathcal{N})$ have identical images in $X_*(A'_G)_{\mathbb{Q}}$. We also note that θ_0 is the unique minimal element of $\Theta(G, \mathcal{N})$.

Proposition 6.3.12 *For each HN-type θ, we have*

$$\overline{\mathcal{F}(G, \mathcal{N})_\theta} \subset \bigcup_{\theta' \geq \theta} \mathcal{F}(G, \mathcal{N})_{\theta'} .$$

For each HN-*vector H, there is a locally closed subvariety $\mathcal{F}(G, \mathcal{N})_H$ of $\mathcal{F}(G, \mathcal{N})$ defined over E, with*

$$\mathcal{F}(G, \mathcal{N})_H(\bar{k}) = \mathcal{F}(G, \mathcal{N})(\bar{k})_H .$$

Furthermore,

$$\overline{\mathcal{F}(G, \mathcal{N})_H} \subset \bigcup_{H' \geq H} \mathcal{F}(G, \mathcal{N})_{H'} .$$

The variety $\mathcal{F}(G, \mathcal{N})_H$ for a given HN-*vector is the scheme-theoretic sum of all* HN-*strata corresponding to* HN-*types with* HN-*vector equal to H:*

$$\mathcal{F}(G, \mathcal{N})_H = \coprod_{\mathrm{HN}(\theta) = H} \mathcal{F}(G, \mathcal{N})_\theta .$$

Proof The argument is the same as in Corollary 2.3.4 for $G = \mathrm{GL}_n$. Again, the union $X = \bigcup_{\theta' \geq \theta} \mathcal{F}(G, \mathcal{N})_{\theta'}$ is constructible which implies for the topological closure $\overline{X}$ the identity $\overline{X} = \bigcup_{x \in X} \overline{\{x\}}$. Let $y \in \overline{\{x\}}$ for some $x \in X$. Let $\rho : G \to \mathrm{GL}(V)$ be any irreducible representation and denote by $\rho_* : \mathcal{F}(G, \mathcal{N}) \to \mathcal{F}(\mathrm{GL}(V), \rho_*(\mathcal{N}))$ the induced morphism. Then $\rho_*(y) \in \overline{\{\rho_*(x)\}}$. By using the valuative criterion (EGA II, Thm. 7.3.8) and Theorem 2.3.1 (a), we conclude from the definition of the ordering on HN-vectors that X is closed. This proves the first statement.

On the other hand we have

$$\mathcal{F}(G, \mathcal{N})_H = \overline{\mathcal{F}(G, \mathcal{N})_H} \setminus \bigcup_{\{\theta' \mid \mathrm{HN}(\theta') > H\}} \mathcal{F}(G, \mathcal{N})_{\theta'} . \tag{3.7}$$

Hence $\mathcal{F}(G, \mathcal{N})_H$ is a locally closed subset. Now Theorem 2.3.1 (b) shows that $\mathcal{F}(G, \mathcal{N})_\theta$ is closed in $\mathcal{F}(G, \mathcal{N})_H$ for any $\theta \in \Theta(G, \mathcal{N})$ with $\mathrm{HN}(\theta) = H$. Hence $\mathcal{F}(G, \mathcal{N})_H$ is the disjoint sum of these closed subsets.

Finally, since H is fixed under Γ_E, we see that $\mathcal{F}(G, \mathcal{N})_H$ comes from a locally closed subvariety of $\mathcal{F}(G, \mathcal{N})$ defined over E. □

Corollary 6.3.13 *A point $x \in \mathcal{F}(G,\mathcal{N})(\bar{k})$ has* HN-*vector H if and only if for all rational representations (V,ρ) of G, the image $\rho_*(x)$ under the map*

$$\mathcal{F}(G,\mathcal{N})(\bar{k}) \to \mathcal{F}(\mathrm{GL}(V),\rho_*(\mathcal{N}))(\bar{k})$$

has HN-*vector $\rho_*(H)$.*

Proof This follows from Proposition 6.3.9 (iii). □

Example 6.3.14 Let us reconsider Example 6.1.13 (i), i.e., $G = \mathrm{GSp}_4$ and $\nu = (x_1, x_2, -x_2, -x_1)$ regular. Since G is split we have $E = k$, $\Delta = \Delta_0 = \{\alpha_1, \alpha_2\}$, where $\alpha_1 = \epsilon_1 - \epsilon_2$ and $\alpha_2 = 2\epsilon_2$. The set of std psgps is $\mathcal{P} = \{B, P_1, P_2, G\}$. Also, W_{P_i} = is generated by the reflection s_{α_i}. A computation shows that there are six HN-type strata $\Theta(\mathcal{N}) = \{\theta_0, \theta_1, \theta_2, \theta_3, \theta_4, \theta_5\}$ given as follows.

θ	$\mathcal{F}(G,\mathcal{N})_\theta$
$\theta_0 = (G,\nu)$	$\mathcal{F}(G,\mathcal{N})^{ss}$
$\theta_1 = (B,\nu)$	$\mathcal{F}(G,\mathcal{N})(k)$
$\theta_2 = (P_2,\nu)$	$\{\mathcal{F} \in \mathcal{F}(G,\mathcal{N}) \mid \mathcal{F}^{x_1}$ *is k-rational,* $\mathcal{F}^{x_2}$ *is not k-rational* $\}$
$\theta_3 = (P_1,\nu)$	$\{\mathcal{F} \in \mathcal{F}(G,\mathcal{N}) \mid \mathcal{F}^{x_2}$ *is k-rational,* $\mathcal{F}^{x_1}$ *is not k-rational* $\}$
$\theta_4 = (P_2,\nu_1)$	$\{\mathcal{F} \in \mathcal{F}(G,\mathcal{N}) \mid \mathcal{F}^{x_2}$ *contains a k-rational line,* $\mathcal{F}^{x_1}$ *and* $\mathcal{F}^{x_2}$ *aré not k-rational* $\}$
$\theta_5 = (P_1,\nu_2)$	$\{\mathcal{F} \in \mathcal{F}(G,\mathcal{N}) \mid \mathcal{F}^{x_1}$ *is contained in a totally isotropic k-rational plane,* $\mathcal{F}^{x_1}$ *and* $\mathcal{F}^{x_2}$ *are not k-rational*$\}$

Here $\nu_1 = (x_2, -x_1, x_1, -x_2)$ and $\nu_2 = (x_1, -x_2, x_2, -x_1)$.

Moreover, the stratifications by HN-types and by HN-vectors coincide in this case. The HN-vectors are as follows.

θ	$\mathrm{HN}(\theta)$
θ_0	$(0,0,0,0)$
θ_1	$(x_1, x_2, -x_2, -x_1)$
θ_2	$(x_1, 0, 0, -x_1)$
θ_3	$\frac{1}{2}((x_1+x_2)^{(2)}, (-(x_1+x_2))^{(2)})$
θ_4	$(x_2, 0, 0, -x_2)$
θ_5	$\frac{1}{2}((x_1-x_2)^{(2)}, (x_2-x_1))^{(2)})$

Notes and References Proposition 6.3.9 and variants of it are discussed in the context of split groups resp. compact Lie groups in several sources, cf. [9], §12, [2], [182], [133], [178]. The notion of an HN-type in the context of G-bundles appears in [9], cf. also [151]. More precisely, the notion of the HN-type used in [151] corresponds to our HN-type, whereas the HN-type in the sense of [9] corresponds to our HN-vector. Analogously to Corollary 6.3.13, Atiyah and

Bott [9] prove that a G-bundle ξ has type μ if and only if the induced bundle $\rho(\xi)$ has type $\rho(\mu)$ for all unitary G-representations ρ of G. Using this property they deduce closure properties of the HN-strata for the moduli space of G-bundles from the case of vector bundles. It turns out that the stratification of the moduli space of G-bundles by HN-types is perfect in the sense of equivariant cohomology with respect to the gauge group.

VII. Cohomology of Period Domains for Reductive Groups

In this chapter we consider the ℓ-adic cohomology of period domains for arbitrary reductive groups. As in the case of GL_n, we compute the Euler–Poincaré characteristic and we count their points using the Grothendieck–Lefschetz fixed point formula.

The structure of this chapter is similar to Chapter III that concerned the case of GL_n. First we formulate the Langlands lemma and then we discuss generalized Steinberg representations. In the second section we then use these ingredients to determine the Euler–Poincaré characteristics of period domains.

In this chapter we use the following notation. Let G be a reductive group over a perfect field k. We fix a minimal parabolic subgroup P_0 and a Levi subgroup M_0 of P_0, both defined over k. Let

$$\mathcal{P} = \{P \subset G \mid P \supset P_0\}$$

be the set of standard parabolic subgroups of G. For every $P \in \mathcal{P}$, we denote as before by M_P the unique Levi subgroup of P containing M_0, and by N_P the unipotent radical of P. Further, we write A_P for the maximal split torus in the center of M_P. In the case $P = P_0$ we also denote the maximal split torus A_{P_0} by S. Let $\Delta_0 = \Delta_{P_0}$ be the set of simple roots of G with respect to $S \subset P_0$. There is a one-to-one correspondence between $\mathcal{P}$ and the set of subsets of Δ_0. To the subset I of Δ_0 corresponds the standard parabolic P_I such that $\Delta_0 \setminus I$ is the set of simple roots occurring in N_{P_I}. We set $M_I = M_{P_I}$ and $N_I = N_{P_I}$. The minimal parabolic subgroup P_0 corresponds to $I = \emptyset$ and the non-proper parabolic G to $I = \Delta_0$. Also $I \subset J$ if and only if $P_I \subset P_J$.

1 The Langlands Lemma and generalized Steinberg representations

We start by formulating the Langlands Lemma for reductive groups. The formulation is a straightforward generalization of that in the case of GL_n, cf. Chapter III, §1.

We first recall some notation already used in previous chapters. For $P \in \mathcal{P}$, let $\mathfrak{a}_P = X_*(A_P)_{\mathbb{Q}}$. For two std psgps $P \subset Q$, we have by Chapter V (2.1), page 131 a direct sum decomposition

$$\mathfrak{a}_P = \mathfrak{a}_Q \oplus \mathfrak{a}_P^Q ,$$

where $\mathfrak{a}_P^Q = \ker(\mathfrak{a}_P \longrightarrow \mathfrak{a}_Q)$. Let

$$\Phi_P = \{\text{characters of } A_P \text{ which occur in Lie } G\} .$$

Hence $\Phi_P \subset (\mathfrak{a}_P^G)^* \subset \mathfrak{a}_P^*$. Since Φ_P is not a root system in general, we define Δ_P as the set of non-trivial restrictions of elements of Δ_{P_0} to A_P ($\subset A_{P_0}$). Then Δ_P is a basis of the subspace $(\mathfrak{a}_P^G)^*$ of $X^*(A_P)_{\mathbb{Q}}$. Then for $P \in \mathcal{P}$, we have

$$\mathfrak{a}_P = \{x \in \mathfrak{a}_{P_0} \mid \alpha(x) = 0 \, , \forall \alpha \in \Delta_0 \setminus \Delta_P\},$$

where we write $\alpha(x)$ instead of $\langle x, \alpha\rangle$.

For $P \in \mathcal{P}$, the projection map

$$\mathfrak{a}_{P_0} \to \mathfrak{a}_P \tag{1.1}$$

is given as follows. Let A'_P be the maximal split torus quotient of P. Then the projection is defined via the composition of the natural homomorphism $X_*(A_{P_0})_{\mathbb{Q}} \to X_*(A'_P)_{\mathbb{Q}}$ and the identification $X_*(A_P)_{\mathbb{Q}} \cong X_*(A'_P)_{\mathbb{Q}}$.

We denote by $(\alpha^\vee)_{\alpha\in\Delta_0}$ the set of simple co-roots, which form a basis of $\mathfrak{a}_{P_0}^G$. We denote by $(\varpi_\alpha)_{\alpha\in\Delta_0} \subset (\mathfrak{a}_{P_0}^G)^*$ its dual basis. For $P \in \mathcal{P}$ and $\alpha \in \Delta_P$, we may define the co-root $\alpha^\vee$ as follows. Let $\beta \in \Delta_0$ with $\beta|A_P = \alpha$. Then $\alpha^\vee$ is defined to be the projection of $\beta^\vee$ onto $\mathfrak{a}_P^G$. Then $(\alpha^\vee)_{\alpha\in\Delta_P}$ gives a basis of $\mathfrak{a}_P^G$.

More generally, if $P \subset Q$ are two std psgps, then P defines the psgp $P \cap M_Q$ of M_Q. Let

$$\Phi_P^Q = \Phi_{P\cap M_Q},$$

and

$$\Delta_P^Q = \Delta_{P\cap M_Q}.$$

Then $\Delta_P^Q \subset \Phi_P^Q$ and Δ_P^Q is a basis of $(\mathfrak{a}_P^Q)^*$. The projection of $(\alpha^\vee)_{\alpha\in\Delta_P^Q}$ onto $\mathfrak{a}_P^Q$ gives a basis of $\mathfrak{a}_P^Q$. We denote its dual basis by $(\varpi_\alpha^Q)_{\alpha\in\Delta_P^Q} \subset (\mathfrak{a}_P^Q)^*$. Then

$$\mathfrak{a}_P^Q = \{x \in \mathfrak{a}_P^G \mid \varpi_\alpha(x) = 0, \forall \alpha \in \Delta_Q\}.$$

As in the case of GL_n, we denote for any pair $P \subset Q$ in $\mathcal{P}$ by

$$\tau_P^Q \text{ (resp. } \widehat{\tau}_P^Q) : \mathfrak{a}_{P_0} \longrightarrow \{0, 1\} \tag{1.2}$$

the characteristic function of the subset

$$\mathfrak{a}_Q + \{x \in \mathfrak{a}_P^Q \mid \alpha(x) > 0, \forall \alpha \in \Delta_P^Q\} + \mathfrak{a}_{P_0}^P$$

(resp.

$$\mathfrak{a}_Q + \{x \in \mathfrak{a}_P^Q \mid \varpi_\alpha^Q(x) > 0, \forall \alpha \in \Delta_P^Q\} + \mathfrak{a}_{P_0}^P\ )$$

of $\mathfrak{a}_{P_0}$. Again, the first set is called the *acute Weyl chamber*, the second set the *obtuse Weyl chamber*. The former is a subset of the latter, as follows from the next lemma.

Lemma 7.1.1 *Let $P \subset Q$ be two std psgps. Let $H \in \mathfrak{a}_P^Q$.*
(i) *Assume for each $\alpha \in \Delta_P^Q$ that $\alpha(H) > 0$ or $\varpi_\alpha^Q(H) > 0$. Then $\varpi_\alpha^Q(H) > 0$ for all $\alpha \in \Delta_P^Q$.*
(ii) *Assume for each $\alpha \in \Delta_P^Q$ that $\alpha(H) \leq 0$ or $\varpi_\alpha^Q(H) \leq 0$. Then $\varpi_\alpha^Q(H) \leq 0$ for all $\alpha \in \Delta_P^Q$.*

Proof See [141], Lemma 3.1. □

For $P \subset Q$, set

$$a_P^Q = \dim \mathfrak{a}_P^Q.$$

Lemma 7.1.2 (Langlands Lemma) *For any std psgps $P \subset R$, we have*

$$\sum_{P \subset Q \subset R} (-1)^{a_P^Q} \tau_P^Q(x) \widehat{\tau}_Q^R(x) = \delta_{P,R}$$

$$\sum_{P \subset Q \subset R} (-1)^{a_P^Q} \widehat{\tau}_P^Q(x) \tau_Q^R(x) = \delta_{P,R}$$

for every $x \in \mathfrak{a}_{P_0}$. Here $\delta_{P,R} = 1$ if $P = R$, and $\delta_{P,R} = 0$ otherwise.

Proof See [4], §6 or [141], Lemma 3.2. □

From now on, let k be a finite field. Then G is quasi-split and P_0 is therefore a Borel subgroup. For psychological reasons, we set $B = P_0$. In the following, we discuss the same kind of representations for $G(k)$ as we did for $\mathrm{GL}_n(k)$ in Chapter III, §2. All representations will be again in vector spaces over a fixed algebraically closed field $\mathbf{C}$ of characteristic zero. For $P \in \mathcal{P}$, we denote by

$$i_P^G = \mathrm{Ind}_P^G \mathbf{1}$$

the induced representation of the trivial representation of P. As in the case of $G = \mathrm{GL}_n$, we call two parabolic subgroups *associate* if their Levi components are conjugate under $G(k)$. The Grothendieck group generated by the irreducible constituents of i_B^G will again be denoted by $K_0(i_B^G)$.

Theorem 7.1.3 (i) *i_P^G is equivalent to $i_{P'}^G$ if and only if P is associate to P'.*
(ii) *The elements $[i_P^G]$, where P ranges over the associate classes in $\mathcal{P}$, are linearly independent in $K_0(i_B^G)$.*

Proof The *if* part of the first statement is classical and due to Harish-Chandra, see [62], 6.1. A proof working over any ring where p is invertible instead of over $\mathbf{C}$ is given in [119]. The *only if* part of course will follow from the second statement.

As for the second statement, let us first note the following consequence of the *if* part: whenever M is a "rational Levi subgroup" (meaning that not only it is rational, but in fact a Levi component of some rational parabolic subgroup) the element $[i_P^G] \in K_0(i_B^G)$ associated with any rational parabolic subgroup P with Levi component M only depends on M and even only on its $G(k)$-conjugacy class $[M]$. Let us denote it by $i_{[M]}$. So, if $\mathcal{M}$ denotes the set of $G(k)$-conjugacy classes of rational Levi subgroups, we have to show that the rank of the submodule of $K_0(i_B^G)$ generated by the $i_{[M]}$ is the cardinality of $\mathcal{M}$.

On $\mathcal{M}$ there is a partial ordering defined by $[M] \leq [M']$ if there is $g \in G(k)$ such that $M \subseteq gM'g^{-1}$. If $s \in G(k)$ is a semi-simple element then, by Proposition 6.5.1 of [38], the subset $\mathcal{M}_s$ consisting of the conjugacy classes $[M] \in \mathcal{M}$ such that some $G(k)$-conjugate of s lies in $[M]$ has a *unique smallest* element $[M](s)$. This defines a map $\mathcal{S} \longrightarrow \mathcal{M}$ from the set of semi-simple conjugacy classes in $G(k)$ to $\mathcal{M}$. Assume that we can prove its *surjectivity*. Then choose for each $[M] \in \mathcal{M}$ a semi-simple element $s_{[M]}$ such that $[M](s_{[M]}) = [M]$, and denote by $\langle i_{[M]}, s_{[M']}\rangle$ the evaluation of the character $i_{[M]}$ on the conjugacy class $s_{[M']}$. By construction we have $\langle i_{[M]}, s_{[M]}\rangle \neq 0$ and $(\langle i_{[M]}, s_{[M']}\rangle \neq 0) \Rightarrow ([M'] \leq [M])$. Hence the associated $|\mathcal{M}| \times |\mathcal{M}|$ matrix with entries $\langle i_{[M]}, s_{[M']}\rangle$ is triangular and therefore invertible, thus implying our second statement.

It remains to show that the map $\mathcal{S} \longrightarrow \mathcal{M}$ is surjective. This amounts to finding *elliptic* semi-simple elements in $M(k)$, i.e., elements not lying in any proper rational parabolic subgroup of M. We state this explicitly in the next proposition. □

Proposition 7.1.4 *Any reductive group G over a finite field k has semi-simple elliptic elements.*

Proof To simplify notations, we will often use the same symbol for the group of k-rational points of G. Let $k = \mathbb{F}_q$.

We begin with a standard reduction argument. First, it is sufficient to prove the statement for the simply connected covering of the derived group of G. The latter is a product of quasi-simple components which are permuted under the action of the Frobenius automorphism F and we may treat each F-orbit separately. Hence assume that F has only one orbit of, say, m quasi-simple factors; then, replacing F by its mth power, k by its extension of degree m, and G by one of the factors, we are reduced to the case where G is absolutely quasi-simple and simply connected.

Here now is the strategy to exhibit elliptic elements. First we will easily construct *elliptic* rational maximal tori, i.e., rational maximal tori which are contained in no rational Levi subgroups. Then we will look for regular k-rational elements in such a torus. Recall that a semi-simple element is called *regular* if it is contained in a unique maximal torus. Hence a k-rational regular element of an elliptic maximal torus is elliptic.

We first review some properties of rational maximal tori. Let T be a *maximally split* maximal torus of G defined over k, and let $W := W(G,T)$ be its Weyl group. The endomorphism F induces an endomorphism of the cocharacter group $Y = X_*(T)$ of T, of the form $F = q \cdot \sigma$, where σ is a finite order automorphism which preserves some basis of the coroot system of T in Y. Note that since G is now assumed to be quasi-simple, the order of σ is in fact at most 3. Now recall from [38], Prop. 3.3.3 that there is a natural bijection between $G(k)$-conjugacy classes of rational maximal tori of G and F-conjugacy classes in W. Concretely, if $w \in W$ lifts to $n_w \in N_G(T)(\bar{k})$, then by Lang's theorem we may choose a $g \in G(\bar{k})$ such that $g^{-1}F(g) = n_w$ and get the desired rational maximal torus by conjugating T by g. We know by [38], Prop. 3.2.2 that the group $({}^gT)(k) = ({}^gT)^F$ of k-rational points of ${}^gT$ can be identified with $Y/(w^{-1}F - 1)Y$, and by [38], Prop. 3.3.6 that the k-rational Weyl group $W(G, {}^gT)^F$ can be identified with $W^{w^{-1}F}$, the F-centralizer of w in W.

Now, if M is a k-rational Levi subgroup containing T, then by [38], Prop. 6.5.3 a k-rational maximal torus of G is contained in a conjugate of M if and only if its associated F-conjugacy class in W meets $W(M,T)$. One particular F-conjugacy class in W is that of *F-Coxeter elements*. By definition, an F-Coxeter element is a product simple reflections, one from each F-conjugacy class. That such elements form a F-conjugacy class follows from [204], Thm. 7.6(i). That this F-conjugacy class does not meet any proper $W(M,T)$ follows from [204], Lemma 7.4 (i).

Let $T_c := {}^gT$ be a k-rational maximal torus in G obtained as above from some F-Coxeter element w in W; by the above discussion, such a torus is elliptic and it remains to prove that it contains regular k-rational elements. Since G is simply connected, for an element in T_c to be regular is equivalent to having trivial centralizer in $W(G,T_c)$. If the element is in $T_c(k)$, then by [58], 5.17, this is also equivalent to having trivial stabilizer in the k-rational Weyl group $W(G,T_c)^F$. We know by [204], Thm. 7.6 (v) that $W(G,T_c)^F$, which is the F-centralizer of w, is cyclic and consists of powers of $w^{-1}\sigma$. Let h_σ be the order of $w^{-1}\sigma$ in $\mathrm{Aut}(Y)$. Since $w^{-1}\sigma$ acts as multiplication by q^{-1} (in additive notation) on $T_c(k) = Y/(w^{-1}F - 1)Y$, it remains only to prove that the order of multiplication by q in $\mathrm{Aut}(T_c(k))$ is exactly h_σ. But by [204], Thm. 7.6 (ii) the characteristic polynomial of $w^{-1}\sigma$ is divisible by the cyclotomic polynomial Φ_{h_σ} of order h_σ, hence the order $|T_c(k)|$ of $T_c(k)$ is divisible by $\Phi_{h_\sigma}(q)$. Moreover, by [18], Thm. V, we know that if $h_\sigma > 2$ and $(q, h_\sigma) \neq (2, 6)$, there exist prime numbers ℓ such that the multiplicative order of q modulo ℓ is precisely h_σ. Such primes of course divide $\Phi_{h_\sigma}(q)$, hence divide $|T_c(k)|$, and it follows that the order of multiplication by q in $\mathrm{Aut}(T_c(k))$ is exactly h_σ, as desired.

It remains to treat the cases $h_\sigma = 2$ and $(q, h_\sigma) = (2, 6)$. The first one is the case of SL_2 and is easy. Indeed, if $N_2 : k_2^\times \longrightarrow k^\times$ denotes the norm map from the degree 2 extension of k, then any choice of a k-basis on k_2 provides an embedding of $\ker N_2$ into $\mathrm{SL}_2(k)$ and the image of any generator of k_2 is a semi-simple elliptic element. The second case, $(q, h_\sigma) = (2, 6)$, consists of finitely many subcases. Namely the base field k is $\mathbb{F}_2$ and the possible types of the group are those with (possibly twisted) Coxeter number 6. The tables of [28] in the untwisted case, resp. of [204], Table 10, page 184 in the twisted case, give the following list: $\mathrm{SL}_6(\mathbb{F}_2)$, $\mathrm{Sp}_6(\mathbb{F}_2) = \mathrm{SO}_7(\mathbb{F}_2)$, $\mathrm{SO}_8(\mathbb{F}_2)$, $\mathrm{SU}_3(\mathbb{F}_2)$, and $\mathrm{SU}_4(\mathbb{F}_2)$. In all these cases one can compute explicitely $T_c(k)$. The results are respectively (and up to isomorphism) $\mathbb{Z}/63\mathbb{Z}$, $\mathbb{Z}/9\mathbb{Z}$, $\mathbb{Z}/3\mathbb{Z} \times \mathbb{Z}/9\mathbb{Z}$, $\mathbb{Z}/3\mathbb{Z}$ and $\mathbb{Z}/9\mathbb{Z}$. So in all cases except $\mathrm{SU}_3(\mathbb{F}_2)$, $T_c(k)$ has elements of order 9 and since the multiplicative order of $2(= q)$ modulo 9 is precisely $6(= h_\sigma)$, such elements are regular, hence elliptic.

Finally, there remains the case of $\mathrm{SU}_3(\mathbb{F}_2)$. Here $T_c(k)$ is the center $Z = \mathbb{F}_4^\times \simeq \mathbb{Z}/3\mathbb{Z}$ and in particular contains no regular element! In fact for the maximally split torus we also have $T(\mathbb{F}_2) = Z$, thus nothing distinguishes rational points of maximally split tori and Coxeter tori! So we need to use another elliptic torus, namely that associated with the longest element w_0 of the Weyl group S_3. Call it T_0. Then $w_0^{-1}\sigma = -\mathrm{id}_Y$ hence $T_0(k) = \mathbb{Z}/3\mathbb{Z} \times \mathbb{Z}/3\mathbb{Z}$. It does not contain any regular element either, but at least any non-central element is elliptic. □

Remark 7.1.5 Note that the above proof also shows that any reductive group over a finite field has k-rational regular semi-simple elements, except if $k = \mathbb{F}_2$ and if the simply connected covering of its derived group has a component SU_3. The computation of the group of rational points of a Coxeter torus shows that it is not always a cyclic group; for classical groups, one can prove that it is cyclic in all cases, except $\mathrm{SO}_{2n}(\mathbb{F}_q)$.

Remark 7.1.6 There is another simple strategy for detecting k-rational elliptic semi-simple elements: from the explicit formula of the order of a finite reductive group, we may look for a prime ℓ dividing the order of the group but not that of any proper parabolic subgroup. Then any element of order ℓ will be elliptic (and also semi-simple since ℓ will have to be prime to q). This strategy might appear simpler but leaves aside the same list of groups as in the strategy we have followed above, with no easy way to recover these exceptions.

Keep the same notation $G, T, W = W(G, T), Y = X_*(T), F = q \cdot \sigma$ as in the above proof and with absolutely simple G. By [209], Thm. 11.16 we have the following formula for the order of $G(k)$:

$$|G(k)| = q^{n_+} \prod_{i=1}^{r} (q^{d_i} - \varepsilon_i) .$$

Here n_+ is the number of positive roots and r is the semi-simple rank of G. It is known, [28], V.5, Thm. 4, that the algebra of W-invariant polynomials on Y is a polynomial algebra of dimension r, and we may choose a set of homogenous generators $\{f_1, \ldots, f_r\}$ which are eigenvectors of σ. Then, in the above formula, d_i denotes the degree of f_i and ε_i its eigenvalue under σ. Let us order the degrees decreasingly. The greatest degree d_1 is also the *Coxeter number h* of W, see [28], V.6, Thm. 1 and Prop. 3, which is also equal to $\sum_\alpha (n_\alpha + 1)$, where $\sum_\alpha n_\alpha \alpha$ is the longest root, see [28], VI.1, Prop. 31. It follows from the latter formula that the Coxeter number of any strict parabolic subgroup of W is less than h. We also note that the order of $T(k)$ is given by $|T(k)| = \prod_{i=1}^{r} (q - \varepsilon_i)$ and that the order of a Levi subgroup divides the product of $|T(k)|$ and of the orders of its various F-simple components.

We will say that a prime number ℓ is *(q, n)-primitive* if the multiplicative order of q modulo ℓ equals n, and *G-primitive* if it divides $|G(k)|$, but not the order of a proper parabolic subgroup.

Consider first the case where G is split. Then all ε_i are equal to 1. It follows from the above discussion that any prime number ℓ which is (q, h)-primitive is G-primitive.

Next consider the case where σ has order 2; we refer to the discussion in section 11 of [209], especially 11.20, for more details on the possibilities which occur. If h is odd, we are in type A_{h-1}, we know that $\varepsilon_1 = -1$, hence any $(q, 2h)$-primitive prime number is also G-primitive. Let us now assume $h = 2h'$. If G has type A_{h-1}, we know that $d_2 = h - 1$ with $\varepsilon_2 = -1$ and one checks that any prime number which is $(q, 2h - 2)$-primitive is G-primitive. If G has type $D_{h'}$, then the general formula specializes to $|G(k)| = q^{n_+}(q^{h'} + 1) \prod_{i<h'} (q^{2i} - 1)$ hence any (q, h)-primitive prime number is G-primitive. If G has type E_6 then the d_j are $2, 5, 6, 8, 9, 12$ and the ε_j are -1 exactly for 5 and 9. Then one checks that any $(q, 18)$-primitive prime number is G-primitive.

Finally let us consider the triality case. So G has type D_4 and σ has order 3. Then the general formula shows that $q^8 + q^4 + 1$ divides $|G(k)|$ and one checks that any $(q, 12)$-primitive prime number is G-primitive.

Thus, invoking again [18], Thm. V for the existence of (q,n)-primitive primes, we get the existence of elliptic elements, except for the same cases as before, namely $\mathrm{SL}_2(k)$, $\mathrm{SL}_6(\mathbb{F}_2)$, $\mathrm{Sp}_6(\mathbb{F}_2) = \mathrm{SO}_7(\mathbb{F}_2)$, $\mathrm{SO}_8(\mathbb{F}_2)$, $\mathrm{SU}_3(\mathbb{F}_2)$, and $\mathrm{SU}_4(\mathbb{F}_2)$. □

Remarks 7.1.7 (i) A statement analogous to point (ii) in Theorem 7.1.3, but for Weyl groups instead of reductive groups, is proved in [203]. It is possible to reduce the case of a reductive group G to that of its Weyl group W using the following two steps (only sketched).

(1) There is an equivalence of categories between the category of $\mathbf{C}[G]$-representations generated by their B-invariants and the category of modules over the intertwining algebra $\mathcal{H}(W)$ of $\mathbf{C}[G/B]$. Moreover for any parabolic $P \supseteq B$, there is a natural embedding $\mathcal{H}(W_{M_P}) \subset \mathcal{H}(W)$ and induction of modules via tensor products corresponds to parabolic induction on representations.

(2) There is a deformation $\mathcal{H}_v(W)$, known as the "generic" Hecke algebra, which is a $\mathbf{C}[v, v^{-1}]$ algebra and specializes to $\mathcal{H}(W)$ for $v = q$ and to $\mathbf{C}[W]$ for $v = 1$. Since both $\mathcal{H}(W)$ and $\mathbf{C}[W]$ are known to be split semi-simple algebras, both specialization morphisms from $K_0(\mathcal{H}_v(W)\otimes\mathbf{C}(v, v^{-1}))$ to either $K_0(\mathcal{H}(W))$ or $K_0(W)$ are isomorphisms. Accordingly there is a "generic" Hecke algebra for W_M which is naturally embedded in $\mathcal{H}_v(W)$ and the embedding specializes to the natural embeddings $\mathcal{H}(W_M) \subset \mathcal{H}(M)$ and $\mathbf{C}[W_M] \subset \mathbf{C}[W]$.

(ii) In general, and in contrast to the case $G = \mathrm{GL}_n$, the subgroup generated by the i_P^G may be *strictly* contained in $K_0(i_B^G)$. Assume for simplicity that G is split and q large enough for the nilpotent cone to be isomorphic to the unipotent variety (as a k-variety). Then we have the following inequalities, where $\mathcal{M}$ has the same meaning as in the above proof:

$$\mathrm{rk}(K_0(i_B^G)) = \mathrm{rk}(K_0(W)) \geq \#\{\text{rational unipotent classes in } G\} \geq \#\mathcal{M}.$$

Indeed, the first inequality comes from the Springer correspondence [205], Thm. 1.13, while the second one follows from Bala–Carter theory [38], Thm. 5.9.5 (ii). Now, when G is not of type A, both inequalities may be strict. For example, for Sp_6, there are 7 classes of Levi subgroups and 8 unipotent classes. Note also that if the subgroup generated by the i_P^G is proper, then, since it contains i_B^G, it is not generated by irreducible representations.

Let $L_0(i_B^G)$ be the subgroup of $K_0(i_B^G)$ generated by the elements $[i_P^G]$ where $P \in \mathcal{P}$. By Theorem 7.1.3 these elements form even a basis of $L_0(i_B^G)$. As in the case of $G = \mathrm{GL}_n$, we give another basis of $L_0(i_B^G)$ via generalized Steinberg representations.

Definition 7.1.8 The *generalized Steinberg representation* associated to $P \in \mathcal{P}$

is

$$v_P^G = i_P^G / \sum_{Q \supsetneq P} i_Q^G.$$

Obviously, v_P^G only depends on the associate class of P. For $P = G$, we have $v_G^G = i_G^G = \mathbf{1}$. The following statements are formulated in the case of the general linear group in Chapter III. They are also valid in the general case. The proofs can be carried over word for word.

Theorem 7.1.9 *Let $I \subset \Delta_0$. Then*

$$v_{P_I}^G = (-1)^{|I|} \sum_{J \supset I} (-1)^{|J|}\, i_{P_J}^G. \tag{1.3}$$

More precisely, the following complex is an acyclic resolution of $v_{P_I}^G$ by G-modules,

$$0 \to i_G^G \to \bigoplus_{\substack{I \subset K \subset \Delta_0 \\ |\Delta_0 \setminus K| = 1}} i_{P_K}^G \to \bigoplus_{\substack{I \subset K \subset \Delta_0 \\ |\Delta_0 \setminus K| = 2}} i_{P_K}^G \to \cdots \to \bigoplus_{\substack{I \subset K \subset \Delta_0 \\ |K \setminus I| = 1}} i_{P_K}^G \to i_{P_I}^G \to v_{P_I}^G \to 0. \tag{1.4}$$

Here the differentials $d_{K',K} : i_{P_{K'}}^G \to i_{P_K}^G$ are defined as follows. For two subsets $K \subset K'$ of Δ_0, we let

$$p_{K',K} : i_{P_{K'}}^G \longrightarrow i_{P_K}^G$$

be the natural homomorphism induced by the surjection $G/P_K \to G/P_{K'}$. For arbitrary subsets $K, K' \subset \Delta_0$, with $|K'| - |K| = 1$ and $K' = \{k_1 < \cdots < k_r\}$, we put

$$d_{K',K} = \begin{cases} (-1)^i p_{K',K} & K' = K \cup \{k_i\} \\ 0 & K \not\subset K' \end{cases}. \qquad \square$$

Let $W_0 \subset W$ be the relative Weyl group of G. For a subset $I \subset \Delta_0$, let $W_{0,I} \subset W_0$ be the subgroup generated by the simple reflections associated to I. Put $i_{W_{0,I}}^{W_0} = \mathrm{Ind}_{W_{0,I}}^{W_0} \mathbf{1}$ and $v_{W_{0,I}}^{W_0} = i_{W_{0,I}}^{W_0} / \sum_{J \supsetneq I} i_{W_{0,J}}^{W_0}$.

Theorem 7.1.10 *The natural complex*

$$0 \to i_{W_0}^{W_0} \to \bigoplus_{\substack{I \subset K \subset \Delta_0 \\ |\Delta_0 \setminus K| = 1}} i_{W_{0,K}}^{W_0} \to \bigoplus_{\substack{I \subset K \subset \Delta_0 \\ |\Delta_0 \setminus K| = 2}} i_{W_{0,K}}^{W_0} \to \cdots \to \bigoplus_{\substack{I \subset K \subset \Delta_0 \\ |K \setminus I| = 1}} i_{W_{0,K}}^{W_0} \to i_{W_{0,I}}^{W_0} \to v_{W_{0,I}}^{W_0} \to 0$$

is acyclic. $\square$

Corollary 7.1.11 *The representation v_B^G is the Steinberg representation. In particular, v_B^G is irreducible.* $\square$

Corollary 7.1.12 *For any* $P \in \mathcal{P}$,

$$i_P^G = \sum_{Q \supset P} v_Q^G .$$

In particular, the generalized Steinberg representations v_P^G, *as* P *ranges over the associate classes in* $\mathcal{P}$, *form a basis of* $L_0(i_B^G)$. □

Notes and References Lemma 7.1.4, asserting the existence of k-rational elliptic semi-simple elements in reductive groups over a finite field, is known to G. Lusztig. However, we are not aware of a proof in the literature. The statement is related to the existence of cuspidal representations of $G(k)$ through Deligne–Lusztig theory [58].

2 The Euler–Poincaré characteristic

Let k be a finite field of characteristic p with q elements. As usual, Γ_k will be the Galois group $\mathrm{Gal}(\bar{k}/k)$ for some fixed algebraic closure $\bar{k}$ of k. Let G be a reductive group over k and let $\mathcal{N}$ be a conjugacy class of $\mathbb{Q}$-1-PS of $G_{\bar{k}}$. Let E be the reflex field of the PD-pair $(G, \mathcal{N})$ and denote by $\Theta(\mathcal{N}) = \Theta(G, \mathcal{N})$ the set of HN-types.

We first describe the cohomology $H_c^*(\mathcal{F}(G, \mathcal{N})) = H_c^*(\mathcal{F}(G, \mathcal{N}) \otimes_E \bar{k}, \bar{\mathbb{Q}}_\ell)$ as a module under $\Gamma_E \times G(k)$. Since k is a finite field, G is quasi-split. We choose a Borel subgroup B and a maximal torus $T \subset B$, both defined over k. By the Kottwitz Lemma 6.1.5, we may choose $\nu \in \mathcal{N}$ defined over E. We may assume that $\nu \in X_*(T)_{\mathbb{Q}}$. Let $W_\nu \subset W$ be the stabilizer of ν in the Weyl group W. The decomposition of $\mathcal{F}(G, \mathcal{N}) \otimes_E \bar{k}$ into Bruhat cells shows that there is no cohomology in odd degrees and that $H_c^{2i}(\mathcal{F}(G, \mathcal{N}))$ has a $\bar{\mathbb{Q}}_\ell$-basis enumerated by the elements of W/W_ν of length i. The action of $G(k)$ is trivial. The action of Γ_E, however, permutes the basis vectors according to the action of Γ_E on W/W_ν. Let $[W/W_\nu]_E$ denote the set of Γ_E-orbits on W/W_ν. For $[w] \in [W/W_\nu]_E$ with representative $w \in W/W_\nu$, let $\Gamma_w \subset \Gamma_E$ be the stabilizer of w. Let

$$\mathrm{ind}_{[w]} = \mathrm{Ind}_{\Gamma_w}^{\Gamma_E} \bar{\mathbb{Q}}_\ell \tag{2.1}$$

be the induced representation of the trivial representation of Γ_w. This is independent, up to equivalence, of the choice of w. We may therefore write

$$H_c^{2i}(\mathcal{F}(G, \mathcal{N})) = \bigoplus_{\substack{[w] \in [W/W_\nu]_E \\ \ell([w]) = i}} \mathrm{ind}_{[w]} \otimes t_E^i . \tag{2.2}$$

Here $\ell([w])$ denotes the minimal length of a representative in W of any coset wW_ν (this is independent of the choice of the representative w of $[w]$). Also t_E

denotes the Tate twist representation of Γ_E. Since $t_E \mid \Gamma_w = t_{E_w}$, it follows that $\mathrm{ind}_{[w]} \otimes t_E^{\ell([w])} = \mathrm{Ind}_{\Gamma_w}^{\Gamma_E}(t_{E_w}^{\ell(w)})$, where $E_w = \bar{k}^{\Gamma_w}$.

For the Euler–Poincaré characteristic we therefore obtain

$$\chi_c(\mathcal{F}(G, \mathcal{N})) = \sum_{[w]\in[W/W_\nu]_E} \mathrm{ind}_{[w]} \cdot t_E^{\ell([w])} . \tag{2.3}$$

We note that if G is split, then $E = k$ and the action of Γ_E on W/W_ν is trivial. Hence this formula is indeed the generalization of the formula Chapter III (3.2), page 83 for $G = \mathrm{GL}_n$.

From the HN-stratification of $\mathcal{F}(G, \mathcal{N})$ in Theorem 6.3.5, we obtain an implicit recursion relation for the Euler–Poincaré characteristic of $\mathcal{F}(G, \mathcal{N})^{ss}$, as follows. From

$$\mathcal{F}(G, \mathcal{N}) \otimes_E \bar{k} = \bigcup_\theta \mathcal{F}(G, \mathcal{N})_\theta ,$$

we get the equality

$$\chi_c(\mathcal{F}(G, \mathcal{N})) = \sum_\theta \chi_c(\mathcal{F}(G, \mathcal{N})_\theta) .$$

Since each stratum is invariant under $G(k)$, this is an equality as virtual $G(k)$-modules. On the other hand, when G is not split, Γ_E acts by permuting the summands.

Let $[\Theta(\mathcal{N})]_E$ denote the set of Γ_E-orbits on $\Theta(\mathcal{N})$. For $[\theta] \in [\Theta(\mathcal{N})]_E$ with representative $\theta \in \Theta(\mathcal{N})$, let $\Gamma_\theta \subset \Gamma_E$ be the stabilizer and $E_\theta = \bar{k}^{\Gamma_\theta}$ the reflex field of the HN-type θ. Denoting by

$$\mathrm{Ind}_{[\theta]} = \mathrm{Ind}_{\Gamma_\theta}^{\Gamma_E}$$

the induction functor, we obtain the following equality in the Grothendieck group of representations of $\Gamma_E \times G(k)$,

$$\chi_c(\mathcal{F}(G, \mathcal{N})) = \sum_{[\theta]\in[\Gamma(\mathcal{N})]_E} \mathrm{Ind}_{[\theta]}\left(\chi_c(\mathcal{F}(G, \mathcal{N})_{[\theta]})\right) .$$

Here $\mathcal{F}(G, \mathcal{N})_{[\theta]} = \mathcal{F}(G, \mathcal{N})_\theta$ for any representative θ of $[\theta]$. As representations of $\Gamma_\theta \times G(k)$ we have from Corollary 6.3.8

$$\chi_c(\mathcal{F}(G, \mathcal{N})_\theta) = \mathrm{Ind}_{P_\theta(k)}^{G(k)}\left(\chi_c(\mathcal{F}(M_\theta, \mathcal{N}_\theta)^{ss}) \otimes t_{E_\theta}^{\ell(w_\theta)}\right) .$$

Summarizing, we have obtained the following equality of virtual $\Gamma_E \times G(k)$-representations,

$$\chi_c(\mathcal{F}(G, \mathcal{N})) = \sum_{[\theta]\in[\Theta(\mathcal{N})]_E} \mathrm{Ind}_{[\theta]}\left(\mathrm{Ind}_{P_\theta(k)}^{G(k)}\left(\chi_c(\mathcal{F}(M_\theta, \mathcal{N}_\theta)^{ss})\right) \otimes t_{E_\theta}^{\ell(w_\theta)}\right) . \tag{2.4}$$

By the discussion after Definition 6.3.2 and (1.1), page 162 we may identify $\Theta(\mathcal{N})$ with

$$\Theta(\mathcal{N}) = \{(Q, w) \mid Q \in \mathcal{P},\ w \in W_Q \setminus W/W_\nu, \tau_Q^G(w\nu) = 1\}\,. \qquad (2.5)$$

Here we set

$$\tau_P^G(w\nu) = \tau_P^G(\rho(w\nu)),$$

where $\rho : X_*(T)_\mathbb{Q} \longrightarrow X_*(S)_\mathbb{Q}$ is defined by the composition of the natural homomorphism $X_*(T) \to X_*(A'_B)$ and the identification $X_*(A'_B)_\mathbb{Q} \cong X_*(A_B)_\mathbb{Q} = X_*(S)_\mathbb{Q}$. If k' is the splitting field of G of degree $n = |k' : k|$ then the map ρ coincides with

$$X_*(T)_\mathbb{Q} \longrightarrow X_*(S)_\mathbb{Q}\,, \qquad \lambda \mapsto \frac{1}{n}\sum\nolimits_{\sigma\in \mathrm{Gal}(k'/k)} {}^\sigma\lambda.$$

The action of Γ_E on the right in (2.5) above is on the second entry and is induced by the action of Γ_E on W. Hence, denoting by $[W_Q \setminus W/W_\nu]_E$ the set of Γ_E-orbits on $W_Q \setminus W/W_\nu$, we can rewrite the equality (2.4), page 171 as

$$\chi_c(\mathcal{F}(G,\mathcal{N})) = \sum_{Q\in\mathcal{P}} \sum_{[w]\in[W_Q\setminus W/W_\nu]_E} \tau_Q^G(w\nu)\cdot t_E^{l([w])}\cdot \mathrm{Ind}_{[w]}\left(\mathrm{Ind}_{Q(k)}^{G(k)}(\chi_c(\mathcal{F}(M_Q, w\nu)^{ss}))\right). \quad (2.6)$$

Here

$$\mathrm{Ind}_{[w]} = \mathrm{Ind}_{\Gamma_w}^{\Gamma_E}$$

denotes the induction functor of Galois modules from Γ_w to Γ_E and $E_w = \bar{k}^{\Gamma_w}$ is the fixed field under the stabilizer of w in Γ.

As in the case of GL_n, we will denote by $[\mathcal{P}]$ the set of associate classes in $\mathcal{P}$. For $P \in \mathcal{P}$, we denote by $[P]$ the image of P in $[\mathcal{P}]$.

Let $K_0(\Gamma_E)$ denote the Grothendieck group of ℓ-adic representations of Γ_E.

Lemma 7.2.1 *In the Grothendieck group of representations of $\Gamma_E \times G(k)$,*

$$\chi_c(\mathcal{F}(G,\mathcal{N})^{ss}) = \sum_{[P]\in[\mathcal{P}]} N_{[P]}^G(\mathcal{N})\cdot i_P^G\,,$$

with uniquely defined coefficients $N_{[P]}^G(\mathcal{N}) \in K_0(\Gamma_E)$. If G is a split reductive group, then $N_{[P]}^G(\mathcal{N}) \in \mathbb{Z}[t_E]$.

Proof We prove this by induction on the dimension of G. Let us first assume that G is split. Then all HN-strata $\mathcal{F}(G,\mathcal{N})_\theta$ are defined over k, and the identity (2.6) above simplifies to

$$\chi_c(\mathcal{F}(G,\mathcal{N})) = \sum_{\theta\in\Theta(\mathcal{N})} \mathrm{Ind}_{P_\theta(k)}^{G(k)}(\chi_c(\mathcal{F}(M_\theta, \mathcal{N}_\theta)^{ss}))\otimes t_E^{\ell(w_\theta)}\,. \qquad (2.7)$$

Here $\chi_c(\mathcal{F}(G,\mathcal{N})^{ss})$ occurs on the RHS as the summand corresponding to $\theta_0 = (G,\mathcal{N})$. For all $\theta \neq \theta_0$ we have $\dim M_\theta < \dim G$ and we deduce from the induction hypothesis and the transitivity of induction that the contribution of θ is of the form $\sum_{P\in\mathcal{P}} n_\theta(P)\cdot i_P^G$, with $n_\theta(P) \in \mathbb{Z}[t_E]$. The LHS is equal to

$$\chi_c(\mathcal{F}(G,\mathcal{N})) = \sum_{w\in W/W_\nu} t_E^{\ell(w)},$$

hence the assertion follows in this case by collecting the coefficients of i_P^G for a given class $[P] \in \mathcal{P}$. The uniqueness follows from Theorem 7.1.3.

When G is not split, the Galois group Γ_E permutes the set of HN-strata, and each single stratum is defined over a finite extension of E. An obvious modification of the previous argument proves the assertion in general. □

The following theorem gives the coefficients $N_{[P]}^G(\mathcal{N})$ explicitly.

Theorem 7.2.2 *For $[P] \in [\mathcal{P}]$, the coefficient $N_{[P]}^G$ is given by*

$$N_{[P]}^G(\mathcal{N}) = (-1)^{a_P^G}\cdot \sum_{P\in[P]}\ \sum_{[w]\in[W/W_\nu]_E} \mathrm{ind}_{[w]}\cdot t_E^{\ell(w)}\cdot \widehat{\tau}_P^G(w\nu)\,. \tag{2.8}$$

Equivalently,

$$\chi_c(\mathcal{F}(G,\mathcal{N})^{ss}) = \sum_{P\in\mathcal{P}}(-1)^{a_P^G}\cdot \sum_{[w]\in[W/W_\nu]_E} \mathrm{ind}_{[w]}\cdot t_E^{\ell(w)}\cdot \widehat{\tau}_P^G(w\nu)\cdot i_P^G\,.$$

Here we note that the value of a_P^G is the same for all $P \in [P]$, and that $\ell(w)$ is independent of the representative w of $[w]$, and so is the value of $\widehat{\tau}_P^G(w\nu) = \widehat{\tau}_P^G(\rho(w\nu))$.

Proof We proceed by induction on the dimension of G. Denote by $n_P^G(\mathcal{N}) = n_P^G(\nu)$ the coefficient of i_P^G in the above expression of $\chi(\mathcal{F}(G,\mathcal{N})^{ss})$,

$$n_P^G(\nu) = (-1)^{a_P^G}\cdot \sum_{[w]\in[W/W_\nu]_E} \mathrm{ind}_{[w]}\cdot t_E^{\ell(w)}\cdot \widehat{\tau}_P^G(w\nu)\,.$$

We have to prove that

$$\chi_c(\mathcal{F}(G,\mathcal{N})^{ss}) = \sum_{P\in\mathcal{P}} n_P^G(\nu)\cdot i_P^G\,.$$

We compare this with the identity (2.6), page 172, which reads

$$\chi_c(\mathcal{F}(G,\mathcal{N})) = \chi_c(\mathcal{F}(G,\mathcal{N})^{ss}) +$$

$$\sum_{Q\neq G}\ \sum_{[w]\in[W_Q\backslash W/W_\nu]_E} \tau_Q^G(w\nu)\cdot t_E^{\ell(w)}\cdot \mathrm{Ind}_{[w]}\left(\mathrm{Ind}_{Q(k)}^{G(k)}\Big(\sum_{P\subset Q} n_P^Q(w\nu) i_P^Q\Big)\right).$$

We need to show the following identity in $K_0(\Gamma_E)$ for all $P \in \mathcal{P}$,

$$\sum_{Q \supset P} \sum_{[w] \in [W_Q \backslash W/W_\nu]_E} \tau_Q^G(w\nu) \cdot t_E^{\ell(w)} \cdot \mathrm{Ind}_{[w]}\left(n_P^Q(w\nu)\right) = \begin{cases} \sum_{[w] \in [W/W_\nu]_E} \mathrm{ind}_{[w]} \cdot t_E^{\ell(w)} & P = G \\ 0 & P \neq G. \end{cases} \tag{2.9}$$

Note that here $n_P^Q(w\nu) \in K_0(\Gamma_{E_w})$. Indeed, once this is shown, it follows that

$$\begin{aligned}
&\sum_{P \in \mathcal{P}} n_P^G(\nu) i_P^G + \sum_{Q \neq G} \sum_{[w] \in [W_Q \backslash W/W_\nu]_E} \tau_Q^G(w\nu) \cdot t_E^{\ell(w)} \, \mathrm{Ind}_{[w]}\left(\mathrm{Ind}_{Q(k)}^{G(k)} \left(\sum_{P \subset Q} n_P^Q(w\nu) i_P^Q \right)\right) \\
&\quad = \sum_{Q} \sum_{[w] \in [W_Q \backslash W/W_\nu]_E} \tau_Q^G(w\nu) \cdot t_E^{\ell(w)} \, \mathrm{Ind}_{[w]}\left(\mathrm{Ind}_{Q(k)}^{G(k)} \left(\sum_{P \subset Q} n_P^Q(w\nu) i_P^Q \right)\right) \\
&\quad = \sum_{P} i_P^G \left(\sum_{Q \supset P} \sum_{[w] \in [W_Q \backslash W/W_\nu]_E} \tau_Q^G(w\nu) t_E^{\ell(w)} \, \mathrm{Ind}_{[w]}\left(n_P^Q(w\nu)\right)\right) \\
&\quad = \left(\sum_{[w] \in [W/W_\nu]_E} \mathrm{ind}_{[w]} \cdot t_E^{\ell(w)} \right) \cdot i_G^G \\
&\quad = \chi_c(\mathcal{F}(G, \mathcal{N})) ,
\end{aligned}$$

as had to be shown.

To prove the identity (2.9) above, we calculate for $P \in \mathcal{P}$, by inserting the values for $n_P^Q(w\nu)$,

$$\begin{aligned}
&\sum_{Q \supset P} \sum_{[w] \in [W_Q \backslash W/W_\nu]_E} \tau_Q^G(w\nu) t_E^{\ell(w)} \, \mathrm{Ind}_{[w]}\left((-1)^{a_P^Q} \cdot \sum_{[w_1] \in [W_Q/(W_Q)_{w\nu}]} \mathrm{Ind}_{\Gamma_{w_1}}^{\Gamma_w} \cdot t_{E_w}^{\ell(w_1)} \cdot \widetilde{\tau}_P^Q(w_1 w\nu) \right) \\
&= \sum_{[\tilde{w}] \in [W/W_\nu]_E} t_E^{\ell(\tilde{w})} \, \mathrm{ind}_{[w]} \cdot \sum_{Q \supset P} (-1)^{a_P^Q} \cdot \tau_Q^G(\tilde{w}\nu) \cdot \widetilde{\tau}_P^Q(\tilde{w}\nu) \\
&= \begin{cases} \sum_{[w] \in [W/W_\nu]_E} \mathrm{ind}_{[w]} \cdot t_E^{\ell(w)} & P = G \\ 0 & P \neq G \, . \end{cases}
\end{aligned}$$

Here we used in the first identity, just as in the proof of Theorem 3.3.4, that an element $\tilde{w}$ in W of shortest length in its class in W/W_ν is the product of the shortest element w in its class in $W_Q \backslash W/W_\nu$ and an element w_1 in W_Q which is of minimal length in its class in $W_Q/(W_Q)_{w\nu}$ in a unique way and that $\ell(\tilde{w}) = \ell(w_1) + \ell(w)$. The uniqueness property implies that for $\tau \in \Gamma_E$ the corresponding decomposition of $\tau(\tilde{w})$ is given by

$$\tau(\tilde{w}) = \tau(w_1) \cdot \tau(w) \, .$$

Hence we may enumerate the Γ_E-orbit of $\tilde{w} \in W/W_\nu$ by first letting w range over its Γ_E-orbit in $W_Q \backslash W/W_\nu$ and secondly letting w_1 range over its Γ_{E_w}-orbit in $W_Q/(W_Q)_{w\nu}$, and this leads to the expression in front of the second sum in the first identity. The second identity follows from the Langlands Lemma 7.1.2. □

As in the GL_n-case we want to express the Euler–Poincaré characteristic of $\mathcal{F}(G,\mathcal{N})$ in terms of generalized Steinberg representations v_P^G, cf. Section 1. Again, by the upper triangular nature of the base change matrix between these two bases of $L_0(i_B^G)$, we deduce from Corollary 7.1.12 that

$$\chi_c(\mathcal{F}(G,\mathcal{N})^{ss}) = \sum\nolimits_{[P]\in[\mathcal{P}]} M_{[P]}^G(\nu)\cdot v_P^G\,, \tag{2.10}$$

with uniquely defined elements $M_{[P]}^G(\nu)\in K_0(\Gamma_E)$.

For $[w]\in[W/W_\nu]_E$, set

$$\Delta(w\nu) := \{\alpha\in\Delta_0 \mid \langle\rho(w\nu),\varpi_\alpha\rangle > 0\}\,.$$

This set is well-defined, i.e., independent of the choice of the representative of $[w]$. Denote by $P(w\nu)$ the unique standard parabolic subgroup P with $\Delta_P = \Delta(w\nu)$, i.e., $\Delta(w\nu)$ is the set of simple roots occurring in the unipotent radical of $P(w\nu)$.

The proof of the following theorem is the same as in Chapter III, except that the Γ_E-coefficients must be changed.

Theorem 7.2.3 *There is the following expression in the Grothendieck group of $\Gamma_E\times G(k)$-modules,*

$$\chi_c(\mathcal{F}(G,\mathcal{N})^{ss}) = \sum_{w\in[W/W_\nu]_E} (-1)^{a^G_{P(w\nu)}}\ \mathrm{ind}_{[w]}\cdot t_E^{\ell(w)}\cdot v^G_{P(w\nu)}\,.$$

In other words, for any $[P]\in[\mathcal{P}]$,

$$M_{[P]}^G(\nu) = (-1)^{a_P^G} \sum_{\{[w]\in[W/W_\nu]_E \mid P(w\nu)\in[P]\}} \mathrm{ind}_{[w]}\, t_E^{\ell(w)}.$$ □

It turns out that Theorem 3.3.10 also has a generalization to arbitrary reductive groups. The cohomology of $\mathcal{F}(G,\mathcal{N})^{ss}$ is given by the following theorem, cf. [170]. Just as with Theorem 3.3.10, we will not give the proof of this theorem, but rather give some comments concerning some of its consequences.

Theorem 7.2.4

$$H_c^*(\mathcal{F}(G,\mathcal{N})^{ss}) = \bigoplus_{[w]\in[W/W_\nu]_E} \mathrm{ind}_{[w]}\cdot t_E^{\ell(w)}\cdot v^G_{P(w\nu)}[-2\ell(w)-|\Delta(w\nu)|]\,.$$

Here the symbol $[-m], m\in\mathbb{N}$, indicates that the corresponding module is shifted into degree m of the graded cohomology ring. □

Variant 7.2.5 In the case of variant $\mathrm{Fil}^k_{\mathbb{F}_1}$ for arbitrary reductive groups, cf. Variants 6.1.12 (ii), we get

$$H_c^*(\mathcal{F}(G,S,\mathcal{N})^{ss}) = \bigoplus_{[w]\in[W/W_\nu]_E} \mathrm{ind}_{[w]} \cdot t_E^{\ell(w)} \cdot v_{W_0(w\nu)}^{W_0}[-2\ell(w) - |\Delta(w\nu)|] .$$

Here $W_0(w\nu) = W_{0,\Delta_0\setminus\Delta_0(w\nu)}$.

Remark 7.2.6 We note that Remarks 3.3.11 (regarding the object in the derived category of $G(k) \times \Gamma_k$-representations defined by the cohomology complex $R\Gamma_c(\mathcal{F}(G,\mathcal{N}))$) and 3.3.18 (regarding the Hodge theory analogue relative to $\mathrm{Fil}_k^{\mathbb{F}_1}$) also have generalizations to the case of an arbitrary reductive group.

Before formulating the next result, we recall that $\mathcal{F}(G,\mathcal{N})^{\mathrm{ss}}$ depends only on the adjoint group G_{ad}, cf. Proposition 6.1.9. Also, if $G = \prod_i G_i$ is the direct product of groups, then the corresponding period domain $\mathcal{F}(G,\mathcal{N})^{ss} = \prod_i \mathcal{F}(G_i,\mathcal{N}_i)^{ss}$ splits into a product of corresponding period domains. Let now G be k-simple and adjoint. Then G is of the form $G = \mathrm{Res}_{k'/k}(G')$ for an absolutely simple group G' over k'. Then $\mathcal{N}$ is of the form $(\mathcal{N}_1,\ldots,\mathcal{N}_n)$ corresponding to the k-embeddings of k' into $\bar{k}$. Here $n = [k' : k]$ and $\mathcal{N}_1,\ldots,\mathcal{N}_n$ are conjugacy classes of G'.

We may deduce from Theorem 7.2.4 the following remarkable vanishing property.

Corollary 7.2.7 *Let G be k-simple and adjoint, and let $\mathcal{N} = (\mathcal{N}_1,\ldots,\mathcal{N}_n)$ be non-trivial, i.e., there is at least one index $1 \le i \le n$ such that $\mathcal{N}_i$ is not defined by a trivial* $\mathbb{Q}$-1-PS. *Let $d = |\Delta_0|$ be the semi-simple k-rank of G. Then*

$$H_c^i(\mathcal{F}(G,\mathcal{N})^{ss}) = 0 \textit{ for } i \textit{ with } 0 \le i \le d-1 ,$$

and

$$H_c^d(\mathcal{F}(G,\mathcal{N})^{ss}) = v_B^G .$$

Furthermore, the top cohomology is

$$H_c^{2\ell(w_0)}(\mathcal{F}(G,\mathcal{N})^{ss}) = t_E^{\ell(w_0)} \cdot i_G^G ,$$

where w_0 is of maximal length in W/W_ν, i.e., $\dim \mathcal{F}(G,\mathcal{N}) = \ell(w_0)$.

Remark 7.2.8 Suppose that in the above situation the conjugacy class $\mathcal{N}$ is trivial. Then the flag variety $\mathcal{F}(G,\mathcal{N})$ is a finite set of points. Thus Corollary 7.2.7 cannot be true in this case.

The proof of Corollary 7.2.7 is similar to the proof of Corollary 3.3.12 which concerns the case of GL_n. In the present case, we have to deal with non-split reductive groups and their relative root systems. We need the following preparation.

Let k' be a Galois splitting field of G of degree n. By fixing an invariant inner product $(\ ,\ )$, we may identify $X^*(S)_{\mathbb{Q}}$ with the subspace of $\mathrm{Gal}(k'/k)$-invariants in $X^*(T)_{\mathbb{Q}}$. This space is orthogonal to the subspace generated by characters vanishing on S, i.e., by the kernel $X^*(T) \longrightarrow X^*(S)$. Via this identification the restriction map is then given by

$$\chi \longrightarrow \frac{1}{n} \sum_{\sigma \in \mathrm{Gal}(k'/k)} {}^{\sigma}\chi. \tag{2.11}$$

Let Δ be the set of simple absolute roots of G with respect to $T \subset B$. The *(relative) root basis* Δ_0 is given by restriction of Δ to S, cf. [26]. More precisely,

$$\Delta_0 = \{\alpha_{|S} \mid \alpha \in \Delta, \alpha_{|S} \neq 0\}.$$

Furthermore, two simple roots with the same non-trivial restriction lie in the same orbit under the action of $\mathrm{Gal}(k'/k)$. The set of *(relative) fundamental weights* $(\omega_\alpha)_{\alpha \in \Delta_0}$ is by definition the dual basis of the co-roots $(\alpha^\vee)_{\alpha \in \Delta_0}$. The co-roots are given via the identification $X_*(S)_{\mathbb{Q}} \cong X^*(S)_{\mathbb{Q}}$ by

$$\alpha^\vee = \frac{2\alpha^*}{(\alpha, \alpha)},$$

where $\alpha^* \in X_*(S)_{\mathbb{Q}}$ corresponds to α, cf. [28], Ch. VI, §1. Every root $\alpha \in \Delta_0$ may be identified by (2.11) above with

$$\frac{1}{n} \sum_{\sigma \in \mathrm{Gal}(k'/k)} {}^{\sigma}\beta \in X^*(T)_{\mathbb{Q}},$$

where $\beta \in \Delta$ is a lift of α. Then $(\alpha, \alpha) = (\beta, \alpha) = \frac{1}{n} \sum_\sigma ({}^\sigma\beta, \beta)$. Thus

$$\alpha^\vee = \frac{2(\frac{1}{n} \sum_\sigma {}^\sigma\beta)^*}{\frac{1}{n} \sum_\sigma ({}^\sigma\beta, \beta)} = \frac{2 \sum_\sigma {}^\sigma(\beta^*)}{\sum_\sigma ({}^\sigma\beta, \beta)}.$$

We obtain

$$\alpha^\vee = c_\alpha \cdot \sum\nolimits_\sigma {}^\sigma(\beta^\vee),$$

where

$$c_\alpha = \frac{(\beta, \beta)}{\sum_\sigma ({}^\sigma\beta, \beta)}.$$

Hence the relative fundamental weights can be expressed through the absolute fundamental weights, as

$$\varpi_\alpha = \frac{n}{c_\alpha} \sum_{\substack{\beta \in \Delta \\ \beta_{|S} = \alpha}} \varpi_\beta. \tag{2.12}$$

In particular, ϖ_α and $\sum_{\substack{\beta\in\Delta\\ \beta_{|S}=\alpha}} \varpi_\beta$ differ at most by a positive rational multiple. Thus we may write

$$\Delta(wv) = \left\{\alpha \in \Delta_0 \mid \left\langle wv, \sum_{\substack{\beta\in\Delta\\ \beta_{|S}=\alpha}} \varpi_\beta\right\rangle > 0\right\}, \tag{2.13}$$

where $\beta \in \Delta$ is an arbitrarily fixed lifting of $\alpha \in \Delta_0$.

Let W^v be the set of representatives of shortest length in their cosets W/W_v.

Lemma 7.2.9 *Let $w \in W^v$. Let $s = s_{\alpha'} \in W$ be a simple reflection associated to a simple root $\alpha' \in \Delta$, such that $sw \in W^v$. Suppose that $\ell(sw) = \ell(w) + 1$. Set $\alpha = \alpha'|S$. Then*

$$\Delta(swv) \setminus \{\alpha\} = \Delta(wv) \setminus \{\alpha\}$$

and

$$\Delta(swv) \subset \Delta(wv).$$

In particular, the sets $\Delta(swv)$ and $\Delta(wv)$ differ at most by one element.

Proof Again the proof is the same as that of Lemma 3.3.13. In fact if G is split then this is immediate. In the general case, let $\beta \in \Delta_0$. We get

$$\left\langle swv, \sum_{\substack{\beta'\in\Delta\\ \beta'_{|S}=\beta}} \varpi_{\beta'}\right\rangle = \left\langle wv, s\left(\sum_{\substack{\beta'\in\Delta\\ \beta'_{|S}=\beta}} \varpi_{\beta'}\right)\right\rangle$$

and

$$s\left(\sum_{\beta'} \varpi_{\beta'}\right) = \begin{cases} \sum_{\beta'} \varpi_{\beta'} & ; \quad \beta \neq \alpha \\ \sum_{\beta'} \varpi_{\beta'} - \alpha' & ; \quad \beta = \alpha \end{cases}$$

([28], VI, §1.10). The first statement follows. The assumption $\ell(sw) = \ell(w) + 1$ is equivalent to the positivity $w^{-1}(\alpha') > 0$ ([38], Prop. 2.2.8). Since v is contained in the closure of the acute Weyl chamber, we deduce that

$$\langle wv, \alpha'\rangle = \langle v, w^{-1}\alpha'\rangle \geq 0\,.$$

Thus, for $\beta = \alpha \in \Delta(swv)$, we get

$$0 < \left\langle swv, \sum_{\beta'} \varpi_{\beta'}\right\rangle = \left\langle wv, \sum_{\beta'} \varpi_{\beta'} - \alpha'\right\rangle = \left\langle wv, \sum_{\beta'} \varpi_{\beta'}\right\rangle - \langle wv, \alpha'\rangle.$$

It follows that

$$\left\langle wv, \sum_{\beta'} \varpi_{\beta'}\right\rangle > \langle wv, \alpha'\rangle \geq 0,$$

hence $\alpha \in \Delta(wv)$. □

Lemma 7.2.10 *Let $w \in W^v$ and let v be non-central. Then*

$$|\Delta_0 \setminus \Delta(wv)| \leq \ell(w).$$

Proof Since ν is not central, we conclude that $\Delta(\nu) = \Delta_0$. In fact, for simplicity let G be absolutely simple adjoint, so that $n = [k' : k] = 1$. Then we may write $\nu = \sum_{\alpha\in\Delta} n_\alpha \varpi^*_\alpha$ with $n_\alpha \geq 0$ for all α. Here $\varpi^*_\alpha \in X_*(T)_{\mathbb{Q}}$ corresponds to ϖ_α via the identification of characters and cocharacters induced by the inner invariant product. Since ν is non-central there is an α such that $n_\alpha > 0$. Hence it suffices to see that $(\varpi_\alpha, \varpi_\beta) > 0$ for all $\alpha, \beta \in \Delta$. But this follows from checking case by case all different root systems, cf. [28], pp. 265–290.

Now the proof of Lemma 3.3.14 can be carried over word for word. □

Proof of Corollary 7.2.7 Applying the previous lemma, we get for all $w \in W^\nu$ the inequality

$$|\Delta(w\nu)| + 2\ell(w) \geq (|\Delta_0| - \ell(w)) + 2\ell(w) = |\Delta_0| + \ell(w) \geq |\Delta_0|.$$

Equality holds in this chain if and only if $w = 1$. The first two assertions follow. For the last statement, we note that $\Delta(w_0\nu) = \emptyset$, since $w_0\nu$ lies in the opposite Weyl chamber and we may apply Lemma 7.1.1 (ii). We note that the fact that the highest cohomology group is one-dimensional can also be deduced from the connectedness of $\mathcal{F}(G, \mathcal{N})^{ss}$. Finally $\mathrm{ind}_{[w]}$ is the trivial representation for $w = 1$, resp. $w = w_0$, since these are the only elements of length 0, resp. of maximal length, and hence are invariant under Γ_E. □

We now give some examples which illustrate the formulas obtained.

Examples 7.2.11 (i) Let $G = \mathrm{GSp}_4$, cf. Example 6.1.13 (i). Since G is split we have $E = k$, $\Delta = \Delta_0$ and $W = W_0$. More precisely, $\Delta = \{\alpha_1, \alpha_2\}$, where $\alpha_1 = \epsilon_1 - \epsilon_2$ and $\alpha_2 = 2\epsilon_2$, $\varpi_{\alpha_1} = \epsilon_1$, $\varpi_{\alpha_2} = \epsilon_1 + \epsilon_2$ and $W = S_2 \ltimes (\mathbb{Z}/2\mathbb{Z})^2$. The set of std psgps is $\mathcal{P} = \{B, P_1, P_2, G\}$, where $P_i = P_{\{\alpha_i\}}$.

Let $\nu = (x_1, x_2, -x_2, -x_1) \in X_*(T)_{\mathbb{Q}}$ with $x_1 > x_2 > 0$. Then $W_\nu = \{1\}$. The sets $\Delta(w\nu)$, $w \in W$, are determined by the following table. Here $s_1 = s_{\alpha_1}$ and $s_2 = s_{\alpha_2}$.

w	1	s_1	s_2	s_1s_2
$w\nu$	$(x_1, x_2, -x_2, -x_1)$	$(x_2, x_1, -x_1, -x_2)$	$(x_1, -x_2, x_2, -x_1)$	$(-x_2, x_1, -x_1, x_2)$
$\Delta(w\nu)$	Δ	Δ	Δ	$\{\alpha_2\}$

w	s_2s_1	$s_1s_2s_1$	$s_2s_1s_2$	$s_1s_2s_1s_2$
$w\nu$	$(x_2, -x_1, x_1, -x_2)$	$(-x_1, x_2, -x_2, x_1)$	$(-x_2, -x_1, x_1, x_2)$	$(-x_1, -x_2, x_2, x_1)$
$\Delta(w\nu)$	$\{\alpha_1\}$	$\emptyset$	$\emptyset$	$\emptyset$

Theorem 7.2.4 gives

$$H_c^0(\mathcal{F}(G,\mathcal{N})^{ss}) = 0 \qquad H_c^5(\mathcal{F}(G,\mathcal{N})^{ss}) = t_k^2 \cdot v_{P_1}^G \oplus t_k^2 \cdot v_{P_2}^G$$

$$H_c^1(\mathcal{F}(G,\mathcal{N})^{ss}) = 0 \qquad H_c^6(\mathcal{F}(G,\mathcal{N})^{ss}) = (t_k^3 \cdot i_G^G)^{\oplus 2}$$

$$H_c^2(\mathcal{F}(G,\mathcal{N})^{ss}) = v_B^G \qquad H_c^7(\mathcal{F}(G,\mathcal{N})^{ss}) = 0$$

$$H_c^3(\mathcal{F}(G,\mathcal{N})^{ss}) = 0 \qquad H_c^8(\mathcal{F}(G,\mathcal{N})^{ss}) = t_k^4 \cdot i_G^G$$

$$H_c^4(\mathcal{F}(G,\mathcal{N})^{ss}) = (t_k \cdot v_B^G)^{\oplus 2} \qquad H_c^i(\mathcal{F}(G,\mathcal{N})^{ss}) = 0 \ \ \forall i > 8\,.$$

(ii) Let $G = \mathrm{U}(3)$ be the unitary group with respect to a quadratic extension k'/k contained in $\bar{k}$. In contrast to Example 6.1.3 (iii), we define $U(3)$ to be

$$U(3)(k) = \{g \in \mathrm{GL}_3(k') \mid gJ\bar{g}^t = J\},$$

where

$$J = \begin{pmatrix} 0 & 0 & 1 \\ 0 & 1 & 0 \\ 1 & 0 & 0 \end{pmatrix}.$$

A maximal torus T is given by

$$T(k) = \{\mathrm{diag}(t_1, t_2, t_3) \mid t_1\bar{t}_3 = 1, t_2\bar{t}_2 = 1\}.$$

A maximal split torus $S \subset T$ is

$$S(k) = \{\mathrm{diag}(s, 1, s^{-1}) \mid s \in k^\times\}.$$

The action of $\mathrm{Gal}(k'/k)$ on $X_*(T)_\mathbb{Q} = \mathbb{Q}^3$ is given by

$$(x_1, x_2, x_3) \mapsto (-x_3, -x_2, -x_1).$$

Thus for $\nu = (x_1, x_2, x_3) \in X_*(T)_\mathbb{Q}$, we get $E = k$ if and only if $x_1 = -x_3$ and $x_2 = 0$. Otherwise, $E = k'$.

We have $W = S_3$, $\Delta = \{\alpha_1, \alpha_2\}$, where $\alpha_1 = \epsilon_1 - \epsilon_2$ and $\alpha_2 = \epsilon_2 - \epsilon_3$, and $\Delta_0 = \{\alpha\}$, where $\alpha = \alpha_1|S = \alpha_2|S$. Thus $\varpi_{\alpha_1} = (\frac{2}{3}, -\frac{1}{3}, -\frac{1}{3})$, $\varpi_{\alpha_2} = (\frac{1}{3}, \frac{1}{3}, -\frac{2}{3})$ and $\varpi_\alpha = \frac{1}{2}(\varpi_{\alpha_1} + \varpi_{\alpha_2}) = (\frac{1}{2}, 0, -\frac{1}{2})$. Furthermore, $\mathcal{P} = \{B, G\}$. We compute.

(a) Let $\nu = (x_1, x_2, x_3)$ with $x_1 > x_2 > x_3$ and $x_1 \neq -x_3$ or $x_2 \neq 0$. Then $W_\nu = \{1\}$, $E = k'$ and

$$\mathcal{F}(G,\mathcal{N})(\bar{k}) = \{0 \subsetneq \mathcal{F}_1 \subsetneq \mathcal{F}_2 \subsetneq \bar{k}^3\}.$$

A computation shows that

$$\mathcal{F}(G,\mathcal{N})(\bar{k})^{ss} = \{\mathcal{F}\in\mathcal{F}(G,\mathcal{N})(\bar{k}) \mid \mathcal{F}_1 \textit{ is not isotropic and there is no isotropic line } U \textit{ with } \mathcal{F}_2 = U^{\perp}\}. \quad (2.14)$$

Since E is also the splitting field of G, the action of Γ_E on W is trivial. Thus $W = [W]_E$ and $\mathrm{ind}_{[w]}$ is the trivial representation for all w. The sets $\Delta(wv)$, $w \in W$, are determined by the following table.

w	1	s_1	s_2	s_1s_2	s_2s_1	$s_1s_2s_1$
wv	(x_1,x_2,x_3)	(x_2,x_1,x_3)	(x_1,x_3,x_2)	(x_3,x_1,x_2)	(x_2,x_3,x_1)	(x_3,x_2,x_1)
$\Delta(wv)$	Δ_0	Δ_0	Δ_0	$\emptyset$	$\emptyset$	$\emptyset$.

Theorem 7.2.4 gives

$$H^0_c(\mathcal{F}(G,\mathcal{N})^{ss}) = 0 \qquad H^4_c(\mathcal{F}(G,\mathcal{N})^{ss}) = (t_E^2\cdot i_G^G)^{\oplus 2}$$

$$H^1_c(\mathcal{F}(G,\mathcal{N})^{ss}) = v_B^G \qquad H^5_c(\mathcal{F}(G,\mathcal{N})^{ss}) = 0$$

$$H^2_c(\mathcal{F}(G,\mathcal{N})^{ss}) = 0 \qquad H^6_c(\mathcal{F}(G,\mathcal{N})^{ss}) = t_E^3\cdot i_G^G$$

$$H^3_c(\mathcal{F}(G,\mathcal{N})^{ss}) = (t_E\cdot v_B^G)^{\oplus 2} \qquad H^i_c(\mathcal{F}(G,\mathcal{N})^{ss}) = 0 \;\; \forall i > 6.$$

(b) Let $v = (x_1,x_2,x_3)$ with $x_1 = -x_3$, $x_2 = 0$ and $x_1 > 0$. Then $W_v = \{1\}$, $E = k$ and the attached period domain has the same shape as (2.14) above. We compute $[W]_\Gamma = \{[1],[s_1],[s_2s_1],[s_1s_2s_1]\}$. The sets $\Delta(wv)$ and Γ_w, $w \in [W]_E$, are determined by the following table.

$[w]$	1	$[s_1]$	$[s_2s_1]$	$[s_1s_2s_1]$
wv	(x_1,x_2,x_3)	(x_2,x_1,x_3),	(x_2,x_3,x_1)	(x_3,x_2,x_1)
$\Delta(wv)$	Δ_0	Δ_0	$\emptyset$	$\emptyset$
Γ_w	Γ_k	$\Gamma_{k'}$	$\Gamma_{k'}$	Γ_k.

Theorem 7.2.4 gives

$$H^0_c(\mathcal{F}(G,\mathcal{N})^{ss}) = 0 \qquad H^4_c(\mathcal{F}(G,\mathcal{N})^{ss}) = \mathrm{ind}_{[s_2s_1]}\, t_k^2 v_G^G$$

$$H^1_c(\mathcal{F}(G,\mathcal{N})^{ss}) = v_B^G \qquad H^5_c(\mathcal{F}(G,\mathcal{N})^{ss}) = 0$$

$$H^2_c(\mathcal{F}(G,\mathcal{N})^{ss}) = 0 \qquad H^6_c(\mathcal{F}(G,\mathcal{N})^{ss}) = t_k^3\cdot v_G^G$$

$$H^3_c(\mathcal{F}(G,\mathcal{N})^{ss}) = \mathrm{ind}_{[s_1]}\cdot t_k\cdot v_B^G \qquad H^i_c(\mathcal{F}(G,\mathcal{N})^{ss}) = 0 \;\; \forall i > 6\,.$$

We finally use as in Chapter III the Grothendieck–Lefschetz fixed point formula from Chapter III (3.3), page 84 to determine the number of points of $\mathcal{F}(G,\mathcal{N})^{ss}$ over any finite extension $\mathbb{F}_{q_E^m}$ of $E = \mathbb{F}_{q_E}$. If $[E : k] = r$, then $q_E = q^r$. We may alternatively use either of the formulas for the Euler–Poincaré characteristic obtained in Theorem 7.2.3 and in Theorem 7.2.2.

We first use the formula for the Euler–Poincaré characteristic given by Theorem 7.2.3. Let $\varphi \in \Gamma_E$ be the topological generator induced by the inverse of the arithmetic Frobenius. Then

$$|\mathcal{F}(G,\mathcal{N})^{ss}(\mathbb{F}_{q_E^m})| = \sum_{w\in[W/W_\nu]_E} (-1)^{a^G_{P(w\nu)}} \cdot \mathrm{trace}(\varphi^m, \mathrm{ind}_{[w]} \cdot t_E^{\ell(w)} \cdot v^G_{P(w\nu)}).$$

If $[\Gamma_E : \Gamma_w] \nmid m$, then we have $\mathrm{trace}(\varphi^m, \mathrm{ind}_{[w]}) = 0$. Otherwise, we get $\mathrm{trace}(\varphi^m, \mathrm{ind}_{[w]}) = \dim \mathrm{ind}_{[w]} = [\Gamma_E : \Gamma_w]$, since $\varphi^m \equiv 1 \bmod \Gamma_w$. Again φ^m acts on $t_E^{\ell(w)}$ by multiplication with $q_E^{m\ell(w)} = q^{|E:k|\cdot m\cdot\ell(w)}$. On the generalized Steinberg representations $v^G_{P(w\nu)}$, $w \in [W/W_\nu]_E$, the action of φ is via the identity. Thus we get

$$\mathrm{trace}(\varphi^m, \mathrm{ind}_{[w]} \cdot t_E^{\ell(w)} \cdot v^G_{P(w\nu)}) = \begin{cases} [\Gamma_E : \Gamma_w] \cdot q_E^{m\ell(w)} \cdot \dim v^G_{P(w\nu)} & \text{if } [\Gamma_E : \Gamma_w] \mid m \\ 0 & \text{otherwise.} \end{cases} \tag{2.15}$$

The dimension of $v^G_{P(w\nu)}$ may be computed by Theorem 7.1.9, as follows. Note that $W_0 = W^{\Gamma_k} \subset W$. Via this embedding we denote for $w \in W_0$, by $\ell(w)$ the length of w considered as an element of W and not of W_0.

Lemma 7.2.12 *Let $I \subset \Delta_0$. Set $Y_I = \{w \in W_0 \mid w\alpha > 0\ \forall\alpha \in I,\ w\alpha < 0\ \forall\alpha \in \Delta_0 \setminus I\}$. Then*

(i) $\dim v^G_{P_I} = \sum_{w\in Y_I} q^{\ell(w)}$;

(ii) $\dim v^{W_0}_{W_{0,I}} = |Y_I|$.

Proof The proof is almost the same as that of Lemma 3.3.19. Let $\Phi^+_{nd} \subset \Phi$ be the set of non-divisible positive roots. For $w \in W_0$, let

$$N_w := \prod_{\substack{\alpha\in\Phi^+_{nd}\\ w\alpha<0}} U_\alpha$$

be the unipotent subgroup of B generated by the (non-divisible) positive root subgroups made negative by w. Using the complex (1.4), page 169 and the *(relative) Bruhat decomposition*

$$(G/P_I)(k) = \bigcup_{w\in W_0^I} N_w(k)wP_I(k)/P_I(k), \tag{2.16}$$

we conclude that the characteristic functions $\chi_{n,w}$ on the sets $nwP_I(k)/P_I(k)$ for

$w \in Y_I, n \in N_w(k)$ form a basis of $v_{P_I}^G$. But N_w is an affine space of dimension $\dim N_w = \ell(w)$. Thus for any k-structure on N_w one has $N_w(k) = q^{\dim N_w}$, cf. [68], Prop. 10.11. The claim follows. □

Remark 7.2.13 If $I = \emptyset$, i.e., $P_I = B$, then $Y_I = \{w_0\}$, where w_0 is the longest Weyl group element in W_0. Thus $\dim v_B^G = q^{\ell(w_0)} = |R_u(B)(k)|$, which coincides with the formula in [38], Cor. 6.4.3., cf. also [46], Prop. 69.9 in the case of a split BN-pair.

We set $Y_{[w]} := Y_{\Delta\setminus\Delta(wv)}$ for $[w] \in [W/W_v]_E$.

Proposition 7.2.14 *We have*

$$|\mathcal{F}(G,\mathcal{N})^{ss}(\mathbb{F}_{q_E^m})| = \sum_{\substack{w\in[W/W_v]_E \\ [\Gamma_E:\Gamma_w]|m}} (-1)^{a^G_{P(wv)}} [\Gamma_E : \Gamma_w] \cdot q_E^{m\ell(w)} \cdot \sum_{y\in Y_{[w]}} q^{\ell(y)}.$$

Proof By Lemma 7.2.12 we have $\dim v_{P_I}^G = \sum_{y\in Y_I} q^{\ell(y)}$. From (2.15), page 182 we see that

$$\mathrm{trace}(\varphi^m, \mathrm{ind}_{[w]} \cdot t_E^{\ell(w)} \cdot v^G_{P(wv)}) = \begin{cases} [\Gamma_E : \Gamma_w] \cdot q_E^{m\ell(w)} \cdot \sum_{y\in Y_{[w]}} q^{\ell(y)} & \text{if } [\Gamma_E : \Gamma_w] \mid m \\ 0 & \text{otherwise.} \end{cases}$$

Thus the claim follows from Theorem 7.2.3. □

Variant 7.2.15 In the context of variant $\mathrm{Fil}^k_{\mathbb{F}_1}$ for arbitrary groups, cf. Variant 6.1.12 (ii), we get

$$|\mathcal{F}(G,S,\mathcal{N})^{ss}(\mathbb{F}_{q_E^m})| = \sum_{\substack{w\in[W/W_v]_E \\ [\Gamma_E:\Gamma_w]|m}} (-1)^{a^{W_0}_{W_0(wv)}} [\Gamma_E : \Gamma_w] \cdot q_E^{m\ell(w)} \cdot |Y_{[w]}|.$$

Now we count the rational points of $\mathcal{F}(G,\mathcal{N})^{ss}$, starting with the formula for the Euler–Poincaré characteristic given by Theorem 7.2.2. We have $\dim i_P^G = |G(k)/P(k)|$ and by the Bruhat decomposition (with the same reasoning as above) $|G(k)/P(k)| = \sum_{w\in W_0/W_{0,P}} q^{\ell(w)}$. It follows that

$$\begin{aligned} |\mathcal{F}(G,\mathcal{N})^{ss}(\mathbb{F}_{q^m})| &= \sum_{P\in\mathcal{P}} (-1)^{a_P^G} \sum_{w'\in W_0/W_{0,P}} q^{\ell(w')} \cdot \sum_{\substack{w\in[W/W_v]_E \\ [\Gamma_E:\Gamma_w]|m}} [\Gamma_E : \Gamma_w] \cdot q_E^{m\ell(w)} \cdot \hat{\tau}_P^G(wv). \\ &= \sum_{P\in\mathcal{P}} (-1)^{a_P^G} \sum_{w'\in W_0/W_{0,P}} q^{\ell(w')} \sum_{\substack{[w]\in[W/W_v]_E \\ [\Gamma_E:\Gamma_w]|m \\ \Delta_P\subset\Delta(wv)}} [\Gamma_E : \Gamma_w] \cdot q_E^{m\ell(w)} \\ &= \sum_{P\in\mathcal{P}} (-1)^{a_P^G} \sum_{w'\in W_0/W_{0,P}} \sum_{\substack{[w]\in[W/W_v]_E \\ [\Gamma_E:\Gamma_w]|m \\ \Delta_P\subset\Delta(wv)}} [\Gamma_E : \Gamma_w] \cdot q_E^{m\ell(w)} q^{\ell(w')}. \qquad (2.17) \end{aligned}$$

Examples 7.2.16 Reconsider Examples 7.2.11.

(i) Let $G = \mathrm{GSp}_4$. We have $E = k$, $W = W_0$, $\Delta = \{\alpha_1, \alpha_2\}$, where $\alpha_1 = \epsilon_1 - \epsilon_2$ and $\alpha_2 = 2\epsilon_2$, and $W = S_2 \ltimes (\mathbb{Z}/2\mathbb{Z})^2$.

Let $\nu = (x_1, x_2, -x_2, -x_1) \in X_*(T)_{\mathbb{Q}}$ with $x_1 > x_2 > 0$. The sets $Y_{[w]}$, $w \in W$, are determined by the following table.

w	1	s_1	s_2	$s_1 s_2$
$Y_{[w]}$	$\{s_1 s_2 s_1 s_2\}$	$\{s_1 s_2 s_1 s_2\}$	$\{s_1 s_2 s_1 s_2\}$	$\{s_2, s_1 s_2, s_2 s_1 s_2\}$
$w\nu$	$(x_1, x_2, -x_2, -x_1)$	$(x_2, x_1, -x_1, -x_2)$	$(x_1, -x_2, x_2, -x_1)$	$(-x_2, x_1, -x_1, x_2)$
$\Delta(w\nu)$	Δ	Δ	Δ	$\{\alpha_2\}$

w	$s_2 s_1$	$s_1 s_2 s_1$	$s_2 s_1 s_2$	$s_1 s_2 s_1 s_2$
$Y_{[w]}$	$\{s_1, s_2 s_1, s_1 s_2 s_1\}$	$\{1\}$	$\{1\}$	$\{1\}$
$w\nu$	$(x_2, -x_1, x_1, -x_2)$	$(-x_1, x_2, -x_2, x_1)$	$(-x_2, -x_1, x_1, x_2)$	$(-x_1, -x_2, x_2, x_1)$
$\Delta(w\nu)$	$\{\alpha_1\}$	$\emptyset$	$\emptyset$	$\emptyset$

By using Proposition 7.2.14 we obtain

$$|\mathcal{F}(G, \mathcal{N})^{ss}(\mathbb{F}_{q^m})| = q^4 + 2q^{m+4} - 2q^{2m}(q + q^2 + q^3) + 2q^{3m} + q^{4m}. \quad (2.18)$$

Note that $[\Gamma_E : \Gamma_w] = 1$ for all w, since G is split. To apply formula (2.17), page 183, we consider the following table.

P	B	P_2	P_1	G
W/W_P	W	$\{1, s_2, s_1 s_2, s_2 s_1 s_2\}$	$\{1, s_1, s_2 s_1, s_1 s_2 s_1\}$	$\{1\}$
Δ_P	Δ	$\{\alpha_1\}$	$\{\alpha_2\}$	$\emptyset$

(2.19)

The contributions of the parabolic subgroups to (2.17), page 183 are as follows.

B	$(1 + 2q + 2q^2 + 2q^3 + q^4)(1 + 2q^m)$
P_1	$-(1 + q + q^2 + q^3)(1 + 2q^m + q^{2m})$
P_2	$-(1 + q + q^2 + q^3)(1 + 2q^m + q^{2m})$
G	$q^{4m} + 2q^{3m} + 2q^{2m} + 2q^m + 1$

Thus we get by applying (2.17), page 183

$$\begin{aligned} |\mathcal{F}(G, \mathcal{N})^{ss}(\mathbb{F}_{q^m})| &= (1 + 2q + 2q^2 + 2q^3 + q^4)(1 + 2q^m) \\ &\quad -2(1 + q + q^2 + q^3)(1 + 2q^m + q^{2m}) \\ &\quad +q^{4m} + 2q^{3m} + 2q^{2m} + 2q^m + 1 \\ &= q^4 + 2q^{m+4} - 2q^{2m}(q + q^2 + q^3) + 2q^{3m} + q^{4m}. \end{aligned}$$

The result coincides with (2.18) above.

(ii) Let $G = U(3)$. Then $W_0 = \{1, s_1 s_2 s_1\}$.

(a) Let $v = (x_1, x_2, x_3)$ with $x_1 > x_2 > x_3$ and $x_1 \neq -x_3$ or $x_2 \neq 0$. Thus $E = k'$ and the action of Γ_E on W is trivial.

The sets Y_w are given by the following table.

w	1	s_1	s_2	$s_1 s_2$	$s_2 s_1$	$s_1 s_2 s_1$
Y_w	$\{s_1 s_2 s_1\}$	$\{s_1 s_2 s_1\}$	$\{s_1 s_2 s_1\}$	$\{1\}$	$\{1\}$	$\{1\}$
$\Delta(wv)$	Δ_0	Δ_0	Δ_0	$\emptyset$	$\emptyset$	$\emptyset$

By using Proposition 7.2.14 we get

$$\begin{aligned} |\mathcal{F}(G, \mathcal{N})^{ss}(\mathbb{F}_{q_E^m})| &= -q^3 - 2q_E^m q^3 + 2q_E^{2m} + q_E^{3m} \\ &= q^{6m} + 2q^{4m} - 2q^{2m+3} - q^3. \end{aligned} \tag{2.20}$$

To apply formula (2.17), page 183, we consider the following tables.

P	B	G
$W_0/W_{0,P}$	W_0	$\{1\}$
Δ_P	Δ_0	$\emptyset$

(2.21)

w	1	s_1	s_2	$s_1 s_2$	$s_2 s_1$	$s_1 s_2 s_1$
$\Delta(wv)$	Δ_0	Δ_0	Δ_0	$\emptyset$	$\emptyset$	$\emptyset$

The contributions of the parabolic subgroups to (2.17), page 183 are as follows.

B	$-(1 + 2q_E^m)(q^3 + 1)$
G	$1 + 2q_E^m + 2q_E^{2m} + q_E^{3m}$

Thus we get by applying (2.17), page 183

$$\begin{aligned} |\mathcal{F}(G, \mathcal{N})^{ss}(\mathbb{F}_{q^m})| &= -(1 + 2q_E^m)(q^3 + 1) + 1 + 2q_E^m + 2q_E^{2m} + q_E^{3m} \\ &= q^{6m} + 2q^{4m} - 2q^{2m+3} - q^3. \end{aligned}$$

The result coincides with (2.20) above.

(b) Let $v = (x_1, x_2, x_3)$ with $x_1 = -x_3$ and $x_2 = 0$. Thus $E = k$. In this case, the sets $Y_{[w]}$ and Γ_w are given by the following table.

w	1	s_1	$s_2 s_1$	$s_1 s_2 s_1$
$Y_{[w]}$	$\{s_1 s_2 s_1\}$	$\{s_1 s_2 s_1\}$	$\{1\}$	$\{1\}$
wv	(x_1, x_2, x_3)	(x_2, x_1, x_3)	(x_2, x_3, x_1)	(x_3, x_2, x_1)
$\Delta(wv)$	Δ_0	Δ_0	$\emptyset$	$\emptyset$
Γ_w	Γ_k	$\Gamma_{k'}$	$\Gamma_{k'}$	Γ_k

By using Proposition 7.2.14 we get

$$|\mathcal{F}(\nu)^{ss}(\mathbb{F}_{q^m})| = \begin{cases} -q^3 + q^{3m} & \text{if } m \text{ is odd} \\ -q^3 + q^{3m} - 2q^{m+3} + 2q^{2m} & \text{if } m \text{ is even.} \end{cases} \tag{2.22}$$

To apply formula (2.17), page 183, we have to consider table (2.21), page 185 again. The contributions of the parabolic subgroups in (2.17), page 183 are given by:

B	$-1 - q^3$
G	$1 + q^{3m}$

if m is odd,

resp.

B	$(-1 - q^3)(1 + 2q^m)$
G	$1 + q^{3m} + 2q^m + 2q^{2m}$

if m is even.

For m odd we get by applying (2.17), page 183

$$\begin{aligned} |\mathcal{F}(G, \mathcal{N})^{ss}(\mathbb{F}_{q^m})| &= (1 + q^{3m}) - (1 + q^3) \\ &= q^{3m} - q^3. \end{aligned}$$

For m even we get

$$\begin{aligned} |\mathcal{F}(G, \mathcal{N})^{ss}(\mathbb{F}_{q^m})| &= (-1 - q^3)(1 + 2q^m) + 1 + q^{3m} + 2q^m + 2q^{2m} \\ &= -q^3 + q^{3m} - 2q^{m+3} + 2q^{2m}. \end{aligned}$$

These formulas coincide with (2.22) above.

Notes and References Theorem 7.2.4 was proved by the second author in [170]. For more remarks on the geometry behind this proof, we refer to Part 4 of this monograph. The non-vanishing statement in Corollary 7.2.7 has some striking consequences that are explained in Chapter XI.

Part 3

Period Domains over p-adic Fields

VIII. Period Domains over p-adic Fields

In this chapter we consider period domains over p-adic fields in the case of GL_n. Because the study of such period domains is motivated by p-adic Hodge theory, we have to enlarge the setting of filtered vector spaces to that of filtered *isocrystals* (also called filtered φ-modules by Fontaine). Moreover, in contrast to the finite field case, p-adic period domains are in general not Zariski-open subsets of their ambiant flag variety, but only *analytic* open subsets. Therefore we have to leave the realm of algebraic geometry for that of rigid-analytic geometry.

In the first section we introduce the tannakian category of isocrystals. Then we define the slope of a filtered isocrystal, which leads to the notion of semi-stability and the HN-filtration. In the second section we define period spaces as open analytic subsets of suitable flag varieties. The proofs are given in the setting of Berkovich's analytic spaces, with comments on the possible modifications to fit in the framework of Huber's adic spaces. In the third section we study the HN-stratification, again in the setting of analytic spaces. In the fourth section we discuss the GIT approach. In the fifth section we look at isocrystals with coefficients.

1 Filtered isocrystals

Let L be a perfect field of characteristic $p > 0$. Denote by $W(L)$ the associated ring of Witt vectors. Recall that $W(L)$ is the unique complete noetherian local ring of characteristic 0 with residue field L, such that the maximal ideal is generated by p, cf. [30]. We denote by

$$K_0 = K_0(L) := \mathrm{Quot}(W(L))$$

the corresponding fraction field and by $\sigma \in \mathrm{Aut}(K_0/\mathbb{Q}_p)$ the Frobenius homomorphism.

Definition 8.1.1 An *isocrystal* over L is a pair $N = (V, \Phi)$ consisting of a finite-dimensional vector space V over K_0 together with a σ-linear bijective endomorphism Φ of V.

The isocrystals over L form a noetherian $\mathbb{Q}_p$-linear tannakian $\otimes$-category $\mathrm{Isoc}(L)$, cf. [63]. Here the tensor product of two isocrystals $N_1 = (V_1, \Phi_1)$, $N_2 = (V_2, \Phi_2)$ is given by

$$(V_1, \Phi_1) \otimes (V_2, \Phi_2) = (V_1 \otimes V_2, \Phi_1 \otimes \Phi_2).$$

Each object $N = (V, \Phi) \in \mathrm{Isoc}(L)$ has a unique *slope decomposition*

$$N = \bigoplus_{\alpha \in \mathbb{Q}} N_\alpha \tag{1.1}$$

into isotypical sub-isocrystals $N_\alpha = (V_\alpha, \Phi)$ of slope $\alpha \in \mathbb{Q}$, cf. [224], Satz 6.22. The latter means that if $\alpha = \frac{r}{s}$ with $(r, s) = 1$ and $s > 0$, then there exists a $W(L)$-lattice $M \subset V_\alpha$ with

$$\Phi^s(M) = p^r M.$$

This slope decomposition is compatible with tensor products in the following sense. If N_1 and N_2 are isotypical of slope α_1 and α_2, then their tensor product $N_1 \otimes N_2$ is isotypical of slope $\alpha_1 + \alpha_2$. It is also compatible with the *base change $\otimes$-functor*

$$\begin{aligned} \varepsilon_L^{L'} : \mathrm{Isoc}(L) &\longrightarrow \mathrm{Isoc}(L') \\ N &\mapsto N_{L'} \end{aligned} \tag{1.2}$$

induced by any extension $L \subset L'$ of perfect fields. Here $N_{L'} = (V \otimes_{K_0} K_0(L'), \Phi')$ where $\Phi' = \Phi \otimes \sigma$.

For $\alpha = \frac{r}{s} \in \mathbb{Q}$ as above, let us denote by

$$E_\alpha^L = (K_0^s, A_\alpha \cdot \sigma)$$

the isocrystal with

$$A_\alpha = \begin{bmatrix} 0 & 1 & & \\ & & \ddots & 1 \\ & p^r & & 0 \end{bmatrix}.$$

This is a simple object of slope α in $\mathrm{Isoc}(L)$. If L contains $\mathbb{F}_{p^s}$, then its endomorphism ring $D_\alpha = \mathrm{End}(E_\alpha^L)$ is "the" $\mathbb{Q}_p$-division algebra with invariant $-\alpha$.

For an algebraically closed field L, it is known [63] that the category $\mathrm{Isoc}(L)$ is semi-simple and the objects E_α^L, for $\alpha \in \mathbb{Q}$, form a set of representatives of isomorphism classes of simple objects in $\mathrm{Isoc}(L)$. If L is not algebraically closed, it will be often convenient to restrict our attention to "split semi-simple" isocrystals in the following sense.

Definition 8.1.2 An isocrystal $N = (V, \Phi)$ over L is called *split semi-simple* if the following equivalent properties hold.

(i) The base change map $\mathrm{End}_{\mathrm{Isoc}(L)}(N) \longrightarrow \mathrm{End}_{\mathrm{Isoc}(L')}(N_{L'})$ is an isomorphism for any (or one) algebraically closed field L' containing L.

(ii) N decomposes as a direct sum of sub-isocrystals of the form E^L_α where $\alpha = \frac{r}{s}$ is such that $\mathbb{F}_{p^s} \subset L$.

(iii) For any slope $\alpha = \frac{r}{s}$ of N, we have $\mathbb{F}_{p^s} \subset L$, and V_α is generated as a K_0-vector space by the $\mathbb{Q}_{p^s}$-vector space $V_{s,\alpha} := \{v \in V \mid \Phi^s(v) = p^{\alpha s} v\}$.

The equivalence of the conditions follows from the theory of isocrystals as presented in [63] or [224].

We remind the reader that evaluation on any column of A_α defines an isomorphism of $\mathbb{Q}_p$-vector spaces $\mathrm{Hom}(E^L_\alpha, N) \xrightarrow{\sim} V_{s,\alpha}$, where the latter space inherits the $\mathbb{Q}_p$-multiplication given by scalar restriction. The full subcategory $\mathrm{Isoc}(L)^{\mathrm{ss}}$ of split semi-simple isocrystals in $\mathrm{Isoc}(L)$ is stable under tensor products and is abelian semi-simple. Here the isocrystals E^L_α, for $\alpha = \frac{r}{s}$ with $\mathbb{F}_{p^s} \subset L$, form a set of representatives of isomorphism classes of simple objects.

Of course, split semi-simple isocrystals can be defined over $\mathbb{Q}_p$, albeit in a non-canonical way. An important fact is that they can be *functorially* defined over some bigger field. More precisely, let s be a positive integer such that $\mathbb{F}_{p^s} \subset L$ and let $\mathrm{Isoc}(L)^{\mathrm{ss}}_s$ denote the abelian $\otimes$-subcategory of $\mathrm{Isoc}(L)^{\mathrm{ss}}$ whose objects have slopes in $\frac{1}{s}\mathbb{Z}$. Let $N = (V, \Phi)$ be an object of $\mathrm{Isoc}(L)^{\mathrm{ss}}_s$ and put

$$V_s := \sum\nolimits_{\alpha \in \frac{1}{s}\mathbb{Z}} V_{s,\alpha}. \tag{1.3}$$

Then V_s is a Φ-stable vector space over $K_0^{\sigma^s} = K_0(L^{\sigma^s}) = \mathbb{Q}_{p^s}$ and $N_s = (V_s, \Phi_{|V_s})$ defines an isocrystal over $L^{\sigma^s} = \mathbb{F}_{p^s}$. Further, the embedding $V_s \hookrightarrow V$ induces an isomorphism

$$V_s \otimes_{K_0^{\sigma^s}} K_0 \xrightarrow{\sim} V.$$

Thus by mapping the isocrystal N over L to its canonical model $N_s = (V_s, \Phi_{|V_s})$ over L^{σ^s}, we get a $\otimes$-equivalence of categories

$$\mathrm{Isoc}(L)^{\mathrm{ss}}_s \xrightarrow{\sim} \mathrm{Isoc}(\mathbb{F}_{p^s})^{\mathrm{ss}}_s$$

which is a quasi-inverse of the base change functor $\varepsilon^L_{\mathbb{F}_{p^s}}$ (1.2), page 190.

Now we are ready to return to the consideration of filtrations.

Definition 8.1.3 Let K/K_0 be a field extension. A *filtered isocrystal* $(N, \mathcal{F})$ over K is an isocrystal $N = (V, \Phi)$ over L together with a $\mathbb{Q}$-filtration $\mathcal{F}$ on the K-vector space V_K.

The filtered isocrystals over K form a $\mathbb{Q}_p$-linear quasi-abelian $\otimes$-category $\mathrm{FilIsoc}_L^K$. The tensor product of two objects $(N_i, \mathcal{F}_i)$, $i = 1, 2$, is given by

$$(N_1, \mathcal{F}_1) \otimes (N_2, \mathcal{F}_2) = (N_1 \otimes N_2, \mathcal{F}_1 \otimes \mathcal{F}_2).$$

Let $N = (V, \Phi)$ be an isocrystal over L and consider its slope decomposition $N = \bigoplus_{\alpha \in \mathbb{Q}} N_\alpha$ into isotypical sub-isocrystals. For $\beta \in \mathbb{Q}$, set

$$\mathcal{F}_0^\beta = \bigoplus_{\alpha \leq -\beta} V_\alpha. \tag{1.4}$$

This is a $\mathbb{Q}$-filtration $\mathcal{F}_0^\bullet$ on V by K_0-vector spaces.

Definition 8.1.4 Let $(N, \mathcal{F}) = (V, \Phi, \mathcal{F}) \in \mathrm{FilIsoc}_L^K$ be a filtered isocrystal over K. We set

$$\deg(N, \mathcal{F}) = \deg(V, \mathcal{F}) + \deg(V, \mathcal{F}_0)$$

and

$$\mu(N, \mathcal{F}) = \mu(V, \mathcal{F}) + \mu(V, \mathcal{F}_0).$$

These invariants are called *degree*, resp. *slope*, of $(N, \mathcal{F})$. We also denote them by $\deg_{\mathcal{F}}(N)$, resp. $\mu_{\mathcal{F}}(N)$.

As in Chapter I, § 1, degrees are additive on short exact sequences and can be computed on the maximal exterior power, while slopes are additive on tensor products and "ultrametric" on short exact sequences, cf. Proposition 1.1.15. Moreover, the statement of Lemma 1.1.12 still holds true in this context, up to replacing the category Fil_k^K by the category $\mathrm{FilIsoc}_L^K$ and the words "underlying k-vector-space" by "underlying isocrystal." Now the natural definition of semi-stability is as follows.

Definition 8.1.5 Let $(N, \mathcal{F}) \in \mathrm{FilIsoc}_L^K$. Then $(N, \mathcal{F})$ is called *semi-stable* if

$$\mu(N', \mathcal{F}') \leq \mu(N, \mathcal{F})$$

for all sub-isocrystals N' of N. Here $\mathcal{F}'$ denotes the filtration on $V' \otimes_{K_0} K$ induced by $\mathcal{F}$.

The statements of Proposition 1.1.20 and its Corollary 1.1.21 still hold true in this context. In particular, the full subcategory of $\mathrm{FilIsoc}_L^K$ formed by semi-stable objects of a fixed slope is abelian. Now, as acknowledged in the Introduction of this monograph, the relevance of all these concepts of semi-stability is in shedding new light on the arithmetically meaningful notion of *weakly admissible* isocrystal due to Fontaine.

Definition 8.1.6 The filtered isocrystal $(N, \mathcal{F})$ is called *weakly admissible* if it is semi-stable and $\mu_{\mathcal{F}}(N) = 0$.

Remark 8.1.7 Let $N = (V, \Phi) \in \mathrm{Isoc}(L)$. Consider the determinant $\det(\Phi)$. Actually this expression is not well-defined since Φ is only σ-linear. Nevertheless, it makes sense to consider the p-order

$$\deg(N) := \mathrm{ord}_p \det(\Phi)$$

of it. Then one checks that

$$-\deg(V, \mathcal{F}_0) = \deg(N).$$

In particular our definition of weak admissibility coincides with the original definition by Fontaine, cf. [80], at least when K is finite and totally ramified over K_0 and the filtration is supported on $\mathbb{Z}$.

One of the cornerstones of Chapter I was the tensor product theorem. The latter still holds true for filtered isocrystals and one ingredient is the following analogue of Lemma 1.2.5, the proof of which is left to the reader.

Lemma 8.1.8 *Let $(N, \mathcal{F}) \in \mathrm{FilIsoc}_L^K$ and let $\mathcal{F}'$ be a $\mathbb{Z}$-filtration of N by sub-isocrystals. Then*

$$\langle \mathcal{F}, \mathcal{F}' \rangle + \langle \mathcal{F}_0, \mathcal{F}' \rangle = \sum_{x \in \mathbb{Z}} (\mu_{\mathcal{F}}(N^x) - \mu_{\mathcal{F}}(N)) \dim(V^x_{\mathcal{F}'}) + \mu_{\mathcal{F}}(N) \mu_{\mathcal{F}'}(V) \dim(V).$$

In particular, if $(N, \mathcal{F})$ is weakly admissible, the RHS is non-positive. □

Now we can outline the proof of the $\otimes$-product theorem.

Theorem 8.1.9 *Let $(N, \mathcal{F}), (N', \mathcal{F}') \in \mathrm{FilIsoc}_L^K$. If $(N, \mathcal{F})$ and $(N', \mathcal{F}')$ are both semi-stable, resp. weakly admissible, then the tensor product $(N, \mathcal{F}) \otimes (N', \mathcal{F}')$ is again semi-stable, resp. weakly admissible.*

Proof We have $\mu((N, \mathcal{F}) \otimes (N', \mathcal{F}')) = \mu(N, \mathcal{F}) + \mu(N', \mathcal{F}')$. Thus by the rescaling argument it suffices to prove the statement for weak admissibility. Then the argument is similar to that of Theorem 1.2.1. Let $S = (W, \Phi) \subset N \otimes N'$ be a sub-isocrystal; we have to show that $\deg_{\mathcal{F} \otimes \mathcal{F}'}(S) \leq 0$. Let us assume the contrary. Then $\deg_{\mathcal{F} \otimes \mathcal{F}'}(W) > 0$ or $\deg_{\mathcal{F}_0 \otimes \mathcal{F}'_0}(W) > 0$, so we may apply Proposition 1.2.9 which says that the function $(\mathcal{G}, \mathcal{G}') \mapsto \deg_{\mathcal{G} \otimes \mathcal{G}'}(W)$ takes its maximum on a unique pair $(\mathcal{G}_0, \mathcal{G}'_0)$ of filtrations of norm equal to 1. Moreover, we have

$$\deg_{\mathcal{F} \otimes \mathcal{F}'}(W) \leq \deg_{\mathcal{G}_0 \otimes \mathcal{G}'_0}(W) \cdot \left[\langle \mathcal{F}, \mathcal{G}_0 \rangle + \langle \mathcal{F}', \mathcal{G}'_0 \rangle \right]$$

and

$$\deg_{\mathcal{F}_0\otimes\mathcal{F}_0'}(W) \leq \deg_{\mathcal{G}_0\otimes\mathcal{G}_0'}(W)\cdot\left[\langle\mathcal{F}_0,\mathcal{G}_0\rangle + \langle\mathcal{F}_0',\mathcal{G}_0'\rangle\right].$$

As in the proof of Theorem 1.2.1, the uniqueness of the pair of filtrations $(\mathcal{G}_0, \mathcal{G}_0')$ implies that these filtrations are defined over K_0. Moreover, since S is an isocrystal, it is stable under Φ. Using the uniqueness argument once again, these filtrations are also Φ-stable. In other words, these are filtrations by sub-isocrystals.

Now, adding the former inequalities yields

$$\deg_{\mathcal{F}\otimes\mathcal{F}'}(S) \leq \deg_{\mathcal{G}_0\otimes\mathcal{G}_0'}(W)\cdot\left[\langle\mathcal{F},\mathcal{G}_0\rangle + \langle\mathcal{F}_0,\mathcal{G}_0\rangle + \langle\mathcal{F}',\mathcal{G}_0'\rangle + \langle\mathcal{F}_0',\mathcal{G}_0'\rangle\right].$$

By Proposition 1.2.9, we know that $\mathcal{G}_0$ and $\mathcal{G}_0'$ are $\mathbb{Q}$-filtrations, so, up to rescaling we may apply Lemma 8.1.8. Since $(N, \mathcal{F})$ and $(N', \mathcal{F}')$ are both weakly admissible we deduce from this lemma that the RHS above is non-positive. Thus we get a contradiction. □

The existence of the HN-filtration also carries over.

Proposition 8.1.10 *Let $(N, \mathcal{F}) \in \mathrm{FilIsoc}_L^K$. Then there is a unique $\mathbb{Q}$-filtration ${}^\bullet N$ of N by sub-isocrystals, such that the graded objects ${}^\alpha\mathrm{gr}(N)$ are semi-stable of slope α. In particular $(N, \mathcal{F})$ is semi-stable if and only if ${}^\bullet N$ has a single jump.*

Proof The proof is the same as the one indicated in Proposition 1.3.1, in the case of the category Fil_k^K. □

As in Chapter I, the canonical filtration ${}^\bullet N$ is functorial and compatible with tensor products. Also, we may again introduce the HN-*vector* and the HN-*polygon* of a filtered isocrystal. We also mention the following functoriality property with respect to L.

Proposition 8.1.11 *Let $L' \supset L$ be a perfect field extension and let $N_{L'} = (V', \Phi')$ be the base change of N in $\mathrm{Isoc}(L')$. If K is an extension of $K_0' = K_0(L')$ and $\mathcal{F}$ is a filtration on $V \otimes_{K_0} K$, then the* HN*-polygon of $(N_{L'}, \mathcal{F})$ lies above the* HN*-polygon of $(N, \mathcal{F})$, with* equality *whenever N is split semi-simple. In the latter case, the* HN*-filtration of $(N_{L'}, \mathcal{F})$ is the base change of the* HN*-filtration of $(N, \mathcal{F})$.*

Proof First we note that $\mu(N, \mathcal{F}) = \mu(N_{L'}, \mathcal{F})$, since base change preserves the isotypical decomposition of any isocrystal. The first assertion on HN-polygons then follows from Corollary 1.3.5. Now if N is split semi-simple, the base change functor induces an isomorphism between the ordered sets of sub-isocrystals of N and $N_{L'}$. This implies the remaining statements. □

Example 8.1.12 The following example illustrates the fact that, if N is not split semi-simple, the HN-polygon can go up under base change.

Let $L = \mathbb{F}_p$ and let $N = (\mathbb{Q}_p^2, \Phi)$ with $\Phi = \begin{pmatrix} 1 & 1 \\ 0 & 1 \end{pmatrix}$. Let $\mathcal{F}^\bullet$ be a filtration with support in $\{0, 1\}$, given by a one-dimensional subspace $\mathcal{F}$ of K^2. Then a little calculation shows that $(N, \mathcal{F})$ is always semi-stable except when $\mathcal{F} = \langle e_1 \rangle$, in which case the HN-vector is equal to $(1, 0)$. Now let $L' = \overline{\mathbb{F}}_p$ and $N' = (K_0'^2, \Phi')$. Then for suitable basis of the form $e_1' = e_1, e_2' = ae_1 + e_2$, the matrix of Φ' is $\Phi' = \begin{pmatrix} 1 & 0 \\ 0 & 1 \end{pmatrix}$. Then an easy calculation shows that $(N', \mathcal{F})$ is always semi-stable, except when $\mathcal{F}$ is of the form $\mathcal{F} = \mathcal{F}_{(a,b)} = \langle ae_1' + be_2' \rangle$ for suitable $a, b \in \mathbb{Q}_p$, in which case the HN-vector is equal to $(1, 0)$. Thus we see that for those subspaces $\mathcal{F}_{(a,b)}$ where $b \neq 0$, the HN-polygon has gone up under base change.

Notes and References The notion of an isocrystal is due to Dieudonné [67]. He classified these objects in the case of an algebraically closed field L. For other proofs see [63], [158], or [224]. The notion of a weakly admissible filtered isocrystal is due to Fontaine [80]. Our definition of a filtered isocrystal is a slight generalization of Fontaine's. In the original definition only finite extensions K/K_0 and $\mathbb{Z}$-filtrations are treated, where in addition the extension K/K_0 is totally ramified. Fontaine introduces a functor from the category of p-adic Galois representations of Gal_K to the category of filtered isocrystals (compare Chapter XI, §4) The image of the crystalline Galois representations are by definition the admissible filtered isocrystals. Fontaine observed that all admissible representations are weakly admissible and conjectured that the converse holds. His conjecture was proved by Colmez and himself in [44]. Other proofs are due to Berger [10], Colmez [43] and Kisin [130]. The fact that the category of crystalline Galois representations is stable under tensor products led Fontaine to his conjecture that the tensor product of weakly admissible filtered isocrystals is again weakly admissible, cf. Theorem 8.1.9.

2 Period domains for GL_n

Let L be a perfect field of characteristic p and let $N = (V, \Phi)$ be an isocrystal over L. Denote by $n = \dim V$ the dimension of V. Let $g : \mathbb{Q} \to \mathbb{Z}_{\geq 0}$ be a multiplicity function with $\sum_x g(x) = n$. Alternatively, let $\nu = (\nu(1)^{(n_1)}, \ldots, \nu(r)^{(n_r)}) \in (\mathbb{Q}^n)_+$ be the attached type with $\mathrm{supp}(g) = \{\nu(1) > \cdots > \nu(r)\}$ and $n_i = g(\nu(i))$.

We consider the associated flag variety $\mathcal{F}(g) = \mathcal{F}(\nu) = \mathcal{F}(V, \nu)$ over K_0 defined in Chapter II.

For any field extension K/K_0, we consider the sets

$$\mathcal{F}(N, \nu)(K)^{\mathrm{ss}} = \{\mathcal{F} \in \mathcal{F}(\nu)(K) \mid (N \otimes_{K_0} K, \mathcal{F}) \text{ is semi-stable }\}$$

resp.

$$\mathcal{F}(N, \nu)(K)^{\mathrm{wa}} = \{\mathcal{F} \in \mathcal{F}(\nu)(K) \mid (N \otimes_{K_0} K, \mathcal{F}) \text{ is weakly admissible }\}$$

of semi-stable resp. weakly admissible points. In accordance with the notation in Part 1, we also sometimes denote these sets by $\mathcal{F}(N, g)(K)^{\mathrm{ss}}$ and $\mathcal{F}(N, g)(K)^{\mathrm{wa}}$. Note that the slope $\mu_{\mathcal{F}}(N)$ is independent of $\mathcal{F} \in \mathcal{F}(\nu)(K)$. It is given by

$$\mu_{\mathcal{F}}(N) = \frac{1}{n}(|\nu| + \deg_{\mathcal{F}_0}(V)).$$

Hence, if $|\nu| \neq -\deg_{\mathcal{F}_0}(V)$, we have $\mathcal{F}(N, \nu)(K)^{\mathrm{wa}} = \emptyset$, while if $|\nu| = -\deg_{\mathcal{F}_0}(V)$, we have $\mathcal{F}(N, \nu)(K)^{\mathrm{wa}} = \mathcal{F}(N, \nu)(K)^{\mathrm{ss}}$.

If $N' \subset N$ is a sub-isocrystal, the locus of points in $\mathcal{F}(\nu)$ where N' violates the semi-stability condition forms a closed subvariety, as in Proposition 2.1.1. But the new feature here is that, in general, there are infinitely many sub-isocrystals (when N is split semi-simple, this occurs as soon as the isocrystal N is not multiplicity-free). So, in this case, the subset

$$\{x \in \mathcal{F}(\nu) \mid \mathcal{F}_x \in \mathcal{F}(N, \nu)(\kappa(x))^{\mathrm{ss}}\}$$

is not Zariski-open in $\mathcal{F}(\nu)$ (here $\mathcal{F}_x$ denotes the stalk at x of the *universal filtration* of type g on V). In other words, the functor $K \mapsto \mathcal{F}(N, \nu)(K)^{\mathrm{ss}}$ is not "represented" by a scheme. Nevertheless, if one restricts this functor to finite extensions, it is "representable" by a rigid-analytic variety. We refer to [183] for a proof of this fact in the classical setting of rigid-analytic geometry, and we give an alternative proof using Berkovich's analytic spaces, cf. [12, 13].

Recall that an analytic space is in particular a locally topologically ringed space. As is customary in Berkovich's papers we will write $\mathscr{H}(x)$ for the completion of the residue field of the local ring at the point x of some analytic space $\mathcal{X}$. This is a non-Archimedean field.

Let $\mathcal{F}(\nu)^{an}$ denote the K_0-analytic space attached to the variety $\mathcal{F}(\nu)$. It is a smooth and compact analytic space. We still write $\mathcal{F}$ for the filtration of type ν on $V \otimes_{K_0} \mathcal{O}_{\mathcal{F}(\nu)^{an}}$ obtained by analytification of the universal filtration over $\mathcal{F}(\nu)$.

Proposition 8.2.1 *Let N be an isocrystal over L and let $\nu \in (\mathbb{Q}^n)_+$.*

(i) *The subset $\{x \in \mathcal{F}(\nu)^{an} \mid \mathcal{F}_x \in \mathcal{F}(N, \nu)(\mathscr{H}(x))^{\mathrm{ss}}\}$ is open in $\mathcal{F}(\nu)^{an}$, hence* *it is the underlying set of a smooth K_0-analytic space which we denote by $\mathcal{F}(N,\nu)^{\mathrm{ss}}$.*

(ii) *If K is a non-Archimedean field over K_0, then $\mathcal{F}(N,\nu)^{\mathrm{ss}}(K) = \mathcal{F}(N,\nu)(K)^{\mathrm{ss}}$.*

(iii) *Let $L' \supset L$ be a perfect field extension and let $N' = (V', \Phi')$ be the base change of N. Let us identify $\mathcal{F}(N',\nu)$ with $\mathcal{F}(N,\nu)\widehat{\otimes}K_0'$. Then $\mathcal{F}(N',\nu)^{\mathrm{ss}}$ is contained in $\mathcal{F}(N,\nu)^{\mathrm{ss}}\widehat{\otimes}K_0'$ with equality if N is split semi-simple.*

Proof For any integer $d \geq 1$, let $\mathbf{Gr}_d(V)$ denote the Grassmannian of d-dimensional subspaces of V. Consider the set

$$\mathcal{S}_d(N) := \mathbf{Gr}_d(V)(K_0)^{\Phi}$$

of all d-dimensional sub-isocrystals of N. We claim that it is a closed, hence a compact, subset of $\mathbf{Gr}_d(V)^{an}$. We will prove this in two steps. We first recall the following facts on the relation between the topological spaces X^{an} and $X(K_0)$ associated with an algebraic variety X over K_0.

- *The topology on $X(K_0)$ is induced from that on X^{an}.* Indeed, for $X = \mathbb{A}^n$, this follows from the definition of X^{an} in [12], 1.5.1. This implies in turn the property for X affine, and then for any X by patching.
- *$X(K_0)$ is closed in X^{an} if K_0 is locally compact, i.e., finite over $\mathbb{Q}_p$.* Indeed, when X is proper, $X(K_0)$ is compact, hence closed in X^{an}. In general we may embed X as an open K_0-subvariety of a proper one, say $\bar{X}$. Then by GAGA [12], 3.4, X^{an} is known to be open in $\bar{X}^{an}$, hence $X(K_0) = \bar{X}(K_0) \cap X^{an}$ is closed in X^{an}.
- *If K_0' is a non-Archimedean field over K_0, the topology on $X(K_0)$ is induced by that on $X(K_0')$, and $X(K_0)$ is closed in $X(K_0')$.* Again this is clear for $X = \mathbb{A}^n$ since K_0 is complete. This implies the property for X affine, and then for general X by patching.

Therefore it is sufficient to prove that the set $\mathcal{S}_d(N)$ is compact for the topology induced by $\mathbf{Gr}_d(V)(K_0)$. We first treat the case where N is split semi-simple. In this case, let s be an integer such that all slopes of N are in $\frac{1}{s}\mathbb{Z}$ and such that $\mathbb{F}_{p^s} \subset L$. Let $N_s = (V_s, \Phi_s)$ be the canonical model of N over $\mathbb{Q}_{p^s}$ defined right before Definition 8.1.3. Then, since all sub-isocrystals of N come from sub-isocrystals of N_s, we have

$$\mathbf{Gr}_d(V)(K_0)^{\Phi} = \mathbf{Gr}_d(V_s)(K_0^{\sigma^s})^{\Phi_s}.$$

Now consider the closed subvariety $\mathbf{Gr}_d'(V_s)$ that parametrizes d-dimensional subspaces W of V_s such that

$$W = \bigoplus_{\alpha}(V_{s,\alpha} \cap W).$$

Then, since $(\Phi_s)^s = p^{s\alpha}$ on $V_{s,\alpha}$, we have $(\Phi_s)^s = 1$ on $\mathbf{Gr}'_d(V_s)$ hence Φ_s defines a Galois descent datum on $\mathbf{Gr}'_d(V_s)$ over $\mathbb{Q}_p$, and $\mathcal{S}_d(N)$ is the set of $\mathbb{Q}_p$-points of the descent of $\mathbf{Gr}'_d(V_s)$ to $\mathbb{Q}_p$ (see [183], p. 27 for more details). In particular it is a closed subset of the compact set $\mathbf{Gr}_d(V_s)(K_0^{\sigma^s})$, hence it is a compact subset of $\mathbf{Gr}_d(V)(K_0)$.

We consider now the case of a general isocrystal N. We can find an extension $L' \supset L$ such that the base change $N' = (V', \Phi')$ of N to $K'_0 = K_0(L')$ is split semi-simple. Then $\mathbf{Gr}_d(V)(K_0)^{\Phi}$ coincides with the intersection $\mathbf{Gr}_d(V)(K_0) \cap \mathbf{Gr}_d(V)(K'_0)^{\Phi'}$ hence it is a closed subset of $\mathbf{Gr}_d(V)(K'_0)^{\Phi'}$ and therefore is compact for the topology induced from $\mathbf{Gr}_d(V)(K_0)$.

Now recall from the proof of Proposition 2.1.1 that on the scheme $\mathbf{Gr}_d(V) \times_{K_0} \mathcal{F}(v)$, the finite range function $(W, \mathcal{F}) \mapsto \mu_{\mathcal{F}}(W)$ is upper semi-continuous. By GAGA [12], 3.4, it follows that the same function remains upper semi-continuous on the analytic space $\mathbf{Gr}_d(V)^{an} \times_{\mathcal{M}(K_0)} \mathcal{F}(v)^{an}$ (here $\mathcal{M}(K_0)$ denotes the analytic space attached to K_0). But since $\mathcal{S}_d(N)$ consists of K_0-rational points of $\mathbf{Gr}_d(V)^{an}$, we have a continuous embedding of $\mathcal{S}_d(N) \times \mathcal{F}(v)^{an}$ (ordinary product of topological spaces) into $\mathbf{Gr}_d(V)^{an} \times_{\mathcal{M}(K_0)} \mathcal{F}(v)^{an}$, which has closed image since $\mathcal{S}_d(N)$ is compact. It follows that the function

$$\begin{array}{rcl} \mathcal{S}_d(N) \times \mathcal{F}(v)^{an} & \longrightarrow & \mathbb{R} \\ (S, x) & \mapsto & \mu_{\mathcal{F}_x}(S \otimes \mathcal{H}(x)) \end{array}$$

is upper semi-continuous for the induced topology on $\mathcal{S}_d(N)$ and the analytic topology on $\mathcal{F}(v)^{an}$ (note that the summand $\mu_{\mathcal{F}_0}(S \otimes \mathcal{H}(x))$ attached to the slope filtration of S is locally constant since it only depends on the isomorphism class of S (and not on x) which, by definition, is locally constant in $\mathcal{S}_d(N)$). Since the function

$$\begin{array}{rcl} \mathcal{F}(v)^{an} & \longrightarrow & \mathbb{R} \\ x & \mapsto & \mu_{\mathcal{F}_x}(N \otimes \mathcal{H}(x)) = \dfrac{1}{n}\Big(\sum_{i \in \mathbb{Q}} ig(i) + \deg_{\mathcal{F}_0}(V)\Big) \end{array}$$

is constant, it follows that the subset

$$\{(S, x) \in \mathcal{S}_d(N) \times \mathcal{F}(v)^{an} \mid \mu_{\mathcal{F}_x}(S \otimes \mathcal{H}(x)) > \mu_{\mathcal{F}_x}(N \otimes \mathcal{H}(x))\}$$

is closed in $\mathcal{S}_d(N) \times \mathcal{F}(v)^{an}$, hence it is compact. Its image $\mathcal{F}_d \subset \mathcal{F}(v)^{an}$ by the second projection is also compact, since this map is continuous. Now, the semi-stable locus considered in statement (i) is the complement of the union of the finitely many $\mathcal{F}_d$ for $d = 1, \ldots, n-1$, hence it is open.

For statement (ii), recall that giving a point in $\mathcal{F}(v)^{an}(K)$ is the same as giving a point $x \in \mathcal{F}(v)^{an}$ together with a continuous morphism of K_0-algebras $\mathcal{H}(x) \longrightarrow K$. The associated filtration on $V \otimes_{K_0} K$ is then obtained by base change from $\mathcal{F}_x$. Since the semi-stability condition is invariant under base change, statement (ii) follows.

Finally, the base change property in the last statement follows from Proposition 8.1.11. □

Definition 8.2.2 The K_0-analytic space $\mathcal{F}(N,\nu)^{\mathrm{ss}}$ is called the *period domain* of (N,ν). If $|\nu| + \deg_{\mathcal{F}_0}(V) = 0$, then we also use the notation $\mathcal{F}(N,\nu)^{\mathrm{wa}}$.

Note that, if we set

$$\nu' := \nu - \frac{|\nu| + \deg_{\mathcal{F}_0}(V)}{n}\mathbf{1}, \tag{2.1}$$

then we have equalities $\mathcal{F}(N,\nu)^{\mathrm{ss}} = \mathcal{F}(N,\nu')^{\mathrm{ss}} = \mathcal{F}(N,\nu')^{\mathrm{wa}}$. Hence period domains for semi-stable filtrations are not more general than period domains for weakly admissible filtrations, but allow more flexibility, e.g. for studying HN-strata.

Remark 8.2.3 Suppose N is split semi-simple and let s be an integer such that all slopes of N are in $\frac{1}{s}\mathbb{Z}$. Then by statement (iii) in Proposition 8.2.1 and the discussion above Definition 8.1.3, the period domain $\mathcal{F}(N,\nu)^{\mathrm{ss}}$ has a canonical model over $\mathbb{Q}_{p^s}$.

We have to compare our analytic period domain with the rigid-analytic one from [183], Prop. 1.36. From [13], 1.6 there is a fully faithful functor $X \mapsto X_0$, from strictly K_0-analytic spaces to rigid spaces over K_0. Note that $\mathcal{F}(N,\nu)^{\mathrm{ss}}$ is strictly K_0-analytic, since it is open in $\mathcal{F}(\nu)^{an}$, which is strictly K_0-analytic (see [13], 1.3).

Proposition 8.2.4 *The rigid space $\mathcal{F}(N,\nu)_0^{\mathrm{ss}}$ is canonically isomorphic to the period domain $\mathcal{F}(N,\nu)^{\mathrm{ss},rig}$ constructed in [183], Prop. 1.36.*

Proof For an (strictly) affinoid K_0-algebra, we denote by $\mathcal{M}(A)$, resp. $\max(A)$, the associated strictly analytic affinoid space, resp. rigid affinoid space. By definition of the Berkovich functor, we have $\mathcal{M}(A)_0 = \max(A)$. Consider the following two set-valued functors on the category of affinoid K_0-algebras.

$$A \mapsto \left\{ \begin{array}{l} \text{filtrations } \mathcal{F}^\bullet \text{ on } A \otimes_{K_0} N \text{ of type } \nu \text{ such that } \forall x \in \mathcal{M}(A), \\ \text{the stalk of } \mathcal{F} \text{ at } x \text{ is a semi-stable filtration of } N \otimes \mathscr{H}(x) \end{array} \right\}$$

and

$$A \mapsto \left\{ \begin{array}{l} \text{filtrations } \mathcal{F}^\bullet \text{ on } A \otimes_{K_0} N \text{ of type } \nu \text{ such that } \forall x \in \max(A), \\ \text{the stalk of } \mathcal{F} \text{ at } x \text{ is a semi-stable filtration of } N \otimes \kappa(x) \end{array} \right\}$$

Here, by filtration we mean a filtration by projective sub-A-modules such that the associated grading is again a projective A-module. By definition of the period domains – analytic or rigid – the first functor is isomorphic to

$$A \mapsto \mathrm{Hom}_{K_0-\mathrm{An}}(\mathcal{M}(A), \mathcal{F}(N,\nu)^{\mathrm{ss}}) = \mathrm{Hom}_{K_0-\mathrm{Rig}}(\max(A), \mathcal{F}(N,\nu)_0^{\mathrm{ss}}),$$

while the second one is isomorphic to

$$A \mapsto \mathrm{Hom}_{K_0-\mathrm{Rig}}(\max(A), \mathcal{F}(N,\nu)^{\mathrm{ss},rig}).$$

Note that for $x \in \max(A)$ we have $\kappa(x) = \mathscr{H}(x)$. In this way we get a morphism $\mathcal{F}(N,\nu)_0^{\mathrm{ss}} \longrightarrow \mathcal{F}(N,\nu)^{\mathrm{ss},rig}$. In order to prove that it is an isomorphism, we have to show that if a filtration of type ν on $A \otimes_{K_0} N$ is semi-stable at each rigid point of $\mathcal{M}(A)$, then it is also semi-stable at each analytic point of $\mathcal{M}(A)$. However, given a sub-isocrystal N' of N, the set of points $x \in \mathcal{M}(A)$ such that $\mu(N', \mathcal{F}_x) > \mu(N, \mathcal{F}_x)$ is *Zariski-closed*. Hence if this set is non-empty, it has to contain a rigid point. □

Variant 8.2.5 (i) Analogous to Variant 1.1.5, we may consider several filtrations $\mathcal{F}_1, \ldots, \mathcal{F}_r$ on V which leads to the category $\mathrm{FilIsoc}_L^r$. Hence the objects of $\mathrm{FilIsoc}_L^r$ are given by tuples $(N, \mathcal{F}_1, \ldots, \mathcal{F}_r)$ where $N \in \mathrm{Isoc}(L)$ and $\mathcal{F}_1, \ldots, \mathcal{F}_r$ are $\mathbb{Q}$-filtrations on V. We define

$$\begin{aligned} \deg(N, \mathcal{F}_1, \ldots, \mathcal{F}_r) &= \deg(V, \mathcal{F}_0) + \sum\nolimits_{i=1}^{r} \deg(V, \mathcal{F}_i) \\ \mu(V, \mathcal{F}_1, \ldots, \mathcal{F}_r) &= \mu(V, \mathcal{F}_0) + \sum\nolimits_{i=1}^{r} \mu(V, \mathcal{F}_i) \end{aligned}$$

and the semi-stability condition in the usual way. Fix an isocrystal N of dimension n and functions $g_i : \mathbb{Q} \to \mathbb{Z}_{\geq 0}$, $i = 1, \ldots, r$, with $\sum_x g_i(x) = n$. Then the subset

$$\{x = (x_1, \ldots, x_r) \in (\mathcal{F}(g_1) \times \cdots \times \mathcal{F}(g_r))^{an} \mid (N \otimes \mathscr{H}(x), \mathcal{F}_{x_1}, \ldots, \mathcal{F}_{x_r}) \text{ is semi-stable}\}$$

is open in $(\mathcal{F}(g_1) \times \cdots \times \mathcal{F}(g_r))^{an}$, hence inherits the structure of a smooth analytic space over K_0, denoted by $\mathcal{F}(N, g_1, \ldots, g_r)^{\mathrm{ss}}$. Then for any non-Archimedean field K over K_0, the set $\mathcal{F}(N, g_1, \ldots, g_r)^{\mathrm{ss}}(K)$ of K-points of this space is the set of r-tuples $(\mathcal{F}_1, \ldots, \mathcal{F}_r) \in \mathcal{F}(g_1)(K) \times \cdots \times \mathcal{F}(g_r)(K)$ for which $(N \otimes K, \mathcal{F}_1, \ldots, \mathcal{F}_r)$ is semi-stable.

(ii) Fontaine's theory of semi-stable p-adic representations suggests the following variant. The category of isocrystals is replaced by the category of triples (V, Φ, N) where (V, Φ) is an isocrystal over L and N is a K_0-linear endomorphism of V satisfying $p\Phi N = N\Phi$. The total slope of a filtered N-isocrystal is by definition that of its underlying isocrystal. Note that the semi-stability condition is generally weaker since subobjects (i.e., N-stable sub-isocrystals) are more scarce. Again, the subset

$$\mathcal{F}(V,\Phi,N,g)^{\mathrm{ss}} := \{x \in \mathcal{F}(\nu)^{an} \mid (V \otimes \mathcal{H}(x), \Phi, N, \mathcal{F}_x) \text{ is semi-stable}\}$$

is open in $\mathcal{F}(\nu)^{an}$. Hence it is the underlying set of a smooth analytic space such that for any non-Archimedean field K over K_0, the set $\mathcal{F}(V,\Phi,N,g)^{\mathrm{ss}}(K)$ is the set of filtrations $\mathcal{F}$ on $V\otimes K$ such that the quadruple $(V,\Phi,N,\mathcal{F})$ is semi-stable.

In contrast to the finite field situation, the semi-stable locus $\mathcal{F}(N,\nu)^{\mathrm{ss}}$ in $\mathcal{F}(\nu)^{an}$ is not stable under the action of the group $\mathrm{Aut}_{K_0}(V)$. But it is stable under the action of the subgroup $\mathrm{Aut}_{\mathrm{Isoc}(L)}(N)$ of *automorphisms of the isocrystal* N. When N is assumed to be split semi-simple, we have the following explicit description. For $\alpha = \frac{r}{s} \in \mathbb{Q}$, let $n_\alpha \geq 0$ be the multiplicity of E^L_α in N, so that $N \cong \bigoplus_\alpha (E^L_\alpha)^{n_\alpha}$ with $n_\alpha \neq 0 \Rightarrow \mathbb{F}_{p^s} \subset L$. Then we get

$$\mathrm{Aut}_{\mathrm{Isoc}(L)}(N) = \prod\nolimits_{\alpha\in\mathbb{Q}} \mathrm{GL}_{n_\alpha}(D_\alpha). \tag{2.2}$$

In particular, $\mathrm{Aut}_{\mathrm{Isoc}(L)}(N)$ is the group of $\mathbb{Q}_p$-rational points of an inner form of a Levi subgroup of $\mathrm{GL}(V)$. In general $\mathrm{Aut}_{\mathrm{Isoc}(L)}(N)$ is the group of $\mathbb{Q}_p$-points of the group functor J on $\mathbb{Q}_p$-algebras defined by

$$J(R) = \{g \in \mathrm{GL}(V \otimes_{\mathbb{Q}_p} R) \mid g\Phi = \Phi g\}$$

for any $\mathbb{Q}_p$-algebra R. By [183], Prop. 1.12, compare also [136], 3.3, this functor is representable by an algebraic group J over $\mathbb{Q}_p$. By the slope decomposition (1.1), page 190, J is of the form $\prod_\alpha J_\alpha$, but it need not be reductive in general. Further, even if J is reductive, it need not be an inner form of a Levi subgroup of $\mathrm{GL}(V)$.

Example 8.2.6 Let $L = \mathbb{F}_p$. In this case an isocrystal $N = (V,\Phi)$ is nothing else but a finite-dimensional $\mathbb{Q}_p$-vector space V together with a $\mathbb{Q}_p$-linear map Φ. Then J can be identified with the centralizer of Φ in $\mathrm{GL}(V)$. This easily yields examples where J is not reductive. Let $N = \bigoplus_\alpha (E^{\mathbb{F}_p}_\alpha)^{n_\alpha}$. Then J is the product of groups J_α, where $J_\alpha = \mathrm{Res}_{L_\alpha|\mathbb{Q}_p} \mathrm{GL}_{n_\alpha}$, where L_α is a totally ramified extension of degree s_α of $\mathbb{Q}_p$, where $\alpha = \frac{r_\alpha}{s_\alpha}$. Hence in this case J is reductive, but not an inner form of a Levi subgroup.

Examples 8.2.7 (i) Let $\nu = (n-1,(-1)^{(n-1)})$ and consider the isocrystal $N = (E_0)^n = (K_0^n, \mathrm{Id}\otimes\sigma)$. Then as in Chapter II, we get $\mathcal{F}(\nu) = \mathbb{P}(V)$, and the corresponding period domain is the *Drinfeld halfspace*

$$\mathcal{F}(N,\nu)^{\mathrm{wa}} = \Omega(V) = \mathbb{P}(V) \setminus \bigcup\nolimits_{H\subsetneq V} \mathbb{P}(H),$$

i.e., the complement of all $\mathbb{Q}_p$-rational hyperplanes in $\mathbb{P}(V)$. Furthermore, in this case we have $J = \mathrm{GL}_n$.

(ii) More generally let $N = (V, \Phi)$ be a trivial isocrystal in the sense that $\Phi = \mathrm{Id} \otimes \sigma$ on $V = K_0^n = \mathbb{Q}_p^n \otimes_{\mathbb{Q}_p} K_0$. Then $J = \mathrm{GL}_n$ and the period domains arising here are of the same type as the period domains in the case of a finite field.

(iii) Let $N \in \mathrm{Isoc}(L)$ be a simple isocrystal, i.e., $N = E_\alpha^L$ with $\alpha = \frac{r}{s} \in \mathbb{Q}$. Let $g : \mathbb{R} \to \mathbb{Z}_{\geq 0}$ be arbitrary with $\sum_x g(x) = s$. Then $\mathcal{F}(N, \nu)^{\mathrm{ss}} = \mathcal{F}(\nu)$ since there are no non-trivial sub-isocrystals. Furthermore, $\mathcal{F}(N, \nu)^{\mathrm{wa}}$ is empty unless $\sum_x x g(x) = r$, in which case it coincides with $\mathcal{F}(N, \nu)^{\mathrm{ss}}$, following a general rule, cf. Theorem 8.2.9 below.

(iv) Let $N = (E_{\frac{1}{2}}^L)^2$ and set $\nu = (1, 1, 0, 0)$, so that $\mathcal{F}(N, \nu) = \mathbf{Gr}_2(K_0^4)$ is the Grassmanian of 2-dimensional subspaces in K_0^4. Then a small computation shows that

$$\mathcal{F}(N, \nu)^{\mathrm{wa}} = \{W \in \mathbf{Gr}_2(K_0^4)^{an} \mid W \textit{ is not an sub-isocrystal of } N\}.$$

(v) Example 8.1.12 illustrates the possibility that the inclusion of $\mathcal{F}(N', \nu)^{\mathrm{ss}}$ in $\mathcal{F}(N, \nu)^{\mathrm{ss}} \widehat{\otimes} K_0'$ in Proposition 8.2.1 (iii) may be strict. Namely for $L = \mathbb{F}_p$, $V = \mathbb{Q}_p^2$, $\Phi = \begin{pmatrix} 1 & 1 \\ 0 & 1 \end{pmatrix}$, $\nu = (1, 0)$, we have $\mathcal{F}(N, \nu)^{\mathrm{ss}} = \mathbb{P}^1 \setminus \{(1, 0)\} = \mathbb{A}^1$. After passing from $L = \mathbb{F}_p$ to $\overline{\mathbb{F}}_p$, we obtain the Drinfeld upper halfspace $\Omega^2 = \mathbb{P}^1 \setminus \mathbb{P}^1(\mathbb{Q}_p)$.

As we have already mentioned above, the set of weakly admissible filtrations may be empty for a trivial reason, namely if $|\nu| + \deg_{\mathcal{F}_0}(V) \neq 0$. In fact there are also more subtle reasons why it might be empty.

Let $N \in \mathrm{Isoc}(L)$ and consider its slope decomposition $N = N_{\alpha(1)} \oplus \cdots \oplus N_{\alpha(t)}$ with $\alpha(1) > \cdots > \alpha(t)$. We set

$$\nu(N) = (\alpha(1)^{(m_1)}, \ldots, \alpha(t)^{(m_t)}) \in (\mathbb{Q}^n)_+ \tag{2.3}$$

where $m_i = \dim N_{\alpha(i)}$.

Definition 8.2.8 The vector $\nu(N) \in (\mathbb{Q})_+^n$ is called the *Newton vector* or *slope vector* of N (it is the type of the opposite filtration to $\mathcal{F}_0$.).

Recall the dominance order $\leq$ on $(\mathbb{Q}^n)_+$, defined right after Theorem 2.3.1.

Theorem 8.2.9 *Let $N \in \mathrm{Isoc}(L)$ of dimension n, and let $\nu \in (\mathbb{Q}^n)_+$. Then we have $\mathcal{F}(N, \nu)^{\mathrm{wa}} \neq \emptyset$ if and only if $\nu \geq \nu(N)$.*

Proof Since the canonical map $\mathcal{F}(\nu)^{an} \longrightarrow \mathcal{F}(\nu)$ is surjective, we see that $\mathcal{F}(N, \nu)^{\mathrm{wa}} \neq \emptyset$ if and only if the locus of points x in the *scheme* $\mathcal{F}(\nu)$ such that $(N, \mathcal{F}_x)$ is weakly admissible is non-empty. We have already seen that this locus need not be Zariski-open, but it is certainly stable by generalization. Hence let

x be the generic point of the scheme $\mathcal{F}(\nu)$. We have $\mathcal{F}(N,\nu)^{\mathrm{wa}} \neq \emptyset$ if and only if $(N, \mathcal{F}_x)$ is semi-stable.

Let us write the type ν in the form $\nu = (\nu_1 \geq \nu_2 \geq \cdots \geq \nu_n)$ with n the dimension of V. Now let $N' \subset N$ be a sub-isocrystal of dimension n'. Then the type ν' of $\mathcal{F}_x$ on N' has to be the "generic one," i.e., $\nu' = (\nu_{n-n'+1} \geq \nu_{n-n'+2} \geq \cdots \geq \nu_n)$. In fact, otherwise the generic point would be contained in a generalized Bruhat cell different from the big cell, a contradiction.

Let us also write $\nu(N)$ in the form $\nu(N) = (\alpha_1 \geq \alpha_2 \geq \cdots \geq \alpha_n)$. Then $\nu(N')$ is of the form $(\alpha_{j_1} \geq \alpha_{j_2} \geq \cdots \geq \alpha_{j_{n'}})$ for some subset $\{j_1, \ldots, j_{n'}\} \subseteq \{1, \ldots, n\}$. As we have already noted, we may assume that $|\nu| = |\nu(N)|$. Then N' violates the semi-stability of $(N, \mathcal{F}_x)$ if and only if $|\nu'| > |\nu(N')|$. In this case we get the inequalities

$$\sum_{i=1}^{n'} \nu_{n-n'+i} > \sum_{i=1}^{n'} \alpha_{j_i} \geq \sum_{i=1}^{n'} \alpha_{n-n'+i}.$$

Since $|\nu| = |\nu(N)|$, this is equivalent to $\sum_{i=1}^{n-n'} \nu_i < \sum_{i=1}^{n-n'} \alpha_i$, which shows that the inequality $\nu \geq \nu(N)$ does not hold.

Conversely, assume that the inequality $\nu \geq \nu(N)$ does not hold. Thus there is an n' with $0 < n' < n$, such that

$$\sum_{i=1}^{n'} \nu_{n-n'+i} > \sum_{i=1}^{n'} \alpha_{n-n'+i}.$$

Let us take n' minimal with this property. Then $\nu_{n-n'+1} > \alpha_{n-n'+1}$. Define $k_0 \in \{1, \ldots, t\}$ by the condition

$$\sum_{k>k_0} m_k < n' \leq \sum_{k \geq k_0} m_k$$

and put $n'' := \sum_{k \geq k_0} m_k$. Then we also have

$$\sum_{i=1}^{n''} \nu_{n-n''+i} > \sum_{i=1}^{n''} \alpha_{n-n''+i}.$$

But this says that the sub-isocrystal $N'' := \bigoplus_{k \geq k_0} N_{\alpha(k)}$ of N violates the semi-stability of $(N, \mathcal{F}_x)$. □

Remark 8.2.10 In [82] it is proved that if L is algebraically closed, then one can find points of $\mathcal{F}(N,\nu)^{\mathrm{wa}}$ over K_0 provided that $\nu \geq \nu(N)$. This statement need not be true any longer if L is not algebraically closed, even when N is split semi-simple, as shown by the Drinfeld halfspace for GL_2, see Example 8.2.7 (i).

As a conclusion, we can decide when the set of semi-stable points is non-empty. Recall that $|\mu| = \mu_1 + \cdots + \mu_n$ for all $\mu \in \mathbb{Q}^n$.

Corollary 8.2.11 *Let $N = (V, \Phi)$ be an isocrystal of dimension n and let $\nu \in (\mathbb{Q}^n)_+$. Then $\mathcal{F}(N, \nu)^{ss} \neq \emptyset$ if and only if*

$$\begin{aligned} \nu_1 + \frac{|\nu(N)|}{n} &\geq \nu(N)_1 + \frac{|\nu|}{n} \\ \nu_1 + \nu_2 + \frac{2|\nu(N)|}{n} &\geq \nu(N)_1 + \nu(N)_2 + \frac{2|\nu|}{n} \\ &\vdots \\ \nu_1 + \cdots + \nu_{n-1} + \frac{(n-1)|\nu(N)|}{n} &\geq \nu(N)_1 + \cdots + \nu(N)_{n-1} + \frac{(n-1)|\nu|}{n}. \end{aligned} \tag{2.4}$$

Proof We reconsider the twisted Hodge vector $\nu' := \nu - \frac{|\nu| + \deg_{\mathcal{F}_0}(V)}{n}\mathbf{1}$, cf. (2.1), page 199. Note that $|\nu(N)| = -\deg_{\mathcal{F}_0}(V)$. As we have already pointed out, we have $\mathcal{F}(N, \nu')^{wa} = \mathcal{F}(N, \nu')^{ss} \cong \mathcal{F}(N, \nu)^{ss}$. By Theorem 8.2.9 we deduce that $\mathcal{F}(N, \nu')^{wa} \neq \emptyset$ if and only if

$$\begin{aligned} \nu'_1 &\geq \nu(N)_1 \\ \nu'_1 + \nu'_2 &\geq \nu(N)_1 + \nu(N)_2 \\ &\vdots \\ \nu'_1 + \cdots + \nu'_{n-1} &\geq \nu(N)_1 + \cdots + \nu(N)_{n-1} \\ |\nu'| &= |\nu(N)|. \end{aligned}$$

But

$$\nu'_1 + \cdots + \nu'_i = \nu_1 + \cdots + \nu_i - \frac{i}{n}(|\nu| - |\nu(N)|),$$

for $1 \leq i \leq n$. The corollary follows. □

Example 8.2.12 Let $N = (V, \Phi)$ be an *isotypical* isocrystal, i.e., $N = (E_\alpha^L)^m$ for $\alpha \in \mathbb{Q}$, $m \in \mathbb{N}$. Then $\mathcal{F}(N, \nu)^{ss} \neq \emptyset$. Furthermore, $\mathcal{F}(N, \nu)^{wa} \neq \emptyset$ if and only if $\dim V \cdot \alpha = |\nu|$.

Notes and References In the general setting, the first p-adic period domains were studied by van der Put and Voskuil [220] in the case of a trivial isocrystal, i.e., all slopes are zero. In [220], formal adic models of the rigid-analytic varieties defined by period domains are constructed in certain (rare) cases, as a generalization of the Drinfeld case, cf. also Chapter XI. The definition given here appears in [183], except that there these spaces are defined as rigid-analytic varieties, whereas here we defined them as analytic spaces. We also mention

[185], [186] for more information on the geometry of period domains. Further, these papers mainly consider the case of an algebraically closed field L. We refer to [183], Prop. 1.36 for the rigid-analytic version of Proposition 8.2.1 and to [183], Prop. 1.34 for a description of an admissible open covering.

Period domains have also canonical counterparts in the language of Huber's adic spaces. We note that the proof of Proposition 8.2.1 can be adapted easily to the adic setting. In [108], Hartl shows that the rigid-analytic period domain from [183] comes from an analytic space via Berkovich's functor, and moreover that this analytic space is an open analytic domain in $\mathcal{F}(\nu)^{an}$. By Proposition 8.2.4, it follows that his analytic space coincides with ours. However, Hartl's method is different since he uses the open admissible covering from [183], 1.34 and associates to each member of this covering an open subspace of the analytic flag space. The resulting (analytic) period domain is then given by pasting these analytic subspaces together.

In [196] Schneider and Stuhler discuss the Drinfeld halfspace on the level of rigid-analytic varieties. They show that it has even the structure of a rigid-analytic Stein space, cf. Chapter XI.

Theorem 8.2.9 is proved in [82] in the case of an algebraically closed field L. It is not known whether $\mathcal{F}(N, \nu)$, if non-empty, has points over a finite extension of $\mathbb{Q}_p$.

Variant 8.2.5 (ii), involving the monodromy operator N, is also considered by Totaro [219]. He proves that the $\otimes$-product theorem also holds in this variant.

3 The Harder–Narasimhan stratification

We continue with the setting of the last section. In particular, we have the isocrystal $N = (V, \Phi)$, where $n = \dim V$ and we have fixed a type $\nu \in (\mathbb{Q}^n)_+$.

Let K/K_0 be a field extension and let $\mathcal{F} \in \mathcal{F}(\nu)(K)$. As in Proposition 8.1.10, $\mathcal{F}$ gives rise to the HN-filtration $({}^{\alpha}N)_{\alpha\in\mathbb{Q}}$ of N by sub-isocrystals ${}^{\alpha}N = ({}^{\alpha}V, \Phi)$, with the characteristic property that $({}^{\alpha}\mathrm{gr}(N) \otimes K, \mathcal{F})$ is semi-stable of slope α. Let g_α be the multiplicity function of the filtration $\mathcal{F}$ on ${}^{\alpha}\mathrm{gr}(N) \otimes K$. Similarly, let h be the *multiplicity function of the slope filtration* $\mathcal{F}_0$ on the isocrystal N, and let h_α be that of ${}^{\alpha}\mathrm{gr}(N)$. Thus in terms of the slope decomposition of N,

$$h(x) = \dim N_x \text{ for all } x \in \mathbb{Q}$$

and $h(x)$ is divisible by $\dim E_x^L$. Ditto for all h_α, $\alpha \in \mathbb{Q}$. Then we get the following equalities:

- $g = \sum_\alpha g_\alpha$ and $h = \sum_\alpha h_\alpha$,
- $\dim({}^{\alpha}\,\mathrm{gr}(N)) = \sum_x g_\alpha(x) = \sum_x h_\alpha(x)$,

• $\alpha = \frac{1}{\dim({}^{\alpha}\mathrm{gr}(N))} \left(\sum_x x(g_\alpha(x) - h_\alpha(x))\right)$.

This leads to the following notion of a HN-*type*.

Definition 8.3.1 Let $N = (V, \Phi) \in \mathrm{Isoc}(L)$ and let $g : \mathbb{Q} \longrightarrow \mathbb{Z}_{\geq 0}$ be a multiplicity function, defining the type ν. An HN-*type* associated to (N, ν) is a pair $\theta = (\underline{h}, \underline{g})$ consisting of tuples $\underline{h} = (h_1, \ldots, h_s)$ and $\underline{g} = (g_1, \ldots, g_s)$ of multiplicity functions $h_i, g_j : \mathbb{Q} \to \mathbb{Z}_{\geq 0}$, which satisfy the following conditions.

(i) For each $i = 1, \ldots, s$, we have $m_i := \sum_x h_i(x) = \sum_x g_i(x) > 0$.

(ii) $\sum_{i=1}^s h_i = h$ and $\sum_{i=1}^s g_i = g$.

(iii) $\frac{1}{m_1} \sum_x x(g_1(x) - h_1(x)) > \frac{1}{m_2} \sum_x x(g_2(x) - h_2(x)) > \cdots > \frac{1}{m_s} \sum_x x(g_s(x) - h_s(x))$.

(iv) For all $i = 1, \ldots, s$ and all $\alpha \in \mathbb{Q}$, the integer $h_i(\alpha)$ is divisible by $\dim E_\alpha^L$.

We denote by $\Theta(N, g)$ or $\Theta(N, \nu)$ the finite set of HN-types associated to the multiplicity function (N, g) or the corresponding type ν. To an element $\theta \in \Theta(N, g)$ we associate its HN-*vector* $\mathrm{HN}(\theta) \in (\mathbb{Q}^n)_+$ by setting

$$\mathrm{HN}(\theta) = \left(\left(\frac{1}{m_1} \sum_x x(g_1(x) - h_1(x))\right)^{(m_1)}, \ldots, \left(\frac{1}{m_s} \sum_x x(g_s(x) - h_s(x))\right)^{(m_s)}\right). \tag{3.1}$$

Let $\mathcal{H}(N, \nu)$ or $\mathcal{H}(N, g)$ be the set of HN-vectors thus obtained.

We denote by θ_0 the distinguished element of $\Theta(N, g)$ consisting of the singleton (h, g). It is the unique element with HN-vector equal to the scalar vector $\frac{|\nu| - |\nu(N)|}{n} \cdot \mathbf{1}$.

Remark 8.3.2 The above definition of an HN-type applies to any pair (N, ν), but is best adapted in the case where N is split semi-simple. If N is not split semi-simple, then we will get a sharper definition of an HN-type in the next chapter, cf. Example 9.6.13 below.

Example 8.3.3 Let $N = (K_0^n, \mathrm{id} \otimes \sigma)$. Then $h(x) = 0$ for all $x \neq 0$ and $h(0) = n$. In this case the additional datum $\underline{h} = (h_1, \ldots, h_s)$ is superfluous. Thus an HN-type coincides with the notion of an HN-type in the case of a finite field in Chapter VI.

As we have seen above, any filtration $\mathcal{F} \in \mathcal{F}(\nu)(K)$ determines via its associated HN-filtration an HN-type $\theta_{\mathcal{F}}$ and hence also an HN-vector $\mathrm{HN}(\theta_{\mathcal{F}})$. Thus we obtain, as in the finite field situation, decompositions into finite disjoint unions

$$\mathcal{F}(\nu)(K) = \dot{\bigcup_\theta} \mathcal{F}(N, \nu)(K)_\theta, \quad \mathcal{F}(\nu)(K) = \dot{\bigcup_H} \mathcal{F}(N, \nu)(K)_H . \tag{3.2}$$

Here θ ranges over all HN-types attached to ν, and H over all HN-vectors. Again the first decomposition is a refinement of the second decomposition and we have

$$\mathcal{F}(N,\nu)(K)^{\mathrm{ss}} = \mathcal{F}(N,\nu)(K)_{\theta_0}.$$

So far our discussion has been only set-theoretical. In order to turn to the analytic properties, we introduce a partial pre-order on $\Theta(N,\nu)$ through the dominance order (as defined below Theorem 2.3.1) on their associated HN-vectors, cf. (3.1), page 206,

$$\theta \le \theta' \iff \mathrm{HN}(\theta) \le \mathrm{HN}(\theta'). \tag{3.3}$$

Note that θ_0 is the unique minimal element of $\Theta(N,\nu)$ provided that $\mathcal{F}(N,\nu)^{\mathrm{ss}} \neq \emptyset$. For the following statement we use the same notation as in Proposition 8.2.1.

Proposition 8.3.4 *The function*

$$\begin{array}{ccc} \mathcal{F}(\nu)^{an} & \longrightarrow & \Theta(N,\nu) \\ x & \mapsto & \theta_{\mathcal{F}_x} \end{array}$$

is upper semi-continuous for the analytic topology on $\mathcal{F}(\nu)^{an}$. This is equivalent to either of the following properties, where we put

$$\mathcal{F}(N,\nu)_\theta := \{x \in \mathcal{F}(\nu)^{an} \mid \mathcal{F}_x \in \mathcal{F}(\nu)(\mathcal{H}(x))_\theta\}$$

for $\theta \in \Theta(N,g)$:

(i) $\dot{\bigcup}_{\theta' \le \theta}\mathcal{F}(N,\nu)_{\theta'}$ *is open in $\mathcal{F}(\nu)^{an}$,*
(ii) $\dot{\bigcup}_{\theta' \ge \theta}\mathcal{F}(N,\nu)_{\theta'}$ *is closed in $\mathcal{F}(\nu)^{an}$.*

In particular, $\mathcal{F}(N,\nu)_{\mathrm{HN}(\theta)} = \dot{\bigcup}_{\theta \le \theta' \le \theta}\mathcal{F}(N,\nu)_{\theta'}$ is locally closed in $\mathcal{F}(\nu)^{an}$.

Proof We will prove statement (i) along the same lines as the proof of Proposition 8.2.1, using the same notation. From the definition of the HN-filtration and Proposition 1.3.4, we have the following criterion to detect whether a point x has HN-type below θ:

$$\theta_{\mathcal{F}_x} \le \theta \text{ if and only if}$$
$$\forall S \subset N, \text{ the point } (\dim S, \deg_{\mathcal{F}_x}(S \otimes \mathcal{H}(x))) \text{ lies below the HN-polygon of } \theta.$$

Here S ranges over sub-isocrystals of N. In other words, if $f_\theta : [0,n] \longrightarrow \mathbb{R}$ is the function whose graph is the HN-polygon of θ, we have:

$$\theta_{\mathcal{F}_x} \le \theta \text{ if and only if } \deg_{\mathcal{F}_x}(S \otimes \mathcal{H}(x)) \le f_\theta(\mathrm{rk}(S)), \quad \forall S \subset N.$$

Now, as in the proof of Proposition 8.2.1, the subset

$$\{(S,x) \in \mathcal{S}_d(N) \times \mathcal{F}(\nu)^{an} \mid \mu_{\mathcal{F}_x}(S \otimes \mathcal{H}(x)) > f_\theta(d)/d\}$$

is closed in $\mathcal{S}_d(N) \times \mathcal{F}(\nu)^{an}$, hence it is compact, hence so is its image $\mathcal{F}_{d,\theta} \subset \mathcal{F}(\nu)^{an}$ by the second projection. But as we have just mentioned, the subset $\{x \in \mathcal{F}(\nu)^{an} \mid \theta_{\mathcal{F}_x} \leq \theta\}$ is the complement of the union of the finitely many $\mathcal{F}_{d,\theta}$'s for $d = 1, \ldots, n-1$, hence it is open. □

In order to study the HN-type strata $\mathcal{F}(N,\nu)_\theta$, we may decompose them further, exactly as in the case of vector spaces. More precisely, adding to the picture the HN-flag associated to a HN-filtration, we have for any extension field K of K_0, the disjoint union

$$\mathcal{F}(N,\nu)(K)_\theta = \dot{\bigcup}_{{}^\bullet N} \mathcal{F}(N,\nu)(K)_{\theta,{}^\bullet N}.$$

Here ${}^\bullet N$ ranges over all flags $0 = N_0 \subset N_1 \subset \cdots \subset N_s = N$ of sub-isocrystals of N such that the multiplicity function of the slope filtration of N_i/N_{i-1} is h_i, where $\theta = (\underline{h}, \underline{g})$. Note that the subset $\mathcal{F}(N,\nu)(K)_\theta$ is stable under the action of the automorphism group $\mathrm{Aut}_{\mathrm{Isoc}(L)}(N) = J(\mathbb{Q}_p)$.

If N is a split semi-simple isocrystal, then J is reductive and any two flags as above are conjugate under $J(\mathbb{Q}_p)$. More precisely, let $P_{\mathrm{GL}(V)}({}^\bullet N)$ be the parabolic subgroup of $\mathrm{GL}(V)$ stabilizing the flag ${}^\bullet N$ and put

$$P_J({}^\bullet N) := P_{\mathrm{GL}(V)}({}^\bullet N) \cap J(\mathbb{Q}_p).$$

This is the group of $\mathbb{Q}_p$-rational points of some parabolic subgroup of J. Then the subset $\mathcal{F}(N,\nu)(K)_{\theta,{}^\bullet N}$ is stable under the action of $P_J({}^\bullet N)$ and we have a bijection

$$\begin{array}{ccc} J(\mathbb{Q}_p) \times^{P_J({}^\bullet N)} \mathcal{F}(N,\nu)(K)_{\theta,{}^\bullet N} & \longrightarrow & \mathcal{F}(N,\nu)(K)_\theta \\ (j,x) & \mapsto & jx \end{array}.$$

For an HN-type $\theta = (\underline{h}, \underline{g})$, define the Weyl group element $w_\theta \in S_n$ to be the one in Corollary 2.3.10 by forgetting $\underline{h}$.

Proposition 8.3.5 *Let $\theta = (\underline{h}, \underline{g})$ be an* HN*-type and fix some flag ${}^\bullet N$ of sub-isocrystals as above. Put*

$$\mathcal{F}(N,\nu)_{\theta,{}^\bullet N} := \{x \in \mathcal{F}(\nu)^{an} \mid (N \otimes \mathcal{H}(x), \mathcal{F}_x) \in \mathcal{F}(N,\nu)(\mathcal{H}(x))_{\theta,{}^\bullet N}\}.$$

Then the following statements hold.

(i) *This is a locally closed K_0-*analytic *subset of $\mathcal{F}(\nu)^{an}$.*

(ii) *If $\mathcal{F}(N,\nu)_{\theta,{}^\bullet N}$ is non-empty, then the map*

$$\pi : \mathcal{F}(N,\nu)_{\theta,{}^\bullet N} \longrightarrow \prod_{i=1}^{s} \mathcal{F}(\mathrm{gr}^i({}^\bullet N), g_i)^{ss}$$

which sends a filtration $\mathcal{F}^\bullet$ to the restrictions $(\mathcal{F}^\bullet | \mathrm{gr}^i({}^\bullet N))_{i=1,\ldots,s}$, is a K_0-analytic $P_J({}^\bullet N)$-equivariant vector bundle of rank $\ell(w_\theta)$.

(iii) *We have*

$$\overline{\mathcal{F}(N,\nu)_{\theta,{}^\bullet N}} \setminus \mathcal{F}(N,\nu)_{\theta,{}^\bullet N} \subset \bigcup\nolimits_{H>HN(\theta)} \mathcal{F}(N,\nu)_H.$$

(iv) *If N is split semi-simple, then the map*

$$J(\mathbb{Q}_p) \times^{P_J({}^\bullet N)} \mathcal{F}(N,\nu)_{\theta,{}^\bullet N} \longrightarrow \mathcal{F}(N,\nu)_\theta$$

is a homeomorphism and $\mathcal{F}(N,\nu)_\theta$ is a closed subset of $\mathcal{F}(N,\nu)_{\mathrm{HN}(\theta)}$.

Proof Here by a locally closed K_0-analytic subset we mean a subset such that every point has an open neighborhood U with two finite sets of analytic functions $\{f_i\}_i, \{g_j\}_j$ on U such that the set in question is defined inside U by the vanishing of the functions f_i and the non-vanishing of the functions g_j.

Let $\mathcal{F}(\nu)_{\underline{g},{}^\bullet N}$ be the locally closed subvariety of $\mathcal{F}(\nu)$ consisting of filtrations $\mathcal{F}^\bullet$ such that $\mathcal{F}^\bullet|\,\mathrm{gr}^i({}^\bullet N) \in \mathcal{F}(\mathrm{gr}^i({}^\bullet N), g_i)$ for $i = 1,\dots,s$. This is a generalized Bruhat cell. In Chapter II (3.12), page 49, we saw that the map

$$\pi : \mathcal{F}(\nu)_{\underline{g},{}^\bullet N} \longrightarrow \prod\nolimits_{i=1}^{s} \mathcal{F}(\mathrm{gr}^i({}^\bullet N), g_i)$$

which sends a flag $\mathcal{F}^\bullet$ to the restrictions $(\mathcal{F}^\bullet|\,\mathrm{gr}^i({}^\bullet N))_{i=1,\dots,s}$, is an algebraic $P_{\mathrm{GL}(V)}({}^\bullet N)$-equivariant vector bundle of K_0-varieties of rank $\ell(w_\theta)$. By analytification we get a K_0-analytic vector bundle

$$\pi^{an} : \mathcal{F}(\nu)^{an}_{\underline{g},{}^\bullet N} \longrightarrow \prod\nolimits_{i=1}^{s} \mathcal{F}(\mathrm{gr}^i({}^\bullet N), g_i)^{an}$$

and by definition, the preimage of $\prod_{i=1}^{s} \mathcal{F}(\mathrm{gr}^i({}^\bullet N), g_i)^{\mathrm{ss}}$ is $\mathcal{F}(N,\nu)_{\theta,{}^\bullet N}$. The latter is thus an open subset of $\mathcal{F}(\nu)^{an}_{\underline{g},{}^\bullet N}$, hence it inherits a K_0-analytic structure satisfying properties (i) and (ii) of the proposition.

Let us admit property (iii) for a while and prove the statements of point (iv). Since $P^J({}^\bullet N)$ preserves $\overline{\mathcal{F}(N,\nu)_{\theta,{}^\bullet N}}$, the action map $J(\mathbb{Q}_p) \times \overline{\mathcal{F}(N,\nu)_{\theta,{}^\bullet N}} \longrightarrow \mathcal{F}(\nu)^{an}$ factors through a map

$$J(\mathbb{Q}_p) \times^{P_J({}^\bullet N)} \overline{\mathcal{F}(N,\nu)_{\theta,{}^\bullet N}} \longrightarrow \mathcal{F}(\nu)^{an}$$

which is continuous since the action of $J(\mathbb{Q}_p)$ on $\mathcal{F}(\nu)^{an}$ is continuous. Under the split semi-simple assumption, we know that J is a reductive group and $P^J({}^\bullet N)$ is a parabolic subgroup, hence the factor set $J(\mathbb{Q}_p)/P^J({}^\bullet N)$ is compact. It follows that the source space of the above map is compact. Thus its image is also compact and coincides therefore with $\overline{\mathcal{F}(N,\nu)_\theta}$. Then property (iii) implies that

$$\overline{\mathcal{F}(N,\nu)_\theta} \setminus \mathcal{F}(N,\nu)_\theta \subset \bigcup_{H>\mathrm{HN}(\theta)} \mathcal{F}(N,\nu)_H.$$

In particular $\mathcal{F}(N,\nu)_\theta$ coincides with $\mathcal{F}(N,\nu)_{\mathrm{HN}(\theta)} \cap \overline{\mathcal{F}(N,\nu)_\theta}$. Hence it is a closed subset of $\mathcal{F}(N,\nu)_{\mathrm{HN}(\theta)}$. Now we consider the map

$$J(\mathbb{Q}_p) \times^{P_J(^\bullet N)} \mathcal{F}(N,\nu)_{\theta,^\bullet N} \longrightarrow \mathcal{F}(N,\nu)_\theta \tag{3.4}$$

of point (iv). It is certainly bijective and continuous, hence it remains to prove that it is closed (in the sense that it takes closed subsets to closed subsets). The target is now known to be locally closed in $\mathcal{F}(\nu)^{an}$, so it is in particular locally compact. Therefore it is sufficient to prove that the above map is proper. Again, since $P^J(^\bullet N)$ is a parabolic subgroup of the reductive group J, there exists by the Iwasawa decomposition [216], 3.3.2 a compact subgroup C of $J(\mathbb{Q}_p)$ such that $C.P^J(^\bullet N) = J(\mathbb{Q}_p)$. Then the action map $C \times \mathcal{F}(N,\nu)_\theta \longrightarrow \mathcal{F}(N,\nu)_\theta$ is certainly proper. Since, by point (iii), $\mathcal{F}(N,\nu)_{\theta,^\bullet N}$ is closed in $\mathcal{F}(N,\nu)_\theta$, the restricted action map $C \times \mathcal{F}(N,\nu)_{\theta,^\bullet N} \longrightarrow \mathcal{F}(N,\nu)_\theta$ also is proper, and thereby so is the map (3.4), page 210.

Now we turn to property (iii). We have already seen at the beginning of the proof that $\mathcal{F}(N,\nu)_{\theta,^\bullet N}$ is the intersection of $\mathcal{F}(N,\nu)_\theta$ and the Bruhat cell $\mathcal{F}(\nu)^{an}_{\underline{g},^\bullet N}$. The closure of this generalized Bruhat cell is a *generalized Schubert variety*. More precisely, it is a disjoint union

$$\overline{\mathcal{F}(\nu)^{an}_{\underline{g},^\bullet N}} = \dot{\bigcup}_{\underline{g}' \leq \underline{g}} \mathcal{F}(\nu)^{an}_{\underline{g}',^\bullet N}$$

of Bruhat cells associated to other multiplicity tuples $\underline{g}' = (g'_1, \ldots, g'_s)$ such that $\sum_x g'_i(x) = m_i$, for any $i = 1, \ldots, s$ and $\sum_i g'_i = g'$. Here the order $\underline{g}' \leq \underline{g}$ is the *Bruhat order* among Bruhat cells. In concrete terms, we have

$$\underline{g}' \leq \underline{g} \Leftrightarrow \left(\sum_{x \geq y} \sum_{j \leq i} g'_j(x) \geq \sum_{x \geq y} \sum_{j \leq i} g_j(x), \quad \forall i = 1, \ldots, s, \ \forall y \in \mathbb{Q} \right). \tag{3.5}$$

Note that we only need the easy implication $\Rightarrow$, which follows from the usual degeneration rule for ranks of intersections of subbundles. The converse implication is in fact more subtle.

Now let z be a point in some $\mathcal{F}(\nu)^{an}_{\underline{g}',^\bullet N}$. The degree of the ith step in the filtration $^\bullet N$ by sub-isocrystals is given by the formula

$$\begin{aligned} \deg(N_i, \mathcal{F}_z) &= \sum_x \sum_{j \leq i} x(g'_j(x) - h_j(x)) \\ &= \sum_{l=1}^{r-1} (x_l - x_{l+1}) \left(\sum_{j \leq i} \sum_{x \geq x_l} g'_j(x) \right) + x_r \left(\sum_{j \leq i} m_j \right) - \sum_x \sum_{j \leq i} x h_j(x). \end{aligned}$$

Here $\mathrm{supp}(g) = \{x_1, \ldots, x_r\}$. On the other hand, the HN-polygon associated with the HN-vector $\mathrm{HN}(\theta)$ is the piecewise linear function which interpolates the origin and the s points $(\dim_{K_0}(N_i), \sum_x \sum_{j \leq i} x(g_j(x) - h_j(x))$. Hence, by the inequalities (3.5) above, we see that the convex polygon hull of the s points $(\dim_{K_0}(N_i), \deg(N_i, \mathcal{F}_z))$ lies *above* the HN-polygon of $\mathrm{HN}(\theta)$ and *strictly above* as soon as $\underline{g}' \neq \underline{g}$. Applying then Corollary 1.3.5 (which clearly holds for isocrystals instead of vector spaces), we get

$$\mathcal{F}(\nu)^{an}_{\underline{g}',\bullet N} \subset \bigcup_{H>\mathrm{HN}(\theta)} \mathcal{F}(N,\nu)_H$$

for any $\underline{g}' < \underline{g}$, and

$$\mathcal{F}(\nu)^{an}_{\underline{g},\bullet N} \setminus (\mathcal{F}(\nu)^{an}_{\underline{g},\bullet N} \cap \mathcal{F}(N,\nu)_{\mathrm{HN}(\theta)}) \subset \bigcup_{H>\mathrm{HN}(\theta)} \mathcal{F}(N,\nu)_H.$$

So it remains only to prove that $\mathcal{F}(\nu)^{an}_{\underline{g},\bullet N} \cap \mathcal{F}(N,\nu)_{\mathrm{HN}(\theta)} = \mathcal{F}(N,\nu)_{\theta,\bullet N}$. But if z is in this intersection, then the piecewise linear function that interpolates the s points $(\dim_{K_0}(N_i), \deg(N_i, \mathcal{F}_z))$ is a convex polygon which coincides with the HN-polygon of $(N, \mathcal{F}_z)$. It follows by immediate induction that each $(N_i/N_{i-1}, \mathcal{F}_z)$ is semi-stable and therefore that ${}^\bullet N$ is the HN-flag of $(N, \mathcal{F}_z)$. □

Remark 8.3.6 The same proof of part (iii) works also in the finite field case. However, in the latter situation we can (and did) use the valuative criterion Theorem 2.3.1, which is not available for analytic spaces. On the other hand, the valuative criterion is available for adic spaces, cf. [121], Cor. 1.3.9. So in this setting, we could apply the proof of the finite field case for proving part (iii).

Remark 8.3.7 If N is not split semi-simple, then it is still true that the HN-type stratum is closed in its associated HN-vector stratum. This will follow from the general case which is studied in Theorem 9.6.8. In Chapter IX we will introduce a sharper stratification and a notion of HN-types which is different from the present one, when N is not split semi-simple.

Corollary 8.3.8 *If N is split semi-simple, then $\mathcal{F}(N,\nu)_\theta$ is locally closed in $\mathcal{F}(\nu)^{an}$. Moreover, for any non-Archimedean field K over K_0, we have*

$$\begin{aligned}\mathcal{F}(N,\nu)(K)_\theta = \mathcal{F}(N,\nu)_\theta(K) :=&\\ \{y \in \mathcal{F}(\nu)^{an}(K) \text{ whose image belongs to } \mathcal{F}(N,\nu)_\theta\}.&\end{aligned}$$

The locally closed subset $\mathcal{F}(N,\nu)_\theta$ is called the HN-*stratum* to θ.

HN-strata are in general not K_0-analytic subsets of $\mathcal{F}(\nu)^{an}$. Here is the simplest example.

Example 8.3.9 Let $N = (K_0^2, \mathrm{id} \otimes \sigma)$ and let $\nu \in (\mathbb{Q}^2)_+$ be regular. Thus we are in the situation of Example 8.2.7 (i) with $\mathcal{F}(N,\nu) = \mathbb{P}^1$. Then we obtain as

HN-stratification $\mathbb{P}^1(K) = (\mathbb{P}^1(K) \setminus \mathbb{P}^1(\mathbb{Q}_p)) \,\dot{\bigcup}\, \mathbb{P}^1(\mathbb{Q}_p)$. But $\mathbb{P}^1(\mathbb{Q}_p)$ is not an analytic subset of $(\mathbb{P}^1)^{an}$.

Remark 8.3.10 We also note that the proofs above are easily adapted to the setting of Huber's adic spaces, but here again the strata are not adic subspaces. Nevertheless, it follows from the isomorphism (3.4), page 210 that $\mathcal{F}(N,\nu)_\theta$ is a *pseudo-adic* subset of $\mathcal{F}(\nu)^{ad}$, i.e., a subset which is locally pro-constructible and convex with respect to the specializing order of points, cf. [121], Def. 1.10.3.

Let us mention the following compatibility with respect to base change. Recall that if $L' \supset L$ is a perfect field extension we have denoted by $N_{L'} = (V', \Phi')$ the corresponding base change of N, cf. (1.2), page 190. We note that by definition we have $\Theta(N,\nu) = \Theta(N_{L'},\nu)$. Further, we may identify $\mathcal{F}(V',\nu)$ with $\mathcal{F}(V,\nu)\widehat{\otimes}K_0'$.

Proposition 8.3.11 *With the above notation, for any* HN-*vector* H,

$$\dot{\bigcup}_{H' \leq H} \mathcal{F}(N_{L'},\nu)_{H'} \subseteq \left(\dot{\bigcup}_{H' \leq H} \mathcal{F}(N,\nu)_{H'}\right)\widehat{\otimes}\, K_0'.$$

If N *is split semi-simple, then*

$$\mathcal{F}(N_{L'},\nu)_\theta = \mathcal{F}(N,\nu)_\theta\widehat{\otimes}K_0'$$

for any $\theta \in \Theta(N,\nu)$.

Proof Indeed, this follows from Proposition 8.1.11. □

Let us now give an example, where we enumerate all HN-types for a given pair (N,ν).

Example 8.3.12 Let $\nu = (\nu_1,\nu_2,\nu_3) \in (\mathbb{Q}^3)_+$ with $\nu_1 > \nu_2 > \nu_3$, and let $N = E_{\frac{1}{2}} \oplus E_0$. Thus $\nu(N) = (\frac{1}{2},\frac{1}{2},0) \in (\mathbb{Q}^3)_+$ and $\mathcal{F}(\nu)$ is the complete flag variety of K_0^3. Further, we have by (2.2), page 201, the identity $J(\mathbb{Q}_p) = D^\times_{\frac{1}{2}} \times \mathbb{Q}_p^\times$.

For any integer $i \in \{1,2,3\}$, let $g_i : \mathbb{Q} \longrightarrow \mathbb{Z}_{\geq 0}$ be the multiplicity function with $g_i(\nu_i) = 1$ and $g_i(x) = 0$ for all $x \neq \nu_i$. For a subset $\{i,j\} \subset \{1,2,3\}$, let $g_{i,j} : \mathbb{Q} \longrightarrow \mathbb{Z}_{\geq 0}$ be the multiplicity function with $g_{i,j}(\nu_i) = g_{i,j}(\nu_j) = 1$ and $g_{i,j}(x) = 0$ for $x \notin \{\nu_i,\nu_j\}$. Further, let $h_1 : \mathbb{Q} \longrightarrow \mathbb{Z}_{\geq 0}$ be the multiplicity function with $h_1(\frac{1}{2}) = 2$ and $h_1(x) = 0$ for $x \neq \frac{1}{2}$ and define $h_2 : \mathbb{Q} \longrightarrow \mathbb{Z}_{\geq 0}$ by $h_2(0) = 1$ and $h_2(x) = 0$ for $x \neq 0$. The non-empty strata are indexed by the following elements of $\Theta = \Theta(N, \nu)$:

$$\begin{aligned}
\theta_0 &= (h, g) \in \Theta \Leftrightarrow 2\nu_1 \geq \frac{1}{2} + \nu_2 + \nu_3 \text{ and } \nu_1 + \nu_2 \geq 1 + 2\nu_3 \\
\theta_1 &= ((h_1, h_2), (g_{1,2}, g_3)) \in \Theta \Leftrightarrow \nu_1 + \nu_2 > 2\nu_3 + 1 \\
\theta_2 &= ((h_1, h_2), (g_{1,3}, g_2)) \in \Theta \Leftrightarrow \nu_1 + \nu_3 > 2\nu_2 + 1 \\
\theta_3 &= ((h_2, h_1), (g_1, g_{2,3})) \in \Theta \\
\theta_4 &= ((h_2, h_1), (g_2, g_{1,3})) \in \Theta \Leftrightarrow 2\nu_2 + 1 > \nu_1 + \nu_3 \\
\theta_5 &= ((h_2, h_1), (g_3, g_{1,2})) \in \Theta \Leftrightarrow 2\nu_3 + 1 > \nu_1 + \nu_2 \,.
\end{aligned}$$

The strict inequalities follow from the requirement that the slopes of the HN-filtration are monotone decreasing. The inequalities for θ_0 are induced by Corollary 8.2.11. The subquotients of the HN-filtration of type $\theta_1, \ldots, \theta_5$ are all isotypic; hence, by Example 8.2.12, the inequalities of Corollary 8.2.11 are automatically satisfied in these cases.

Notes and References More generally than Theorem 8.2.9, one may ask for which HN-vectors the corresponding HN-stratum is non-empty. This question is raised in [82]. An answer to this problem is given in [173], even in the refined form of determining the set of HN-*types* with non-empty corresponding stratum, in the case that L is algebraically closed.

The relation of period domains to generalized Bruhat cells and a description of their closures appear in [169].

4 The relation to GIT

Next we want to describe the relation of period domains to GIT in the p-adic setting. Let L be a perfect field. Fix a pair (N, ν) consisting of a *split semi-simple* isocrystal $N = (V, \Phi)$ of dimension n over L and a vector $\nu \in (\mathbb{Q}^n)_+$. Write $\nu = (\nu(1)^{(n_1)}, \ldots, \nu(r)^{(n_r)})$ with $\nu(1) > \nu(2) > \cdots > \nu(r)$ and $n_1, \ldots, n_r \in \mathbb{N}$. Denote by $\mathcal{F}(\nu)$ the corresponding flag variety over $\mathbb{Q}_p$. Let $\nu(N) = (\alpha_1^{(\dim N_{\alpha_1})}, \ldots, \alpha_t^{(\dim N_{\alpha_t})}) \in (\mathbb{Q}^n)_+$ be the Newton vector as in (2.3), page 202. Set

$$\nu_0 = (-\alpha_t^{(\dim N_{\alpha_t})}, \ldots, -\alpha_1^{(\dim N_{\alpha_1})}) \in (\mathbb{Q}^n)_+.$$

Let $\mathcal{F}_0 = \mathcal{F}(\nu_0)$ be the algebraic variety parametrizing all filtrations of type ν_0 on V. By construction we have $\mathcal{F}_0 \in \mathcal{F}_0(K_0)$, where $\mathcal{F}_0$ was defined in (1.4), page 192. Consider the closed embedding

$$i : \mathcal{F}(\nu) \hookrightarrow \mathcal{F}(\nu) \times \mathcal{F}_0,$$

given by the identity on the first factor and by the K_0-rational point $\mathcal{F}_0$ of $\mathcal{F}_0$ on the second factor.

Let $\mathcal{L}(\nu)$ be the homogeneous $\mathbb{Q}$-line bundle on $\mathcal{F}(\nu)$ of Chapter II (2.4), page 38. Further, let $\mathcal{L}_0 := \mathcal{L}(\nu_0)$ be the ample $\mathbb{Q}$-line bundle on the flag variety $\mathcal{F}_0$ defined by the same procedure applied to ν_0. We let

$$\mathcal{L} := i^*(\mathcal{L}(\nu) \boxtimes \mathcal{L}_0) \tag{4.1}$$

be the induced $\mathbb{Q}$-line bundle on $\mathcal{F}$. We consider $\mathcal{L}$ as a J_{K_0}-equivariant line bundle. Its fiber in a point $\mathcal{F}$ is given by

$$\bigotimes_i \det(\mathrm{gr}_{\mathcal{F}}^{\nu(i)}(V))^{\otimes - \nu(i)} \otimes \bigotimes_i \det(\mathrm{gr}_{\mathcal{F}_0}^{-\alpha_i}(V))^{\otimes \alpha_i}.$$

Let K/K_0 be a field extension and let $x \in \mathcal{F}(\nu)(K)$. For any 1-PS $\lambda : \mathbb{G}_m \to J$, let $\mu^{\mathcal{L}}(x, \lambda)$ be the slope of x with respect to λ and the $\mathbb{Q}$-line bundle $\mathcal{L}$. The p-adic analogue of Theorem 2.2.3 is given as follows.

Theorem 8.4.1 *Let K be a field extension of K_0. Let $\mathcal{F}$ be $\mathbb{Q}$-filtration on V_K which corresponds to a point $x \in \mathcal{F}(\nu)(K)$. Then*

(i) *$(N, \mathcal{F})$ is semi-stable, i.e., $x \in \mathcal{F}(N, \nu)^{\mathrm{ss}}(K)$ if and only if $\mu^{\mathcal{L}}(x, \lambda) \geq 0$ for all* 1-PS *λ of J defined over $\mathbb{Q}_p$ which factor through* SL(V).

(ii) *$(N, \mathcal{F})$ is weakly admissible, i.e., $x \in \mathcal{F}(N, \nu)^{\mathrm{wa}}(K)$ if and only if $\mu^{\mathcal{L}}(x, \lambda) \geq 0$ for all* 1-PS *λ of J defined over $\mathbb{Q}_p$.*

The proof is the same as that of Theorem 2.2.3. It uses the following lemma. Let λ be a 1-PS of J defined over $\mathbb{Q}_p$. Then we get a $\mathbb{Z}$-grading of N by sub-isocrystals,

$$N = \bigoplus_\alpha N_\alpha^\lambda.$$

This gives a descending filtration $\mathcal{F}_\lambda$ on N by sub-isocrystals. Conversely, any filtration on N by sub-isocrystals arises in this way, cf. also Proposition 9.3.7 below.

Lemma 8.4.2 *Let K be a field extension of K_0. Let $[\mathcal{F}] \in \mathcal{F}(\nu)(K)$ and let $\lambda \in X_*(J)$. Let $\mathcal{F}_\lambda$ be the induced filtration on N. Then*

$$\mu^{\mathcal{L}}([\mathcal{F}], \lambda) = -\big(\langle \mathcal{F}, \mathcal{F}_\lambda \rangle + \langle \mathcal{F}_0, \mathcal{F}_\lambda \rangle\big).$$

Proof The proof is the same as the conjunction of Lemmas 2.2.1 and 2.2.2. □

Proof of Theorem 8.4.1 Part (i) is proved in the same way as Theorem 2.2.3. Thus we need only to prove part (ii). Here we have to show that $\mu^{\mathcal{L}}(x, \lambda) \geq 0$ for all 1-PS λ of J defined over $\mathbb{Q}_p$ which factor through the centre $Z(\mathrm{GL}(V))$ of $\mathrm{GL}(V)$ if and only if $\mu_{\mathcal{F}_x}(N) = 0$. For this, we note that these 1-PS correspond exactly to the trivial filtrations $\mathcal{F}^i_\lambda = N$ for $i \leq j$ and $\mathcal{F}^i_\lambda = (0)$ for $i > j$ for some integer $j \in \mathbb{Z}$. Then we get, by using Lemma 8.4.2 and Lemma 8.1.8, that $\mu^{\mathcal{L}}(x, \lambda) = j \cdot \deg_{\mathcal{F}_x}(N)$. The statement follows at once. □

Notes and References The proof of Theorem 8.4.1 is modeled on Totaro's proof [219]. A similar criterion holds for the variants considered in Variant 8.2.5, cf. Theorem 2.2.7.

5 Isocrystals with coefficients

In this section we outline a variant of the previous results where we replace the base field $\mathbb{Q}_p$ by a finite extension F.

Definition 8.5.1 Let L be a perfect field of characteristic p. An isocrystal over L *with coefficients* in F is a pair (N, κ) consisting of an isocrystal N over L and a morphism of $\mathbb{Q}_p$-algebras $\kappa : F \longrightarrow \mathrm{End}(N)$.

The isocrystals over L with coefficients in F form in an obvious way an F-linear abelian category $\mathrm{Isoc}(L)_F$. The tensor product of (N, κ) and (N', κ') is defined as the quotient of $N \otimes_{\mathbb{Q}_p} N'$ by the $\mathbb{Q}_p$-linear span of elements of the form $\kappa(x)(v) \otimes v' - v \otimes \kappa'(x)(v')$ for $x \in F$, $v \in V$ and $v' \in V'$, equipped with the embedding of F induced indifferently by κ or κ'. In this way the category $\mathrm{Isoc}(L)_F$ becomes a tannakian category over F.

Now we present another way of looking at these objects. Let $q_F = p^{f_F}$ be the cardinality of the residue field of F. Further, let p^f be the cardinality of the residue field of the σ^{f_F}-fixed subfield $K_0^{\sigma^{f_F}}$ of K_0, so that in general f divides f_F. Note that $f = f_F$ if L is algebraically closed. Then the tensor product algebra $K_0 \otimes_{\mathbb{Q}_p} F$ decomposes as a product of fields

$$K_0 \otimes_{\mathbb{Q}_p} F = \prod_\iota K_0 \otimes_{K_0^{\sigma^f}, \iota} F,$$

where ι runs over all embeddings of the subfield $K_0^{\sigma^f}$ of σ^f-fixed points in K_0 into F. Moreover, the action of $\sigma \otimes \mathrm{id}$ on the LHS permutes the set of such embeddings. Let us fix an embedding ι_0 and define

$$K_F = K_0(L)_F := K_0 \otimes_{K_0^{\sigma^f}, \iota_0} F. \tag{5.1}$$

This is an unramified extension of F, equipped with the automorphism σ_F induced by $\sigma^f \otimes \mathrm{id}$.

Definition 8.5.2 A σ_F-K_F-*space* is a pair (V_F, Φ_F) consisting of a finite-dimensional vector space V_F over K_F together with a σ_F-linear automorphism Φ_F of V_F.

The σ_F-K_F-spaces form a tannakian category σ_F-K_F-Sp over F and we will now show that this tannakian category is equivalent to $\mathrm{Isoc}(L)_F$. This requires further notation.

Notation 8.5.3 • We denote by ε_0 the idempotent of $K_0 \otimes_{\mathbb{Q}_p} F$ associated with the embedding ι_0 defining K_F. Thus we have $K_F = \varepsilon_0(K_0 \otimes_{\mathbb{Q}_p} F)$.

• We denote by τ_0 the composition $K_F \xrightarrow{\sim} \varepsilon_0(K_0 \otimes_{\mathbb{Q}_p} F) \hookrightarrow K_0 \otimes_{\mathbb{Q}_p} F$. This is a *non-unital* morphism of rings.

• We obtain a diagonal *unital* embedding

$$\tau = (\tau_0, (\sigma \otimes \mathrm{id}) \circ \tau_0, \ldots, (\sigma \otimes \mathrm{id})^{f-1} \circ \tau_0) : K_F \hookrightarrow K_0 \otimes_{\mathbb{Q}_p} F.$$

In the last item we use the decomposition

$$K_0 \otimes_{\mathbb{Q}_p} F = \prod_{i \in \mathbb{Z}/f\mathbb{Z}} K_F^{(i)}.$$

where $K_F^{(i)}$ is the field $K_0 \otimes_{K_0^{\sigma^f}, \iota_0 \circ \sigma^i} F$, which is also cut out by the idempotent

$$\varepsilon_i := (\sigma \otimes \mathrm{id})^i(\varepsilon_0).$$

The action of $\sigma \otimes \mathrm{id}$ on $K_0 \otimes_{\mathbb{Q}_p} F$ induces F-linear and K_0-σ-linear field isomorphisms $K_F^{(i)} \xrightarrow{\sim} K_F^{(i+1)}$ which will be again denoted by σ. Their f-fold composition yields the automorphism σ_F of K_F. We note that

$$(\sigma \otimes \mathrm{id}) \circ \tau = (\sigma_F \times \mathrm{id}_{K_F^{(1)}} \times \cdots \times \mathrm{id}_{K_F^{(f-1)}}) \circ \tau.$$

We define a functor

$$\begin{array}{rccc} \varepsilon_{0,*} : & \mathrm{Isoc}(L)_F & \longrightarrow & \sigma_F\text{-}K_F\text{-Sp} \\ & (V, \Phi, \kappa) & \mapsto & (\varepsilon_0 V, \Phi^f) \end{array} . \tag{5.2}$$

In the RHS, the datum κ is implicitly used for the definition of the K_F-vector space $\varepsilon_0 V$. Moreover, since Φ^f is $K_0^{\sigma^f}$-linear, it certainly leaves $\varepsilon_0 V$ stable.

Conversely we define a functor

$$\begin{array}{rccl} \tau_* : & \sigma_F\text{-}K_F\text{-Sp} & \longrightarrow & \mathrm{Isoc}(L)_F \\ & (V_F, \Phi_F) & \mapsto & (V_F \otimes_{K_F,\tau} (K_0 \otimes_{\mathbb{Q}_p} F), (\Phi_F, \mathrm{id}, \dots, \mathrm{id}), \kappa) \end{array} . \quad (5.3)$$

In the RHS,the notation $(\Phi_F, \mathrm{id}, \dots, \mathrm{id})$ refers to the product decomposition $V_F \otimes_{K_F,\tau} (K_0 \otimes_{\mathbb{Q}_p} F) = \prod_{i \in \mathbb{Z}/f\mathbb{Z}} V_F \otimes_{K_F} K_F^{(i)}$ and κ is given by the obvious action of F. The proof of the following lemma contains another point of view on these functors.

Lemma 8.5.4 *The above functors induce inverse equivalences of tannakian categories between* σ_F-K_F-Sp *and* $\mathrm{Isoc}(L)_F$.

Proof Of course this can be checked directly. But it is also instructive to remark that this is an instance of Morita equivalence.

Let us denote by $K_F\langle\Phi_F\rangle$ the skew Laurent polynomial ring in one variable Φ_F with relations $\Phi_F\lambda = {}^{\sigma_F}\lambda\Phi_F$ for $\lambda \in K_F$. Then the category σ_F-K_F-Sp can be identified with the category of left modules over this ring which are finite-dimensional as K_F-vector spaces. We note that the tensor structure on σ_F-K_F-Sp is induced by the comultiplication which takes Φ_F to $\Phi_F \otimes \Phi_F$. Further, the dual object is given by the σ-linear antipode which takes Φ_F to Φ_F^{-1}.

Similarly the category $\mathrm{Isoc}(L)_F$ is equivalent to the category of left modules over the skew Laurent polynomial ring $(K_0 \otimes_{\mathbb{Q}_p} F)\langle\Phi\rangle$ with relations $\Phi\lambda = {}^{\sigma}\lambda\Phi$, for $\lambda \in K_0 \otimes F$ which are finite-dimensional over K_0. We note that $\Phi^i \cdot \varepsilon_0 = \varepsilon_{\bar{i}} \cdot \Phi^i$ for any integer $i \in \mathbb{Z}$, where $\bar{i} \in \mathbb{Z}/f\mathbb{Z}$ is i modulo f, and where $\varepsilon_{\bar{i}}$ is the idempotent introduced earlier. Define

$$P_{0,F} := (K_0 \otimes_{\mathbb{Q}_p} F)\langle\Phi\rangle\varepsilon_0 .$$

This is a projective monogeneous left $(K_0 \otimes_{\mathbb{Q}_p} F)\langle\Phi\rangle$-module. Its endomorphism ring $(\varepsilon_0(K_0 \otimes_{\mathbb{Q}_p} F)\langle\Phi\rangle\varepsilon_0)^{\mathrm{opp}}$ is identified with $K_F\langle\Phi_F\rangle^{\mathrm{opp}}$ via the ring isomorphism

$$\begin{array}{rcl} K_F\langle\Phi_F\rangle & \longrightarrow & \varepsilon_0(K_0 \otimes_{\mathbb{Q}_p} F)\langle\Phi\rangle\varepsilon_0 \\ \sum_n \lambda_n \Phi_F^n & \mapsto & \sum_n \tau_0(\lambda_n)\Phi^{fn} \cdot \varepsilon_0 \end{array} .$$

In this way $P_{0,F}$ appears as a finite free right $K_F\langle\Phi_F\rangle$-module of rank f. More precisely, we have

$$P_{0,F} = \bigoplus_{i=0}^{f-1} \Phi^i \cdot \tau_0(K_F)\langle\Phi^f\rangle.$$

Further, since $\varepsilon_i = \Phi^i\varepsilon_0\Phi^{-i}$, we have $(K_0 \otimes_{\mathbb{Q}_p} F)\langle\Phi\rangle\varepsilon_0(K_0 \otimes_{\mathbb{Q}_p} F)\langle\Phi\rangle = (K_0 \otimes_{\mathbb{Q}_p}$

$F)\langle\Phi\rangle$. Hence we may apply Morita's theorem [46]. It says that the functors $\mathrm{Hom}_{(K_0\otimes F)\langle\Phi\rangle}(P_{0,F},-)$ and $P_{0,F}\otimes_{K_F\langle\Phi_F\rangle}-$ are inverse equivalences of categories.

However, these functors are nothing other than $\varepsilon_{0,*}$ and τ_*, respectively. For $\varepsilon_{0,*}$ the assertion is clear, while for τ_* it follows from the following alternative description

$$P_{0,F} = \bigoplus_{i=0}^{f-1} K_F^{(i)} \cdot \Phi^{-i}\langle\Phi^f\rangle,$$

which exhibits $P_{0,F}$ as a free $K_0 \otimes_{\mathbb{Q}_p} F$-module of rank 1. We leave it to the reader to check that these equivalences preserve the tensor structures. □

Let us fix some uniformizer ϖ in F. To any rational number $\alpha = \frac{r}{s} \in \mathbb{Q}$, we associate the σ-L-space

$$E_\alpha^{L,F} := (K_F^s, A_\alpha^\varpi \cdot \sigma_F),$$

where the matrix A_α^ϖ has the same shape as the matrix A_α above in Definition 8.1.2, with p replaced by ϖ. If L is algebraically closed, then the category σ_F-K_F-Sp is semi-simple, the spaces $E_\alpha^{L,F}$, $\alpha \in \mathbb{Q}$, form a set of representatives of isomorphism classes of simple objects, and the endomorphism algebra of $E_\alpha^{L,F}$ is "the" central division algebra of invariant $-\alpha$ over F [63].

If L is only perfect, then we still have a slope decomposition and we may define split semi-simple objects as in the case of isocrystals. The Newton degree $\deg(N_F)$ of a σ_F-K_F-space $N_F = (V_F, \Phi_F)$ is then defined as the negative of the degree of the slope filtration of N_F, compare Remark 8.1.7. As in the case of isocrystals, we have the formula

$$\deg N_F = \mathrm{ord}_F(\det(\Phi_F|V_F)).$$

Let us compare this with the degree of the corresponding isocrystal $N_0 = (V_0, \Phi_0, \kappa) \in \mathrm{Isoc}(L)_F$. By the explicit formula in (5.3), page 217 we have

$$\deg N_0 = \mathrm{ord}_p(\det(\Phi_F|V_{F|K_0})),$$

where, on the RHS, V_F is considered as a K_0-vector space. However, we have

$$\mathrm{ord}_p(\det(\Phi_F|V_{F|K_0}) = \mathrm{ord}_p(N_{K_F|K_0}\det(\Phi_F|V_F)) = \mathrm{ord}_F(\det(\Phi_F)),$$

and therefore we get $\deg N_0 = \deg N_F$. It follows that

$$\mu(N_0) = \frac{1}{[F:\mathbb{Q}_p]}\mu(N_F). \tag{5.4}$$

Now if $\mathcal{F}_F$ is a filtration on $V_F \otimes_{K_F} K$ for some extension field K of K_F, one defines the total slope

$$\mu(N_F, \mathcal{F}_F) := \mu_{\mathcal{F}_F}(V_F) - \mu(N_F)$$

as in the case of isocrystals. We shall associate to $\mathcal{F}_F$ a filtration $\mathcal{F}_0$ on $V_0 \otimes_{K_0} K$.

First, since K contains F we have a decomposition of the algebra $F \otimes_{\mathbb{Q}_p} K$ as a product $K \times K'$, where K' is a product of fields. Denote by ξ_0 the corresponding idempotent of $F \otimes_{\mathbb{Q}_p} K$. We note that it is finer than ε_0, i.e., we have $\xi_0 \varepsilon_0 = \xi_0$. Moreover, we have a canonical identification of K-vector spaces $\xi_0(V_0 \otimes_{K_0} K) \simeq V_F \otimes_{K_F} K$. Then we define the filtration $\mathcal{F}_0$ on $V_0 \otimes_{K_0} K$ as

$$\mathcal{F}_0 = \mathcal{F}_F \oplus \mathcal{F}^{\text{triv}} \text{ on } \quad \xi_0(V_0 \otimes_{K_0} K) \oplus (1 - \xi_0)(V_0 \otimes_{K_0} K),$$

i.e., the direct sum of $\mathcal{F}_F$ on the summand $\xi_0(V_0 \otimes_{K_0} K)$ and the trivial filtration on the complement $(1 - \xi_0)(V_0 \otimes_{K_0} K)$. Then we clearly have $\deg_{\mathcal{F}_0}(V_0) = \deg_{\mathcal{F}_F}(V_F)$ and therefore we get

$$\mu(N_0, \mathcal{F}_0) = \frac{1}{[F : \mathbb{Q}_p]} \mu(N_F, \mathcal{F}_F). \tag{5.5}$$

Of course we define the semi-stability of filtered σ_F-K_F-spaces as usual. Then the reader can check that all the results on period domains and HN-stratification for filtered isocrystals have obvious analogues for filtered σ_F-K_F-spaces. However, we do not give any more details here since we will see in the next chapter, Section 8, that these period domains also have an interpretation as period domains for "isocrystals with G-structure," for a reductive group over $\mathbb{Q}_p$ of the form $G = \mathrm{Res}_{F|\mathbb{Q}_p}(\mathrm{GL}_{n,F})$.

One ingredient of this re-interpretation will be the following lemma.

Lemma 8.5.5 *The filtered σ_F-K_F-space $(N_F, \mathcal{F}_F)$ is semi-stable if and only if $(N_0, \mathcal{F}_0)$ is semi-stable as an ordinary filtered isocrystal (without coefficients).*

Proof First of all we note that, for any subobject N_F' of N_F, the restriction of $\mathcal{F}_0$ to the corresponding subobject N_0' of N_0 (given by the equivalence in Lemma 8.5.4 between σ_F-K_F-spaces and isocrystals with coefficients in F) is the filtration $\mathcal{F}_0'$ associated to the restriction $\mathcal{F}_F'$ of $\mathcal{F}_F$ to N_F'. It follows that the pair $(N_F, \mathcal{F}_F)$ is semi-stable if and only if the pair $(N_0, \mathcal{F}_0)$ is semi-stable as a filtered isocrystal *with coefficients in* F.

Since any "sub-isocrystal with coefficients" of N_0 is in particular a sub-isocrystal, we see that the "if" part of the statement of the lemma is clear. Conversely, assume that $(N_0, \mathcal{F}_0)$ is semi-stable as an isocrystal with coefficients in F. Since Φ is F-linear and since the filtration $\mathcal{F}_0$ is F-stable (due to its definition), for any $x \in F$ and any sub-isocrystal N_0' of N_0, the image $x(N_0')$ is again a sub-isocrystal and we have $\mu(N_0', \mathcal{F}_0) = \mu(x(N_0'), \mathcal{F}_0)$. In particular, the sub-isocrystal of N_0 which has maximal slope with respect to $\mathcal{F}$ (i.e., the first step of the HN-filtration) is F-stable. By our semi-stability hypothesis, it coincides with N_0. Therefore $(N_0, \mathcal{F}_0)$ is a semi-stable filtered isocrystal. □

Notes and References The notion of a σ_F-K_F-space appears in Kottwitz [135]. Kottwitz states (without proof) the content of Lemma 8.5.4. In [183] period domains are defined in the context of σ_F-K_F-spaces, but it is already mentioned there that this variant does not yield a new theory since it is realized by considering isocrystals with G-structure, cf. the next chapter.

IX. Period Domains for p-adic Reductive Groups

In this chapter we consider period domains for arbitrary reductive groups G over $\mathbb{Q}_p$. Recall from the Introduction that the motivation comes from the arithmetic applications, which may involve more general reductive groups than GL_n, such as unitary or symplectic groups. Again we follow the Tannaka formalism already used in Part 2. We consider a conjugacy class $\mathcal{N}$ of $\mathbb{Q}$-1-PS of $G_{\overline{\mathbb{Q}}_p}$ which gives rise to a flag variety $\mathcal{F}(G, \mathcal{N})$ over the reflex field $E = E(G, \mathcal{N})$ of $(G, \mathcal{N})$. The additional data of an isocrystal is now replaced by a G-isocrystal. However, to ensure stability under the HN-formalism, we have to broaden the context a little, by replacing the traditional G-isocrystals by "augmented group schemes over Isoc(L)," which are groups over K_0 endowed with a Weil descent datum to $\mathbb{Q}_p$ and some "augmentation".

As in the previous chapter, we will consider period domains as analytic spaces rather than rigid-analytic varieties.

1 C-isocrystals and group schemes over Isoc(L)

Let $K_0 = K_0(L)$ for a perfect field L of characteristic p.

Definition 9.1.1 Let C be a Tannaka category over $\mathbb{Q}_p$. A *C-isocrystal* over L is an exact faithful and $\mathbb{Q}_p$-linear tensor functor $N : C \longrightarrow \mathrm{Isoc}(L)$ from C into the category of isocrystals over L.

By their definition as functors between categories, C-isocrystals form a category C-Isoc(L), with natural transformations as morphisms. By Proposition 4.1.10 all morphisms in this category are invertible, i.e., C-Isoc(L) is simply a groupoid.

For our purposes, the most important case is when $C = \mathrm{Rep}_{\mathbb{Q}_p}(G)$ for some algebraic group G over $\mathbb{Q}_p$. In this case C comes with a fiber functor ω^G to

$\mathrm{Vec}_{\mathbb{Q}_p}$ and on the other hand, $\mathrm{Isoc}(L)$ comes with a natural fiber functor ω_{iso} : $\mathrm{Isoc}(L) \longrightarrow \mathrm{Vec}_{K_0}$.

Definition 9.1.2 Let G be a linear algebraic group over $\mathbb{Q}_p$. A *G-isocrystal* is a $\mathrm{Rep}_{\mathbb{Q}_p}(G)$-isocrystal $N : \mathrm{Rep}_{\mathbb{Q}_p}(G) \longrightarrow \mathrm{Isoc}(L)$ which commutes (strictly) with fiber functors. More precisely, we ask that $\omega_{\mathrm{iso}} \circ N = \omega^G \otimes_{\mathbb{Q}_p} K_0$.

We denote by G-$\mathrm{Isoc}(L)$ the corresponding full subcategory of the category $\mathrm{Rep}_{\mathbb{Q}_p}(G)$-$\mathrm{Isoc}(L)$, which is a groupoid as well.

We can produce G-isocrystals as follows. Let $b \in G(K_0)$. For a finite-dimensional rational representation V of G, we put

$$N_b(V) := \left(V \otimes_{\mathbb{Q}_p} K_0, b(\mathrm{id}_V \otimes \sigma)\right).$$

For a homomorphism $f : V_1 \to V_2$ of G-representations, let

$$N_b(f) := f \otimes \mathrm{id}_{K_0} : N_b(V_1) \to N_b(V_2)$$

be the induced morphism. This construction yields a G-isocrystal N_b. Moreover, any $g \in G(K_0)$ defines an isomorphism $N_b \longrightarrow N_{gb\sigma(g)^{-1}}$ in an obvious way.

Definition 9.1.3 Let G be a linear algebraic group over $\mathbb{Q}_p$. We denote by $\mathcal{B}(G, L)$ the category whose objects are elements b in $G(K_0)$ and whose morphisms $b \longrightarrow b'$ are elements g in $G(K_0)$ such that $b' = gb\sigma(g)^{-1}$.

Clearly, the category $\mathcal{B}(G, L)$ is a groupoid. Hence we have a functor of categories (groupoids)

$$\mathcal{B}(G, L) \longrightarrow G\text{-}\mathrm{Isoc}(L). \tag{1.1}$$

Lemma 9.1.4 *The functor* (1.1) *above is an* isomorphism *of categories (not merely an equivalence).*

Proof Let b, b' be elements in $G(K_0)$. By composing with the forgetful functor ω_{iso}, a given isomorphism $N_b \longrightarrow N_{b'}$ induces an automorphism of the fiber functor $\omega^G \otimes K_0$. By Tannaka duality, Theorem 4.1.18, such an isomorphism is given by the action of an element $g \in G(K_0)$, which then has to σ-conjugate b into b'. Hence the functor is full. Moreover, since any $g \in G(K_0)$ acts non-trivially on the fiber functor $\omega^G \otimes K_0$, the functor is also faithful.

Now let us start with a G-isocrystal N. Since $\omega_{\mathrm{iso}} \circ N = \omega^G \otimes K_0$, we may write $N(V) = (V_{K_0}, \Phi_V)$ for $V \in \mathrm{Rep}_{\mathbb{Q}_p}(G)$. Then the collection of $\Phi_V \circ (\mathrm{id} \otimes \sigma^{-1})$ for $V \in \mathrm{Rep}_{\mathbb{Q}_p}(G)$ defines an automorphism of the fiber functor $\omega^G \otimes K_0$, hence it is given by the action of some $b \in G(K_0)$, i.e., $N = N_b$. Thus the functor (1.1) is surjective. Finally, $N_b = N_{b'}$ if and only if $\rho(b) = \rho(b')$ for any representation ρ of G. But this implies that $b = b'$, again by Tannaka duality. □

In particular, if V is a $\mathbb{Q}_p$-vector space, then a GL(V)-isocrystal is the same as the structure of an isocrystal on $V \otimes K_0$. Indeed any σ-linear bijective map Φ on $V \otimes K_0$ has a unique factorization $\Phi = b(\mathrm{id}_V \otimes \sigma)$ for some $b \in \mathrm{GL}(V)(K_0)$.

On the other hand, the following lemma clarifies the difference between G-isocrystals and $\mathrm{Rep}_{\mathbb{Q}_p}(G)$-isocrystals.

Lemma 9.1.5 *The obvious functor G-Isoc(L) $\longrightarrow$ $\mathrm{Rep}_{\mathbb{Q}_p}(G)$-Isoc($L$) is fully faithful. It is an equivalence if and only if $H^1(K_0, G)$ is trivial.*

Proof By definition G-Isoc(L) is a full subcategory of $\mathrm{Rep}_{\mathbb{Q}_p}(G)$-Isoc($L$), so we only have to deal with the essential surjectivity. Now, if $H^1(K_0, G) = \{1\}$, it follows from Theorem 4.1.23 that any two fiber functors on $\mathrm{Rep}_{\mathbb{Q}_p}(G)$ over K_0 are isomorphic. In particular the composition of any $\mathrm{Rep}_{\mathbb{Q}_p}(G)$-isocrystal N with ω_{iso} is isomorphic to $\omega^G \otimes K_0$. Any isomorphism $\omega_{\mathrm{iso}} \circ N \xrightarrow{\sim} \omega^G \otimes K_0$ induces an isomorphism from N to a G-isocrystal by the previous lemma. □

Remark 9.1.6 Here are some examples where $H^1(K_0, G)$ is trivial.

(i) If L is algebraically closed and G is connected (Steinberg's theorem), cf. [208], Thm. 1.9.

(ii) If $G = \mathrm{GL}_n$ (Hilbert 90).

(iii) If G is semi-simple simply connected and L is finite (Kneser [132], Bruhat–Tits [36]).

In order to define and study period domains, we could have restricted our attention to G-isocrystals for some reductive group G over $\mathbb{Q}_p$. However, when we deal with HN-filtrations, we will see that more general C-isocrystals show up naturally. In order to have a more concrete theory, we now look for a more explicit description of C-isocrystals.

Notation 9.1.7 To the automorphism σ of K_0, one associates base change and restriction functors in two related settings: linear algebra over K_0 and algebraic geometry over Spec K_0. Here are our conventions about these functors.

• For any vector space (resp. algebra) V over K_0, we set $\sigma_* V := V \otimes_{K_0,\sigma} K_0$ (*base change*) and we denote by $\sigma^* V$ the vector space V with K_0 acting through σ (*restriction*). Of course we have a canonical isomorphism $(\sigma^{-1})_* V \xrightarrow{\sim} \sigma^* V$. Moreover, giving a σ-linear map $V \longrightarrow W$ is the same as giving a linear map $V \longrightarrow \sigma^* W$, resp. a linear map $\sigma_* V \longrightarrow W$.

• For a K_0-scheme $X \longrightarrow \operatorname{Spec} K_0$, we denote by $\sigma_* X$ the K_0-scheme whose underlying scheme is X but whose K_0-structure is the composition $X \longrightarrow \operatorname{Spec} K_0 \xrightarrow{\sigma^*} \operatorname{Spec} K_0$. We denote by $\sigma^* X$ the base change of X along σ. Hence, if X is affine, then we have $O_{\sigma_* X} = \sigma^* O_X$ and $O_{\sigma^* X} = \sigma_* O_X$. We also see that for two affine schemes X, Y over K_0, a K_0-morphism $\sigma_* X \longrightarrow Y$ is the same as a σ-linear morphism of rings $O_Y \longrightarrow O_X$.

• We denote by φ_σ the canonical isomorphism $\sigma_* \operatorname{Spec} K_0 \xrightarrow{\sim} \operatorname{Spec} K_0$ given by $\sigma : K_0 \xrightarrow{\sim} \sigma^* K_0$.

Definition 9.1.8 (i) An *affine group scheme over* Isoc(L) is a pair (G, φ) consisting of an affine group scheme G over K_0 and an isomorphism $\varphi : \sigma_* G \xrightarrow{\sim} G$ of affine groups over K_0, such that the algebra O_G is an increasing union of finite-dimensional φ-stable subspaces.

(ii) A *morphism* $\rho : (G, \varphi) \longrightarrow (G', \varphi')$ *of affine group schemes over* Isoc(L) is a morphism $\rho : G \longrightarrow G'$ of K_0-groups such that $\rho \circ \varphi = \varphi' \circ \sigma_* \rho$.

We will denote by GS(L) the category of affine group schemes over Isoc(L). The datum φ in the definition above is called a *Weil descent datum* on G. More precisely, in [183], Def. 3.45, a Weil descent datum is an isomorphism

$$\alpha : G \longrightarrow G^\sigma,$$

where G^σ is the functor which associates to a scheme X over $\operatorname{Spec} K_0$ the set $G^\sigma(X) = G(\sigma_* X)$. The isomorphism φ in Definition 9.1.8 corresponds to α under the identification

$$\operatorname{Hom}(\sigma_*(G), G) \simeq \operatorname{Hom}(G, G^\sigma).$$

Remark 9.1.9 Our terminology here is in accordance with *Deligne's affine algebraic geometry over tannakian categories*, see [54], 5.4, and [55], 7.5. More precisely, an affine group scheme over Isoc(L) is nothing else but a Isoc(L)-group scheme, i.e., an object which is dual to an Hopf-algebra object in the category of Ind-isocrystals.

Example 9.1.10 If (V, Φ_V) is an isocrystal over L, then the morphism

$$\varphi_V : \sigma_* \operatorname{GL}(V) = \operatorname{GL}(\sigma_* V) \longrightarrow \operatorname{GL}(V)$$

which maps α to $\Phi_V \alpha \Phi_V^{-1} \in \operatorname{GL}(V)$ is a Weil descent datum which satisfies the required condition. Therefore we get an affine group scheme $(\operatorname{GL}(V), \varphi_V)$ over Isoc(L).

Example 9.1.11 For any scheme X over $\mathbb{Q}_p$, we have a canonical isomorphism

$$\sigma_*(X \times_{\operatorname{Spec} \mathbb{Q}_p} \operatorname{Spec} K_0) = X \times_{\operatorname{Spec} \mathbb{Q}_p} \sigma_* \operatorname{Spec} K_0,$$

whence an isomorphism $\mathrm{id}_X \times \varphi_\sigma : \sigma_*(X_{K_0}) \xrightarrow{\sim} X_{K_0}$ (see the third item in Notation 9.1.7). If G is an affine group scheme over $\mathbb{Q}_p$, we thus get an affine group scheme $(G_{K_0}, \varphi) = (G_{K_0}, \mathrm{id}_G \times \varphi_\sigma)$ over Isoc(L). In this case, the Weil descent datum is called *effective*, cf. [183], 3.4.6.

Let ω_{iso} denote the natural fiber functor from Isoc(L) to Vec_{K_0} and let $\sigma_*\omega_{\mathrm{iso}}$ be its base change along the automorphism σ of K_0, which again is a fiber functor to Vec_{K_0}. Thus

$$\sigma_*\omega_{\mathrm{iso}}(V, \Phi_V) = V \otimes_{K_0,\sigma} K_0.$$

Then we have an isomorphism $\sigma_*\omega_{\mathrm{iso}} \xrightarrow{\sim} \omega_{\mathrm{iso}}$ of fiber functors (given by the σ-linear bijective maps). Let

$$D_L := \underline{\mathrm{Aut}}^{\otimes}(\omega_{\mathrm{iso}})$$

be the group scheme over K_0 corresponding to the fiber functor ω_{iso} of Isoc(L) over K_0. Then the isomorphism $\sigma_*\omega_{\mathrm{iso}} \xrightarrow{\sim} \omega_{\mathrm{iso}}$ induces a Weil descent datum

$$\varphi_L : \sigma_* D_L \xrightarrow{\sim} D_L$$

on D_L, which thus becomes an affine group scheme over Isoc(L). Moreover, by definition, every isocrystal (V, Φ_V) is endowed with a canonical and functorial morphism of group schemes over Isoc(L)

$$\nu_{V,\Phi_V} : (D_L, \varphi_L) \longrightarrow (\mathrm{GL}(V), \varphi_V). \tag{1.2}$$

Remark 9.1.12 The group scheme (D_L, φ_L) over Isoc(L) thus obtained is the *fundamental group scheme* of the tannakian category Isoc(L) in the sense of [54], §6, and [55], §8. Further, the above action ν_{V,Φ_V} on the object (V, Φ_V) is the canonical action of the fundamental group scheme on this object.

Remark 9.1.13 Since the slope decomposition of isocrystals defines a grading of ω_{iso}, we have by Corollary 4.2.3 a morphism of K_0-groups $\mathbb{D}_{K_0} \xrightarrow{sl} D_L$. If we equip $\mathbb{D}_{K_0}$ with the descent datum $\mathrm{id} \times \varphi_\sigma$ associated with its $\mathbb{Q}_p$-model $\mathbb{D}_{\mathbb{Q}_p}$, then we get a morphism of group schemes over Isoc(L):

$$sl : (\mathbb{D}_{K_0}, \mathrm{id} \times \varphi_\sigma) \longrightarrow (D_L, \varphi_L). \tag{1.3}$$

By definition, the restriction to $\mathbb{D}_{K_0}$ of the canonical action of D_L on an isocrystal (V, Φ_V) is via the character α on the summand (V_α, Φ_V), for all $\alpha \in \mathbb{Q}$.

We can make the pair (D_L, φ_L) explicit in a couple of important cases.

Examples 9.1.14 (i) If L is algebraically closed, then Dieudonné theory says that sl is an isomorphism, cf. [193], Thm. VI.3.3.2.
(ii) If $L = \mathbb{F}_p$, then the Jordan decomposition of matrices shows that $D_{\mathbb{F}_p}$ is the product of $\mathbb{G}_a$ with the multiplicative group associated to the $\mathrm{Gal}(\overline{\mathbb{Q}}_p/\mathbb{Q}_p)$-module $\overline{\mathbb{Q}}_p^{\times}$ (see [193], VI.3.4.1). Moreover, sl is given by the valuation map $\overline{\mathbb{Q}}_p^{\times} \longrightarrow \mathbb{Q}$. Alternatively we may consider $D_{\mathbb{F}_p}$ as the "hull" of the abstract group $\mathbb{Z}$ in the category of group schemes over $\mathbb{Q}_p$, i.e., the unique group scheme over $\mathbb{Q}_p$ which represents the functor $G \mapsto G(\mathbb{Q}_p)$. In this way we see that the canonical action (1.2), page 225 determines the isocrystal structure, namely $\Phi_V = \nu_{V,\Phi_V}(\epsilon) \in \mathrm{GL}(V)$, where $\epsilon \in D_{\mathbb{F}_p}(\mathbb{Q}_p)$ denotes the universal element.
(iii) If $L = \mathbb{F}_{p^r}$, then the discussion in [193], VI.3.4.2 shows that $D_{\mathbb{F}_{p^r}}$ is the product of $\mathbb{G}_a$ with the multiplicative group associated to the $\mathrm{Gal}(\overline{\mathbb{Q}}_p/\mathbb{Q}_{p^r})$-module $\overline{\mathbb{Q}}_p^{\times}/\mu_r$. The Weil descent datum $\varphi_{\mathbb{F}_{p^r}}$ is induced by the natural model over $\mathbb{Q}_p$. Again sl is given by the valuation map. As in the case $L = \mathbb{F}_p$, we note that $D_{\mathbb{F}_{p^r}}$ represents the functor $G \mapsto G(\mathbb{Q}_{p^r})$ in the category of group schemes over $\mathbb{Q}_{p^r}$. But this time the canonical action (1.2), page 225 does not determine completely the isocrystal structure. In fact, we merely have the equation $\Phi_V^r = \nu_{V,\Phi_V}(\epsilon)$ in $\mathrm{GL}(V)$.

Observe that in each of these cases, D_L is abelian. Indeed this holds true if and only if the Galois group of L is abelian, by [193], VI.3.2.3.1.

Let (G, φ) be a group scheme over $\mathrm{Isoc}(L)$. By definition, the algebra O_G is a union of (finite-dimensional) sub-isocrystals, hence it is endowed with a canonical action of (D_L, φ_L) as in (1.2), page 225. This action preserves the Hopf algebra structure: it is an action of D_L on G compatible with descent data.

Definition 9.1.15 An *augmentation* of an affine group scheme (G, φ) over $\mathrm{Isoc}(L)$ is a morphism of group schemes

$$\nu : (D_L, \varphi_L) \longrightarrow (G, \varphi)$$

over $\mathrm{Isoc}(L)$ such that the canonical action of D_L on G above is given by ν via the inner action of G on itself. The triple $\mathbb{G} = (G, \varphi, \nu)$ is called an *augmented group scheme* over $\mathrm{Isoc}(L)$.

Augmented group schemes form a category $\mathrm{AGS}(L)$. Here morphisms are morphisms of group schemes which are compatible with the respective augmentations. We will sometimes write "a.g.s." for "augmented group scheme." Note that the "a" stands for "augmented" and not for "affine".

Examples 9.1.16 (i) The triple $(D_L, \varphi_L, \mathrm{id})$ is tautologically an augmented group scheme over Isoc(L).

(ii) For any (V, Φ_V) in Isoc(L), the morphism ν_{V,Φ_V} of (1.2), page 225 is tautologically an augmentation of $(\mathrm{GL}(V), \varphi_V)$. We will also denote by

$$\mathbb{GL}(V, \Phi_V) = (\mathrm{GL}(V), \varphi_V, \nu_{V,\Phi_V})$$

the a.g.s. thus obtained.

(iii) According to the description of $D_{\mathbb{F}_{p^r}}$ in Examples 9.1.14 (ii) and (iii), an augmentation of a group scheme (G, φ) over Isoc($\mathbb{F}_{p^r}$) is the same as an element $g \in G(\mathbb{Q}_{p^r})$ such that

(a) $\varphi^{(r)} = \mathrm{Int}(g)$ in Aut(G). Here $\varphi^{(r)}$ is the automorphism $\varphi \circ \sigma_* \varphi \circ \cdots \circ \sigma_*^{r-1}\varphi : \sigma_*^r G = G \longrightarrow G$.

(b) $\varphi \circ \sigma_* g = g \circ \varphi_\sigma$ in $G(K_0)$. Here g is considered as a section Spec $K_0 \longrightarrow G$ and φ_σ is defined in Notation 9.1.7.

Remark 9.1.17 Let ν and ν' be two augmentations of a group scheme (G, φ) over Isoc(L). Then, denoting by ρ_{ad} the canonical projection $G \longrightarrow G_{\mathrm{ad}}$, we have $\rho_{\mathrm{ad}} \circ \nu = \rho_{\mathrm{ad}} \circ \nu'$. Therefore there is a unique central augmentation ζ of the affine subgroup scheme $(Z(G), \varphi)$ over Isoc(L) such that $\nu' = \zeta \cdot \nu$.

Definition 9.1.18 (i) A *representation* of an augmented group scheme $\mathbb{G} = (G, \varphi, \nu)$ over Isoc(L) is a triple (V, Φ_V, ρ), where (V, Φ_V) is an isocrystal over L and

$$\rho : \mathbb{G} \longrightarrow \mathbb{GL}(V, \Phi_V)$$

is a morphism of augmented group schemes.

(ii) A *morphism* $f : (V, \Phi_V, \rho_V) \longrightarrow (W, \Phi_W, \rho_W)$ of $\mathbb{G}$-representations is a homomorphism of isocrystals $(V, \Phi_V) \longrightarrow (W, \Phi_W)$ such that the underlying map $V \longrightarrow W$ is a homomorphism of G-representations.

Concretely, a morphism ρ as above is the same as a structure of O_G-comodule on V which is compatible with the σ-linear maps φ and Φ_V, and such that the O_{D_L}-comodule structure induced by ν on V is the canonical one. A morphism of representations is then a morphism of isocrystals that is compatible with these comodule structures. We thus obtain a $\mathbb{Q}_p$-linear category $\mathrm{Rep}_{\mathbb{Q}_p}(\mathbb{G})$. Further, the tensor structure on Isoc(L) induces a tensor structure on this category, and one checks rather formally the following fact.

Lemma 9.1.19 *The category* $\mathrm{Rep}_{\mathbb{Q}_p}(\mathbb{G})$ *is a Tannaka category over* $\mathbb{Q}_p$. □

Moreover, by construction, this category comes equipped with a $\mathrm{Rep}_{\mathbb{Q}_p}(\mathbb{G})$-isocrystal $N^{\mathbb{G}}$

$$N^{\mathbb{G}} : \mathrm{Rep}_{\mathbb{Q}_p}(\mathbb{G}) \longrightarrow \mathrm{Isoc}(L) ,$$

which simply takes (V, Φ_V, ρ) to (V, Φ_V). We also put

$$\omega^{\mathbb{G}} := \omega_{\mathrm{iso}} \circ N^{\mathbb{G}},$$

which is a K_0-valued fiber functor on $\mathrm{Rep}_{\mathbb{Q}_p}(\mathbb{G})$. In the proof of the Proposition below, we will exhibit a canonical isomorphism of K_0-groups $G \xrightarrow{\sim} \underline{\mathrm{Aut}}^{\otimes}(\omega^{\mathbb{G}})$.

Let Tan be the category of all tannakian categories over $\mathbb{Q}_p$ with exact tensor functors as morphisms. Let $\mathrm{Tan}_{/\,\mathrm{Isoc}(L)}$ be the category of tannakian categories "over" Isoc(L). Thus, objects are pairs (C, N) consisting of a Tannaka category C over $\mathbb{Q}_p$ endowed with a C-isocrystal $N : C \longrightarrow \mathrm{Isoc}(L)$. A morphism $(C, N) \longrightarrow (C', N')$ in $\mathrm{Tan}_{/\,\mathrm{Isoc}(L)}$ is an exact tensor functor $C \xrightarrow{\alpha} C'$ such that $N' \circ \alpha = N$. We have just defined a contravariant functor

$$\begin{aligned} \tau : \mathrm{AGS}(L) &\longrightarrow \mathrm{Tan}_{/\,\mathrm{Isoc}(L)} \\ \mathbb{G} &\mapsto (\mathrm{Rep}_{\mathbb{Q}_p}(\mathbb{G}), N^{\mathbb{G}}). \end{aligned} \tag{1.4}$$

Now we are going to define a functor in the opposite direction. So let $(C, N) \in \mathrm{Tan}_{/\,\mathrm{Isoc}(L)}$. The $\otimes$-functor

$$\omega := \omega_{\mathrm{iso}} \circ N$$

is a fiber functor from C to Vec_{K_0}. Via base change with the automorphism σ of K_0 we get another fiber functor $\sigma_*\omega$ and the fact that ω comes from N yields an isomorphism of fiber functors $\varphi : \sigma_*\omega \xrightarrow{\sim} \omega$. Therefore, the automorphism K_0-group scheme

$$G := \underline{\mathrm{Aut}}^{\otimes}(\omega)$$

of ω comes equipped with a Weil descent datum φ. Moreover, the morphism of K_0-groups

$$\nu : D_L = \underline{\mathrm{Aut}}^{\otimes}(\omega_{\mathrm{iso}}) \longrightarrow G = \underline{\mathrm{Aut}}^{\otimes}(\omega)$$

induced by N is compatible with the Weil descent data φ and φ_D. In this way we get an a.g.s. (G, φ, ν), and hence a functor

$$\begin{aligned} \eta : \mathrm{Tan}_{/\,\mathrm{Isoc}(L)} &\longrightarrow \mathrm{AGS}(L) \\ (C, N) &\mapsto (\underline{\mathrm{Aut}}^{\otimes}(\omega), \varphi, \nu). \end{aligned} \tag{1.5}$$

Proposition 9.1.20 *The contravariant functors τ and η are quasi-inverse equivalences of categories.*

Proof Let us start with a pair $(C, N) \in \mathrm{Tan}_{/\,\mathrm{Isoc}(L)}$ and let $(G, \varphi, \nu) = \eta(C, N)$ denote the associated a.g.s. From the definition of $G = \underline{\mathrm{Aut}}^{\otimes}(\omega_{\mathrm{iso}} \circ N)$, we have the following $\otimes$-functor factorization of $\omega_{\mathrm{iso}} \circ N$:

$$\begin{array}{ccc} C & \xrightarrow{\omega_{\mathrm{iso}} \circ N} & \mathrm{Vec}_{K_0} \\ \tilde{\omega} \big\downarrow & \nearrow_{\omega^G} & \\ \mathrm{Rep}_{K_0}(G) & & \end{array} .$$

From the definition of φ and ν, this factorization in fact comes from the following one:

$$\begin{array}{ccc} C & \xrightarrow{N} & \mathrm{Isoc}(L) \\ \tilde{N} \big\downarrow & \nearrow_{N^{\mathbb{G}}} & \\ \mathrm{Rep}_{\mathbb{Q}_p}(\mathbb{G}) & & \end{array} .$$

We claim that $\tilde{N}$ is an equivalence of tannakian categories over $\mathbb{Q}_p$. Indeed, this is the content of Deligne's generalization of (half of) Tannaka duality in Proposition 6.5 of [54] (see also Theorem 8.17 in [55]), once we have noticed that the pair (G, φ) is the group scheme over Isoc(L) which would be denoted by $N\pi(C)$ in Deligne's notation (the "image" by N of the fundamental group scheme of C). In fact, this follows from [55], 8.15. The augmentation ν is the canonical morphism (6.4.2) of [54] $\pi(\mathrm{Isoc}(L)) \longrightarrow N\pi(C)$ (see also (8.15.2) in [55]). This proves that the composite $\tau \circ \eta$ is isomorphic to the identity functor of $\mathrm{Tan}_{/\,\mathrm{Isoc}(L)}$.

Conversely, start with an a.g.s. (G, φ, ν) over Isoc(L) and denote by $\omega^{\varphi,\nu}$ the functor on $\mathrm{Rep}_{\mathbb{Q}_p}(\mathbb{G})$ which forgets φ and ν. We have a diagram

$$\begin{array}{ccc} \mathrm{Rep}_{\mathbb{Q}_p}(\mathbb{G}) & \xrightarrow{N^{\mathbb{G}}} & \mathrm{Isoc}(L) \\ \omega^{\varphi,\nu} \big\downarrow & \searrow^{\omega^{\mathbb{G}}} & \big\downarrow \omega_{\mathrm{iso}} \\ \mathrm{Rep}_{K_0}(G) & \xrightarrow[\omega^G]{} & \mathrm{Vec}_{K_0} \end{array}$$

whence a morphism of K_0-group schemes $G = \underline{\mathrm{Aut}}^{\otimes}(\omega^G) \longrightarrow \underline{\mathrm{Aut}}^{\otimes}(\omega^{\mathbb{G}})$, which is compatible with the respective Weil descent data and morphisms ν, for formal reasons. Then we claim:

The above morphism $G \longrightarrow \underline{\mathrm{Aut}}^{\otimes}(\omega^{\mathbb{G}})$ *is an isomorphism.* (1.6)

This will prove that the composite $\eta \circ \tau$ is isomorphic to the identity of AGS(L).

We first reformulate (1.6) above. There is a functorial base change $C \mapsto C_{K_0}$

from tannakian categories over $\mathbb{Q}_p$ to tannakian categories over K_0, cf. Chapter IV (1.5), page 114. An exact $\otimes$-functor $C \longrightarrow C'$ from a tannakian category over $\mathbb{Q}_p$ to another one over K_0 induces a functor $C_{K_0} \longrightarrow C$ of tannakian categories over K_0. In this setting, the claim (1.6) is equivalent to the claim:

$$\text{The } \otimes\text{-functor } \mathrm{Rep}_{\mathbb{Q}_p}(\mathbb{G})_{K_0} \xrightarrow{\omega_0^{\varphi,\nu}} \mathrm{Rep}_{K_0}(G) \text{ induced by } \omega^{\varphi,\nu} \text{ is an equivalence of Tannaka categories over } K_0. \tag{1.7}$$

Let us first consider the category $\mathrm{Rep}_\varphi(G)$ of representations of the group scheme (G, φ) over $\mathrm{Isoc}(L)$, in the sense of [54], 5.7. Such representations may be defined in an elementary way as we did for representations of (G, φ, ν) but without any condition on augmentations. By Tannaka duality the fiber functor ω_{iso} identifies $\mathrm{Isoc}(L)_{K_0}$ with $\mathrm{Rep}_{K_0}(D_L)$. It therefore identifies $\mathrm{Rep}_\varphi(G)_{K_0}$ with the category $\mathrm{Rep}_{D_L}(G)$ of representations of the group scheme $\omega_{\mathrm{iso}}(G, \varphi)$ over $\mathrm{Rep}_{K_0}(D_L)$. The latter group scheme is simply given by the K_0-Hopf algebra O_G endowed with the canonical action of D_L. Concretely, objects in $\mathrm{Rep}_{D_L}(G)$ are rational representations (V, ρ) of D_L equipped with a D_L-equivariant action $G \longrightarrow \mathrm{Aut}(V)$. In other words, $\mathrm{Rep}_{D_L}(G)$ (and therefore $\mathrm{Rep}_\varphi(G)_{K_0}$) can be identified with $\mathrm{Rep}_{K_0}(G \rtimes D_L)$.

Now let us add the condition on augmentations. As above the fiber functor ω_{iso} identifies $\mathrm{Rep}_{\mathbb{Q}_p}(\mathbb{G})_{K_0}$ with the category $\mathrm{Rep}_{D_L,\nu}(G)$ of representations of $\omega_{\mathrm{iso}}(G, \varphi)$ which are compatible with ν. Concretely, objects in $\mathrm{Rep}_{D_L,\nu}(G)$ are rational representations (V, ρ) of D_L equipped with a D_L-equivariant action $G \xrightarrow{\mu} \mathrm{Aut}(V)$ such that $\mu \circ \nu = \rho$. But these are nothing more than representations of G. All in all, we have shown that ω_{iso} identifies $\mathrm{Rep}_{\mathbb{Q}_p}(\mathbb{G})_{K_0}$ with $\mathrm{Rep}_{K_0}(G)$, as desired. □

Remark 9.1.21 Using Deligne's language of algebraic geometry over tannakian categories, we see that the notions of augmented group schemes and their representations make sense over any Tannaka category $\mathcal{T}$. Indeed an *a.g.s. over* $\mathcal{T}$ is a pair $(\mathcal{G}, \nu)$ consisting of a group scheme $\mathcal{G}$ over $\mathcal{T}$ endowed with a morphism $\pi(\mathcal{T}) \longrightarrow \mathcal{G}$ which induces the canonical action of $\pi(\mathcal{T})$ on $\mathcal{G}$ via inner action. A representation is a pair (X, ρ) consisting of an object $X \in \mathcal{T}$ with an action ρ of $\mathcal{G}$ whose restriction to $\pi(\mathcal{T})$ is the canonical one. As in the above proposition, an exact $\otimes$-functor $C \xrightarrow{u} \mathcal{T}$ from another Tannaka category (over the same field) yields an a.g.s. $\pi(\mathcal{T}) \longrightarrow u\pi(C)$ and this defines an equivalence of categories between the category of such $\otimes$-functors and the category of a.g.s. over $\mathcal{T}$.

Example 9.1.22 Let G be an algebraic group over $\mathbb{Q}_p$ and let $b \in G(K_0)$. We denote by

$$\mathbb{G}_b = (G_{K_0}, \varphi_b, \nu_b)$$

the a.g.s. associated by the previous proposition with the G-isocrystal N_b : $\mathrm{Rep}_{\mathbb{Q}_p}(G) \longrightarrow \mathrm{Isoc}(L)$. We have $\varphi_b = \mathrm{Int}(b) \circ (\mathrm{id}_G \times \varphi_\sigma)$. The functor $(V, \rho) \mapsto (V_{K_0}, \rho(b)(\mathrm{id}_{K_0} \otimes \sigma), \rho_{K_0})$ induces a tensor equivalence

$$\iota_b : \mathrm{Rep}_{\mathbb{Q}_p}(G) \xrightarrow{\sim} \mathrm{Rep}_{\mathbb{Q}_p}(\mathbb{G}_b)$$

such that $N_b = N^{\mathbb{G}_b} \circ \iota_b$. Moreover, conjugation by an element $g \in G(K_0)$ induces an isomorphism of a.g.s. $\mathbb{G}_b \xrightarrow{\sim} \mathbb{G}_{gb\sigma(g)^{-1}}$.

Conversely, we may ask when an a.g.s. $\mathbb{G}$ comes from a G-isocrystal. Obviously this is equivalent to the conjunction of the two following properties.

(i) $\mathrm{Rep}_{\mathbb{Q}_p}(\mathbb{G})$ has a fiber functor ω over $\mathbb{Q}_p$.

(ii) The difference $\alpha \in H^1(K_0, G)$ between the fiber functors $\omega_{\mathrm{iso}} \circ N^{\mathbb{G}}$ and ω_{K_0} comes from an element of $H^1(\mathbb{Q}_p, \underline{\mathrm{Aut}}^{\otimes}(\omega))$.

Note that condition (ii) is always satisfied if $H^1(K_0, G)$ is trivial, in particular if G is connected and L is algebraically closed by Steinberg's theorem.

Example 9.1.23 It is instructive to look at the special case where L is a finite field.

(i) Assume first $L = \mathbb{F}_p$. In this case, conditions (i) and (ii) of the previous example are satisfied, so that any a.g.s. is of the form $\mathbb{G}_b$. In fact, we already know from Example 9.1.16 (iii) that the category of a.g.s. over $\mathrm{Isoc}(\mathbb{F}_p)$ is equivalent to that of pairs (G, g) consisting of a group scheme over $\mathbb{Q}_p$ and a rational element of G. Explicitly we have $\varphi = \mathrm{Int}(g)$ and $g = \nu(\epsilon)$ (recall from Example 9.1.14 (ii) that $D_{\mathbb{F}_p}$ is the "hull" of $\mathbb{Z}$).

(ii) Consider more generally the case $L = \mathbb{F}_{p^r}$. Here an a.g.s. need not be of the form $\mathbb{G}_b$. However, Example 9.1.16 (iii) says that an a.g.s. $\mathbb{G} = (G, \varphi, \nu)$ over $\mathbb{F}_{p^r}$ is the same as a triple (G, φ, g) with φ a Weil descent datum and $g \in G(\mathbb{Q}_{p^r})$ such that $\varphi \circ \sigma_* g = g \circ \varphi_\sigma$ and $\varphi^{(r)} = \mathrm{Int}(g)$. In this language, a morphism $(G, \varphi) \xrightarrow{\rho} (\mathrm{GL}(V), \varphi_V)$ is a representation of $\mathbb{G}$ iff we have $\Phi_V^r = \rho(g)$.

Remark 9.1.24 By [55] we know that a Tannaka category C over $\mathbb{Q}_p$ is isomorphic to the representation category $\mathrm{Rep}_{\mathbb{Q}_p}(\underline{\mathcal{G}})$ of an affine gerb $\underline{\mathcal{G}}$ (in the sense of Giraud) over $\mathbb{Q}_p$. Moreover, denoting by $\underline{\mathcal{D}}_L$ the gerb corresponding to $\mathrm{Isoc}(L)$, the functor $N : C \longrightarrow \mathrm{Isoc}(L)$ is given by a morphism of gerbs $\underline{\mathcal{D}}_L \longrightarrow \underline{\mathcal{G}}$. Let us sketch the connection to augmented group schemes over $\mathrm{Isoc}(L)$ in the language of Galois gerbs.

By Proposition 9.1.20 we know that an a.g.s. over $\mathrm{Isoc}(L)$ is the same as an affine gerb $\underline{\mathcal{G}}$ over $\mathbb{Q}_p$ which is equipped with an exact faithful $\mathbb{Q}_p$-linear tensor functor to Isoc(L). If $C = \mathrm{Rep}_{\mathbb{Q}_p}(\underline{\mathcal{G}})$ is an algebraic Tannaka category (i.e., $\underline{\mathcal{G}}$ is an algebraic gerb), we have the more concrete notion of a *Galois gerb* over $\mathbb{Q}_p$ [144] at our disposal. In fact, as is explained in [144], these two different notions of a gerb are essentially equivalent (see the summary below). The restriction to the algebraic situation is not serious, since each affine gerb is an inverse limit of algebraic gerbes. On the other hand, any group scheme over Isoc(L) is an inverse limit of algebraic group schemes over Isoc(L) (the same proof as for ordinary affine group schemes in [62], Cor. II.2.7 applies), cf. the proof of Proposition 9.2.28 below.

Recall that a Galois gerb $\mathcal{G}$ over $\mathbb{Q}_p$ is an extension of topological groups,

$$1 \longrightarrow H(\overline{\mathbb{Q}}_p) \longrightarrow \mathcal{G} \longrightarrow \Gamma_{\mathbb{Q}_p} \longrightarrow 1\,,$$

where H is an algebraic group over $\overline{\mathbb{Q}}_p$ and $H(\overline{\mathbb{Q}}_p)$ (the *kernel* of the Galois gerb) is equipped with the discrete topology. It is required that there exist a section $\Gamma_K \longrightarrow \mathcal{G}, \tau \mapsto g_\tau$, for some finite Galois extension K of $\mathbb{Q}_p$, such that the maps $H(\overline{\mathbb{Q}}_p) \longrightarrow H(\overline{\mathbb{Q}}_p), h \mapsto g_\tau h g_\tau^{-1}$, are induced by a K-rational group structure G of H (i.e., $\mathcal{G}$ is neutralized by K). We obtain then an extension

$$1 \longrightarrow G(K) \longrightarrow \mathcal{G}_K \longrightarrow \mathrm{Gal}(K/\mathbb{Q}_p) \longrightarrow 1\,,$$

which gives rise, via pull-back by the map $\Gamma_{\mathbb{Q}_p} \longrightarrow \mathrm{Gal}(K/\mathbb{Q}_p)$ and via push-out by $G(K) \longrightarrow H(\overline{\mathbb{Q}}_p)$, to the Galois gerb with which we started. A *morphism of Galois gerbes* over $\mathbb{Q}_p$ is a homomorphism of group extensions which induces on $\Gamma_{\mathbb{Q}_p}$ the identity. It is required that the restriction to the kernels is induced by a homomorphism of algebraic groups over $\overline{\mathbb{Q}}_p$ and which is compatible with the splittings over a sufficiently large finite extension of $\mathbb{Q}_p$.

Let C be an algebraic Tannaka category over $\mathbb{Q}_p$, together with a fiber functor $\omega : C \longrightarrow \mathrm{Vec}_K$ for some finite Galois extension K of $\mathbb{Q}_p$. The attached Galois gerb is defined as follows, cf. [144, 136]. For any element $\tau \in \mathrm{Gal}(K/\mathbb{Q}_p)$, one sets

$$(\mathcal{G}_K)_\tau = \mathrm{Isom}^{\otimes}(\tau_*\omega, \omega).$$

Then $\mathcal{G}_K = \coprod_{\tau \in \mathrm{Gal}(K/\mathbb{Q}_p)} (\mathcal{G}_K)_\tau$ defines a Galois gerb with the following group law. Let $g \in (\mathcal{G}_K)_\tau$ and $g' \in (\mathcal{G}_K)_{\tau'}$. Then $g \cdot g' := g \circ \tau(g') \in (\mathcal{G}_K)_{\tau\cdot\tau'}$, where $\tau(g') \in \mathrm{Isom}^{\otimes}((\tau\cdot\tau')_*\omega, \tau_*\omega)$ is the induced twisted element. The homomorphism $\mathcal{G}_K \longrightarrow \mathrm{Gal}(K/\mathbb{Q}_p)$ is the obvious one, whereas H is given by the algebraic group scheme $\underline{\mathrm{Isom}}^{\otimes}(\omega, \omega)$ over K.

Conversely, one recovers the pair (C, ω) from $\mathcal{G}$ by considering the representation category of $\mathcal{G}$ together with its canonical fiber functor. Here a *representation* of a Galois gerb $\mathcal{G}$ is a discrete semilinear algebraic action ρ of $\mathcal{G}$

on a finite-dimensional $\overline{\mathbb{Q}}_p$-vector space V. "Algebraic" means that the restriction to H is an algebraic representation. "Discrete" means that the stabilizer of any $v \in V$ is an open subgroup of $\mathcal{G}$. The resulting category Rep($\mathcal{G}$) of all $\mathcal{G}$-representations is tannakian over $\mathbb{Q}_p$.

By the above correspondence and the existence of a fiber functor of an algebraic gerb over a finite extension of $\mathbb{Q}_p$ (cf. the proposition on p. 225 in [62]), we conclude that an algebraic gerb $\underline{\mathcal{G}}$ over $\mathbb{Q}_p$ which admits a fiber functor over K_0 corresponds to a Galois gerb $\mathcal{G}$ over $\mathbb{Q}_p$ which is neutralized by a finite *unramified* extension of $\mathbb{Q}_p$ contained in K_0.

Now let us suppose that $L = \overline{L}$ is algebraically closed. Let $K \subset K_0$ be a subfield which is of finite degree s over $\mathbb{Q}_p$, i.e., $K = \mathbb{Q}_{p^s}$. Let $\mathrm{Isoc}(L)^{\mathrm{ss}}_s \subset \mathrm{Isoc}(L)$ be the algebraic tannakian subcategory defined in Chapter VIII, §1, i.e., the subcategory consisting of split semi-simple isocrystals having slopes in $\frac{1}{s}\mathbb{Z}$. Then the map $(V, \Phi_V) \longrightarrow V_s$ given by Chapter VIII (1.3), page 191 defines a fiber functor $\omega^K_{\mathrm{iso}} : \mathrm{Isoc}(L)^{\mathrm{ss}}_s \longrightarrow \mathrm{Vec}_K$. By the construction above and the description of split semi-simple isocrystals (cf. Proposition 9.1.31 below), we get an extension of groups

$$1 \longrightarrow \mathbb{G}_m(K) \longrightarrow \mathcal{D}^0_K \longrightarrow \mathrm{Gal}(K/\mathbb{Q}_p) \longrightarrow 1$$

or, equivalently, a Galois gerb over $\mathbb{Q}_p$ with a neutralization over K,

$$1 \longrightarrow \mathbb{G}_m(\overline{\mathbb{Q}}_p) \longrightarrow \mathcal{D}_K \longrightarrow \Gamma_{\mathbb{Q}_p} \longrightarrow 1.$$

The gerb $\mathcal{D}_K$ is the Galois gerb of $\mathrm{Isoc}(L)^{\mathrm{ss}}_s$. We have a canonical element $\phi_K \in (\mathcal{D}_K)_\sigma$ which corresponds to the isomorphism $\sigma_*\omega^K_{\mathrm{iso}} \xrightarrow{\sim} \omega^K_{\mathrm{iso}}$ given by the σ-linear maps Φ_V. One has the relation $\phi^s_K = p$ in $\mathbb{G}_K(K)$. By forming the projective limit over all such gerbs with varying K, we get an extension of groups

$$1 \longrightarrow \mathbb{D}(\overline{\mathbb{Q}}_p) \longrightarrow \mathcal{D} \longrightarrow \Gamma_{\mathbb{Q}_p} \longrightarrow 1,$$

where $\mathbb{D}$ is the torus over $\mathbb{Q}_p$ with character group $\mathbb{Q}$.

Now let (C, N) be an algebraic Tannaka category corresponding to the Galois gerb $\mathcal{G}$, together with an exact faithful $\mathbb{Q}_p$-linear tensor functor $N : C \longrightarrow \mathrm{Isoc}(L)$. Let $\omega = \omega_{\mathrm{iso}} \circ N$ be the induced fiber functor as in the construction before Proposition 9.1.20. The functor N gives rise by Tannaka duality to a morphism of groups $\nu' : \mathcal{D} \longrightarrow \mathcal{G}$ which factorizes over some Galois gerb $\mathcal{D}_K$. Let (G, φ, ν) be the a.g.s. corresponding to (C, N). Then G is defined by the kernel of $\mathcal{G}$ (note that $\mathcal{G}$ is neutralized over a subextension of K_0). The isomorphism $\sigma_*\omega \xrightarrow{\sim} \omega$ (which gives rise to φ) induced by $\sigma_*\omega_{\mathrm{iso}} \xrightarrow{\sim} \omega_{\mathrm{iso}}$ corresponds just to the image $\nu'(\phi_K) \in \mathcal{G}_\sigma$. Finally, the homomorphism $\nu : D_L = \mathbb{D}_{K_0} \longrightarrow G_{K_0}$ is nothing else but the restriction of ν' to the kernels.

Conversely, an a.g.s. (G, φ, ν) over Isoc(L) corresponding to (C, N) defines ν' : $\mathcal{D} \longrightarrow \mathcal{G}$ by simply inverting the above rules.

Thus we get an equivalence of categories between the category of algebraic a.g.s. over Isoc(L) and the category of Galois gerbs $\mathcal{G}$ over $\mathbb{Q}_p$ together with a morphism of Galois gerbs $\mathcal{D}_K \longrightarrow \mathcal{G}$ for some finite Galois extension K of $\mathbb{Q}_p$ contained in K_0.

If L is no longer algebraically closed, one can imitate the above construction, once one writes down the affine gerb attached to the category Isoc(L) as a projective limit of algebraic gerbs. By the same process one gets an extension of groups

$$1 \longrightarrow D_L(\overline{\mathbb{Q}}_p) \longrightarrow \mathcal{D}_L \longrightarrow \Gamma_{\mathbb{Q}_p} \longrightarrow 1$$

and an equivalence of the corresponding categories as described above.

The general theory of tannakian categories leads to the following criteria.

Corollary 9.1.25 *Let* $\mathbb{G} = (G, \varphi, \nu)$ *be an augmented group scheme over* Isoc(L).

(i) $\mathrm{Rep}_{\mathbb{Q}_p}(\mathbb{G})$ *has a tensor generator if and only if G is an algebraic group over* K_0.

(ii) $\mathrm{Rep}_{\mathbb{Q}_p}(\mathbb{G})$ *is a semi-simple category if and only if G is pro-reductive.*

Proof As a corollary of the proof of Proposition 9.1.20, we know that G is the automorphism group scheme of the K_0-valued fiber functor $\omega_{\mathrm{iso}} \circ N^{\mathbb{G}}$ on $\mathrm{Rep}_{\mathbb{Q}_p}(\mathbb{G})$. Therefore, point (i) follows from [193], III.3.3.1.1 and point (ii) follows from [193], III.3.3.3.b, cf. Proposition 4.1.21. □

Definition 9.1.26 An a.g.s. $\mathbb{G} = (G, \varphi, \nu)$ over Isoc(L) is called *reductive* if the K_0-group G is reductive.

Let $\mathbb{G} = (G, \varphi, \nu)$ be a reductive a.g.s. over Isoc(L), and let P be a K_0-parabolic subgroup of G which is stable under the Weil descent datum φ. This simply means that φ restricts to an isomorphism $\sigma_* P \longrightarrow P$. Thus we have an embedding $(P, \varphi) \hookrightarrow (G, \varphi)$ of group schemes over Isoc(L) which is automatically equivariant for the canonical actions. Now, on G this action is given by the inner action of (D_L, φ_L) via the augmentation ν. Therefore we see that this inner action preserves P. Since P is its own normalizer in G, it follows that ν factors through P. We thus get an a.g.s

$$\mathbb{P} = (P, \varphi, \nu)$$

over Isoc(L). We note that φ induces a Weil descent datum φ_M on the Levi quotient M of P and that the composition ν_M of ν with the projection $P \longrightarrow M$

is an augmentation of (M, φ_M). Therefore we get an a.g.s.

$$\mathbb{M} = (M, \varphi_M, \nu_M)$$

over Isoc(L).

Definition 9.1.27 The a.g.s $\mathbb{P}$ above is called the *parabolic subgroup* of $\mathbb{G}$ associated to the φ-invariant psgp P of G. We call $\mathbb{M}$ its *Levi factor*.

Remark 9.1.28 These constructions are the main reason why we work with a.g.s. over Isoc(L) rather than with G-isocrystals attached to $\mathbb{Q}_p$-groups G. Indeed, let G be a $\mathbb{Q}_p$-group and let $b \in G(K_0)$. Then a psgp of G_{K_0} which is stable under the Weil descent datum $\varphi_b = \mathrm{Int}(b) \circ (\mathrm{id}_G \otimes \varphi_\sigma)$ need not be defined over $\mathbb{Q}_p$ and (equivalently) b need not belong to $P(K_0)$ (think of the extreme case when G is anisotropic). In other words, although P defines a psgp $\mathbb{P}$ of $\mathbb{G}_b$, this psgp need not be of the form $\mathbb{P}_b$ for some $\mathbb{Q}_p$-psgp P of G. This causes troubles when dealing with the HN-formalism.

Construction 9.1.29 *Extension of the field L.* Let $L' \supset L$ be a perfect field extension of L and let $K_0' := K_0(L')$. Consider the exact faithful $\otimes$-functor

$$\varepsilon_L^{L'} : \mathrm{Isoc}(L) \longrightarrow \mathrm{Isoc}(L')$$

of Tannaka categories over $\mathbb{Q}_p$ defined in Chapter VIII (1.2), page 190. Composition with $\varepsilon_L^{L'}$ gives in turn a functor

$$\begin{array}{ccc} \mathrm{Tan}_{/\mathrm{Isoc}(L)} & \longrightarrow & \mathrm{Tan}_{/\mathrm{Isoc}(L')} \\ (C, N) & \mapsto & (C, N' := \varepsilon_L^{L'} \circ N) \end{array}$$

and thus a functor

$$\begin{array}{ccc} \mathrm{AGS}(L) & \longrightarrow & \mathrm{AGS}(L') \\ \mathbb{G} & \mapsto & \mathbb{G}' \end{array}$$

such that the square

$$\begin{array}{ccc} \mathrm{Rep}_{\mathbb{Q}_p}(\mathbb{G}) & \xrightarrow{\sim} & \mathrm{Rep}_{\mathbb{Q}_p}(\mathbb{G}') \\ {\scriptstyle N^{\mathbb{G}}}\downarrow & & \downarrow{\scriptstyle N^{\mathbb{G}'}} \\ \mathrm{Isoc}_L & \xrightarrow[\varepsilon_L^{L'}]{} & \mathrm{Isoc}_{L'} \end{array}$$

is commutative. Here, the upper horizontal functor is given by base change of isocrystals and is an equivalence of Tannaka categories over $\mathbb{Q}_p$ (since both categories can be identified with the original C).

Explicitly, the functor $\mathbb{G} \mapsto \mathbb{G}'$ maps a triple (G, φ, ν) to the triple (G', φ', ν') with $G' = G \times_{K_0} K_0'$, $\varphi' = \varphi \times \sigma^*$, and $\nu' = \nu_{K_0'} \circ \nu_L^{L'}$, where

$$\nu_L^{L'} : (D_{L'}, \varphi_{L'}) \longrightarrow (D_L \times_{K_0} K_0', \varphi_L \times \sigma^*) \tag{1.8}$$

is the morphism induced by the functor $\varepsilon_L^{L'}$ in the sense of Chapter IV (1.6), page 114. We note that if L' is algebraically closed, $\nu_L^{L'}$ can be identified with the base change of sl from K_0 to K_0', see (1.3), page 225. If $L = \mathbb{F}_p$ and $L' = \mathbb{F}_{p^r}$, both sides of (1.8) above represent the functor $G \mapsto G(\mathbb{Q}_{p^r})$ on the category of affine group schemes over $\mathbb{Q}_{p^r}$ and the map (1.8) is the r-power map $G(\mathbb{Q}_{p^r}) \longrightarrow G(\mathbb{Q}_{p^r})$.

In the previous chapter, we introduced the full tannakian subcategory $\mathrm{Isoc}(L)^{\mathrm{ss}}$ of $\mathrm{Isoc}(L)$ whose objects are the *split semi-simple* isocrystals, cf. Definition 8.1.2. There we have observed how much nicer the theory of period domains behaves for such isocrystals, in particular, with regard to extending the field L. This suggests the following definition.

Definition 9.1.30 Let C be a Tannaka category over $\mathbb{Q}_p$. A C-isocrystal is *split semi-simple* if it factors through $\mathrm{Isoc}(L)^{\mathrm{ss}}$.

The embedding of $\mathrm{Isoc}(L)^{\mathrm{ss}}$ in $\mathrm{Isoc}(L)$ is an $\mathrm{Isoc}(L)^{\mathrm{ss}}$-isocrystal hence defines an augmentation

$$(D_L, \varphi_L) \xrightarrow{\nu_L^{\mathrm{ss}}} (D_L^{\mathrm{ss}}, \varphi_L^{\mathrm{ss}}),$$

where D_L^{ss} is the K_0-group of automorphisms of ω_{iso} restricted to $\mathrm{Isoc}(L)^{\mathrm{ss}}$. By definition of split semi-simple isocrystals, we have the following description of $(D_L^{\mathrm{ss}}, \varphi_L^{\mathrm{ss}})$. Let us define a subgroup Λ_L of $\mathbb{Q}$ by

$$\Lambda_L := \sum_{\mathbb{F}_{p^s} \subset L} \frac{1}{s}\mathbb{Z}.$$

Proposition 9.1.31 (i) *D_L^{ss} is the multiplicative split K_0-group* $\mathrm{Diag}_{K_0}(\Lambda_L)$ *whose group of characters is Λ_L, and φ_L^{ss} is the natural descent datum (to a split multiplicative group).*

(ii) *The composition $\nu_L^{\mathrm{ss}} \circ sl$ is induced by the inclusion $\Lambda_L \subset \mathbb{Q}$.*

(iii) *A C-isocrystal N is split semi-simple if and only if the augmentation ν in the associated a.g.s. $(\mathbb{G}, \varphi, \nu)$ factors through D_L^{ss}.*

The proof is rather formal and is left to the reader. In the case of G-isocrystals for a $\mathbb{Q}_p$-group G, this notion of split semi-simplicity of course still makes sense. It is related to the notion of a *decent* σ-conjugacy class as follows.

Let G be an affine algebraic group over $\mathbb{Q}_p$ and let $b \in G(K_0)$. To simplify the notation, we let

$$\tilde{\nu}_b := \nu_b \circ sl : \mathbb{D}_{K_0} \longrightarrow G_{K_0}$$

be the $\mathbb{Q}$-1-PS which gives the Newton slope grading on N_b. Now, since G is algebraic, there always exists a positive integer s, such that $s\tilde{\nu}_b$ factors through $\mathbb{G}_{m\,K_0}$. Here, we consider $\mathbb{G}_m$ as a quotient of $\mathbb{D}$ via the inclusion $\mathbb{Z} \hookrightarrow \mathbb{Q}$ of their character groups.

Definition 9.1.32 Let G be an algebraic group over $\mathbb{Q}_p$ and let s be a positive integer such that $\mathbb{F}_{p^s} \subset L$. An element $b \in G(K_0)$ is called *s-decent* if

(i) $s\tilde{\nu}_b : \mathbb{D}_{K_0} \longrightarrow G_{K_0}$ factors through $\mathbb{G}_{m\,K_0}$,
(ii) the following equality (*decency equation*) holds in $G(K_0) \rtimes \sigma^{\mathbb{Z}}$:

$$(b\sigma)^s = s\tilde{\nu}_b(p)\sigma^s.$$

The element b is called *decent* if it is s-decent for some integer s. Finally, a σ-conjugacy class in $B(G, L)$ is called decent if it contains a decent element.

We draw the reader's attention to the condition $\mathbb{F}_{p^s} \subset L$. In the GL_n-case, a similar condition was imposed to distinguish between "semi-simple" and "split semi-simple" isocrystals. Obviously, an element that is s-decent is also s'-decent for any multiple s' of s, provided that $\mathbb{F}_{p^{s'}} \subset L$. Also we note that if b is s-decent, then $\tilde{\nu}_b$ is defined over $K_0^{\sigma^s} = \mathbb{Q}_{p^s}$ and $b \in G(K_0^{\sigma^s})$, see [183], Cor. 1.9.

Lemma 9.1.33 *If the σ-conjugacy class of b is decent, then the G-isocrystal N_b factors through* $\mathrm{Isoc}(L)^{\mathrm{ss}}$. *The converse is true if G is connected and L contains $\overline{\mathbb{F}}_p$, or if $G = \mathrm{GL}_n$ or if G is semi-simple connected and simply connected.*

Proof We first show that if b is s-decent, then N_b factors through $\mathrm{Isoc}(L)^{\mathrm{ss}}$. Since $b \in G(K_0^{\sigma^s}) = G(\mathbb{Q}_{p^s})$ and since split semi-simple isocrystals remain split semi-simple when extending the perfect field L, we may assume $K_0 = \mathbb{Q}_{p^s}$. Let (V, ρ) be a rational representation of G. Then the decency equation implies that $(\rho(b)\sigma)^s = s\rho_*(\tilde{\nu}_b)(p)$ on $N_b(V) = V \otimes \mathbb{Q}_{p^s}$. Decomposing $N_b(V) = \sum_{\alpha \in \frac{1}{s}\mathbb{Z}} N_b(V)_\alpha$, this gives $(\rho(b)\sigma)^s = p^{s\alpha}$ on $N_b(V)_\alpha$. Hence $N_b(V)$ is split semi-simple, by Definition 8.1.2 (iii).

Conversely, suppose that N_b factors through $\mathrm{Isoc}(L)^{\mathrm{ss}}$. Then there is some integer $s \in \mathbb{N}$, such that $\mathbb{F}_{p^s} \subset L$ and $s\tilde{\nu}_b$ factors through $\mathbb{G}_m$. By the discussion above Definition 8.1.3, N_b factors through a unique G-isocrystal $N_{b,s}$: $\mathrm{Rep}_{\mathbb{Q}_p}(G) \longrightarrow \mathrm{Isoc}(\mathbb{F}_{p^s})$. Namely we have

$$N_{b,s}(V) = \{x \in N_b(V) \mid \rho_*(\tilde{\nu}_b)(p^s)x = (\rho(b)\sigma)^s x\}.$$

Now, assume that $G = \mathrm{GL}_n$ or G is simply connected. Then by Remark 9.1.6, $H^1(\mathbb{Q}_{p^s}, G)$ is trivial. Thus by Lemma 9.1.5, $N_{b,s}$ is of the form $N_{b'}$ for some $b' \in G(\mathbb{Q}_{p^s})$ in the σ-conjugacy class of b. Moreover, for any rational representation (V, ρ) we have $(\rho(b')\sigma)^s = s\rho_*(\tilde{\nu}_{b'})(p)$ on $V \otimes \mathbb{Q}_{p^s}$ because $N_{b'}(V)$ is split semi-simple. This implies that $(b'\sigma)^s = s\tilde{\nu}_{b'}(p)\sigma^s$, i.e., that b' is s-decent. Finally assume that L contains $\overline{\mathbb{F}}_p$. Then we may increase s for the divisibility order so that the class of the fiber functor $\omega_{\mathrm{iso}} \circ N_{b,s}$ in $H^1(\mathbb{Q}_{p^s}, G)$ becomes trivial (thanks to Steinberg's theorem in Remark 9.1.6), in which case we argue as above. □

Remarks 9.1.34 Assume that G is connected.

(i) If L is algebraically closed, then any σ-conjugacy class is decent. This is a consequence of the above lemma.

(ii) If $L' \supset L$ is a pair of algebraically closed fields, then the induced map $B(G, L) \to B(G, L')$ is a bijection. This also follows from the above lemma and Lemma 9.1.5, see also [183], 1.16.

Notes and References The notion of an G-isocrystal (or isocrystal with G-structure) comes from the paper of Kottwitz [135]. Strictly speaking, Kottwitz introduces the set $\mathcal{B}(G)$ of σ-conjugacy classes in $\mathcal{B}(G, L)$, for an algebraically closed field L. He observes that for $G = \mathrm{GL}_n$, the set $\mathcal{B}(G)$ equals the set of isomorphism classes of isocrystals of dimension n over L. The set $\mathcal{B}(G)$ is useful for studying points mod p of Shimura varieties.

In [182] the notion of an isocrystal with G-structure is extended from perfect fields L to general schemes in characteristic p, following an idea of J. de Jong. The main result of [182] is a generalization to arbitrary reductive groups of the specialization theorem of Grothendieck and Katz (the Newton polygon of isocrystals decreases under specialization and its end point is constant).

We note that in the literature the slope homomorphism is generally denoted by ν_b. The reason is that most often L is assumed algebraically closed, and in this case, there is indeed no need to make any difference between ν_b and $\tilde{\nu}_b$, since sl is an isomorphism.

The notion of decency originates in [135], 4.3, and was formally introduced in [183], 1.8.

We note that our definition of a G-isocrystal differs from the traditional one, in which the compatibility with fiber functors is omitted, cf. [182]. Indeed, for a connected reductive group and when L is algebraically closed, both definitions coincide by Steinberg's theorem. Kottwitz [136] defines for non-connected groups a larger set $\mathbf{B}(G)$ which contains $\mathcal{B}(G)$ (for L algebraically closed). He shows that the set $\mathbf{B}(G)$ parametrizes isomorphism classes of $\mathrm{Rep}(G)$-isocrystals.

2 Filtrations on $\omega^{\mathbb{G}}$

In this section, we adapt the concept from Chapter V of a filtration of the natural fiber functor on Rep(G) to the setting of augmented groups schemes over Isoc(L). It turns out that, once the initial definitions are properly modified, the arguments are formally the same.

Let $\mathbb{G} := (G, \varphi, \nu)$ be an augmented group scheme over Isoc(L). As usual, we denote by A'_G the maximal K_0-split quotient torus of G. It comes equipped with a Weil descent datum which in turn induces an automorphism, still denoted by φ, of $X^*(A'_G)$. Explicitly, this automorphism is defined by

$$\chi \mapsto {}^{\varphi}\chi := (\mathrm{id}_{\mathbb{G}_m} \times \varphi_\sigma) \circ \sigma_* \chi \circ \varphi^{-1}, \quad \forall \chi : A'_G \longrightarrow \mathbb{G}_{m,K_0}.$$

Similarly, the maximal K_0-split central torus A_G of G is stable under the Weil descent datum φ and this descent datum is given by an automorphism φ of $X_*(A_G)$. Explicitly, it is given by

$$\lambda \mapsto {}^{\varphi}\lambda := \varphi \circ \sigma_* \lambda \circ (\mathrm{id}_{\mathbb{G}_m} \times \varphi_\sigma^{-1}), \quad \forall \lambda : \mathbb{G}_{m,K_0} \longrightarrow A_G.$$

Definition 9.2.1 We denote by

- $A'_{\mathbb{G}}$ the quotient torus of A'_G defined by $X^*(A'_{\mathbb{G}}) = X^*(A'_G)^{\varphi}$
- $A_{\mathbb{G}}$ the subtorus of A_G defined by $X_*(A_{\mathbb{G}}) = X_*(A_G)^{\varphi}$.

We note that $A'_{\mathbb{G}}$ and $A_{\mathbb{G}}$ only depend on the underlying group scheme (G, φ) over Isoc(L). Sometimes we denote them by $A'_{G,\varphi}$ and $A_{G,\varphi}$.

Lemma 9.2.2 *Assume that G is reductive. Then the natural morphism $A_{\mathbb{G}} \longrightarrow A'_{\mathbb{G}}$ is an isogeny.*

Proof Since G is reductive, we know that $A_G \longrightarrow A'_G$ is an isogeny. Hence the statement follows from the fact that the automorphism φ has *finite order* on both $X_*(A_G)$ and $X^*(A'_G)$. Indeed, this finiteness property is due to the requirement that $\mathcal{O}_G$ be an increasing union of finite-dimensional φ-stable subspaces in Definition 9.1.8. □

We are interested in $\mathbb{Q}$-filtrations of the fiber functor $\omega^{\mathbb{G}}$ of $\mathrm{Rep}_{\mathbb{Q}_p}(\mathbb{G})$ over an extension field K of K_0, as in Variant 4.2.9 (iii). The set of such filtrations is denoted by $\mathrm{Fil}_K(\omega^{\mathbb{G}})$. We note that any filtration of ω^G over K induces a filtration of $\omega^{\mathbb{G}}$ over K by composition with the forgetful functor $\omega^{\varphi,\nu} : \mathrm{Rep}_{\mathbb{Q}_p}(\mathbb{G}) \longrightarrow \mathrm{Rep}_{K_0}(G)$. Conversely any filtration of $\omega^{\mathbb{G}}$ over K factors through $\mathrm{Rep}_{K_0}(G)$, due to claim (1.7), page 230 in the proof of Proposition 9.1.20. Therefore we get a canonical bijection

$$\mathrm{Fil}_K(\omega^G) \xrightarrow{\sim} \mathrm{Fil}_K(\omega^{\mathbb{G}}), \tag{2.1}$$

and we will make no difference between both kinds of filtrations in the sequel.

With these adjustments, we may repeat Definition 5.1.1 and thus associate to any $\mathcal{F} \in \mathrm{Fil}_K(\omega^{\mathbb{G}})$ its degree

$$\deg(\mathcal{F}) \in X_*(A'_{\mathbb{G}})_{\mathbb{Q}} \tag{2.2}$$

and, provided that G is reductive, its slope

$$\mu(\mathcal{F}) \in X_*(A_{\mathbb{G}})_{\mathbb{Q}} \tag{2.3}$$

by applying Lemma 9.2.2. This applies in particular to the Newton filtration

$$\mathcal{F}_0 \in \mathrm{Fil}_{K_0}(\omega^{\mathbb{G}})$$

associated to the $\mathrm{Rep}_{\mathbb{Q}_p}(\mathbb{G})$-isocrystal $N^{\mathbb{G}}$. Following our convention in Chapter VIII (1.4), page 192, this is the filtration induced by the *opposite* of the $\mathbb{Q}$-1-PS $\nu \circ sl$, where sl is defined in (1.3), page 225.

Definition 9.2.3 Let $\mathbb{G} = (G, \varphi, \nu)$ be an augmented group scheme over $\mathrm{Isoc}(L)$ and let $\mathcal{F} \in \mathrm{Fil}_K(\omega^G)$ for an extension field K of K_0. We define the *degree*

$$\deg(\mathbb{G}, \mathcal{F}) := \deg(\mathcal{F}_0) + \deg(\mathcal{F}) \ \in X_*(A'_{\mathbb{G}})_{\mathbb{Q}}$$

and, if G is reductive, the *slope*

$$\mu(\mathbb{G}, \mathcal{F}) := \mu(\mathcal{F}_0) + \mu(\mathcal{F}) \ \in X_*(A_{\mathbb{G}})_{\mathbb{Q}}.$$

Remark 9.2.4 Let $\mathbb{G} = \mathbb{GL}(V, \Phi_V)$ for some $(V, \Phi) \in \mathrm{Isoc}(L)$. By definition the Weil descent datum φ_V on $\mathrm{GL}(V)$ induces the trivial one on $A'_{\mathrm{GL}(V)} = \mathbb{G}_m$. Therefore, the determinant induces an isomorphism $X_*(A'_{\mathbb{G}})_{\mathbb{Q}} \xrightarrow{\sim} \mathbb{Q}$. As in Remark 5.1.2, under this isomorphism $\deg(\mathbb{GL}(V, \Phi_V), \mathcal{F})$ corresponds to $\deg(V, \Phi_V, \mathcal{F}^\bullet V)$. Similarly, φ_V induces the trivial descent datum on $A_{\mathrm{GL}(V)} = \mathbb{G}_m$, so that the action of scalars induces an isomorphism $\mathbb{Q} \xrightarrow{\sim} X_*(A_{\mathbb{G}})_{\mathbb{Q}}$. The latter isomorphism identifies $\mu(V, \Phi_V, \mathcal{F}^\bullet V)$ with $\mu(\mathbb{GL}(V, \Phi_V), \mathcal{F})$. This ensures compatibility with Chapter VIII.

Example 9.2.5 Let G be an algebraic group over $\mathbb{Q}_p$ and let $b \in G(K_0)$. Recall that b gives rise to a G-isocrystal. A pair $(b, \mathcal{F})$ is called a *filtered G-isocrystal* over K. Let $\mathbb{G}_b := (G_{K_0}, \varphi_b, \nu_b)$ be the associated a.g.s. Since $\varphi_b = \mathrm{Int}(b) \circ (\mathrm{id}_G \times \varphi_\sigma)$, we have $A'_{\mathbb{G}_b} = A'_{G_{K_0}, \mathrm{id} \times \varphi_\sigma} = A'_G$, where A'_G is the maximal $\mathbb{Q}_p$-split quotient torus of G. Likewise, we have $A_{\mathbb{G}_b} = A_{G_{K_0}, \mathrm{id} \times \varphi_\sigma} = A_G$. In this setting, we will also write

$$\deg(b, \mathcal{F}) \in X_*(A'_G)_{\mathbb{Q}}, \ \text{resp. } \mu(b, \mathcal{F}) \in X_*(A_G)_{\mathbb{Q}}$$

instead of $\deg(\mathbb{G}_b, \mathcal{F})$, resp. $\mu(\mathbb{G}_b, \mathcal{F})$ if G is reductive. If $G = \mathrm{GL}(V)$ for some

$\mathbb{Q}_p$-vector space V, then $\deg(b, \mathcal{F})$ corresponds to $\deg((V, b(\mathrm{id}_V \otimes \sigma)), \mathcal{F}^\bullet V)$ via the identification $X_*(A'_{\mathrm{GL}(V)})_{\mathbb{Q}} \xrightarrow{\sim} \mathbb{Q}$ given by the determinant. Likewise for $\mu(b, \mathcal{F})$.

Degrees and slopes are functorial with respect to morphisms of augmented group schemes, for the same formal reason as in Chapter V. The functorial property for a morphism $\rho : \mathbb{G} \longrightarrow \mathbb{G}'$ is expressed in terms of the push-forward map

$$\rho_* : X_*(A'_{\mathbb{G}})_{\mathbb{Q}} \longrightarrow X_*(A'_{\mathbb{G}'})_{\mathbb{Q}},$$

which is defined for any morphism of the underlying group schemes over $\mathrm{Isoc}(L)$. Further, $\rho_* : \mathrm{Fil}_K(\omega^G) \longrightarrow \mathrm{Fil}_K(\omega^{G'})$ is defined for any morphism of the underlying group schemes over K_0. However, here there is also a more subtle functoriality property.

Definition 9.2.6 Let $\mathbb{G} = (G, \varphi, \nu)$ and $\mathbb{G}' = (G', \varphi', \nu')$ be two a.g.s. over $\mathrm{Isoc}(L)$. A *weak morphism* $\rho : \mathbb{G} \dashrightarrow \mathbb{G}'$ is a morphism of the underlying group schemes over $\mathrm{Isoc}(L)$ such that $\rho \circ \nu$ is an augmentation of (G', φ').

Let $\rho : \mathbb{G} \dashrightarrow \mathbb{G}'$ be a weak morphism of augmented group schemes over $\mathrm{Isoc}(L)$. We denote by

$$\tilde{\rho}_* : \mathrm{Fil}_K(\omega^{\mathbb{G}}) \longrightarrow \mathrm{Fil}_K(\omega^{\mathbb{G}'}) \tag{2.4}$$

the map which takes a filtration $\mathcal{F}$ to the filtration $(\zeta_\rho \circ sl) \cdot \rho_*(\mathcal{F})$, meaning $\rho_*\mathcal{F}$ shifted by the central $\mathbb{Q}$-1-PS $\zeta_\rho \circ sl$. Here ζ_ρ is the unique augmentation of $(Z(G'), \varphi')$ such that $\nu' = \zeta_\rho \cdot (\rho \circ \nu)$, as in Remark 9.1.17.

We note that a weak morphism always factors as the product of a genuine morphism of a.g.s. followed by a weak *iso*morphism. Moreover, a weak isomorphism is nothing but an isomorphism of the underlying group schemes over $\mathrm{Isoc}(L)$.

The reason why we renormalize the push-forward of filtrations in (2.4) above comes from the following observation: the Newton filtrations $\mathcal{F}'_0$ on $\omega^{G'}$ and $\mathcal{F}_0$ on ω^G are related by the equality

$$\rho_*\mathcal{F}_0 = (\zeta_\rho \circ sl) \cdot \mathcal{F}'_0$$

where the RHS denotes $\mathcal{F}'_0$ shifted by the central $\mathbb{Q}$-1-PS $\zeta_\rho \circ sl$.

Using this observation, and repeating the arguments in Chapter V, we get the same type of functorialities as for genuine morphisms of a.g.s.

Definition 9.2.7 A weak morphism $\rho : \mathbb{G} \dashrightarrow \mathbb{G}'$ is called *homogeneous* if $\rho(A_{\mathbb{G}}) \subset A_{\mathbb{G}'}$.

Proposition 9.2.8 *Let $\rho : \mathbb{G} \dashrightarrow \mathbb{G}'$ be a weak morphism of augmented group schemes over* Isoc(L). *Then*

$$\deg(\mathbb{G}', \tilde{\rho}_* \mathcal{F}) = \rho_* \deg(\mathbb{G}, \mathcal{F}).$$

Furthermore, if ρ is homogeneous and assuming that G and G' are reductive, then we also have

$$\mu(\mathbb{G}', \tilde{\rho}_* \mathcal{F}) = \rho_* \mu(\mathbb{G}, \mathcal{F}).$$

□

Two particular cases deserve attention. The case when ρ is a genuine morphism of augmented group schemes is illustrated by the following example.

Example 9.2.9 Let G, G' be two algebraic $\mathbb{Q}_p$-groups and let $(b, \mathcal{F})$ be a filtered G-isocrystal. Then a morphism $\rho : G \longrightarrow G'$ of $\mathbb{Q}_p$-groups induces a morphism of a.g.s. $\mathbb{G}_b \longrightarrow \mathbb{G}'_{\rho(b)}$ and the functoriality property becomes

$$\deg(\rho(b), \rho_* \mathcal{F}) = \rho_* \deg(b, \mathcal{F}). \tag{2.5}$$

Further, if ρ is *homogeneous* and G, G' are reductive, then

$$\mu(\rho(b), \rho_* \mathcal{F}) = \rho_* \mu(b, \mathcal{F}). \tag{2.6}$$

The second particular case is when ρ is a *weak isomorphism*, i.e., an isomorphism of the underlying group schemes. This is illustrated in the following example of "endoscopic" functoriality.

Example 9.2.10 Again let G, G' be algebraic groups over $\mathbb{Q}_p$ and $b \in G(K_0)$. Assume an *inner twisting* $\rho : G_{K_0} \longrightarrow G'_{K_0}$ is given. This means that ρ is an isomorphism of K_0-groups and that there is some $\gamma_{\text{ad}} \in G'_{\text{ad}}(K_0)$ such that the following equality of morphisms from $\sigma_*(G_{K_0})$ to G'_{K_0} holds:

$$\rho \circ (\text{id}_G \times \varphi_\sigma) = \text{Int}(\gamma_{\text{ad}}) \circ (\text{id}_{G'} \times \varphi_\sigma) \circ \sigma_* \rho.$$

Assume further that γ_{ad} lifts to $G'(K_0)$ and choose such a lift γ. Then ρ induces a *weak* isomorphism $\mathbb{G}_b \dashrightarrow \mathbb{G}'_{\rho(b)\gamma}$. Therefore, setting $b' := \rho(b)\gamma$ and $\mathcal{F}' := \tilde{\rho}_* \mathcal{F}$ we get equalities

$$\deg(b, \mathcal{F}) = \deg(b', \mathcal{F}'), \quad \text{and} \quad \mu(b, \mathcal{F}) = \mu(b', \mathcal{F}'),$$

where for the second one we are assuming that G is reductive. This example will turn out to be useful in reducing the study of filtered G-isocrystals to the case where G is quasi-split.

Let us explain now the relation with a possible external definition of degrees and slopes. Let C be a Tannaka category over $\mathbb{Q}_p$. We denote by A'_C the $\mathbb{Q}_p$-split torus whose group of characters $X^*(A'_C)$ is the group of isomorphism classes of invertible objects in C with product induced by the tensor product of C. Now, let N be a C-isocrystal, and let $\mathcal{F}$ be a filtration on $\omega_{\text{iso}} \circ N$ over K. Then for any object X in C we have a filtered isocrystal $(N(X), \mathcal{F}^\bullet N(X))$, whence a function

$$X \mapsto \deg(N(X), \mathcal{F}^\bullet N(X))$$

on isomorphism classes of objects in C. Restricting to invertible objects we get an element

$$\deg(N, \mathcal{F}) \in X_*(A'_C)_{\mathbb{Q}}. \tag{2.7}$$

Then for any object X in C, we have

$$\deg(N(X), \mathcal{F}^\bullet N(X)) = \langle \deg(N, \mathcal{F}), \textstyle\bigwedge^{\max} X \rangle.$$

Now let $\mathbb{G} = (G, \varphi, \nu)$ be the a.g.s. over $\mathrm{Isoc}(L)$ associated with (C, N) by Proposition 9.1.20. So we may identify C with $\mathrm{Rep}_{\mathbb{Q}_p}(\mathbb{G})$. Any invertible object in C defines a character of G over K_0 which is φ-invariant, and hence which factors through $A'_{\mathbb{G}}$. In this way we get a morphism

$$X^*(A'_C) \longrightarrow X^*(A'_{\mathbb{G}}).$$

Proposition 9.2.11 (i) *The above morphism is injective and has torsion cokernel. In particular it induces a bijection* $X^*(A'_C)_{\mathbb{Q}} \xrightarrow{\sim} X^*(A'_{\mathbb{G}})_{\mathbb{Q}}$.
(ii) *Its adjoint maps* $\deg(\mathbb{G}, \mathcal{F})$ *to* $\deg(N, \mathcal{F})$. *Equivalently, we have the formula*

$$\deg(V, \Phi_V, \mathcal{F}^\bullet V) = \langle \deg(\mathbb{G}, \mathcal{F}), \textstyle\bigwedge^{\max} V \rangle\,,$$

for any $(V, \Phi_V, \rho) \in \mathrm{Rep}_{\mathbb{Q}_p}(\mathbb{G})$.
(iii) *When G is reductive we have the formula*

$$\mu(V, \Phi_V, \mathcal{F}^\bullet V) = \langle \mu(\mathbb{G}, \mathcal{F}), \zeta_{V,\varphi_V} \rangle\,,$$

for any homogeneous $(V, \Phi_V, \rho) \in \mathrm{Rep}_{\mathbb{Q}_p}(\mathbb{G})$. *Here ζ_{V,φ_V} denotes the central character of the homogeneous representation (V, Φ_V), which is defined as the composition* $A_{\mathbb{G}} \longrightarrow A_{\mathbb{GL}(V,\Phi_V)} \xrightarrow{\sim} \mathbb{G}_m$.

Proof Note that statements (ii) and (iii) are independent of the first one. Statement (ii) is a consequence of the functorial property given by Proposition 9.2.8 and of the compatibility with the previous chapter in the case $\mathbb{GL}(V, \Phi_V)$, see Remark 9.2.4. The same holds for statement (iii).

Let us turn to injectivity in statement (i). Let (V, Φ_V) be a one-dimensional isocrystal over L equipped with a trivial action $\mathbb{G} \longrightarrow \mathbb{GL}(V, \Phi_V)$. We have to show that (V, Φ_V) is isomorphic to the unit isocrystal. But the canonical action of (D_L, φ_L) on (V, Φ_V) factors through $\mathbb{G}$, hence is trivial. Therefore the claim follows from [55], Cor. 8.18.

Let us now study the cokernel in statement (i). We start with a character $\chi : (G,\varphi) \longrightarrow (\mathbb{G}_{m,K_0}, \mathrm{id}\otimes\varphi_\sigma)$. The composition $\chi\circ\nu$ is a character of (D_L,φ_L). Since $D_L = \underline{\mathrm{Aut}}(\omega_{\mathrm{iso}})$, there is an isocrystal (V,Φ_V) such that the character $\chi\circ\nu$ occurs in the D_L-module V (compare [55], Lemme 5.9 (ii)). Then the $(\chi\circ\nu)$-isotypical part of V is Φ_V-stable, whence an isocrystal (W,Φ_W) on which the canonical action of D_L is via the character $\chi\circ\nu$. In other words, we have a factorization

$$\nu_{W,\Phi_W} : (D_L,\varphi_L) \xrightarrow{\nu} (G,\varphi) \xrightarrow{\chi} (\mathbb{G}_{m,K_0}, \mathrm{id}\times\varphi_\sigma) \xrightarrow{\iota} \mathbb{GL}(W,\Phi_W),$$

with ι the natural isomorphism between $(\mathbb{G}_{m,K_0}, \mathrm{id}\times\varphi_\sigma)$ and $A_{\mathbb{GL}(W,\Phi_W)}$. In particular, this makes (W,Φ_W) into an object of $\mathrm{Rep}_{\mathbb{Q}_p}(\mathbb{G})$. It is not one-dimensional, but its highest exterior power is, and the associated character is a multiple of χ. □

Let us mention the following trivial but important particular case.

Remark 9.2.12 If $\mathbb{G}$ has anisotropic center (meaning that $A_{\mathbb{G}} = \{1\}$), then for any representation $(V,\Phi_V,\rho) \in \mathrm{Rep}_{\mathbb{Q}_p}(\mathbb{G})$, we have $\mu(V,\Phi_V,\mathcal{F}^\bullet V) = 0$.

Example 9.2.13 Let G be a $\mathbb{Q}_p$-group. In the case of a filtered G-isocrystal $(b,\mathcal{F})$ the proposition above yields the formulas

$$\deg(N_b(V),\mathcal{F}^\bullet V) = \langle \deg(b,\mathcal{F}), \textstyle\bigwedge^{\max} V\rangle \tag{2.8}$$

for any $(V,\rho) \in \mathrm{Rep}_{\mathbb{Q}_p}(G)$ and

$$\mu(N_b(V),\mathcal{F}^\bullet V) = \langle \mu(b,\mathcal{F}), \zeta_V\rangle \tag{2.9}$$

for any *homogeneous* $(V,\rho) \in \mathrm{Rep}_{\mathbb{Q}_p}(G)$ with central character ζ_V. Thus we see that if G has anisotropic center, then for any rational representation $(V,\rho) \in \mathrm{Rep}_{\mathbb{Q}_p}(G)$, we have $\mu(N_b(V),\mathcal{F}^\bullet V) = 0$.

Having slopes at our disposal, we now define semi-stability as in Chapter V, and we define weak admissibility by adding a vanishing slope condition. We note that, for an a.g.s. $\mathbb{G} = (G,\varphi,\nu)$ over $\mathrm{Isoc}(L)$, the Lie algebra $\mathrm{Lie}(G)$ is endowed with a σ-linear bijective map $\mathrm{Lie}(\varphi)$. We denote by

$$\mathrm{Lie}\,\mathbb{G} = (\mathrm{Lie}(G), \mathrm{Lie}(\varphi))$$

the isocrystal thus obtained. The adjoint action of G on $\mathrm{Lie}(G)$ is compatible with descent data and augmentations, so that $\mathrm{Lie}\,\mathbb{G}$ is in fact a representation of $\mathbb{G}$.

Definition 9.2.14 Let $\mathbb{G} := (G,\varphi,\nu)$ be an a.g.s. over $\mathrm{Isoc}(L)$ and let $\mathcal{F} \in$

$\mathrm{Fil}_K(\omega^G)$. Then $\mathcal{F}$ is called *semi-stable* if $(\mathrm{Lie}\,\mathbb{G}, \mathcal{F}^\bullet\,\mathrm{Lie}(G))$ is a semi-stable object of $\mathrm{FilIsoc}_L^K$. It is called *weakly admissible* if furthermore $\mu(\mathbb{G}, \mathcal{F}) = 0$.

If we want to keep track of the a.g.s. $\mathbb{G}$, we also say that the pair $(\mathbb{G}, \mathcal{F})$ is semi-stable or weakly admissible.

Remark 9.2.15 Let $\mathbb{G} = \mathbb{GL}(V, \Phi_V)$ for some isocrystal (V, Φ_V). Then the isocrystal Lie $\mathbb{GL}(V, \Phi_V)$ is isomorphic to the tensor product $(V, \Phi_V) \otimes (V, \Phi_V)^*$ and the filtration $\mathcal{F}^\bullet\,\mathrm{Lie}\,G$ is the tensor product $\mathcal{F}^\bullet V \otimes \mathcal{F}^\bullet V^*$. Therefore, using the tensor product theorem as in Example 5.2.2, we see that $(\mathbb{GL}(V, \Phi_V), \mathcal{F})$ is semi-stable if and only if the corresponding filtered isocrystal $(V, \Phi_V, \mathcal{F}^\bullet V)$ is semi-stable in the sense of the previous chapter. The same holds for weak admissibility, by Remark 5.1.2.

Example 9.2.16 Let us consider the setting of Example 9.2.5. So we start with a $\mathbb{Q}_p$-group G and a filtered G-isocrystal $(b, \mathcal{F})$. We say that $(b, \mathcal{F})$ is semi-stable if $\mathcal{F}$ is semi-stable with respect to the associated a.g.s. $\mathbb{G}_b$. This means that $(N_b(\mathrm{Lie}\,G), \mathcal{F}^\bullet\,\mathrm{Lie}\,G)$ is semi-stable. Assume further that $G = \mathrm{GL}(V)$ for some $\mathbb{Q}_p$-vector space V. Then $(b, \mathcal{F})$ is semi-stable if and only if the filtered isocrystal $(V \otimes K_0, b \otimes \sigma, \mathcal{F}^\bullet V)$ is semi-stable. The same holds for weak admissibility.

As in Chapter V we want to prove functorial properties of semi-stability (resp. weak admissibility) and to give both internal and external characterizations. The most powerful tool for this is the HN-filtration. We need some preparations.

Definition 9.2.17 Let C be a Tannaka category over $\mathbb{Q}_p$ and let N be a C-isocrystal. A *$\mathbb{Q}$-filtration* of N is a tensor functor ${}^\bullet N : C \longrightarrow \mathrm{FilIsoc}_L^{K_0}$ such that

(i) $\omega_{\mathrm{fil}} \circ {}^\bullet N = N$, where $\omega_{\mathrm{fil}} : \mathrm{FilIsoc}_L^{K_0} \longrightarrow \mathrm{Isoc}(L)$ forgets the filtration,
(ii) $\omega_{\mathrm{iso}} \circ {}^\bullet N$ is a $\mathbb{Q}$-filtration of the fiber functor $\omega_{\mathrm{iso}} \circ N$ over K_0, in the sense of Definition 4.2.6.

Let $\mathbb{G} = (G, \varphi, \nu)$ be the a.g.s. over $\mathrm{Isoc}(L)$ associated with (C, N), cf. Proposition 9.1.20. We can describe a filtration of $N = N^{\mathbb{G}}$ purely in terms of $\mathbb{G}$. First we note that a filtration ${}^\bullet N^{\mathbb{G}}$ of $N^{\mathbb{G}}$ is uniquely determined by $\omega_{\mathrm{iso}} \circ {}^\bullet N^{\mathbb{G}}$, which is a filtration on the K_0-valued fiber functor $\omega^{\mathbb{G}}$ on $C = \mathrm{Rep}_{\mathbb{Q}_p}(\mathbb{G})$. Conversely, let ${}^\bullet\omega^{\mathbb{G}}$ be any filtration of $\omega^{\mathbb{G}}$ over K_0. It induces a filtration $\sigma_*({}^\bullet\omega^{\mathbb{G}})$ of the fiber functor $\sigma_*\omega^{\mathbb{G}}$ on $\mathrm{Rep}_{\mathbb{Q}_p}(\mathbb{G})$ deduced from $\omega^{\mathbb{G}}$ by base change along σ. The latter is equipped with an isomorphism

$$\omega^\varphi : \sigma_*\omega^{\mathbb{G}} \xrightarrow{\sim} \omega^{\mathbb{G}}. \tag{2.10}$$

Now the filtration ${}^\bullet\omega^{\mathbb{G}}$ comes from a filtration of N (i.e., is of the form ${}^\bullet\omega^{\mathbb{G}} = \omega_{\mathrm{iso}} \circ {}^\bullet N^{\mathbb{G}}$) if and only if ω^{φ} respects the filtrations, i.e., induces an isomorphism $\sigma_*({}^\bullet\omega^{\mathbb{G}}) \xrightarrow{\sim} {}^\bullet\omega^{\mathbb{G}}$.

In terms of splittings, let $\lambda : \mathbb{D}_{K_0} \longrightarrow G$ be a $\mathbb{Q}$-1-PS of G. It defines a filtration of ω^G hence a filtration of $\omega^{\mathbb{G}}$, see (2.1), page 239. This filtration extends to a filtration of $N^{\mathbb{G}}$ if and only if the cocharacter

$$ {}^{\varphi}\lambda := \varphi \circ \sigma_*\lambda \circ (\mathrm{id}_{\mathbb{D}} \times \varphi_{\sigma}^{-1}) \tag{2.11} $$

is par-equivalent to λ. In particular, the K_0-psgp $P_{{}^\bullet N^{\mathbb{G}}}$ of G associated to a filtration ${}^\bullet N^{\mathbb{G}}$ is stable under the Weil descent datum φ. Thus it defines a parabolic subgroup $\mathbb{P}_{{}^\bullet N^{\mathbb{G}}}$ of $\mathbb{G}$, in the sense of Definition 9.1.27.

From the definition of filtrations in terms of functors, we can push filtrations forward. In terms of a.g.s., given a morphism $\rho : \mathbb{G} \longrightarrow \mathbb{G}'$ and a filtration ${}^\bullet\omega^{\mathbb{G}}$ of $\omega^{\mathbb{G}}$ which is preserved by ω^{φ} as in (2.10), page 245 above, the filtration $\rho_*({}^\bullet\omega_{\mathbb{G}})$ of $\omega^{\mathbb{G}'}$ is again preserved by $\omega^{\varphi'}$. This defines the desired push-forward

$$ {}^\bullet N^{\mathbb{G}} \mapsto \rho_*({}^\bullet N^{\mathbb{G}}). $$

Moreover this argument only uses the compatibility of ρ and φ, therefore it applies to a *weak* morphism ρ as in Definition 9.2.6, and we also get a push-forward map ${}^\bullet N^{\mathbb{G}} \mapsto \rho_*({}^\bullet N^{\mathbb{G}})$ in this case.

Now we turn to HN-filtrations. The following result is proved exactly as in the characteristic 0 case of the proof of Theorem 5.3.1.

Theorem 9.2.18 *Let C be a semi-simple Tannaka category over $\mathbb{Q}_p$ and let N be a C-isocrystal. Let $\mathcal{F}$ be a filtration of $\omega_{\mathrm{iso}} \circ N$ over K. Then there is a unique $\mathbb{Q}$-filtration ${}^\bullet N_{\mathcal{F}}$ of N, such that for any $X \in C$, the induced filtration ${}^\bullet N_{\mathcal{F}}(X)$ on $N(X)$ is the Harder–Narasimhan filtration of the filtered isocrystal $(N(X), \mathcal{F}^\bullet N(X))$.* □

Definition 9.2.19 The filtration ${}^\bullet N_{\mathcal{F}} = {}^\bullet N_{\mathcal{F}}^{\mathbb{G}}$ is called the HN-*filtration* of $\mathcal{F}$. Its associated psgp $\mathbb{P}_{\mathrm{HN}} = \mathbb{P}_{{}^\bullet N_{\mathcal{F}}^{\mathbb{G}}} = (P_{\mathrm{HN}}, \varphi, \nu)$ is called the HN-*parabolic subgroup* of $\mathcal{F}$.

As in Remark 5.3.2, HN-filtrations are tautologically functorial. This means that for any morphism $\rho : \mathbb{G} \longrightarrow \mathbb{G}'$ of a.g.s., we have

$$ {}^\bullet N_{\rho_*\mathcal{F}}^{\mathbb{G}'} = \rho_*({}^\bullet N_{\mathcal{F}}^{\mathbb{G}}). \tag{2.12} $$

In fact, HN-filtrations are also functorial with respect to weak morphisms.

Proposition 9.2.20 *Let $\rho : \mathbb{G} \dashrightarrow \mathbb{G}'$ be a weak morphism of a.g.s. over* $\mathrm{Isoc}(L)$

and let $\mathcal{F} \in \mathrm{Fil}_K(\omega^{\mathbb{G}})$. *Then we have*

$$^\bullet N^{\mathbb{G}'}_{\tilde{\rho}_*\mathcal{F}} = \rho_*(^\bullet N^{\mathbb{G}}_{\mathcal{F}}).$$

Proof By functoriality it suffices to treat the case $\mathbb{G}' = \mathbb{GL}(V, \Phi_V)$ for some isocrystal (V, Φ_V). In this case $(Z(G'), \varphi') = (\mathbb{G}_{m,K_0}, \mathrm{id}_{\mathbb{G}_m} \times \varphi_\sigma)$, so the $\mathbb{Q}$-1-PS ζ_ρ of Definition 9.2.6 is an element of $X^*(A'_{D_L,\varphi_L})$. By Proposition 9.2.11 (i), or rather its proof, there is an isocrystal (W, Φ_W) such that the following factorization holds:

$$\nu_{W,\Phi_W} : (D_L, \varphi_L) \xrightarrow{\zeta_\rho^{-1}} (\mathbb{G}_{m,K_0}, \mathrm{id}_{\mathbb{G}_m} \times \varphi_\sigma) \longrightarrow \mathbb{GL}(W, \Phi_W)\,,$$

where the right map identifies $(\mathbb{G}_{m,K_0}, \mathrm{id}_{\mathbb{G}_m} \times \varphi_\sigma)$ with $A_{\mathbb{GL}(W,\Phi_W)}$. Therefore the composition

$$\rho' : (G, \varphi) \xrightarrow{\rho} \mathbb{GL}(V, \Phi_V) \longrightarrow \mathbb{GL}(V \otimes W, \Phi_V \otimes \Phi_W)$$

is compatible with augmentations and thus makes $(V \otimes W, \Phi_V \otimes \Phi_W)$ into an object of $\mathrm{Rep}_{\mathbb{Q}_p}(\mathbb{G})$, to which we can apply (2.12), page 246. This says that the HN-filtration of $(V \otimes W, \Phi_V \otimes \Phi_W, \rho'_*(\mathcal{F}))$ is $\rho'_*(^\bullet N^{\mathbb{G}}_{\mathcal{F}})$. Now the link beween HN-filtrations of $(V \otimes W, \Phi_V \otimes \Phi_W, \rho'_*(\mathcal{F}))$ and $(V, \Phi_V, \rho_*(\mathcal{F}))$ is given by

$$^\alpha(V \otimes W) = (^{\alpha-\mu(W,\Phi_W)}V) \otimes W, \ \forall \alpha \in \mathbb{Q}.$$

Here $\mu(W, \Phi_W)$ is the slope of (W, Φ_W). By construction it coincides with $\zeta_\rho \circ sl$ viewed as a rational element via $X^*(\mathbb{D}) \xrightarrow{\sim} \mathbb{Q}$. Therefore the HN-filtration of $(V, \Phi_V, \rho_*(\mathcal{F}))$ is $\rho_*(^\bullet N^{\mathbb{G}}_{\mathcal{F}})$ shifted by $-(\zeta_\rho \circ sl)$. Hence the HN-filtration of $(V, \Phi_V, \tilde{\rho}_*(\mathcal{F}))$ is $\rho_*(^\bullet N^{\mathbb{G}}_{\mathcal{F}})$. □

Now we state an analogue of Corollary 5.3.3, whose proof was rather formal and is easily adapted here.

Proposition 9.2.21 *Let* $\mathbb{G}$ *be a reductive augmented group scheme over* Isoc(L) *and let* $\mathcal{F} \in \mathrm{Fil}_K(\omega^{\mathbb{G}})$. *The following properties are equivalent.*

(i) $\mathcal{F}$ *is semi-stable.*
(ii) *The* HN*-filtration* $^\bullet N^{\mathbb{G}}_{\mathcal{F}}$ *is split by* $\mu(\mathbb{G}, \mathcal{F}) : \mathbb{D}_{K_0} \longrightarrow A_{\mathbb{G}} \subset G$.
(iii) *The* HN*-psgp* P_{HN} *is equal to* G. □

These properties of HN-filtrations, although rather formal, imply strong properties of semi-stability.

Corollary 9.2.22 *Let* $\mathbb{G}$ *be reductive. A pair* $(\mathbb{G}, \mathcal{F})$ *is semi-stable if and only if the adjoint pair* $(\mathbb{G}_{\mathrm{ad}}, \mathcal{F}_{\mathrm{ad}})$ *is semi-stable.* □

We note that for an adjoint a.g.s. (more generally an a.g.s. with anisotropic center), semi-stability is equivalent to weak admissibility because of Remark 9.2.12. Hence the above result is generally not true for "weakly admissible."

We now turn to functorial properties of semi-stability and weak admissibility.

Definition 9.2.23 A morphism $\rho : (G, \varphi) \longrightarrow (G', \varphi')$ of group schemes over $\mathrm{Isoc}(L)$ is called *injective* if the underlying morphism of K_0-groups is a closed immersion.

Corollary 9.2.24 *Let* $\rho : \mathbb{G} \dashrightarrow \mathbb{G}'$ *be a weak morphism of reductive augmented group schemes over* $\mathrm{Isoc}(L)$ *and let* $\mathcal{F} \in \mathrm{Fil}_K(\omega^{\mathbb{G}})$.

(i) *If* ρ *is injective, then "*$\tilde{\rho}_*\mathcal{F}$ *semi-stable" implies "*$\mathcal{F}$ *semi-stable."*

(ii) *If* ρ *is homogeneous, then "*$\mathcal{F}$ *semi-stable" implies "*$\tilde{\rho}_*\mathcal{F}$ *semi-stable."*

Proof Part (ii) is a direct consequence of (2.12), page 246 and the equivalence between items (i) and (ii) in Proposition 9.2.21.

Statement (i) is proved exactly as its analogue, Proposition 5.2.6 (i). □

Concerning weak admissibility, it turns out that we can drop the assumption "homogeneous" in (ii).

Corollary 9.2.25 *Let* $\rho : \mathbb{G} \dashrightarrow \mathbb{G}'$ *be a weak morphism of reductive augmented group schemes over* $\mathrm{Isoc}(L)$ *and let* $\mathcal{F} \in \mathrm{Fil}_K(\omega^{\mathbb{G}})$.

(i) *If* ρ *is injective, then "*$\tilde{\rho}_*\mathcal{F}$ *weakly admissible" implies "*$\mathcal{F}$ *weakly admissible."*

(ii) *"*$\mathcal{F}$ *weakly admissible" implies "*$\tilde{\rho}_*\mathcal{F}$ *weakly admissible."*

Proof (i) We already know that "$\tilde{\rho}_*\mathcal{F}$ semi-stable implies $\mathcal{F}$ semi-stable" by Corollary 9.2.24. Hence by Proposition 9.2.21 (ii), together with Proposition 9.2.20, we get $\mu(\mathbb{G}', \rho_*\mathcal{F}) = \rho \circ \mu(\mathbb{G}, \mathcal{F})$, which implies that $\mu(\mathbb{G}, \mathcal{F}) = 0$ since ρ is injective.

(ii) From the equivalence between items (i) and (ii) of Proposition 9.2.21, we deduce that a pair $(\mathbb{G}, \mathcal{F})$ is weakly admissible if and only if its HN-filtration is trivial (i.e., split by the trivial cocharacter). Hence statement (ii) follows immediately from the functoriality (2.12), page 246 of HN-filtrations. □

As in Chapter V, these functorial properties together with Remark 9.2.15 may serve as external definitions of semi-stability and weak admissibility. For semi-stability, we get an obvious analogue of Corollary 5.2.7 which we omit here. For weak admissibility, we state it in the context of filtered G-isocrystals, since it shows that our initial definition coincides with the traditional definition, as in [183].

Corollary 9.2.26 *Let G be a reductive $\mathbb{Q}_p$-group and let $(b,\mathcal{F})$ be a filtered G-isocrystal over K. Then $(b,\mathcal{F})$ is weakly admissible if and only if for any $(V,\rho) \in \mathrm{Rep}_{\mathbb{Q}_p}(G)$, the filtered isocrystal $(N_b(V),\mathcal{F}^\bullet V)$ is weakly admissible. In fact, it suffices to test this on one faithful representation (V,ρ).* □

In other words, a weakly admissible filtered G-isocrystal is a tensor functor from $\mathrm{Rep}_{\mathbb{Q}_p}(G)$ to the tannakian category of weakly admissible filtered isocrystals, which commutes with fiber functors.

Let us illustrate the above functorial properties in the case of inner twisting.

Example 9.2.27 Let $\rho : G_{K_0} \longrightarrow G'_{K_0}$ be an inner twisting between two reductive $\mathbb{Q}_p$-groups, and let $\gamma, b' \in G'(K_0)$ be as in Example 9.2.10. Let $(b,\mathcal{F})$ be a filtered G-isocrystal and let $(b',\mathcal{F}')$ the corresponding "inner twist" filtered G'-isocrystal, see Example 9.2.10. Then Proposition 9.2.20 says that

$$^\bullet N^G_{b',\mathcal{F}'} = \rho_*({}^\bullet N^G_{b,\mathcal{F}}).$$

In particular, $(b,\mathcal{F})$ is semi-stable if and only if $(b',\mathcal{F}')$ is semi-stable.

We have defined the HN-filtration in a tannakian, purely external way. We now turn to a purely internal characterization of this filtration, analogous to Proposition 5.3.5. It will imply, as in Chapter V, a purely internal characterization of semi-stability, analogous to that of Proposition 5.2.9.

Let $\mathbb{G} = (G,\varphi,\nu)$ be a reductive augmented group scheme over K_0 and let ${}^\bullet N^{\mathbb{G}}$ be a filtration of $N^{\mathbb{G}}$, with associated parabolic subgroup $\mathbb{P} = (P_{^\bullet N^{\mathbb{G}}},\varphi,\nu)$. We denote by $\mathbb{M} = (M,\varphi_M,\nu_M)$ the Levi quotient of $\mathbb{P}$, cf. Definition 9.1.27. Let

$$\overline{\lambda} : \mathbb{D}_{K_0} \longrightarrow M$$

be the central cocharacter associated with $\omega_{\mathrm{iso}} \circ {}^\bullet N$ as in Theorem 4.2.13 (c). Then $\overline{\lambda}$ is defined over K_0 and φ-stable, hence it defines an element $\overline{\lambda} \in X_*(A_{\mathbb{M}})_{\mathbb{Q}}$. Now, let further $\mathcal{F} \in \mathrm{Fil}_K(\omega^{\mathbb{G}})$. We denote by $\mathcal{F}_{\mathbb{P}} \in \mathrm{Fil}_K(\omega^{\mathbb{P}})$ the unique filtration on ω^P over K which induces $\mathcal{F}$, as in Proposition 4.2.17. Let $\mathcal{F}_{\mathbb{M}} \in \mathrm{Fil}_K(\omega^{\mathbb{M}})$ be its push-forward to $\mathbb{M}$.

Proposition 9.2.28 *Let $\mathbb{G}$ be a reductive augmented group scheme over* Isoc(L), *let $\mathcal{F} \in \mathrm{Fil}_K(\omega^{\mathbb{G}})$ and let ${}^\bullet N^{\mathbb{G}}$ be a filtration of $N^{\mathbb{G}}$. Then, using the above notation, we have ${}^\bullet N^{\mathbb{G}} = {}^\bullet N^{\mathbb{G}}_{\mathcal{F}}$ if and only if $\mathcal{F}_{\mathbb{M}}$ is semi-stable of slope $\mu(\mathbb{M},\mathcal{F}_{\mathbb{M}}) = \overline{\lambda}$.*

Proof There is no difficulty in adapting the proof of Proposition 5.3.5 to the setting of augmented group schemes over Isoc(L). The strategy is the same. Indeed, first we consider the case of $\mathbb{G} = \mathbb{GL}(V, \Phi_V)$. Then we treat the general case using functoriality. At this point, one just has to check that there indeed exist faithful representations of $\mathbb{G}$.

In order to check this, we first choose a finite-dimensional G-stable subspace W of O_G which is faithful as a representation of G. Since O_G is an increasing union of φ-stable subspaces, there is a finite-dimensional φ-stable subspace W' containing W. It does not need to be G-stable, but its G-saturate V is again finite-dimensional and still φ-stable. Then $(V, \varphi_{|V})$ is a faithful representation of $\mathbb{G}$. □

Remark 9.2.29 Most of this foundational section could have been developed exclusively in the traditional setting of G-isocrystals, except for the above result. Indeed, assuming $\mathbb{G}$ to be of the form $\mathbb{G}_b$ for some $b \in G(K_0)$ then P is stable under $\mathrm{Int}(b) \circ (\mathrm{id} \times \varphi_\sigma)$ but this does not imply that P is defined over $\mathbb{Q}_p$ (i.e., stable under $\mathrm{id} \times \varphi_\sigma$), or that $b \in P(K_0)$ (equivalently, that P is stable under $\mathrm{Int}(b)$). With some hypothesis, such as, e.g., G quasi-split, one can show that $\mathbb{P}$ comes from a P'-isocrystal, for a suitable psgp P' of G. But such an isocrystal is not even well defined up to isomorphism.

We now derive the internal characterization of semi-stability. Again $\mathbb{G} = (G, \varphi, \nu)$ is an a.g.s. over Isoc(L). Let $\mathbb{P} = (P, \varphi, \nu)$ be a parabolic subgroup of $\mathbb{G}$. So P is stable under the Weil descent datum φ. Therefore its unipotent radical U_P is also stable under φ, and

$$\mathfrak{u}_P := \mathrm{Lie}\, U_P$$

defines a representation of $\mathbb{P}$. In turn its determinant $\bigwedge^{\max} \mathfrak{u}_P$ defines a character of $A'_{\mathbb{P}}$.

Corollary 9.2.30 *Let $\mathcal{F} \in \mathrm{Fil}_K(\omega^{\mathbb{G}})$. Then $\mathcal{F}$ is semi-stable if and only if for any maximal parabolic subgroup $\mathbb{P}$ of $\mathbb{G}$, we have*

$$\langle \deg(\mathbb{P}, \mathcal{F}_P), \textstyle\bigwedge^{\max} \mathfrak{u}_P \rangle \leq 0.$$

Here, as usual, $\mathcal{F}_P$ denotes the unique filtration on ω^P over K which induces $\mathcal{F}$, as in Proposition 4.2.17. We note that $\mathbb{P}$ may be a maximal psgp of $\mathbb{G}$ without P being maximal in G. The corollary is derived from Proposition 9.2.28 exactly as in Chapter V.

We end this section with a discussion on the effects of a change of the field L. So let $L' \supset L$ be a perfect field extension and let us take up the notation of Construction 9.1.29. Let $\mathbb{G}$ be an a.g.s. over Isoc(L) with "base change"

$\mathbb{G}'$ over $\mathrm{Isoc}(L')$ and let $\mathcal{F} \in \mathrm{Fil}_K(\omega^{\mathbb{G}})$ for some extension K of $K_0' = K_0(L')$. Thus $\mathcal{F}$ defines also an element $\mathcal{F}' \in \mathrm{Fil}_K(\omega^{\mathbb{G}'})$. In this case we have canonical identifications $A'_{\mathbb{G}'} \xrightarrow{\sim} A'_{\mathbb{G}}$ and $A_{\mathbb{G}} \xrightarrow{\sim} A_{\mathbb{G}'}$ which respectively take $\deg(\mathbb{G}', \mathcal{F})$ to $\deg(\mathbb{G}, \mathcal{F})$ and $\mu(\mathbb{G}, \mathcal{F})$ to $\mu(\mathbb{G}', \mathcal{F}')$ (if $\mathbb{G}$ is reductive).

Proposition 9.2.31 *Assume that $(\mathbb{G}', \mathcal{F}')$ is semi-stable. Then $(\mathbb{G}, \mathcal{F})$ is semi-stable. If $N^{\mathbb{G}}$ is split semi-simple (cf. Definition 9.1.30), then the converse holds.* □

More generally, we will see later that the "type" of the HN-filtration of $(\mathbb{G}, \mathcal{F})$ (called its HN-vector) lies below that of $(\mathbb{G}', \mathcal{F}')$ for some suitable dominance order, with equality whenever $N^{\mathbb{G}}$ is split semi-simple.

Notes and References The notion of a weakly admissible filtered isocrystal for arbitrary groups comes from [183]. More precisely, [183] considers filtered G-isocrystals $(b, \mathcal{F})$ as in Example 9.2.16. Such a pair is called weakly admissible if the external weakly admissible condition is satisfied, i.e., if for any (or one faithful) representation the induced filtered isocrystal is weakly admissible in the sense of Chapter VIII.

3 Automorphism groups and decency

Let C be a Tannaka category over $\mathbb{Q}_p$ and let $N : C \longrightarrow \mathrm{Isoc}(L)$ be a C-isocrystal. We associate to N its *automorphism group functor*

$$J := \underline{\mathrm{Aut}}^{\otimes}(N)$$

on $\mathbb{Q}_p$-algebras, which is representable by an affine group scheme over $\mathbb{Q}_p$, cf. Proposition 4.1.24. It is compatible with base change so that J_{K_0} is the automorphism group functor associated to the induced tensor functor $N_{K_0} : C_{K_0} \longrightarrow \mathrm{Isoc}(L)_{K_0}$ between tannakian categories over K_0. Let $G := \underline{\mathrm{Aut}}^{\otimes}(\omega_{\mathrm{iso}} \circ N)$ be the automorphism group of the fiber functor $\omega_{\mathrm{iso}} \circ N : C \longrightarrow \mathrm{Vec}_{K_0}$. Then there is a commutative diagram of $\otimes$-functors between tannakian categories over K_0,

$$\begin{array}{ccc} C_{K_0} & \xrightarrow{N_{K_0}} & \mathrm{Isoc}(L)_{K_0} \\ \downarrow & & \downarrow \\ \mathrm{Rep}_{K_0}(G) & \xrightarrow[\omega^{\nu}]{} & \mathrm{Rep}_{K_0}(D_L), \end{array}$$

where the vertical functors are equivalences and $\nu : D_L = \underline{\mathrm{Aut}}^{\otimes}(\omega_{\mathrm{iso}}) \longrightarrow G$ is the canonical map. Therefore we get an isomorphism

$$J_{K_0} \xrightarrow{\sim} \underline{\mathrm{Aut}}^{\otimes}(\omega^{\nu}) = G_{\nu}, \tag{3.1}$$

where G_ν is the centralizer of ν in G, cf. Proposition 4.1.24.

In more explicit terms, for any $\mathbb{Q}_p$-algebra R we have

$$\begin{aligned} J(R) = &\ \{g \in G(R \otimes_{\mathbb{Q}_p} K_0) \mid \varphi \circ \sigma_* g = g \circ (\mathrm{id}_R \times \varphi_\sigma)\} \\ = &\ \{g \in G_\nu(R \otimes_{\mathbb{Q}_p} K_0) \mid \varphi \circ \sigma_* g = g \circ (\mathrm{id}_R \times \varphi_\sigma)\}. \end{aligned} \tag{3.2}$$

Here φ is the Weil descent datum on G associated to N, so that $\mathbb{G} = (G, \varphi, \nu)$ is the a.g.s. over $\mathrm{Isoc}(L)$ associated by Proposition 9.1.20 to (C, N). In fact, the relation between φ and ν shows that φ stabilizes the centralizer G_ν, i.e., defines a Weil descent datum on G_ν, and (3.1), page 251 means that this Weil descent datum is effective and J is the descended group over $\mathbb{Q}_p$. In particular, if L is a finite field, $\varphi_{|G_\nu}$ extends to a genuine Galois descent datum on G_ν with respect to the extension $K_0/\mathbb{Q}_p$. Finally we note that the embedding $J_{K_0} \hookrightarrow G$ is induced by (3.2) above and the multiplication map $K_0 \otimes_{\mathbb{Q}_p} K_0 \longrightarrow K_0$.

Definition 9.3.1 The C-isocrystal N, or the corresponding a.g.s. $\mathbb{G}$, is called *basic* if ν factors through the center of G, i.e., if $G = G_\nu$.

Thus in the basic case, the Weil descent datum φ on G is effective and J is the descended group, a $\mathbb{Q}_p$-form of G. In other words the map (3.1), page 251 defines a weak isomorphism

$$\mathbb{J}_1 \overset{\sim}{\dashrightarrow} \mathbb{G}, \tag{3.3}$$

where

$$\mathbb{J}_1 = (J_{K_0}, \mathrm{id}_J \times \varphi_\sigma, 1)$$

is the a.g.s. associated to the trivial J-isocrystal, i.e., to the element $1 \in J(K_0)$. We will see later how this fact reduces the study of period domains for basic isocrystals to that for trivial isocrystals.

In the non-basic case, we still have a weak morphism $\mathbb{J}_1 \dashrightarrow \mathbb{G}$, if we assume that *the Galois group of L is commutative*. Indeed in this case D_L is a commutative group scheme, cf. Examples 9.1.14. Hence ν factors through G_ν. Therefore we get an a.g.s.

$$\mathbb{G}_\nu := (G_\nu, \varphi, \nu)$$

together with an embedding $\mathbb{G}_\nu \hookrightarrow \mathbb{G}$. The map (3.1), page 251 defines a *weak isomorphism* $\mathbb{J}_1 \overset{\sim}{\dashrightarrow} \mathbb{G}_\nu$, whence a diagram

$$\mathbb{J}_1 \overset{\sim}{\dashrightarrow} \mathbb{G}_\nu \hookrightarrow \mathbb{G}. \tag{3.4}$$

Example 9.3.2 Let $\mathbb{G} = \mathbb{GL}(V, \Phi_V)$ for some isocrystal (V, φ_V). The new J coincides with that defined in Chapter VIII (2.2), page 201. We pointed out at the time that when (V, Φ_V) is split semi-simple, the group J is an inner form of a Levi factor of $\mathrm{GL}(V)$. The diagram (3.4) above is what remains from this observation in the most general situation. Further, in this case, the a.g.s. $\mathbb{GL}(V, \Phi_V)$ is basic if and only if (V, Φ_V) is isotypic.

Example 9.3.3 Let G be defined over $\mathbb{Q}_p$ and let $b \in G(K_0)$. The automorphism group J_b of $(\mathrm{Rep}_{\mathbb{Q}_p}(G), N_b)$ is given by

$$J_b(R) = \{g \in G(R \otimes_{\mathbb{Q}_p} K_0) \mid g(b\sigma) = (b\sigma)g\}$$

for any $\mathbb{Q}_p$-algebra R. Moreover, conjugation by some $g \in G(K_0)$ induces an isomorphism

$$J_b \xrightarrow{\sim} J_{gb\sigma(g)^{-1}}$$

over $\mathbb{Q}_p$. In general, $J_b = J$ is not a $\mathbb{Q}_p$-subgroup of G. However, we will discuss below how properties of b yield information on J_b and on the diagram (3.4), page 252 above. We also note that if b is decent (see Definition 9.1.32), then N_b is basic in our sense iff b is basic in the sense of Kottwitz [135], §5.

Remark 9.3.4 Here are some negative properties of J.
(i) Assume G is reductive. In general J does not need to be reductive. By [206], II, 4.1, cf. also [38], Thm. 3.5.3, its neutral component J^o is reductive if ν factors through the multiplicative part of D_L; conversely, if ν does not factor through the multiplicative part of D_L, then the center of J^o contains unipotent elements. On the other hand, ν factors through the multiplicative part of D_L if and only if N factors through the subcategory of semi-simple isocrystals in $\mathrm{Isoc}(L)$ (those which are direct sums of simple isocrystals).
(ii) In general the formation of J is not compatible with changes of the perfect field L. More precisely, let $L' \supset L$ be a perfect field extension of L, and let us take up the notation of Construction 9.1.29. Denote by J' the automorphism group of $N' = N \circ \varepsilon_L^{L'}$. Then we formally have a morphism of $\mathbb{Q}_p$-groups $J \longrightarrow J'$. But it need not be an isomorphism. This was already clear by Example 8.2.6 in the case $L = \mathbb{F}_p$, $L' = \overline{\mathbb{F}}_p$ and $\mathbb{G} = \mathbb{GL}(V, \Phi_V)$.

Proposition 9.3.5 *Let N be a split semi-simple C-isocrystal with associated a.g.s. $\mathbb{G} = (G, \varphi, \nu)$.*

(i) *If G is connected and reductive, then J is connected and reductive. More precisely, $J_{K_0} = G_\nu$ is the centralizer of the torus $\nu \circ sl(\mathbb{D}_{K_0})$, hence is a Levi subgroup of G.*

(ii) *If L' is a perfect field extension of L, then the canonical morphism $J \longrightarrow J'$ in Remark 9.3.4 (ii) is an isomorphism.* □

For an a.g.s. $\mathbb{G}$ we will often denote by $J^{\mathbb{G}}$ the automorphism group of

$N^{\mathbb{G}}$. By definition, any morphism $\rho : \mathbb{G} \longrightarrow \mathbb{G}'$ induces a $\mathbb{Q}_p$-morphism $\rho_J : J^{\mathbb{G}} \longrightarrow J^{\mathbb{G}'}$. Moreover, ρ_J is a closed immersion if ρ is.

Proposition 9.3.6 *Let $\mathbb{G}$ be a connected reductive a.g.s. such that $J^{\mathbb{G}}$ is also connected reductive and let $\mathbb{P}$ be a psgp of $\mathbb{G}$. Then $J^{\mathbb{P}}$ is a $\mathbb{Q}_p$-psgp of $J^{\mathbb{G}}$.*

Proof By base change and property (3.1), page 251, it suffices to prove that P_ν is a K_0-psgp of G_ν. Since $J^{\mathbb{G}}$ is assumed to be reductive, $\nu : D_L \longrightarrow P$ factors through the multiplicative part of D_L, cf. Remark 9.3.4 (i). Therefore ν factors through a Levi subgroup M of P. We may choose a cocharacter $\lambda : \mathbb{G}_{m,K_0} \longrightarrow G_{K_0}$ such that $P = P_\lambda$ and $M = M_\lambda$. In particular, λ factors through the center of M and thus it centralizes ν, i.e., it factors through G_ν. Therefore P_ν may be identified with the K_0-psgp of G_ν defined by λ. □

We note that, when $J^{\mathbb{G}}$ is not reductive, then $J^{\mathbb{P}}$ is not necessarily a parabolic subgroup of $J^{\mathbb{G}}$. This already occurs for $L = \mathbb{F}_p$ and $\mathbb{G} = \mathbb{GL}(V, \Phi_V)$ for a three-dimensional unipotent isocrystal.

There is also a relation between filtrations of the natural fiber functor ω^J and filtrations of $N^{\mathbb{G}}$, in the sense of Definition 9.2.17. Indeed, the canonical embedding (3.1), page 251 of $J^{\mathbb{G}}_{K_0}$ into G induces a push-forward map $\mathrm{Fil}_{K_0}(\omega^J) \longrightarrow \mathrm{Fil}_{K_0}(\omega^G)$. If we start with a filtration of ω^J over $\mathbb{Q}_p$, then we can split it by a $\mathbb{Q}$-1-PS $\lambda_J : \mathbb{D} \longrightarrow J$ defined over $\mathbb{Q}_p$. The composed $\mathbb{Q}$-1-PS $\lambda_G = \mathrm{incl} \circ \lambda_{J,K_0} : \mathbb{D}_{K_0} \longrightarrow J_{K_0} \hookrightarrow G$ is φ-stable and thus defines a filtration of $N^{\mathbb{G}}$. In other words, the above push-forward map induces a map

$$\mathrm{Fil}_{\mathbb{Q}_p}(\omega^J) \longrightarrow \mathrm{Fil}(N^{\mathbb{G}}) \tag{3.5}$$

from filtrations of ω^J over $\mathbb{Q}_p$ to filtrations of $N^{\mathbb{G}}$.

Proposition 9.3.7 *Assume that $\mathbb{G}$ is reductive and that $N^{\mathbb{G}}$ is a split semi-simple isocrystal. Then the map*

$$\mathrm{Fil}_{\mathbb{Q}_p}(\omega^J) \longrightarrow \mathrm{Fil}(N^{\mathbb{G}})$$

is a bijection.

Proof Let us first prove injectivity. This only requires that $J^{\mathbb{G}}$ be reductive and connected. Moreover, it is enough to prove that the push-forward map $\mathrm{Fil}_{K_0}(\omega^J) \longrightarrow \mathrm{Fil}_{K_0}(\omega^G)$ is injective. Let λ, λ' be two $\mathbb{Q}$-1-PS of J over K_0 which induce the same filtration on ω^G. Then the psgps $P^J_\lambda = G_\nu \cap P^G_\lambda$ and $P^J_{\lambda'} = G_\nu \cap P^G_{\lambda'}$ of J_{K_0} coincide. Therefore, up to par-equivalence (in J) we may assume that λ, λ' factor through the same maximal torus of G_ν. But since they are par-equivalent in G they have to coincide. In particular they induce the same filtration on $\omega^J_{K_0}$.

Now we turn to surjectivity. We first reduce to the case where L is a finite field. Indeed, since G is algebraic over K_0, the category $\mathrm{Rep}_{\mathbb{Q}_p}(\mathbb{G})$ has a tensor generator, see Corollary 9.1.25. Hence $N^{\mathbb{G}}$ factors through the subcategory of split semi-simple isocrystals whose slopes belong to $\frac{1}{s}\mathbb{Z}$ for some integer s. By the discussion following Definition 8.1.2, this tannakian subcategory has a fiber functor over $\mathbb{Q}_{p^s}$. Therefore we may assume $L = \mathbb{F}_{p^s}$.

Now let us start with a filtration ${}^\bullet N$ of $N^{\mathbb{G}}$. What we have to show is that it can be split by a φ-stable $\mathbb{Q}$-1-PS. Let λ be any splitting of $\omega_{\mathrm{iso}} \circ {}^\bullet N$. Since ${}^\bullet N$ defines a filtration of the slope grading functor, the $\mathbb{Q}$-1-PS ${}^{\nu(p^s)}\lambda$ is par-equivalent to λ. Here we recall that, since $N^{\mathbb{G}}$ is split semi-simple, ν factors as in Proposition 9.1.31, and this makes the evaluation $\nu(p^s)$ meaningful. Since $\nu(p^s)$ is semi-simple, it is conjugate under $P_\lambda(\mathbb{Q}_{p^s})$ to an element of the centralizer $Z(\lambda)(\mathbb{Q}_{p^s})$. In other words, there is a $\mathbb{Q}$-1-PS λ' par-equivalent to λ and centralized by $\nu(p^s)$. Such a λ' factors through G_ν. Since the filtration defined by λ' comes from a filtration of $N^{\mathbb{G}}$, the $\mathbb{Q}$-1-PS ${}^\varphi\lambda'$ is par-equivalent (in G) to λ'. By the same argument as above regarding injectivity, λ' and ${}^\varphi\lambda'$ are then also par-equivalent in G_ν. In particular the parabolic subgroup $P_{\lambda'} \cap G_\nu$ is stable under the Galois descent datum defined by φ on G_ν, hence it descends to a parabolic subgroup $P^J_{\lambda'}$ of J. Moreover there is some element u in the unipotent radical $U^J_{\lambda'}(\mathbb{Q}_{p^s})$ such that ${}^\varphi\lambda' = {}^u\lambda'$. Since the centralizer of λ' in $U^J_{\lambda'}$ is trivial, the map $\sigma \mapsto u$ defines a Galois 1-cocycle which can be trivialized since $U^J_{\lambda'}$ is unipotent. So we may write $u = {}^\varphi v^{-1} v$ for some $v \in U^J_{\lambda'}$ and, the cocharacter $\lambda_G := {}^v\lambda'$ has the required property ${}^\varphi\lambda_G = \lambda_G$. □

We now focus our attention on G-isocrystals, especially to decent ones. In this situation, one can be more precise about the nature of J, as in [183]. This is the object of the following couple of statements.

Lemma 9.3.8 *Let G be a reductive group over $\mathbb{Q}_p$ and let $b \in G(K_0)$ be an s-decent element, as in Definition 9.1.32. Recall the notation $\tilde{\nu}_b = \nu_b \circ sl$, and write J_b for $J^{\mathbb{G}_b}$. Then $J_{b,\mathbb{Q}_{p^s}}$ is the centralizer $G_{\tilde{\nu}_b}$ of $\tilde{\nu}_b$ in $G_{\mathbb{Q}_{p^s}}$. If furthermore $\tilde{\nu}_b$ is defined over $\mathbb{Q}_p$ (hence so is $G_{\tilde{\nu}_b}$), then J_b is an* inner *form of $G_{\tilde{\nu}_b}$.*

Proof Under the decent hypothesis, there is no need to differentiate between ν_b and $\tilde{\nu}_b$, in view of Proposition 9.1.31. In particular $G_{\nu_b,K_0} = G_{\tilde{\nu}_b,K_0}$. Therefore, by (3.1), page 251, we have

$$J_b(R) = \{g \in G_{\tilde{\nu}_b}(R \otimes K_0) \mid g(b\sigma) = (b\sigma)g\}.$$

In particular for $g \in J_b(R)$, we have $g(b\sigma)^s = (b\sigma)^s g$, hence by the decency equation we get ${}^{\sigma^s}g = g$. Thus

$$J_b(R) = \{g \in G_{\tilde{\nu}_b}(R \otimes \mathbb{Q}_{p^s}) \mid g(b\sigma) = (b\sigma)g\}.$$

Since the element $b\sigma$ centralizes $\tilde{\nu}_b$ by definition, conjugation by b induces an isomorphism

$$G_{\tilde{\nu}_b} \times_{\mathbb{Q}_{p^s},\sigma} \mathbb{Q}_{p^s} \xrightarrow{\sim} G_{\tilde{\nu}_b}$$

and the decency equation says that this is a descent datum with respect to the Galois covering $\operatorname{Spec} \mathbb{Q}_{p^s} \longrightarrow \operatorname{Spec} \mathbb{Q}_p$. From its definition, J_b is the descent of $G_{\tilde{\nu}_b}$ defined by this cocycle. In particular $J_{b,\mathbb{Q}_{p^s}} = G_{\tilde{\nu}_b}$.

Now assume further that $\tilde{\nu}_b$ is defined over $\mathbb{Q}_p$, i.e., that it is σ-invariant. Since it is also centralized by $b\sigma$, it is likewise centralized by b. Hence $b \in G_{\tilde{\nu}_b}(\mathbb{Q}_{p^s})$ and the decency equation says that the map induced by $\sigma \mapsto b$ is a 1-cocycle from $\operatorname{Gal}(\mathbb{Q}_{p^s}/\mathbb{Q}_p)$ to the adjoint group $G_{\tilde{\nu}_b,\mathrm{ad}}(\mathbb{Q}_{p^s})$. Thus J_b is an inner form of $G_{\tilde{\nu}_b}$. □

Corollary 9.3.9 *Let $b \in G(K_0)$ with decent σ-conjugacy class. If G is reductive and quasi-split over $\mathbb{Q}_p$, then J_b is isomorphic to an inner form of a Levi subgroup of G.*

Proof By Lemma 9.3.8, it is enough to show that b is σ-conjugate to a decent element b' such that $\nu_{b'}$ is defined over $\mathbb{Q}_p$. We may start assuming that b is s-decent. Since $b\sigma$ centralizes ν_b, the $G(\mathbb{Q}_{p^s})$-conjugacy class of ν_b is σ-stable, hence it is defined over $\mathbb{Q}_p$. If G is reductive and quasi-split, then the Kottwitz Lemma 6.1.5 shows that this conjugacy class has a $\mathbb{Q}_p$-rational point, say $g\nu_b g^{-1}$ for some $g \in G(\mathbb{Q}_{p^s})$. Then $gb\sigma(g)^{-1}$ is also decent, satisfying the same decency equation as b. □

In general, this result no longer holds for a non-quasi-split group, simply because there are not enough rational Levi subgroups. We will give a counterexample below, see Example 9.3.13. However, we recall that any (connected) reductive group over $\mathbb{Q}_p$ is an inner form of a uniquely determined quasi-split one. As an important special case of Lemma 9.3.8, we have a convenient way of constructing inner forms.

Corollary 9.3.10 *Let (G, b) be as in Lemma 9.3.8. If b is basic, then J_b is an inner form of G.*

Proof Indeed, $\tilde{\nu}_b$ is centralized by b and $b\sigma$, hence also by σ. This means that it is defined over $\mathbb{Q}_p$, so we may apply Lemma 9.3.8. □

Conversely, one may ask when an inner form G' of G is isomorphic to some J_b for a basic $b \in G(K_0)$.

Lemma 9.3.11 *Let G, G' be two inner-equivalent groups over $\mathbb{Q}_p$. Assume that G'_{K_0} is isomorphic to G_{K_0}, and that G and G' are adjoint. Then there is a decent and basic element $b \in G(K_0)$ such that J_b isomorphic to G' over $\mathbb{Q}_p$.*

Proof The first assumption means that K_0 contains a sufficiently large $\mathbb{Q}_{p^s}$ such that G and G' are isomorphic over $\mathbb{Q}_{p^s}$. Let $\rho : G_{\mathbb{Q}_{p^s}} \xrightarrow{\sim} G'_{\mathbb{Q}_{p^s}}$ be an isomorphism. Then the automorphism ${}^{\sigma}\rho^{-1} \circ \rho$ of $G_{\mathbb{Q}_{p^s}}$ has to be inner. Hence it is given by conjugation by an element $b \in G(\mathbb{Q}_{p^s})$, since G is adjoint. In other words the map $\sigma \mapsto b$ defines a 1-cocycle for the Galois extension $\mathbb{Q}_{p^s}$ of $\mathbb{Q}_p$. However, this property is equivalent to b being decent and basic. □

We note that the first assumption is always satisfied if the map

$$H^1(\mathbb{Q}_p, G_{\mathrm{ad}}) \longrightarrow H^1(K_0, G_{\mathrm{ad}})$$

is trivial. In particular, this is the case if L is algebraically closed and G_{ad} is connected. For non-adjoint reductive groups G, however, this condition is not sufficient because the map $G(K_0) \longrightarrow G_{\mathrm{ad}}(K_0)$ need not be surjective. For example, if G is simply connected semi-simple, the only basic σ-conjugacy class is the trivial one, so the only G' of the form J_b is G itself (cf. [135], Prop. 5.4). Fortunately, the adjoint case will be sufficient for our applications to period domains in the next sections.

The following result allows us to transfer G-isocrystals from G to an inner form of J_b. It is a sharpening of Examples 9.2.10 and 9.2.27.

Lemma 9.3.12 *Let G' be an inner form of G. Assume $G' = J_\gamma$ for some basic element $\gamma \in G(K_0)$, thus $G'_{K_0} = G_{K_0}$ and the rational structures are related by $\sigma' = \mathrm{Int}(\gamma) \circ \sigma$. Then the map $b \mapsto b' := b\gamma^{-1}$ yields an equivalence of categories $\mathcal{B}(G, L) \xrightarrow{\sim} \mathcal{B}(G', L)$, takes decent elements to decent elements, and we have $J'_{b'} = J_b$ and $\tilde{\nu}_{b'} = \tilde{\nu}_b \tilde{\nu}_\gamma^{-1}$. In particular, the identity map of G_{K_0} induces a weak isomorphism $\mathbb{G}_b \overset{\sim}{\dashrightarrow} \mathbb{G}'_{b'}$.*

Proof Our map sends elements $g \in G(K_0)$ to $g\gamma^{-1}$ and $g^{-1}b\sigma(g)$ to $g^{-1}b\sigma(g)\gamma^{-1} = g^{-1}(b\gamma^{-1})\sigma'(g)$. Thus it defines a functor, which is an equivalence of categories with inverse map $b' \mapsto b\gamma$. If γ is s_γ-decent and b is s-decent, then b' is

s'-decent for $s' = \mathrm{lcm}(s, s_\gamma)$. Moreover, for any $\mathbb{Q}_p$-algebra, we have

$$\begin{aligned} J'_{b'}(R) = & \{g \in G(K_0) \mid g^{-1}(b\gamma^{-1})\sigma'(g) = b'\} \\ = & \{g \in G(K_0) \mid g^{-1}b\sigma(g)\gamma^{-1} = b\gamma^{-1}\} \\ = & J_b(R). \end{aligned}$$

In order to prove the statement on the cocharacters $\tilde{\nu}$, we may restrict our attention to adjoint groups and tori, thanks to the isogeny $G \longrightarrow G_{\mathrm{ad}} \times G_{\mathrm{ab}}$. For adjoint groups, the equivalence $\mathcal{B}(G, L) \xrightarrow{\sim} \mathcal{B}(G', L)$ between G-isocrystals and G'-isocrystals can be understood directly via the equivalence $\mathrm{Rep}_{\mathbb{Q}_p}(G) \xrightarrow{\sim} \mathrm{Rep}_{\mathbb{Q}_p}(G')$ which sends (V, ρ) to (V', ρ') where $V' := (V \otimes_{\mathbb{Q}_p} K_0)^{\rho(\gamma)\sigma}$. Then we have

$$N_b^G(V) = (V \otimes K_0, \rho(b)\sigma) = (V' \otimes K_0, \rho(b')\sigma') = N_{b'}^{G'}(V')$$

and we deduce that $\tilde{\nu}_b = \tilde{\nu}_{b'}$.

As for tori, it is enough to test the desired identity on $\mathbb{Q}_p$-rational characters χ. Since $\tilde{\nu}_b, \tilde{\nu}_{b'}$, and $\tilde{\nu}_\gamma$ are $\mathbb{Q}_p$-rational $\mathbb{Q}$-1-PS, the assertion follows from the identities $\langle \tilde{\nu}_b, \chi \rangle = \mathrm{ord}_p(\chi(b))$, resp. $\langle \tilde{\nu}_{b'}, \chi \rangle = \mathrm{ord}_p(\chi(b'))$. The last statement was already explained in Example 9.2.10. □

It follows from Lemmas 9.3.12 and 9.3.11 and Corollary 9.3.9 that if K_0 is large enough and G is adjoint, then a group J_b is always an inner form of some Levi subgroup of the quasi-split inner form of G. However, using Lemma 9.3.12, we can produce a pair (G, b) with b decent and such that J_b is not an inner form of some Levi subgroup of G.

Example 9.3.13 Start with a non-isotypic isocrystal, hence we consider a pair (GL_n, b') with $J_{b'}$ an inner form of a proper Levi subgroup of GL_n. Then twist this pair by some basic element $\gamma \in \mathrm{GL}_n(K_0)$ such that $G := J_\gamma$ is anisotropic modulo center. We get a pair (G, b) such that J_b is not an inner form of some Levi subgroup of G.

Notes and References The notion of a basic isocrystal is due to Kottwitz [135]. He proves that for a connected *quasi-split* group, each element in $B(G)$ is the class of a basic element in some Levi subgroup M, which can in fact be taken to be *G-regular* in M (i.e., the centralizer of ν_b is M). In [136] Kottwitz gives a description of $B(H)$ for any connected inner form H of a quasi-split group G. This is done (roughly speaking) by studying the union $B_s(G)$ of all such $B(H)$. The set $B_s(G)$ allows then a similar description as above in the quasi-split case.

A somewhat less precise, but sometimes useful, description of $B(G)$, also valid for any connected reductive group over $\mathbb{Q}_p$, is given in [182].

4 Structure of $X_*(G)_{\mathbb{Q}}/G$ for a reductive a.g.s.

As before, L is a perfect field of characteristic p and $K_0 = W(L)_{\mathbb{Q}}$. We fix an algebraic closure $\overline{K}_0$ of K_0 and denote by $\Gamma_0 := \mathrm{Gal}(\overline{K}_0/K_0)$ its associated Galois group. Recall that for any algebraic group G over K_0 we write

$$X_*(G)_{\mathbb{Q}} := \mathrm{Hom}_{\overline{K}_0-\mathrm{gr}}(\mathbb{D}_{\overline{K}_0}, G_{\overline{K}_0}),$$

and we let $X_*(G)_{\mathbb{Q}}/G$ be the set of conjugacy classes of $\mathbb{Q}$-1-PS of G over $\overline{K}_0$. We have already seen in Chapter VI that if G is reductive, then the set $X_*(G)_{\mathbb{Q}}/G$ has a rich structure. It is an ordered divisible monoid equipped with a smooth, linear and order-preserving, action of Γ_0.

In this section we gather some properties of $X_*(G)_{\mathbb{Q}}/G$, especially the dominance order, when G is part of an augmented group scheme $\mathbb{G} = (G, \varphi, \nu)$ over $\mathrm{Isoc}(L)$. In this case, the Weil descent datum $\varphi : \sigma_*G \xrightarrow{\sim} G$ allows one to extend the action of Γ_0 on $X_*(G)_{\mathbb{Q}}/G$ to the group

$$\widetilde{\Gamma}_0 := \{\gamma \in \mathrm{Aut}(\overline{K}_0) \mid \gamma_{|K_0} \in \sigma^{\mathbb{Z}}\}$$

as follows. If $\sigma|K_0$ has infinite order, then we define for $\gamma \in \widetilde{\Gamma}_0$ the integer $c = c(\gamma) \in \mathbb{Z}$ by $\gamma_{|K_0} = \sigma^c$. Otherwise, we let $c(\gamma)$ be the smallest positive integer c such that $\gamma_{|K_0} = \sigma^c$. For a $\mathbb{Q}$-1-PS $\lambda : \mathbb{D}_{\overline{K}_0} \longrightarrow G_{\overline{K}_0}$, the composition $\lambda \circ (\mathrm{id}_{\mathbb{D}} \times \gamma^{-1})$ defines a $\mathbb{Q}$-1-PS of the group

$$^{\sigma^{c(\gamma)}}G := \begin{cases} \sigma_*(\sigma_*(\ldots(\sigma_*G)) & (c(\gamma) \text{ times}) \;:\; c(\gamma) \geq 0, \\ (\sigma^{-1})_*((\sigma^{-1})_*(\ldots((\sigma^{-1})_*G)) & (-c(\gamma) \text{ times}) \;:\; c(\gamma) < 0. \end{cases}$$

Composing with

$$\varphi^{(c(\gamma))} := \begin{cases} \varphi \circ {}^{\sigma^1}\varphi \circ \cdots \circ {}^{\sigma^{c(\gamma)-1}}\varphi & : \; c(\gamma) \geq 0 \\ {}^{\sigma^{-1}}(\varphi^{-1}) \circ \cdots \circ {}^{\sigma^{-c(\gamma)}}(\varphi^{-1}) & : \; c(\gamma) < 0 \end{cases}$$

we get a $\mathbb{Q}$-1-PS ${}^{\gamma}\lambda$ of G. If σ has infinite order on K_0, then we have $c(\gamma\gamma') = c(\gamma) + c(\gamma')$ for any two $\gamma, \gamma' \in \widetilde{\Gamma}_0$, and therefore the map

$$(\gamma, \lambda) \mapsto {}^{\gamma}\lambda = \varphi^{(c(\gamma))} \circ \lambda \circ (\mathrm{id}_{\mathbb{D}} \times \gamma^{-1}) \tag{4.1}$$

defines an action of $\widetilde{\Gamma}_0$ on $X_*(G)_{\mathbb{Q}}$. If σ has finite order r on K_0, the above map is generally not an action of $\widetilde{\Gamma}_0$ on $X_*(G)_{\mathbb{Q}}$. Nevertheless, since $\varphi^{(r)}$ is an inner automorphism of G (by Example 9.1.16 (iii)), it defines an action $(\gamma, \mathcal{N}) \mapsto {}^{\gamma}\mathcal{N}$ on $X_*(G)_{\mathbb{Q}}/G$.

Lemma 9.4.1 *The extended action of $\widetilde{\Gamma}_0$ on $X_*(G)_{\mathbb{Q}}/G$ factors through a finite quotient and is compatible with the ordered monoid structure.*

Proof Recall first how the ordered monoid structure is defined on $X_*(G)_{\mathbb{Q}}/G$.

Let $T \subset B$ be a Borel pair in $G_{\overline{K}_0}$. The inclusion $T \subset G_{\overline{K}_0}$ induces a bijection $X_*(T)_{\mathbb{Q}}/W \xrightarrow{\sim} X_*(G)_{\mathbb{Q}}/G$. The closed Weyl chamber $\overline{C}_{T,B} \subset X_*(T)_{\mathbb{Q}}$ is a fundamental domain for the action of W on $X_*(T)_{\mathbb{Q}}$, whence a bijection $\overline{C}_{T,B} \xrightarrow{\sim} X_*(G)_{\mathbb{Q}}/G$. This closed Weyl chamber is a divisible monoid and is endowed with the dominance order attached to B. This defines the structure of an ordered monoid on $X_*(G)_{\mathbb{Q}}/G$, which is easily shown not to depend on the choice of the Borel pair $T \subset B$.

Now we extend the L-action of Γ_0 on the Weyl chamber $\overline{C}_{T,B}$ (cf. Chapter VI (2.1), page 144) to an action

$$\widetilde{\Gamma}_0 \times \overline{C}_{T,B} \longrightarrow \overline{C}_{T,B}, \quad (\gamma, x) \mapsto \gamma \cdot x.$$

Namely, let $\iota_T : T = \operatorname{Spec} \overline{K}_0[X^*(T)] \longrightarrow G$ be the embedding of T in G. Then, for $\gamma \in \widetilde{\Gamma}_0$, the morphism $\iota_T \circ (\operatorname{Spec} \mathrm{id}_{X^*(T)} \otimes \gamma^{-1})$ defines a maximal torus of ${}^{\sigma^{c(\gamma)}}G$, hence, composing with $\varphi^{(c(\gamma))}$ we get a maximal torus ${}^{\gamma}T$ in G, together with a canonical isomorphism $X_*(T) \xrightarrow{\sim} X_*({}^{\gamma}T)$. In the same way we define a Borel subgroup ${}^{\gamma}B$ of G, containing ${}^{\gamma}T$. Then there is an element $g_\gamma \in G(\overline{K}_0)$ such that $\mathrm{Int}(g_\gamma)$ takes $({}^{\gamma}T, {}^{\gamma}B)$ to (T, B), whence a linear automorphism $\gamma\cdot$ of $X_*(T)_{\mathbb{Q}}$, which preserves $\overline{C}_{T,B}$ and is compatible with the dominance order associated to B. Since the class $g_\gamma T(\overline{K}_0)$ is uniquely determined, this defines the desired action, at least when σ has infinite order in K_0. Once again, this also defines an action when σ has finite order r, since $\varphi^{(r)}$ is inner.

Now it follows from the definitions that the action we have just defined is compatible with the one in (4.1), page 259, via the isomorphism $\overline{C}_{T,B} \xrightarrow{\sim} X_*(G)_{\mathbb{Q}}/G$.

It remains to prove that this action factors through a finite quotient. By the quasi-isogeny $G \longrightarrow G_{\mathrm{ad}} \times G_{\mathrm{ab}}$ we may consider separately the case of a semi-simple group and that of a torus. In the semi-simple case, the finiteness follows from the finiteness of the automorphism group of the root datum of G. In the case when $G = T$ is a torus, the finiteness was already noticed in the proof of Lemma 9.2.2 (use that φ acts by an automorphism of finite order on $X_*(T)$). □

As an interesting consequence of the finiteness of the $\widetilde{\Gamma}_0$-action, we get a projection

$$p_{\widetilde{\Gamma}_0} : X_*(G)_{\mathbb{Q}}/G \longrightarrow (X_*(G)_{\mathbb{Q}}/G)^{\widetilde{\Gamma}_0},$$

obtained by averaging the $\widetilde{\Gamma}_0$-action for the monoid structure. Note that $(X_*(G)_{\mathbb{Q}}/G)^{\widetilde{\Gamma}_0}$ inherits the structure of an ordered monoid. By compatibility between the dominance order, the monoid structure, and the $\widetilde{\Gamma}_0$-action, $p_{\widetilde{\Gamma}_0}$ is order-preserving, i.e.,

$$\mathcal{N} \leq \mathcal{N}' \Rightarrow p_{\widetilde{\Gamma}_0}\mathcal{N} \leq p_{\widetilde{\Gamma}_0}\mathcal{N}'.$$

Similarly we have the order-preserving projection

$$p_{\Gamma_0} : X_*(G)_{\mathbb{Q}}/G \longrightarrow (X_*(G)_{\mathbb{Q}}/G)^{\Gamma_0}$$

on the Γ_0-invariants.

The next proposition is an "external" characterization of the dominance order, similar to but different from that given in Chapter VI. For a rational representation $(\overline{V},\overline{\rho})$ of $G_{\overline{K}_0}$, we recall the identification $X_*(\mathrm{GL}(\overline{V}))_{\mathbb{Q}}/\mathrm{GL}(\overline{V}) = (\mathbb{Q}^{\dim\overline{\rho}})_+$.

Proposition 9.4.2 *Let $\mathbb{G} = (G,\varphi,\nu)$ be an a.g.s. over* Isoc(L). *There are finite subsets $\mathcal{I}_{\overline{K}_0} \subset \mathrm{Rep}_{\overline{K}_0}(G)$, $\mathcal{I}_{K_0} \subset \mathrm{Rep}_{K_0}(G)$ and $\mathcal{I}_{\mathbb{Q}_p} \subset \mathrm{Rep}_{\mathbb{Q}_p}(\mathbb{G})$, such that for any two elements $\mathcal{N},\mathcal{N}'$ in $X_*(G)_{\mathbb{Q}}/G$, the following statements hold.*

(i) *$\mathcal{N} \le \mathcal{N}'$ if and only if one of the following two properties hold:*

  (a) *$\forall(\overline{V},\overline{\rho}) \in \mathcal{I}_{\overline{K}_0}$, $\overline{\rho}_*\mathcal{N} \le \overline{\rho}_*\mathcal{N}'$ (usual order on $(\mathbb{Q}^{\dim\overline{\rho}})_+$).*

  (b) *$\forall(\overline{V},\overline{\rho}) \in \mathcal{I}_{\overline{K}_0}$, $\max(\overline{\rho}_*\mathcal{N}) \le \max(\overline{\rho}_*\mathcal{N}')$.*

(ii) *$p_{\Gamma_0}\mathcal{N} \le p_{\Gamma_0}\mathcal{N}'$ if and only if one of the following two properties hold:*

  (a) *$\forall(V,\rho) \in \mathcal{I}_{K_0}$, $\rho_*\mathcal{N} \le \rho_*\mathcal{N}'$ (usual order on $(\mathbb{Q}^{\dim\rho})_+$).*

  (b) *$\forall(V,\rho) \in \mathcal{I}_{K_0}$, $\max(\rho_*\mathcal{N}) \le \max(\rho_*\mathcal{N}')$.*

(iii) *$p_{\overline{\Gamma}_0}\mathcal{N} \le p_{\overline{\Gamma}_0}\mathcal{N}'$ if and only if one of the following two properties hold:*

  (a) *$\forall(V,\Phi_V,\rho) \in \mathcal{I}_{\mathbb{Q}_p}$, $\rho_*\mathcal{N} \le \rho_*\mathcal{N}'$ (usual order on $(\mathbb{Q}^{\dim\rho})_+$).*

  (b) *$\forall(V,\Phi_V,\rho) \in \mathcal{I}_{\mathbb{Q}_p}$, $\max(\rho_*\mathcal{N}) \le \max(\rho_*\mathcal{N}')$.*

Remark 9.4.3 We could have prolonged the list of conditions by adding the conditions in (i) that the inequalities in (a), resp. in (b) hold for all $(\overline{V},\overline{\rho})$ in $\mathrm{Rep}_{\overline{K}_0}(G)$. Similarly in (ii), relative to $\mathrm{Rep}_{K_0}(G)$, and in (iii) relative to $\mathrm{Rep}_{\mathbb{Q}_p}(G)$.

In Chapter VI we have dealt with a certain "relative" dominance order (3.5), page 155 on the subset $X_*(G)^{\Gamma_0}/G(K_0)$ of $(X_*(G)_{\mathbb{Q}}/G)^{\Gamma_0}$. It was defined in terms of *relative* root data, but the external characterizations of point (ii) above and Proposition 6.3.9 show that this order is nothing but the restriction of the "absolute" dominance order which is considered here. Note that the inclusion $X_*(G)^{\Gamma_0}/G(K_0) \subset (X_*(G)_{\mathbb{Q}}/G)^{\Gamma_0}$ is an equality iff G is quasi-split, by the Kottwitz Lemma 6.1.5.

Proof We first note that in all three points (i), (ii) and (iii), the equivalence of the properties (a) and (b) follows from the fact that for any $\overline{\rho} \in \mathrm{Rep}_{\overline{K}_0}(G)$, we have

$$\overline{\rho}_*\mathcal{N} \leq \overline{\rho}_*\mathcal{N}' \Leftrightarrow \left(\begin{array}{l} \max((\bigwedge^i \overline{\rho})_*\mathcal{N}) \leq \max((\bigwedge^i \overline{\rho})_*\mathcal{N}'),\ \forall i = 1,\ldots,\dim\rho \\ \text{and} \\ \max((\bigwedge^{\dim\rho} \overline{\rho}^\vee)_*\mathcal{N}) \leq \max((\bigwedge^{\dim\rho} \overline{\rho}^\vee)_*\mathcal{N}') \end{array} \right),$$

cf. Remark 6.3.11.

Point (i) is a particular case of Proposition 6.3.9, Lemma 6.3.10 and Remark 6.3.11.

To prove point (ii), recall how the Galois group Γ_0 acts on rational representations over $\overline{K}_0$. Let $\gamma \in \Gamma_0$. We use similar pieces of notation as in Notation 9.1.7. Namely, we have functors γ_* and γ^* on vector spaces over $\overline{K}_0$, resp. on schemes over $\operatorname{Spec}\overline{K}_0$, and we denote by $\varphi_\gamma : \gamma_* \operatorname{Spec}\overline{K}_0 \xrightarrow{\sim} \operatorname{Spec}\overline{K}_0$ the isomorphism of $\overline{K}_0$-schemes induced by γ. Then, given $(\overline{V},\overline{\rho}) \in \operatorname{Rep}_{\overline{K}_0}(G)$, we set ${}^\gamma\overline{V} := \gamma_*\overline{V}$ and

$${}^\gamma\overline{\rho} : G_{\overline{K}_0} \overset{\mathrm{id}_G\times\varphi_\gamma^{-1}}{\longrightarrow} \gamma_*(G_{\overline{K}_0}) \overset{\gamma_*(\overline{\rho})}{\longrightarrow} \gamma_*(\mathrm{GL}(\overline{V})) = \mathrm{GL}({}^\gamma\overline{V}).$$

Then the pair $({}^\gamma\overline{V}, {}^\gamma\overline{\rho})$ is a rational representation of $G_{\overline{K}_0}$. In this way we get an action of Γ_0 on the category $\operatorname{Rep}_{\overline{K}_0}(G)$. We note that this action does not need to be smooth (i.e., does not factor through a finite quotient of Γ_0), but the induced action on the set of *isomorphism classes* of representations is smooth. It follows from the definitions that in $(\mathbb{Q}^{\dim\overline{V}})_+$ we have ${}^\gamma\overline{\rho}_*\,{}^\gamma\mathcal{N} = \overline{\rho}_*\mathcal{N}$ for any $\mathcal{N} \in X_*(G)_\mathbb{Q}/G$. In particular, if $\overline{\rho}$ is Γ_0-invariant, e.g., if it is defined over K_0, we have $\overline{\rho}_*(p_{\Gamma_0}\mathcal{N}) = \overline{\rho}_*\mathcal{N}$. The "only if" part of item (ii) therefore follows from the "only if" part of item (i) (if we define $\mathcal{I}_{K_0}$ as below).

To prove the "if" part, we use the formalism of Weil restriction of scalars. Indeed, let us pick a model of $(\overline{V},\overline{\rho})$ over a finite subextension K of K_0 inside $\overline{K}_0$, and let us denote this model by the same symbol $(\overline{V},\overline{\rho})$. Define its "restriction of scalars to K_0" to be the following linear representation

$$\rho : G \overset{\text{can}}{\longrightarrow} \operatorname{Res}_{K|K_0}(G_K) \overset{\operatorname{Res}(\overline{\rho})}{\longrightarrow} \operatorname{Res}_{K|K_0}(\mathrm{GL}(\overline{V})) \overset{\text{can}}{\longrightarrow} \mathrm{GL}(V)$$

where V is $\overline{V}$, viewed as a K_0-vector space. Then $\rho_{\overline{K}_0}$ is isomorphic to a direct sum of Galois twists ${}^\gamma\overline{\rho}$ of $\overline{\rho}$. For each such Galois twist, we have ${}^\gamma\overline{\rho}_*(p_{\Gamma_0}\mathcal{N}) = \overline{\rho}_*(p_{\Gamma_0}\mathcal{N})$. Thus $\max(\rho_*(p_{\Gamma_0}\mathcal{N})) = \max(\overline{\rho}_*(p_{\Gamma_0}\mathcal{N}))$. Therefore, denoting by $\mathcal{I}_{K_0}$ the set of "restrictions of scalars to K_0" of the elements in $\mathcal{I}_{\overline{K}_0}$, we see that property (ii)(b) implies $p_{\Gamma_0}\mathcal{N} \leq p_{\Gamma_0}\mathcal{N}'$. Moreover, the equivalence between properties (ii)(a) and (ii)(b) was already noted above.

We now turn to the proof of point (iii). We will follow the same strategy as for point (ii). We first extend the former action of Γ_0 on $\operatorname{Rep}_{\overline{K}_0}(G)$ to $\widetilde{\Gamma}_0$. Let $(\overline{V},\overline{\rho}) \in \operatorname{Rep}_{\overline{K}_0}(G)$ and $\gamma \in \widetilde{\Gamma}_0$, and let us keep the same notation as above.

Then we define a linear representation on ${}^{\gamma}\overline{V} := \gamma_*\overline{V}$ by setting

$$^{\gamma}\overline{\rho} : G_{\overline{K}_0} \overset{\varphi^{(-c(\gamma))}\times\varphi_{\overline{\gamma}}^{-1}}{\longrightarrow} \sigma_*^{c(\gamma)}G \times_{\operatorname{Spec} K_0} \gamma_* \operatorname{Spec} \overline{K}_0 = $$
$$\gamma_*(G_{\overline{K}_0}) \overset{\gamma_*(\overline{\rho})}{\longrightarrow} \gamma_*(\mathrm{GL}(\overline{V})) = \mathrm{GL}({}^{\gamma}\overline{V}).$$

If σ has infinite order on K_0, then this defines an action of $\widetilde{\Gamma}_0$ on $\mathrm{Rep}_{\overline{K}_0}(G)$. If σ has finite order r, this does not need to define such an action. But since $\varphi^{(r)}$ is an inner automorphism, at least this defines an action of $\widetilde{\Gamma}_0$ on the set of isomorphism classes of representations. In either case, this action on isomorphism classes factors through a finite quotient of $\widetilde{\Gamma}_0$. As above, we have ${}^{\gamma}\overline{\rho}_* {}^{\gamma}\mathcal{N} = \overline{\rho}_*\mathcal{N} \in (\mathbb{Q}^{\dim\overline{\rho}})_+$ for any $\mathcal{N} \in X_*(G)_{\mathbb{Q}}/G$. Therefore if $\overline{\rho}$ is $\widetilde{\Gamma}_0$-invariant, we have $\overline{\rho}_*(p_{\widetilde{\Gamma}_0}\mathcal{N}) = \overline{\rho}_*\mathcal{N}$. This applies in particular to any $\overline{\rho}$ coming from a representation (V, Φ_V, ρ) of $\mathbb{G}$. Indeed, in this case, the $\overline{K}_0$-linear isomorphism

$$\Phi_V^{c(\gamma)} \otimes \gamma : {}^{\gamma}\overline{V} = \sigma_*^{c(\gamma)} V \otimes_{K_0} \gamma_*\overline{K}_0 \overset{\sim}{\longrightarrow} V \otimes_{K_0} \overline{K}_0 = \overline{V}$$

intertwines $\overline{\rho}$ and ${}^{\gamma}\overline{\rho}$. Therefore, the "only if" part of item (iii) follows from the "only if" part of item (i) (if we define $\mathcal{I}_{\mathbb{Q}_p}$ as below).

To prove the "if part," we need an analogue of the formalism of Weil restriction of scalars, going from $\mathrm{Rep}_{K_0}(G)$ to $\mathrm{Rep}_{\mathbb{Q}_p}(\mathbb{G})$. As a general convention, for $(V, \rho) \in \mathrm{Rep}_{K_0}(G)$ we put

$$^{\varphi}\rho := \sigma_*\rho \circ \varphi^{-1}.$$

This is a K_0-rational representation of G on the K_0-vector space σ_*V. We also define inductively

$$^{\varphi^{n+1}}\rho := {}^{\varphi}({}^{\varphi^n}\rho)$$

for any integer $n \geq 0$. We start with the particular case where σ has finite order.

Lemma 9.4.4 *Let $L = \mathbb{F}_{p^r}$ for some positive integer r. Then the forgetful functor* $\mathrm{Rep}_{\mathbb{Q}_p}(\mathbb{G}) \longrightarrow \mathrm{Rep}_{K_0}(G)$ *has a right adjoint, denoted by* $(V, \rho) \mapsto (\tilde{V}, \Phi_{\tilde{V}}, \tilde{\rho})$. *Moreover, in* $\mathrm{Rep}_{K_0}(G)$ *we have* $\tilde{\rho} = (\rho \oplus {}^{\varphi}\rho \oplus \cdots \oplus {}^{\varphi^{r-1}}\rho)$.

Proof By Example 9.1.16 (iii), we know that the a.g.s. $\mathbb{G}$ over $\mathrm{Isoc}(L)$ can be considered as a triple (G, φ, g) with $g \in G(\mathbb{Q}_{p^r})$ such that $\varphi^{(r)} = \mathrm{Int}(g)$. In particular, for any $(V, \rho) \in \mathrm{Rep}_{K_0}(G)$, we have ${}^{\varphi^r}\rho = \rho \circ \mathrm{Int}(g^{-1}) = \mathrm{Int}(\rho(g^{-1})) \circ \rho$. Then we set

$$\tilde{\rho} := \rho \oplus {}^{\varphi}\rho \oplus \cdots \oplus {}^{\varphi^{r-1}}\rho$$

which is a rational representation of G on the $\mathbb{Q}_{p^r}$-vector space

$$\tilde{V} := V \oplus \sigma_*V \oplus \cdots \oplus \sigma_*^{r-1}V.$$

On this vector space we define a Frobenius

$$\begin{array}{rcll}\Phi_{\tilde V}: & \sigma_*\tilde V = \sigma_* V\oplus\cdots\oplus\sigma_*^{r-1}V\oplus V & \longrightarrow & \tilde V\\ & (v_1,\ldots,v_r) & \mapsto & (\rho(g)v_r, v_1,\ldots,v_{r-1})\end{array}.$$

The associated Weil descent datum $\varphi_{\tilde V}:\sigma_*\,\mathrm{GL}(\tilde V)\longrightarrow\mathrm{GL}(\tilde V)$ preserves the Levi subgroup $\mathrm{GL}(V)\times\mathrm{GL}(\sigma_*V)\times\cdots\times\mathrm{GL}(\sigma_*^{r-1}V)$ and its restriction to this subgroup is given by

$$\begin{array}{rcl}\varphi_{\tilde V}:\ \mathrm{GL}(\sigma_*V)\times\cdots\times\mathrm{GL}(\sigma_*^{r-1}V)\times\mathrm{GL}(V) & \longrightarrow & \\ & & \mathrm{GL}(V)\times\mathrm{GL}(\sigma_*V)\times\cdots\times\mathrm{GL}(\sigma_*^{r-1}V)\\ (h_1,\ldots,h_r) & \mapsto & (\mathrm{Int}(\rho(g))h_r, h_1,\ldots,h_{r-1}).\end{array}$$

Therefore, the following equation holds:

$$\tilde\rho=\varphi_{\tilde V}\circ{}^{\varphi}\tilde\rho.$$

In other words, $\tilde\rho$ is a morphism $(G,\varphi)\longrightarrow(\mathrm{GL}(\tilde V),\varphi_{\tilde V})$ of group schemes over $\mathrm{Isoc}(L)$. On the other hand, we have

$$\Phi_{\tilde V}^{(r)}=\rho(g)\oplus\sigma_*\rho({}^{\sigma}g)\oplus\cdots\oplus\sigma_*^{r-1}\rho({}^{\sigma^{r-1}}g)$$

in $\mathrm{GL}(\tilde V)(K_0)$, where

$${}^{\sigma}g=\sigma_*g\circ\varphi_\sigma\in(\sigma_*G)(K_0).$$

However, condition (b) in Example 9.1.16 (iii) says that this also equals $\rho(g)\oplus{}^{\varphi}\rho(g)\oplus\cdots\oplus{}^{\varphi^{r-1}}\rho(g)=\tilde\rho(g)$. Therefore, as in Example 9.1.23 (ii), the triple $(\tilde V,\Phi_{\tilde V},\tilde\rho)$ is a representation of $\mathbb{G}$. This defines an exact "restriction" functor $\rho\mapsto\tilde\rho$ from $\mathrm{Rep}_{\mathbb{Q}_{p^r}}(G)$ to $\mathrm{Rep}_{\mathbb{Q}_p}(\mathbb{G})$ which has the last property mentioned in the lemma. It remains to define the adjunction morphisms. They are given respectively by the first projection

$$\tilde\rho=(\rho\oplus{}^{\varphi}\rho\oplus\cdots\oplus{}^{\varphi^{r-1}}\rho)\twoheadrightarrow\rho$$

for any $(V,\rho)\in\mathrm{Rep}_{K_0}(G)$, and the diagonal embedding

$$\rho\hookrightarrow\tilde\rho=(\rho\oplus{}^{\varphi}\rho\oplus\cdots\oplus{}^{\varphi^{r-1}}\rho)$$

given by $\mathrm{id}\oplus\Phi_V^{-1}\oplus\cdots\oplus\Phi_V^{1-r}$, for any $(V,\Phi,\rho)\in\mathrm{Rep}_{\mathbb{Q}_p}(\mathbb{G})$. We let the reader check that these indeed define adjunctions between forgetful and restriction functors. □

Now let us see what remains in the case where L is a general perfect field.

Corollary 9.4.5 *Let $(V,\rho)\in\mathrm{Rep}_{K_0}(G)$. Then there exists $(\tilde V,\Phi_{\tilde V},\tilde\rho)\in\mathrm{Rep}_{\mathbb{Q}_p}(\mathbb{G})$ and a positive integer r such that $\tilde\rho_{\overline{K}_C}\simeq(\rho_{\overline{K}_0}\oplus({}^{\varphi}\rho)_{\overline{K}_0}\oplus\cdots\oplus({}^{\varphi^{r-1}}\rho)_{\overline{K}_0})$ in $\mathrm{Rep}_{\overline{K}_0}(G)$.*

Proof Since G is algebraic over K_0, we know by Corollary 9.1.25 that $\mathrm{Rep}_{\mathbb{Q}_p}(\mathbb{G})$ has a tensor generator, say $(V_t, \Phi_{V_t}, \rho_t)$. Let L' be an extension of L such that the base change $(V'_t, \Phi_{V'_t}, \rho'_t)$ to $K'_0 = K_0(L')$ is split and semi-simple. Then the base change $\mathbb{G}'$ defined in Construction 9.1.29 has the property that $N^{\mathbb{G}'}$ factors through $\mathrm{Isoc}(L')^{\mathrm{ss}}$. Moreover $(V'_t, \Phi_{V'_t}, \rho'_t)$ has a canonical model over some $\mathbb{Q}_{p^r}$, see Definition 8.1.2 and the comment following it; thus $\mathbb{G}'$ is the base change of an a.g.s. $\mathbb{G}'_r$ over $\mathrm{Isoc}(\mathbb{F}_{p^r})$.

Now, let us start with some representation $(V, \rho) \in \mathrm{Rep}_{K_0}(G)$ as in the corollary. Let (V', ρ') be its base change in $\mathrm{Rep}_{K'_0}(G')$. After maybe enlarging r, we may assume that (V', ρ') comes from a representation (V'_r, ρ'_r) in $\mathrm{Rep}_{\mathbb{Q}_{p^r}}(G'_r)$. So we can define its restriction $(\tilde{V}'_r, \Phi_{\tilde{V}'_r}, \tilde{\rho}'_r)$ in $\mathrm{Rep}_{\mathbb{Q}_p}(\mathbb{G}'_r)$ as in the previous lemma. Base changing to L' we get an object $(\tilde{V}', \Phi_{\tilde{V}'}, \tilde{\rho}') \in \mathrm{Rep}_{\mathbb{Q}_p}(\mathbb{G}')$ such that $\tilde{\rho}' \simeq \rho' \oplus {}^{\varphi}\rho' \oplus \cdots \oplus {}^{\varphi^{r-1}}\rho'$ in $\mathrm{Rep}_{K'_0}(G')$. Recall from Construction 9.1.29 that the base change functor $\mathrm{Rep}_{\mathbb{Q}_p}(\mathbb{G}) \longrightarrow \mathrm{Rep}_{\mathbb{Q}_p}(\mathbb{G}')$ is an equivalence of categories. Therefore we may choose a preimage $(\tilde{V}, \Phi_{\tilde{V}}, \tilde{\rho}) \in \mathrm{Rep}_{\mathbb{Q}_p}(\mathbb{G})$ of $(\tilde{V}', \Phi_{\tilde{V}'}, \tilde{\rho}')$. This satisfies the required properties. □

Continuation of the proof of Proposition 9.4.2 For any representation $(V, \rho) \in \mathcal{I}_{K_0}$, we choose $(\tilde{V}, \Phi_{\tilde{V}}, \tilde{\rho}) \in \mathrm{Rep}_{\mathbb{Q}_p}(\mathbb{G})$ as in the above corollary. We get a finite subset $\mathcal{I}_{\mathbb{Q}_p}$ of $\mathrm{Rep}_{\mathbb{Q}_p}(\mathbb{G})$. Since we have ${}^{\varphi}\rho(p_{\overline{\Gamma}_0}\mathcal{N}) = \rho(p_{\overline{\Gamma}_0}\mathcal{N})$ in $(\mathbb{Q}^{\dim \rho})_+$, we also have

$$\max(\tilde{\rho}_*(p_{\overline{\Gamma}_0}\mathcal{N})) = \max(\rho_*(p_{\overline{\Gamma}_0}\mathcal{N})).$$

Therefore, we see that property (iii)(b) implies $p_{\overline{\Gamma}_0}\mathcal{N} \le p_{\overline{\Gamma}_0}\mathcal{N}'$. Again, the equivalence between properties (iii)(a) and (iii)(b) was explained in the beginning of the proof. □

We now aim at an "internal" characterization of the dominance order on $X_*(G)_{\mathbb{Q}}/G$. For any psgp $P \subset G_{\overline{K}_0}$, we have a commutative diagram

$$\begin{array}{ccc} X_*(P)_{\mathbb{Q}}/P & \xrightarrow{\iota_{P\subset G,*}} & X_*(G)_{\mathbb{Q}}/G \\ {\scriptstyle \deg_{\overline{K}_0}}\big\downarrow & & \big\downarrow{\scriptstyle \deg_{\overline{K}_0}} \\ X_*(A'_P)_{\mathbb{Q}} & \longrightarrow & X_*(A'_{G_{\overline{K}_0}})_{\mathbb{Q}}\,. \end{array} \tag{4.2}$$

Here A'_P, resp. $A'_{G_{\overline{K}_0}}$, are the maximal quotient tori *over* $\overline{K}_0$ of P, resp. G. The upper horizontal map is finite to one and each fiber carries the so-called Bruhat order. Recall the projective variety $\mathcal{F}(G, \mathcal{N})$ over $\overline{K}_0$ attached to $\mathcal{N} \in X_*(G)_{\mathbb{Q}}/G$ and its locally closed P-homogeneous subvariety $\mathcal{F}(P, \mathcal{N}_P)$ (*a generalized Bruhat cell*) attached to $\mathcal{N}_P \in \iota^{-1}_{P\subset G,*}(\mathcal{N})$, cf. the discussion following Theorem 6.3.5. In particular, the $\overline{K}_0$-valued points of $\mathcal{F}(G, \mathcal{N})$ correspond to the filtrations $\mathcal{F}$ of ω^G of type $\mathcal{N}$ over $\overline{K}_0$, and the $\overline{K}_0$-valued points of

$\mathcal{F}(P, \mathcal{N}_P)$ correspond to the filtrations $\mathcal{F}$ of ω^G such that the induced filtration $\mathcal{F}_P$ of ω^P is of type $\mathcal{N}_P$.

Definition 9.4.6 Let $\mathcal{N} \in X_*(G)_{\mathbb{Q}}/G$. The *Bruhat order* on $\iota^{-1}_{P\subset G,*}(\mathcal{N})$ is defined by

$$\mathcal{N}_P \preceq \mathcal{N}'_P \Leftrightarrow \mathcal{F}(P, \mathcal{N}_P) \subseteq \overline{\mathcal{F}(P, \mathcal{N}'_P)}.$$

Note that after fixing some std psgp Q corresponding to $\mathcal{N}$, we have an identification of $\iota^{-1}_{P\subset G,*}(\mathcal{N})$ with the set of double cosets $W_P\backslash W/W_Q$ as partially ordered sets. Thus the set $\iota^{-1}_{P\subset G,*}(\mathcal{N})$ always has a greatest element, whose associated generalized Bruhat cell is open, and a lowest element, whose associated generalized Bruhat cell is closed.

Definition 9.4.7 An element $\mathcal{N}_P \in X_*(P)_{\mathbb{Q}}/P$ is called *dominant*, resp. *generic*, if the generalized Bruhat cell $\mathcal{F}(P, \mathcal{N}_P)$ is closed, resp. open, in $\mathcal{F}(G, \iota_{P\subset G,*}\mathcal{N}_P)$.

Example 9.4.8 Assume that $P = B$ is a Borel subgroup and choose a maximal torus $T \subset B$. Then we have a bijection

$$\begin{array}{ccc} X_*(T)_{\mathbb{Q}} & \xrightarrow{\sim} & X_*(B)_{\mathbb{Q}}/B \\ \lambda & \mapsto & \mathcal{N}_\lambda \,. \end{array}$$

It is well-known that $\mathcal{N}_\lambda$ is dominant if and only if $\lambda \in \overline{C}_{T,B}$, while $\mathcal{N}_\lambda$ is generic if and only if $\lambda \in -\overline{C}_{T,B}$. Furthermore, let $\lambda \in \overline{C}_{T,B}$. Then we have an identification $G/P_\lambda \xrightarrow{\sim} \mathcal{F}(G, \iota_{B\subset G,*}\mathcal{N}_\lambda)$.

For more general psgps, here are some characterizations of dominant and generic elements.

Lemma 9.4.9 *Let P be a psgp of $G_{\overline{K}_0}$ and let $\mathcal{N}_P \in X_*(P)_{\mathbb{Q}}/P$. The following properties are equivalent.*

(i) *$\mathcal{N}_P$ is dominant, resp. generic.*
(ii) *For every maximal psgp $Q \supset P$, the push-forward $\mathcal{N}_Q := \iota_{P\subset Q,*}\mathcal{N}_P$ is dominant, resp. generic.*
(iii) *For any Borel subgroup $B \subset P$, the pull-back set $\iota^{-1}_{B\subset P,*}(\mathcal{N}_P) \subset X_*(B)_{\mathbb{Q}}/B$ contains a dominant, resp. a generic element.*

Proof Let Q be any psgp containing P, put $\mathcal{N}_Q := \iota_{P\subset Q,*}\mathcal{N}_P$ and $\mathcal{N} := \iota_{P\subset G,*}\mathcal{N}_P$. Since $\mathcal{F}(Q, \mathcal{N}_Q)$ contains $\mathcal{F}(P, \mathcal{N}_P)$, we see that $\mathcal{N}_P$ generic implies $\mathcal{N}_Q$ generic. On the other hand, $\mathcal{F}(Q, \mathcal{N}_Q)$ is the image of the action map $Q \times^P \mathcal{F}(P, \mathcal{N}_P) \longrightarrow \mathcal{F}(G, \mathcal{N})$. If $\mathcal{N}_P$ is dominant, then with $\mathcal{F}(P, \mathcal{N}_P)$ also

$Q \times^P \mathcal{F}(P, \mathcal{N}_P)$ is proper, and therefore $\mathcal{F}(Q, \mathcal{N}_Q)$ is closed in $\mathcal{F}(G, \mathcal{N})$, i.e., $\mathcal{N}_Q$ is dominant. This proves the implications (iii) $\Rightarrow$ (i) $\Rightarrow$ (ii).

Let us now prove (i) $\Rightarrow$ (iii). Start with a dominant, resp. generic, $\mathcal{N}_P \in X_*(P)_{\mathbb{Q}}/P$ and choose a Borel subgroup $B \subset P$. Let $\{\mathcal{N}_{B,1}, \ldots, \mathcal{N}_{B,r}\}$ be the inverse image of $\mathcal{N}_P$ by $\iota_{B\subset P,*}$. Then we have $\mathcal{F}(P, \mathcal{N}_P) = \bigsqcup_i \mathcal{F}(B, \mathcal{N}_{B,i})$. Since this union is closed, resp. open in $\mathcal{F}(G, \mathcal{N})$, it contains the closed, resp. open, B-orbit in $\mathcal{F}(G, \mathcal{N})$. Therefore $\iota^{-1}_{B\subset P,*}(\mathcal{N}_P)$ contains a (unique) dominant, resp. generic element.

We now prove (ii) $\Rightarrow$ (iii). So we fix a Borel pair $T \subset B \subset P$ and a cocharacter $\lambda \in X_*(T)_{\mathbb{Q}}$ such that $\mathcal{N}_P = i_{B\subset P,*}\mathcal{N}_\lambda$ (notation of Example 9.4.8). We denote by $\lambda^+ \in \overline{C}_{T,B}$ the unique dominant element in the W_G-orbit of λ. Applying the implication (i) $\Rightarrow$ (iii) to maximal psgps Q containing P and Example 9.4.8 to B, the hypothesis (ii) in the dominant case gives

$$\lambda \in W_Q \cdot \lambda^+ \text{ for each maximal } Q \supset P,$$

where W_Q is the Weyl group of the Levi component of Q which contains T. In this setting, statement (iii) in the dominant case is equivalent to the condition $\lambda \in W_P \cdot \lambda^+$. So let w_P be the unique element of minimal length in $\{w \in W_G \mid w \cdot \lambda^+ = \lambda\}$ and choose elements $w_Q \in W_Q$ for each $Q \supset P$ maximal, such that $w_Q \cdot \lambda^+ = \lambda$. For each Q, we have $\ell(w_Q) = \ell(w_P) + \ell(w_P^{-1}w_Q)$, thus w_P admits a minimal expression which is extracted from a minimal expression for w_Q, and therefore $w_P \in W_Q$. Hence $w_P \in \bigcap_{Q\supset P} W_Q = W_P$. This proves (ii) $\Rightarrow$ (iii) in the dominant case. The generic case is proved in the same way, using the generic representative λ^- instead of λ^+. □

We now want to relate these notions to an avatar of the *dominance order on* $X_*(A'_P)_{\mathbb{Q}}$.

Definition 9.4.10 A non-zero element $\alpha^\vee$ in $X_*(A'_P)_{\mathbb{Q}}$ is called *a simple co-root* if it is the push-forward of a simple co-root in $X_*(T)_{\mathbb{Q}}$ for some Borel pair $T \subset B \subset P$. For any $\lambda_P, \lambda'_P \in X_*(A'_P)_{\mathbb{Q}}$ we set

$$\lambda_P \leq \lambda'_P \text{ iff } \lambda'_P - \lambda_P \text{ is a non-negative sum of simple co-roots in } X_*(A'_P)_{\mathbb{Q}}.$$

Remark 9.4.11 For $Q \supset P$, denote by λ_Q the push-forward of λ_P by the natural morphism $A'_P \longrightarrow A'_Q$. Then we have

$$\lambda_P \leq \lambda'_P \Leftrightarrow \lambda_Q \leq \lambda'_Q \text{ for all maximal psgp } Q \supset P.$$

Moreover, if P is maximal, then we have

$$\lambda_P \leq \lambda'_P \Leftrightarrow \langle \lambda_P, \textstyle\bigwedge^{\max} \mathfrak{u}_P \rangle \leq \langle \lambda'_P, \textstyle\bigwedge^{\max} \mathfrak{u}_P \rangle \text{ and } \lambda_G = \lambda'_G.$$

Proposition 9.4.12 *Let P be a psgp of $G_{\overline{K}_0}$ and let $\mathcal{N}_P^1, \mathcal{N}_P^2 \in X_*(P)_{\mathbb{Q}}/P$ inducing the same $\mathcal{N} \in X_*(G)_{\mathbb{Q}}/G$. Consider the following properties.*

(i) *$\mathcal{N}_P^1 < \mathcal{N}_P^2$ for the Bruhat order on $\iota_{P\subset G,*}^{-1}(\mathcal{N})$.*

(ii) *$\deg_{\overline{K}_0}(\mathcal{N}_P^1) > \deg_{\overline{K}_0}(\mathcal{N}_P^2)$ for the dominance order in $X_*(A'_P)_{\mathbb{Q}}$.*

Then we have (i) $\Rightarrow$ (ii).

Proof Let $T \subset B \subset P$ be a Borel pair and let M denote the Levi component of P that contains T. Also, let W and W_M denote the respective Weyl groups in G and M. Denote by $\lambda^+ \in \overline{C}_{T,B} \subset X_*(T)_{\mathbb{Q}}$ the unique dominant element such that $\iota_{B\subset G,*}(\mathcal{N}_{\lambda^+}) = \mathcal{N}$. Then the map

$$\begin{array}{ccc} W \cdot \lambda^+ \cap \overline{C}_{T,B\cap M} & \longrightarrow & \iota_{P\subset G,*}^{-1}(\mathcal{N}) \\ \lambda & \mapsto & \iota_{B\subset P,*}(\mathcal{N}_\lambda) \end{array}$$

is a bijection. By Stembridge [211], Prop. 1.5, the Bruhat order on the LHS is characterized as the transitive closure of the set of relations

$$\begin{aligned} \lambda < s_\beta \cdot \lambda, \quad & \forall \lambda \in W \cdot \lambda^+ \cap \overline{C}_{T,B\cap M}, \\ & \forall \beta \in \{\beta \in \Phi(T,B) \mid s_\beta \cdot \lambda \in (W \cdot \lambda^+ \cap \overline{C}_{T,B\cap M}), \langle \lambda, \beta \rangle > 0\}. \end{aligned}$$

Note that for such λ, β, the root β does not belong to $\Phi(T, B \cap M)$ and therefore the push-forward of β to $X_*(A'_P)_*$ is non-zero. Since we have $\lambda - s_\beta \cdot \lambda = \langle \lambda, \beta \rangle \beta$, it follows that $\deg_{\overline{K}_0}(\iota_{B\subset P,*}(\mathcal{N}_\lambda)) > \deg_{\overline{K}_0}(\iota_{B\subset P,*}(\mathcal{N}_{s_\beta \cdot \lambda}))$. Taking the transitive closure we get the statement of the proposition. □

Corollary 9.4.13 *Let P be a psgp of $G_{\overline{K}_0}$ and $\mathcal{N}_P \in X_*(P)_{\mathbb{Q}}/P$. The following properties are equivalent.*

(i) *$\mathcal{N}_P$ is dominant, resp. generic.*

(ii) *$\deg_{\overline{K}_0}(\mathcal{N}_P)$ is the maximal, resp. minimal, element in*

$$\{\deg_{\overline{K}_0}(\mathcal{N}'_P) \mid \mathcal{N}'_P \in X_*(P)_{\mathbb{Q}}/P,\ \iota_{P\subset G,*}(\mathcal{N}_P) = \iota_{P\subset G,*}(\mathcal{N}'_P)\}. \qquad \square$$

Remark 9.4.14 As a particular case of the previous corollary, if P is maximal, then we see that $\mathcal{N}_P$ is dominant, resp. generic, if and only if the function $\mathcal{N}'_P \mapsto \langle \deg(\mathcal{N}'_P), \bigwedge^{\max} \mathfrak{u}_P \rangle$ on the fiber $i_{P\subset G,*}^{-1}(i_{P\subset G,*}(\mathcal{N}_P))$ takes its maximum, resp. its minimum, at $\mathcal{N}_P$.

The following variant of Proposition 9.4.12 will be used in the study of the HN-stratification.

Corollary 9.4.15 *We take up the setting of Proposition 9.4.12 and assume further that P is part of a psgp $\mathbb{P}$ of $\mathbb{G}$, cf. Definition 9.1.27. Define a dominance* *order on $X_*(A'_{\mathbb{P}})_{\mathbb{Q}}$ by the same process as in Definition 9.4.10. Then again*

$$\mathcal{N}_P^1 < \mathcal{N}_P^2 \Rightarrow \deg(\mathcal{N}_P^1) > \deg(\mathcal{N}_P^2).$$

Here deg *is the usual degree, as in Definition 9.2.3.*

Proof This follows from the fact that the action of $\widetilde{\Gamma}_0$ on $X_*(A'_{P_{\overline{K}_0}})_{\mathbb{Q}}$ preserves the set of simple co-roots, so that the projection $X_*(A'_{P_{\overline{K}_0}})_{\mathbb{Q}} \longrightarrow X_*(A'_{\mathbb{P}})_{\mathbb{Q}}$ is a strictly order-preserving map. □

We now get back to the dominance order on $X_*(G)_{\mathbb{Q}}/G$. By definition (as recalled in the proof of Lemma 9.4.1), for $\mathcal{N}, \mathcal{N}' \in X_*(G)_{\mathbb{Q}}/G$ we have

$$\mathcal{N} \leq \mathcal{N}' \Leftrightarrow \left(\deg_{\overline{K}_0} \mathcal{N}_B^{\mathrm{dom}} \leq \deg_{\overline{K}_0} \mathcal{N}'^{\mathrm{dom}}_B \text{ for any (or one) Borel } B \subset G_{\overline{K}_0}\right).$$

Here $\mathcal{N}_B^{\mathrm{dom}}$ is the unique dominant element in $\iota_{B\subset P,*}^{-1}(\mathcal{N})$ and the inequality in the LHS sits in $X_*(A'_B)_{\mathbb{Q}}$. Therefore, using Remark 9.4.11 and Lemma 9.4.9, we get the following "internal" characterization of the dominance order.

Proposition 9.4.16 *Let $\mathcal{N}, \mathcal{N}' \in X_*(G)_{\mathbb{Q}}/G$ such that $\deg_{\overline{K}_0} \mathcal{N} = \deg_{\overline{K}_0} \mathcal{N}'$ in $X_*(A'_{G_{\overline{K}_0}})_{\mathbb{Q}}$. Then the following properties are equivalent.*

(i) $\mathcal{N} \leq \mathcal{N}'$.

(ii) *For every maximal psgp $P \subset G_{\overline{K}_0}$, we have*

$$\left\langle \deg_{\overline{K}_0}(\mathcal{N}_P^{\mathrm{dom}}), \textstyle\bigwedge^{\max} \mathfrak{u}_P \right\rangle \leq \left\langle \deg_{\overline{K}_0}(\mathcal{N}'^{\mathrm{dom}}_P), \textstyle\bigwedge^{\max} \mathfrak{u}_P \right\rangle,$$

where $\mathcal{N}_P^{\mathrm{dom}}, \mathcal{N}'^{\mathrm{dom}}_P$ denote the dominant representatives of $\mathcal{N}, \mathcal{N}'$ in $X_*(P)_{\mathbb{Q}}/P$.

(iii) *For every maximal pgsp $P \subset G_{\overline{K}_0}$, we have*

$$\left\langle \deg_{\overline{K}_0}(\mathcal{N}_P^{\mathrm{gen}}), \textstyle\bigwedge^{\max} \mathfrak{u}_P \right\rangle \geq \left\langle \deg_{\overline{K}_0}(\mathcal{N}'^{\mathrm{gen}}_P), \textstyle\bigwedge^{\max} \mathfrak{u}_P \right\rangle,$$

where we have denoted by $\mathcal{N}_P^{\mathrm{gen}}, \mathcal{N}'^{\mathrm{gen}}_P$ the generic representatives of $\mathcal{N}, \mathcal{N}'$ in $X_*(P)_{\mathbb{Q}}/P$. □

Because of its similarities with Corollary 9.2.30, this characterization will be useful, in the next section, to test that the period domains are not empty. However, we will also need to study the interplay with the averaging operators p_{Γ_0} and $p_{\tilde{\Gamma}_0}$.

In the next corollary, we consider only K_0-rational psgps of G. In this setting A'_P denotes again the maximal K_0-*split* quotient torus of P and $\deg_{K_0}$ is the push-forward map $X_*(P)_{\mathbb{Q}} \longrightarrow X_*(A'_P)_{\mathbb{Q}}$.

Corollary 9.4.17 *Let $\mathcal{N}, \mathcal{N}' \in X_*(G)_{\mathbb{Q}}/G$ such that $\deg_{K_0} \mathcal{N} = \deg_{K_0} \mathcal{N}' \in X_*(A'_G)_{\mathbb{Q}}$. Consider the following properties (we use the same notation as in the previous proposition).*

(i) $p_{\Gamma_0}\mathcal{N} \leq p_{\Gamma_0}\mathcal{N}'$.

(ii) *For every (maximal) K_0-psgp $P \subset G$,*

$$\left\langle \deg_{K_0}(\mathcal{N}_P^{\mathrm{dom}}), \wedge^{\max} \mathfrak{u}_P \right\rangle \leq \left\langle \deg_{K_0}(\mathcal{N}'^{\mathrm{dom}}_P), \wedge^{\max} \mathfrak{u}_P \right\rangle.$$

(iii) *For every (maximal) K_0-psgp $P \subset G$,*

$$\left\langle \deg_{K_0}(\mathcal{N}_P^{\mathrm{gen}}), \wedge^{\max} \mathfrak{u}_P \right\rangle \geq \left\langle \deg_{K_0}(\mathcal{N}'^{\mathrm{gen}}_P), \wedge^{\max} \mathfrak{u}_P \right\rangle.$$

Then (i) $\Rightarrow$ (ii) $\Leftrightarrow$ (iii). *The implication* (ii) $\Rightarrow$ (i) *holds if $p_{\Gamma_0}\mathcal{N}$ and $p_{\Gamma_0}\mathcal{N}'$ lie in $X_*(G)^{\Gamma_0}/G(K_0)$. In particular it always holds if G is quasi-split over K_0.*

Proof The equivalence (ii) $\Leftrightarrow$ (iii) is easy. The implication (i) $\Rightarrow$ (ii) follows from the previous proposition and the following equalities:

$$\deg_{K_0}(\mathcal{N}_P^{\mathrm{dom}}) = p_{\Gamma_0}(\deg_{\overline{K}_0}(\mathcal{N}_P^{\mathrm{dom}})) = \deg_{\overline{K}_0}(p_{\Gamma_0}\mathcal{N}_P^{\mathrm{dom}}) = \deg_{\overline{K}_0}((p_{\Gamma_0}\mathcal{N})_P^{\mathrm{dom}}).$$

In the first equality we abuse notation a little, by identifying $X_*(A'_{P_{\overline{K}_0}})_{\mathbb{Q}}^{\Gamma_0}$ with $X_*(A'_P)_{\mathbb{Q}}$. The second equality follows from the (obvious) Γ_0-equivariance of $\deg_{\overline{K}_0}$. The third equality follows from the observation that, for any $\gamma \in \Gamma_0$, we have $({}^{\gamma}\mathcal{N})_P^{\mathrm{dom}} = {}^{\gamma}(\mathcal{N}_P^{\mathrm{dom}})$, as follows readily from the definition.

Assume now that $p_{\Gamma_0}\mathcal{N}$ and $p_{\Gamma_0}\mathcal{N}'$ lie in $X_*(G)^{\Gamma_0}/G(K_0)$ and let P_0 be a minimal parabolic subgroup in G. Then property (ii) and the above equalities imply

$$\deg_{K_0}((p_{\Gamma_0}\mathcal{N})_{P_0}) \leq \deg_{K_0}((p_{\Gamma_0}\mathcal{N}')_{P_0})$$

for the dominance order in $X_*(A'_{P_0})$ defined by the same process as in Definition 9.4.10. But this order coincides with the relative order defined in Chapter VI (3.5), page 155. Therefore the last statement follows from Remark 9.4.3. □

Finally, we turn to the case where $\mathcal{N}, \mathcal{N}'$ are assumed to be $\tilde{\Gamma}_0$-invariant. In the next Corollary, we consider psgps of $\mathbb{G}$. In this setting $A'_{\mathbb{P}}$ and the degree deg are defined as in Definition 9.2.3.

Corollary 9.4.18 *Let $\mathcal{N}, \mathcal{N}' \in X_*(G)_{\mathbb{Q}}/G$ such that $\deg \mathcal{N} = \deg \mathcal{N}' \in X_*(A'_{\mathbb{G}})_{\mathbb{Q}}$. Consider the following properties (we use the same notation as above).*

(i) $p_{\tilde{\Gamma}_0}\mathcal{N} \leq p_{\tilde{\Gamma}_0}\mathcal{N}'$.

(ii) *For every (maximal) psgp $\mathbb{P}$ of $\mathbb{G}$,*

$$\left\langle \deg(\mathcal{N}_P^{\mathrm{dom}}), \wedge^{\max} \mathfrak{u}_P \right\rangle \leq \left\langle \deg(\mathcal{N}'^{\mathrm{dom}}_P), \wedge^{\max} \mathfrak{u}_P \right\rangle.$$

(iii) *For every (maximal) psgp* $\mathbb{P}$ *of* $\mathbb{G}$,

$$\left\langle \deg(\mathcal{N}_P^{\mathrm{gen}}), \bigwedge^{\max} \mathfrak{u}_P \right\rangle \geq \left\langle \deg(\mathcal{N}'^{\mathrm{gen}}_P), \bigwedge^{\max} \mathfrak{u}_P \right\rangle.$$

We have (i) $\Rightarrow$ (ii) $\Leftrightarrow$ (iii).

Proof The proof is the same as in the corollary above. □

Notes and References The comparison between the Bruhat and the dominance order started with Ehresmann [72]. He observed that for root systems of type A both orders are the same. In [66], it is proved that the converse of Proposition 9.4.12 holds true (at least for $P = B$) if and only if the simple components of G have rank less than 3 or are of type A. Fomin and Zelevinsky [79] used the relation between these two orders in order to describe Schubert cells by the vanishing of Plücker coordinates.

5 Period domains

We keep the notation of the previous sections. In particular, L is a perfect field of characteristic $p > 0$ and $K_0 = W(L)_{\mathbb{Q}}$. We have fixed an algebraic closure $\overline{K}_0$ of K_0 with Galois group Γ_0. Also $X_*(G)_{\mathbb{Q}}/G$ is the Γ_0-set of conjugacy classes of $\mathbb{Q}$-1-PS of G over $\overline{K}_0$. Similarly to Chapter VI, we will associate to any *PD-pair* a *period domain*.

Definition 9.5.1 A *PD-pair* is a pair $(\mathbb{G}, \mathcal{N})$ consisting of a reductive augmented group scheme $\mathbb{G} = (G, \varphi, \nu)$ over $\mathrm{Isoc}(L)$ and a conjugacy class of $\mathbb{Q}$-1-PS $\mathcal{N} \in X_*(G)_{\mathbb{Q}}/G$. The *Shimura field* or *reflex field* of the PD-pair $(\mathbb{G}, \mathcal{N})$ is the fixed subfield $\breve{K}_0 = \breve{K}_0(\mathbb{G}, \mathcal{N})$ of $\overline{K}_0$ under the stabilizer of $\mathcal{N}$ in Γ_0. This is a finite extension of K_0.

In Chapter VI we saw that the conjugacy class $\mathcal{N}$ defines a flag variety $\breve{\mathcal{F}}(G, \mathcal{N})$ over $\breve{K}_0$. Recall from Remark 6.1.6 that for any extension $K/\breve{K}_0$, the set $\breve{\mathcal{F}}(G, \mathcal{N})(K)$ parametrizes the $\mathbb{Q}$-filtrations of ω^G over K which are "of type" $\mathcal{N}$. Given a PD-pair $(\mathbb{G}, \mathcal{N})$ as above, we define, using the concepts of "semi-stability" resp. "weak admissibility" from Definition 9.2.14,

$$\breve{\mathcal{F}}(\mathbb{G}, \mathcal{N})(K)^{\mathrm{ss}} := \{\mathcal{F} \in \breve{\mathcal{F}}(G, \mathcal{N})(K) \mid \mathcal{F} \text{ is semi-stable w.r.t. } \mathbb{G}\},$$

and

$$\breve{\mathcal{F}}(\mathbb{G}, \mathcal{N})(K)^{\mathrm{wa}} := \{\mathcal{F} \in \breve{\mathcal{F}}(G, \mathcal{N})(K) \mid \mathcal{F} \text{ is weakly admissible w.r.t. } \mathbb{G}\}.$$

However, we note that $\deg(\mathbb{G}, \mathcal{F})$ is independent of $\mathcal{F} \in \breve{\mathcal{F}}(G, \mathcal{N})(K)$. Namely, we have

$$\deg(\mathbb{G}, \mathcal{F}) = \deg(\mathbb{G}, \mathcal{N}) := \deg(\mathcal{F}_0) + \deg(\mathcal{N}).$$

Therefore we have $\breve{\mathcal{F}}(\mathbb{G},\mathcal{N})(K)^{\mathrm{wa}} = \breve{\mathcal{F}}(\mathbb{G},\mathcal{N})(K)^{\mathrm{ss}}$ if $\deg(\mathbb{G},\mathcal{N}) = 0$ and $\breve{\mathcal{F}}(\mathbb{G},\mathcal{N})(K)^{\mathrm{wa}} = \emptyset$, otherwise.

Example 9.5.2 Let $N = (V, \Phi_V)$ be an isocrystal over L and let $\nu \in (\mathbb{Q}^{\dim V})_+$. As in Example 6.1.1, ν corresponds to some conjugacy class $\mathcal{N}$ for $G = \mathrm{GL}(V)$, so that $\breve{\mathcal{F}}(\mathrm{GL}(V),\mathcal{N}) = \mathcal{F}(V,\nu)$. What is now denoted $\breve{\mathcal{F}}(\mathbb{GL}(V,\Phi_V),\mathcal{N})(K)^{\mathrm{ss}}$ is nothing else but what was denoted $\mathcal{F}(N,\nu)(K)^{\mathrm{ss}}$ in the previous chapter.

Let $(\mathbb{G},\mathcal{N})$ be a PD-pair and let $\rho : \mathbb{G} \dashrightarrow \mathbb{G}'$ be a *weak* morphism of a.g.s. over Isoc(L), cf. Definition 9.2.6. By the definition of $\mathcal{N}' = \tilde{\rho}_*(\mathcal{N})$, we have an inclusion of reflex fields $\breve{K}_0(\mathbb{G}',\mathcal{N}') \subset \breve{K}_0(\mathbb{G},\mathcal{N})$, cf. Definition 9.2.6. Further, it induces a *normalized* morphism

$$\tilde{\rho}_* : \breve{\mathcal{F}}(G,\mathcal{N}) \longrightarrow \breve{\mathcal{F}}(G',\mathcal{N}')_{\breve{K}_0(G,\mathcal{N})}.$$

Corollary 9.2.24 says that, if ρ is injective, then we have

$$\breve{\mathcal{F}}(\mathbb{G},\mathcal{N})(K)^{\mathrm{ss}} \supset \tilde{\rho}_*^{-1}(\breve{\mathcal{F}}(\mathbb{G}',\mathcal{N}')(K)^{\mathrm{ss}}), \tag{5.1}$$

while if ρ is homogeneous, then we have

$$\breve{\mathcal{F}}(\mathbb{G},\mathcal{N})(K)^{\mathrm{ss}} \subset \tilde{\rho}_*^{-1}(\breve{\mathcal{F}}(\mathbb{G}',\mathcal{N}')(K)^{\mathrm{ss}}) \tag{5.2}$$

for any extension K of $\breve{K}_0(\mathbb{G},\mathcal{N})$.

Let us now turn to analytic properties.

Proposition 9.5.3 *Let $(\mathbb{G},\mathcal{N})$ be a PD-pair and let $\breve{\mathcal{F}}(G,\mathcal{N})^{an}$ be the $\breve{K}_0$-analytic space associated with $\breve{\mathcal{F}}(G,\mathcal{N})$. Then the set*

$$\breve{\mathcal{F}}(\mathbb{G},\mathcal{N})^{\mathrm{ss}} := \left\{x \in \breve{\mathcal{F}}(G,\mathcal{N})^{an} \mid \mathcal{F}_x \in \breve{\mathcal{F}}(\mathbb{G},\mathcal{N})(\mathscr{H}(x))^{\mathrm{ss}}\right\}$$

is open in $\breve{\mathcal{F}}(G,\mathcal{N})^{an}$, hence has a natural structure of a $\breve{K}_0$-analytic space. Moreover, the following statements hold.

(i) $\breve{\mathcal{F}}(\mathbb{G},\mathcal{N})^{\mathrm{ss}}(K) = \breve{\mathcal{F}}(\mathbb{G},\mathcal{N})(K)^{\mathrm{ss}}$ *for any complete extension field K of $\breve{K}_0$.*

(ii) *For any faithful homogeneous representation (V,Φ_V,ρ) of $\mathbb{G}$, the open subset $\breve{\mathcal{F}}(\mathbb{G},\mathcal{N})^{\mathrm{ss}}$ is the preimage of $\breve{\mathcal{F}}(\mathbb{GL}(V,\Phi_V),\rho_*(\mathcal{N}))^{\mathrm{ss}}\hat{\otimes}_{K_0}\breve{K}_0$ under ρ_*^{an}.*

(iii) *For any weak isomorphism $\rho : \mathbb{G} \dashrightarrow \mathbb{G}'$, the normalized isomorphism $\tilde{\rho}_*^{an} : \breve{\mathcal{F}}(\mathbb{G},\mathcal{N}) \xrightarrow{\sim} \breve{\mathcal{F}}(\mathbb{G}',\tilde{\rho}_*\mathcal{N})$ takes $\breve{\mathcal{F}}(\mathbb{G},\mathcal{N})^{\mathrm{ss}}$ to $\breve{\mathcal{F}}(\mathbb{G}',\tilde{\rho}_*\mathcal{N})^{\mathrm{ss}}$. In particular, the action of the p-adic group $J^{\mathbb{G}}(\mathbb{Q}_p) = \underline{\mathrm{Aut}}^{\otimes}(N^{\mathbb{G}})$ on $\breve{\mathcal{F}}(\mathbb{G},\mathcal{N})$ preserves $\breve{\mathcal{F}}(\mathbb{G},\mathcal{N})^{\mathrm{ss}}$.*

(iv) *The adjoint quotient $\rho_{\mathrm{ad}} : \mathbb{G} \longrightarrow \mathbb{G}_{\mathrm{ad}}$ induces an isomorphism of $\breve{K}_0$-analytic spaces $\rho_{\mathrm{ad}}^{an} : \breve{\mathcal{F}}(\mathbb{G},\mathcal{N})^{\mathrm{ss}} \xrightarrow{\sim} \breve{\mathcal{F}}(\mathbb{G}_{\mathrm{ad}},\mathcal{N}_{\mathrm{ad}})^{\mathrm{ss}}$.*

(v) *Let $L' \supset L$ be a perfect field extension, and let us use the notation of Proposition 9.2.31. Then, the corresponding period domain $\breve{\mathcal{F}}(\mathbb{G}',\mathcal{N})^{\mathrm{ss}}$ over*

$\breve{K}_0'$ is contained in $\breve{\mathcal{F}}(\mathbb{G},\mathcal{N})^{ss}\widehat{\otimes}_{\breve{K}_0}\breve{K}_0'$, with equality if $N^{\mathbb{G}}$ is split semi-simple, in the sense of Definition 9.1.30.

Proof Statements (ii) and (iii) follow from (5.1), page 272 and (5.2), page 272 above. Statement (iv) is due to Proposition 9.2.22 and allows us to reduce the openness statement to the case of adjoint groups. Since such groups admit faithful representations (see the proof of Proposition 9.2.28) and, since these are also homogeneous for trivial reasons, the openness statement follows from (ii) and statement (i) in Proposition 8.2.1. The base change property (v) follows from the corresponding statement of the same proposition together with Proposition 9.2.31. □

Definition 9.5.4 The $\breve{K}_0$-analytic space $\breve{\mathcal{F}}(\mathbb{G},\mathcal{N})^{ss}$ is called the *period domain* associated to the PD-pair $(\mathbb{G},\mathcal{N})$. If $\deg(\mathbb{G},\mathcal{N}) = 0$, then we denote it also by $\breve{\mathcal{F}}(\mathbb{G},\mathcal{N})^{wa}$.

We note that all properties except (iv) in the above proposition hold with ^{wa} instead of ^{ss}.

Example 9.5.5 Let G be a reductive group over $\mathbb{Q}_p$ and let $b \in G(K_0)$. Following a traditional notation, we may also write $\breve{\mathcal{F}}(G,b,\mathcal{N})^{ss}$ instead of $\breve{\mathcal{F}}(\mathbb{G}_b,\mathcal{N})^{ss}$. Let us translate some statements of Proposition 9.5.3 in this setting.

(i) $\breve{\mathcal{F}}(G,b,\mathcal{N})^{ss}$ is a $J_b(\mathbb{Q}_p)$-invariant open analytic subset of $\mathcal{F}(G,\mathcal{N})^{an}_{\breve{K}_0}$.

(ii) If b is s-decent, then, by the base change property (v), $\breve{\mathcal{F}}(G,b,\mathcal{N})^{ss}$ has a canonical model over $\breve{\mathbb{Q}}_{p^s} = E(G,\mathcal{N}).\mathbb{Q}_{p^s}$.

(iii) Let $\rho : G_{K_0} \longrightarrow G'_{K_0}$ be an inner twisting of reductive groups over $\mathbb{Q}_p$ and assume given an element γ as in Example 9.2.10. Recall that ρ induces a weak isomorphism $\mathbb{G}_b \dashrightarrow \mathbb{G}'_{\rho(b)\gamma}$. Therefore, applying statement (iii) of Proposition 9.5.3 we see that the normalized isomorphism $\breve{\mathcal{F}}(G_{K_0},\mathcal{N}) \xrightarrow{\sim} \breve{\mathcal{F}}(G'_{K_0},\mathcal{N}')$ takes $\breve{\mathcal{F}}(G,b,\mathcal{N})^{ss}$ to $\breve{\mathcal{F}}(G',b',\mathcal{N}')^{ss}$. Here $\mathcal{N}' = \tilde{\rho}_*(\mathcal{N})$.

We note that in order to study period domains we may restrict ourselves to adjoint groups because of statement (iv) of the last proposition. In this case, if L is "large enough" (e.g. algebraically closed) an element γ as above always exists for any inner twisting $\rho : G_{K_0} \longrightarrow G'_{K_0}$ by Lemma 9.3.11.

Remark 9.5.6 The last statement about inner twisting invariance can be used to lower the field of definition of a period domain. For example, take $G = \mathrm{GL}_n$ and start with a basic (i.e., isotypical) isocrystal of slope $0 < \frac{r}{s} < 1$. The construction of period domains in the previous chapter gives a model over $\mathbb{Q}_{p^s}$. In the setting of the previous chapter, such a GL_n-isocrystal is obtained for some basic element $b \in \mathrm{GL}_n(K_0)$ satisfying the decency equation in Definition 9.1.32. By Example 9.5.5 (iii), we have $\breve{\mathcal{F}}(G, b, \mathcal{N})^{ss} = \breve{\mathcal{F}}(J_b, 1, \mathcal{N}')^{ss}$, so our period domain is also associated to the *trivial* J_b-isocrystal hence has a natural model over $\mathbb{Q}_p$. For example, in the case $(r, s) = (1, n)$ and where $\mathcal{N}$ corresponds to the projective space $\mathbb{P}^{n-1}_{\mathbb{Q}_{p^s}}$, the period domain over $\mathbb{Q}_{p^s}$ coincides with $\mathbb{P}^{n-1}_{\mathbb{Q}_{p^s}}$. The rational structure we find over $\mathbb{Q}_p$ in this way is the Severi–Brauer variety attached to the division algebra with invariant $1/n$.

The following example realizes the dual version of the Drinfeld upper half-space by starting with a central divison algebra, cf. [183], 1.44.

Example 9.5.7 Let $n \in \mathbb{N}$ and let D be a central division algebra of invariant $-1/n$ over $\mathbb{Q}_p$. Let $\mathbb{Q}_{p^n} \subset D$ be an unramified extension of degree n of $\mathbb{Q}_p$ and denote by $\Pi \in D$ a uniformizer. Let $\tau \in \mathrm{Gal}(\mathbb{Q}_{p^n}/\mathbb{Q}_p)$ be the Frobenius homomorphism. Thus we may write $D = \mathbb{Q}_{p^n}[\Pi]$ with relations $\Pi^n = p$ and $\Pi x = \tau(x)\Pi,\ x \in \mathbb{Q}_{p^n}$.

Let $G = (D^{\mathrm{opp}})^\times$ and set $b = \Pi^{-1} \in G(K_0)$, where $K_0 = K_0(\mathbb{F}_{p^n})$. Then $V = D$ is a faithful G-representation and the isocrystal $N = N_b(V)$ is split semi-simple and isotypical of slope $-\frac{1}{n}$. Since G is an inner form of GL_n there is an isomorphism $G_{K_0} \cong \mathrm{GL}_{n,K_0}$. Let $\mathcal{N}$ be the conjugacy class of 1-PS of G given by the Hodge vector $\nu = (0^{(n-1)}, -1) \in \mathbb{Z}^n$ under this identification. Then $\mathcal{F}(G, \mathcal{N})$ parametrizes D-stable $\mathbb{Q}_p$-rational subspaces of codimension n in V and the reflex field $E(G, \mathcal{N})$ is equal to $\mathbb{Q}_p$.

Fix an embedding $\epsilon : \mathbb{Q}_{p^n} \hookrightarrow K_0$. The decomposition $\mathbb{Q}_{p^n} \otimes_{\mathbb{Q}_p} K_0 = \bigoplus_{\mathbb{Z}/n\mathbb{Z}} K_0$ gives rise to a decomposition

$$N = \bigoplus_{i \in \mathbb{Z}/n\mathbb{Z}} N_i$$

where $N_i = \{v \in N \mid fv = \epsilon(\tau^i(f))v,\ f \in \mathbb{Q}_{p^n}\}$. The operator Φ is of degree 1, whereas Π is of degree -1. Thus the endomorphism $\Phi_0 := \Phi \circ \Pi$ is of degree 0 and induces on N_0 a τ-linear bijective endomorphism of slope equal to 0. We get a $\mathbb{Q}_p$-rational structure on N_0 by setting $V_0 := N_0^{\Phi_0}$.

The projection map $V \longrightarrow V_0$ induces an isomorphism of varieties

$$\mathcal{F}(G, \mathcal{N}) \xrightarrow{\sim} \mathcal{F}(\mathrm{GL}_n, \nu) = \mathbf{Gr}_{n-1}(V_0).$$

Under this identification one verifies (see [183], Prop. 1.45) that $\breve{\mathcal{F}}(G, b, \mathcal{N})^{wa}$ is equal to $\check{\Omega}(V_0)$, where $\check{\Omega}(V_0) = \check{\mathbb{P}}(V) \setminus \bigcup_{H \subsetneq V_0} \check{\mathbb{P}}(V/H)$ is the dual Drinfeld space parametrizing hyperplanes not containing any $\mathbb{Q}_p$-rational subspace. Further, one checks that $J = J_b$ is identified with $\mathrm{GL}(V_0)$.

Variant 9.5.8 Let $\mathbb{G}$ be a reductive a.g.s. over $\mathrm{Isoc}(L)$. Let $\mathcal{N} = (\mathcal{N}_1, \ldots, \mathcal{N}_r)$ be a tuple of r conjugacy classes of $\mathbb{Q}$-1-PS of G. Let $\breve{K}_0 = \breve{K}_0(\mathbb{G}, \mathcal{N}) =$

$\breve{K}_0(\mathbb{G}, \mathcal{N}_1) \cdot \cdots \cdot \breve{K}_0(\mathbb{G}, \mathcal{N}_r)$ be the composite of the individual reflex fields. Then we may consider the product of flag varieties

$$\breve{\mathcal{F}}(G, \mathcal{N}) = \prod_{i=1}^{r} \breve{\mathcal{F}}(G, \mathcal{N}_i)_{\breve{K}_0},$$

which is defined over $\breve{K}_0$. For every faithful homogeneous representation (V, Φ_V, ρ) of $\mathbb{G}$, let $\rho_*(\mathcal{N}) = (\rho_*(\mathcal{N}_1), \ldots, \rho_*(\mathcal{N}_r))$. The resulting morphism

$$\breve{\mathcal{F}}(G, \mathcal{N}) \longrightarrow \breve{\mathcal{F}}(\mathbb{GL}(V, \Phi_V), \rho_*(\mathcal{N}))_{\breve{K}_0}$$

is a closed immersion and, for every field extension K of $\breve{K}_0$, the intersections

$$\breve{\mathcal{F}}(\mathbb{G}, \mathcal{N})(K)^{\mathrm{ss}} = \breve{\mathcal{F}}(G, \mathcal{N})(K) \cap \breve{\mathcal{F}}(\mathbb{GL}(V, \Phi_V), \rho_*(\mathcal{N}))(K)^{\mathrm{ss}}$$
$$\breve{\mathcal{F}}(\mathbb{G}, \mathcal{N})(K)^{\mathrm{wa}} = \breve{\mathcal{F}}(G, \mathcal{N})(K) \cap \breve{\mathcal{F}}(\mathbb{GL}(V, \Phi_V), \rho_*(\mathcal{N}))(K)^{\mathrm{wa}}$$

are the sets of semi-stable, resp. weakly admissible $\mathcal{F}$ in $\breve{\mathcal{F}}(G, \mathcal{N})(K)$.

Statements analogous to those of Proposition 9.5.3 also hold in this setting.

Remark 9.5.9 If $b = 1$ *and* G is quasi-split, then the period domains "look like" those in the finite field situation, except for their analytic nature. If $b = 1$ but G is not quasi-split, period domains generally have no analogue in the finite field case. For example, if G is anisotropic modulo center, they coincide with the full ambient flag variety. Note that $J = G$ if $b = 1$.

We proceed with a generalization of Theorem 8.2.9, i.e., a criterion for the period domains to be non-empty. Recall that the set $X_*(G)_{\mathbb{Q}}/G$ is an *ordered, uniquely divisible monoid*, and that the natural action of Γ_0 extends to a linear and order-preserving action of the group $\widetilde{\Gamma}_0 = \{\gamma \in \mathrm{Aut}(\overline{K}_0) \mid \gamma_{|K_0} \in \sigma^{\mathbb{Z}}\}$. Further we have a projection map $p_{\widetilde{\Gamma}_0} : X_*(G)_{\mathbb{Q}}/G \longrightarrow (X_*(G)_{\mathbb{Q}}/G)^{\Gamma_0}$. We note that, by (1.3), page 225, the conjugacy class of the slope $\mathbb{Q}$-1-PS $\nu \circ sl$ of $\mathbb{G}$ is defined over $\mathbb{Q}_p$.

Theorem 9.5.10 *Let $(\mathbb{G}, \mathcal{N})$ be a PD-pair. Denote by $\mathcal{N}_{\mathbb{G}} \in (X_*(G)_{\mathbb{Q}}/G)^{\widetilde{\Gamma}_0}$ the conjugacy class of the slope $\mathbb{Q}$-1-PS $\nu \circ sl$ of $\mathbb{G}$. Then $\breve{\mathcal{F}}(\mathbb{G}, \mathcal{N})^{\mathrm{wa}} \neq \emptyset$ if and only if $p_{\widetilde{\Gamma}_0}\mathcal{N} \geq \mathcal{N}_{\mathbb{G}}$. More generally, $\breve{\mathcal{F}}(\mathbb{G}, \mathcal{N})^{\mathrm{ss}} \neq \emptyset$ if and only if $p_{\widetilde{\Gamma}_0}\mathcal{N}_{\mathrm{ad}} \geq \mathcal{N}_{\mathbb{G}_{\mathrm{ad}}}$.*

Proof We first note that the second statement about the semi-stable locus follows from the first one. In fact, consider the weakly admissible locus of $(\mathbb{G}_{\mathrm{ad}}, \mathcal{N}_{\mathrm{ad}})$ and apply property (iv) of Proposition 9.5.3, and the fact that for adjoint groups the total slope is trivial, see Remark 9.2.12. Conversely, the second statement implies the first one, since we have

$$p_{\widetilde{\Gamma}_0}\mathcal{N} \geq \mathcal{N}_{\mathbb{G}} \;\Leftrightarrow\; p_{\widetilde{\Gamma}_0}\mathcal{N}_{\mathrm{ad}} \geq \mathcal{N}_{\mathbb{G}_{\mathrm{ad}}} \text{ and } \deg(\mathcal{N}) = \deg(\mathcal{N}_{\mathbb{G}}),$$

as follows from Corollary 9.4.18. In what follows we thus may, and will, assume that $\deg(\mathcal{N}) = \deg(\mathcal{N}_{\mathbb{G}})$, so that semi-stability and weak admissibility coincide.

Let us first assume that $\breve{\mathcal{F}}(\mathbb{G}, \mathcal{N})^{\mathrm{wa}} \neq \emptyset$. Then, by Theorem 8.2.9 and Corollary 9.2.25 (ii), for all $\rho \in \mathrm{Rep}_{\mathbb{Q}_p}(\mathbb{G})$, we have $\rho_*\mathcal{N} \geq \rho_*\mathcal{N}_{\mathbb{G}}$. By Proposition 9.4.2 (iii), since $\mathcal{N}_{\mathbb{G}}$ is $\widetilde{\Gamma}_0$-invariant, it follows that $p_{\widetilde{\Gamma}_0}\mathcal{N} \geq \mathcal{N}_{\mathbb{G}}$.

Conversely, assume that $p_{\widetilde{\Gamma}_0}\mathcal{N} \geq \mathcal{N}_{\mathbb{G}}$. Thus we may apply Corollary 9.4.18, which says that for every maximal parabolic subgroup $\mathbb{P}$ of $\mathbb{G}$, we have

$$\left\langle \deg \mathcal{N}_P^{\mathrm{gen}}, \bigwedge\nolimits^{\max} \mathfrak{u}_P \right\rangle \leq \left\langle \deg \mathcal{N}_{\mathbb{G},P}^{\mathrm{gen}}, \bigwedge\nolimits^{\max} \mathfrak{u}_P \right\rangle, \tag{5.3}$$

where $\mathcal{N}_P^{\mathrm{gen}}, \mathcal{N}_{\mathbb{G},P}{}^{\mathrm{gen}} \in X_*(P)_{\mathbb{Q}}/P$ are the generic lifts of $\mathcal{N}$ and $\mathcal{N}_{\mathbb{G}}$, in the sense of Definition 9.4.7. Now, let x be the generic point of the scheme $\breve{\mathcal{F}}(G, \mathcal{N})$. As in the proof of Theorem 8.2.9, it suffices to prove that the pair $(\mathbb{G}, \mathcal{F}_x)$ is weakly admissible. Since x is generic, the filtration $\mathcal{F}_{x,P}$ induced by $\mathcal{F}_x$ on any $\mathbb{P}$ as above is "generic," meaning that the P-conjugacy class of $\mathcal{F}_{x,P}$ is $\mathcal{N}_P^{\mathrm{gen}}$. In particular, we have

$$\left\langle \deg(\mathcal{F}_{x,P}), \bigwedge\nolimits^{\max} \mathfrak{u}_P \right\rangle = \left\langle \deg \mathcal{N}_P^{\mathrm{gen}}, \bigwedge\nolimits^{\max} \mathfrak{u}_P \right\rangle.$$

On the other hand, we have for the slope filtration $\mathcal{F}_0$ and its lift $\mathcal{F}_{0,P}$ to ω^P,

$$\left\langle \deg(\mathcal{F}_{0,P}), \bigwedge\nolimits^{\max} \mathfrak{u}_P \right\rangle = -\left\langle \mathcal{N}_{\mathbb{P}}, \bigwedge\nolimits^{\max} \mathfrak{u}_P \right\rangle \leq -\left\langle \mathcal{N}_{\mathbb{G},P}^{\mathrm{gen}}, \bigwedge\nolimits^{\max} \mathfrak{u}_P \right\rangle.$$

Here $\mathcal{N}_{\mathbb{P}} \in X_*(P)_{\mathbb{Q}}/P$ is the P-conjugacy class of $\nu \circ sl$ (recall that ν factors through P) and the inequality on the right follows from the definition of the generic lift, see Remark 9.4.14. Plugging this into the inequality (5.3) above, we conclude by Corollary 9.2.30 that $(\mathbb{G}, \mathcal{F}_x)$ is semi-stable. □

Notes and References Period domains for arbitrary reductive groups appear in [183] (using rigid-analytic varieties). The case of a split reductive group and $b = 1$ is considered in [220], where van der Put and Voskuil use the GIT-approach, in order to define semi-stable points (cf. Section 7 below). The analogue of Theorem 9.5.3 (i) is proved there by choosing a faithful representation of G. The same holds true for Hartl's proof [108] of the openness of period domains.

The result of Theorem 9.5.10 generalizes [82], Thm. 3. In [82] only G-isocrystals with G quasi-split and L algebraically closed are considered. In contrast to [82], we use the internal characterization of semi-stability in addition to the external one. We also note that, as in the case of Theorem 8.2.9, one can produce $\breve{K}_0$-valued points under the assumption that L is algebraically closed, but such "rational" points need not exist in general.

6 The Harder–Narasimhan stratification

We continue with the setting and notation of the previous section. In particular, we fix a PD-pair $(\mathbb{G}, \mathcal{N})$ consisting of a reductive augmented group scheme $\mathbb{G} = (G, \varphi, \nu)$ over Isoc(L) and a conjugacy class $\mathcal{N} \in X_*(G)_{\mathbb{Q}}/G$ of $\mathbb{Q}$-1-PS of $G_{\overline{K}_0}$.

Our aim here is to introduce a suitable set of HN-*types* associated with the PD-pair $(\mathbb{G}, \mathcal{N})$ and study the associated HN-stratification on $\breve{\mathcal{F}}(G, \mathcal{N})^{an}$, in a way which is a mixture of what we did for isocrystals in the previous chapter, and what we did in Chapter VI for reductive groups over finite fields. We will start by studying the decomposition according to HN-*vectors*, because it is much easier.

Definition 9.6.1 Let $\mathcal{F} \in \mathrm{Fil}_K(\omega^G)$ for some extension K of K_0. We define its HN-*vector* $H_{\mathcal{F}} \in (X_*(G)_{\mathbb{Q}}/G)^{\tilde{\Gamma}_0}$ as the conjugacy class of any $\mathbb{Q}$-1-PS which splits the HN-filtration ${}^\bullet N_{\mathcal{F}}^{\mathbb{G}}$ of $\mathcal{F}$.

We note that a $\mathbb{Q}$-1-PS as in this definition is defined over K_0, so that we also have $H_{\mathcal{F}} \in X_*(G)_{\mathbb{Q}}^{\Gamma_0}/G(K_0)$. In general, a representative of $H_{\mathcal{F}}$ is (in contrast to its $G(K_0)$-conjugacy class) not φ-stable.

Fixing an extension K of $\breve{K}_0$, we get a disjoint union

$$\breve{\mathcal{F}}(G, \mathcal{N})(K) = \dot{\bigcup}_{H \in (X_*(G)_{\mathbb{Q}}/G)^{\tilde{\Gamma}_0}} \breve{\mathcal{F}}(\mathbb{G}, \mathcal{N})(K)_H \tag{6.1}$$

according to the HN-vector of the filtration. We claim that the dominance partial order on $(X_*(G)_{\mathbb{Q}}/G)^{\tilde{\Gamma}_0}$ is functorial for weak morphisms. Indeed, let $\rho : \mathbb{G} \dashrightarrow \mathbb{G}'$ be a weak morphism of a.g.s. It certainly induces a map

$$(X_*(G)_{\mathbb{Q}}/G)^{\tilde{\Gamma}_0} \longrightarrow (X_*(G')_{\mathbb{Q}}/G')^{\tilde{\Gamma}_0}.$$

If ρ is a genuine morphism, then the external characterization in Proposition 9.4.2 (iii) shows that this map is order-preserving. The same holds true for any weak morphism, by an argument similar to the proof of Proposition 9.2.20. We also have a "normalized" push-forward morphism

$$\tilde{\rho}_* : \breve{\mathcal{F}}(G, \mathcal{N}) \longrightarrow \breve{\mathcal{F}}(G', \rho_*\mathcal{N})_{\breve{K}_0},$$

and by Proposition 9.2.20, we have

$$\tilde{\rho}_*(\breve{\mathcal{F}}(\mathbb{G}, \mathcal{N})(K)_H) \subset \breve{\mathcal{F}}(\mathbb{G}', \rho_*\mathcal{N})(K)_{\rho_* H}$$

for any $H \in (X_*(G)_{\mathbb{Q}}/G)^{\tilde{\Gamma}_0}$ and for any extension field K of $\breve{K}_0$, and even

$$\breve{\mathcal{F}}(\mathbb{G}, \mathcal{N})(K)_H = \tilde{\rho}_*^{-1}(\breve{\mathcal{F}}(\mathbb{G}', \rho_*\mathcal{N})(K)_{\rho_* H}), \tag{6.2}$$

if ρ is injective. In particular, the p-adic group $J^{\mathbb{G}}(\mathbb{Q}_p)$ preserves the sets

$\breve{\mathcal{F}}(\mathbb{G},\mathcal{N})(K)_H$. Finally, we note that for $\mathbb{G} = \mathbb{GL}(V,\Phi_V)$, these notions of HN-vectors and stratification according to HN-vectors coincide with those of the previous chapter, via the usual identification $X_*(\mathrm{GL}(V))_{\mathbb{Q}}/\,\mathrm{GL}(V) \cong (\mathbb{Q}^{\dim V})_+$.

The following result generalizes both Proposition 8.3.4 and Proposition 9.5.3.

Theorem 9.6.2 *Let $\breve{\mathcal{F}}(G,\mathcal{N})^{an}$ be the $\breve{K}_0$-analytic space associated with $\breve{\mathcal{F}}(G,\mathcal{N})$. The function*

$$\begin{array}{ccc} \breve{\mathcal{F}}(G,\mathcal{N})^{an} & \longrightarrow & (X_*(G)_{\mathbb{Q}}/G)^{\breve{\Gamma}_0} \\ x & \mapsto & H_{\mathcal{F}_x} \end{array}$$

is upper semi-continuous for the analytic topology on $\breve{\mathcal{F}}(G,\mathcal{N})^{an}$ and the dominance order on HN*-vectors. In particular, the subset*

$$\breve{\mathcal{F}}(\mathbb{G},\mathcal{N})_H := \left\{x \in \breve{\mathcal{F}}(G,\mathcal{N})^{an} \mid H_{\mathcal{F}_x} = H\right\}$$

is locally closed in $\breve{\mathcal{F}}(G,\mathcal{N})^{an}$.

Proof As in Chapter VI, the proof is by reduction to the GL_n-case, which is given by Proposition 8.3.4. Let $\mathcal{I}_{\mathbb{Q}_p} = \{(V_i,\Phi_{V_i},\rho_i)\}$ be the finite set of representations of $\mathbb{G}$ of Proposition 9.4.2 (iii). Put $\rho := \prod_i \rho_i : \mathbb{G} \longrightarrow \mathbb{G}' := \prod_i \mathbb{GL}(V_i,\Phi_{V_i})$. Then we have:

(i) ρ is an embedding;

(ii) $x \le x' \Leftrightarrow \rho_* x \le \rho_* x'$, for all $x, x' \in (X_*(G)_{\mathbb{Q}}/G)^{\breve{\Gamma}_0}$.

Then we have a morphism $\rho_*^{an} : \breve{\mathcal{F}}(G,\mathcal{N})^{an} \longrightarrow \breve{\mathcal{F}}(G',\rho_*\mathcal{N})^{an}$ of analytic spaces over K_0 and, by (6.2), page 277 above, we get

$$\breve{\mathcal{F}}(\mathbb{G},\mathcal{N})_H = \rho_*^{an^{-1}}(\breve{\mathcal{F}}(\mathbb{G}',\rho_*\mathcal{N})_{\rho_* H}) \tag{6.3}$$

for any $H \in (X_*(G)_{\mathbb{Q}}/G)^{\breve{\Gamma}_0}$. Hence the semi-continuity statement follows from Proposition 8.3.4. □

Remark 9.6.3 There is currently no known description of the set $\mathcal{H}(\mathbb{G},\mathcal{N})$ of HN-vectors which actually contribute to the stratification $\breve{\mathcal{F}}(G,\mathcal{N})^{an} = \dot{\bigcup}_H \breve{\mathcal{F}}(\mathbb{G},\mathcal{N})_H$, although in principle a criterion can be worked out from Theorem 9.5.10 above and Theorem 9.6.8 (ii) below. We will see later that this set is finite, cf. Proposition 9.6.15. Note that it is mapped into the singleton $\{\deg(\mathbb{G},\mathcal{N})\}$ via the canonical map $(X_*(G)_{\mathbb{Q}}/G)^{\breve{\Gamma}_0} \longrightarrow X_*(A'_{\mathbb{G}})_{\mathbb{Q}}$. Also we have $\breve{\mathcal{F}}(\mathbb{G},\mathcal{N})^{ss} = \breve{\mathcal{F}}(\mathbb{G},\mathcal{N})_{H_{\mathbb{G},\mathcal{N}}}$, where $H_{\mathbb{G},\mathcal{N}}$ denotes the image of $\mu(\mathbb{G},\mathcal{N})$ under the map $X_*(A_{\mathbb{G}})_{\mathbb{Q}} \longrightarrow (X_*(G)_{\mathbb{Q}}/G)^{\breve{\Gamma}_0}$. If the semi-stable locus is non-empty, then $H_{\mathbb{G},\mathcal{N}}$ is the unique minimal element of $\mathcal{H}(\mathbb{G},\mathcal{N})$.

Next we turn to the notion of an HN-*type* and the associated stratification. As in Chapter VI, we start with the definition of HN-pairs. Recall the notion of a psgp $\mathbb{P}$ in $\mathbb{G}$ from Definition 9.1.27. For a conjugacy class of filtrations $\mathcal{N}_P \in X_*(P)/P$, we define the degree $\deg(\mathbb{P}, \mathcal{N}_P) \in X_*(A'_{\mathbb{P}})_{\mathbb{Q}}$ as in Definition 9.2.3.

Definition 9.6.4 An HN-*pair* for $(\mathbb{G}, \mathcal{N})$ is a pair $(\mathbb{P}, \mathcal{N}_P)$ consisting of a psgp $\mathbb{P}$ of $\mathbb{G}$ and a conjugacy class $\mathcal{N}_P \in X_*(P)_{\mathbb{Q}}/P$ which satisfy the following conditions.

(i) Under the inclusion $P \subset G$ the conjugacy class $\mathcal{N}_P$ is mapped to the conjugacy class $\mathcal{N}$.
(ii) For any (or equivalently, one) lifting $\lambda : \mathbb{D}_{K_0} \longrightarrow P$ of $\deg(\mathbb{P}, \mathcal{N}_P)$ to a $\mathbb{Q}$-1-PS of P, we have the equality $P = P_\lambda$. (Here P_λ is the K_0-psgp of G associated to λ.)

Let $J^{\mathbb{G}} = \underline{\mathrm{Aut}}^{\otimes}(N^{\mathbb{G}})$ be the automorphism $\mathbb{Q}_p$-group scheme of $N^{\mathbb{G}}$, as defined in Section 3. By the explicit description (3.2), page 252, we see that conjugation by any $j \in J^{\mathbb{G}}(\mathbb{Q}_p) \subset G(K_0)$ takes one HN-pair to another.

Definition 9.6.5 (i) A HN-*type* associated to $(\mathbb{G}, \mathcal{N})$ is a $J^{\mathbb{G}}(\mathbb{Q}_p)$-conjugacy class of HN-*pairs* $(\mathbb{P}, \mathcal{N}_P)$.
(ii) The HN-*vector* of an HN-type θ is the $G(K_0)$-conjugacy class $H_\theta \in (X_*(G)_{\mathbb{Q}}/G)^{\check{\Gamma}_0}$ of any $\mathbb{Q}$-1-PS λ as in Definition 9.6.4 (ii) for any HN-pair $(\mathbb{P}, \mathcal{N}_P)$ in θ.

We denote by $\Theta(\mathbb{G}, \mathcal{N})$ the set of HN-types associated to $(\mathbb{G}, \mathcal{N})$ and by $\theta_{\mathbb{G},\mathcal{N}}$ the distinguished element $\theta_{\mathbb{G},\mathcal{N}} = (\mathbb{G}, \mathcal{N})$. Let us record some functorial properties of these sets.

First, any weak morphism $\rho : \mathbb{G} \dashrightarrow \mathbb{G}'$ of augmented reductive groups over $\mathrm{Isoc}(L)$ induces a map

$$\rho_* : \Theta(\mathbb{G}, \mathcal{N}) \longrightarrow \Theta(\mathbb{G}', \tilde{\rho}_*(\mathcal{N})). \tag{6.4}$$

Explicitly, let $(\mathbb{P}, \mathcal{N}_P)$ be an HN-pair for $(\mathbb{G}, \mathcal{N})$ and let λ be a $\mathbb{Q}$-1-PS over K_0 as in condition (ii) of Definition 9.6.4. Since λ is par-equivalent to ${}^{\varphi}\lambda$ (see (2.11), page 246), its push-forward $\rho_*\lambda$ is par-equivalent to ${}^{\varphi'}\rho_*\lambda$, which means that the K_0-psgp $P'_{\rho_*\lambda}$ in G' is φ'-stable. By Proposition 9.2.8, the pair $(\mathbb{P}'_{\rho_*(\lambda)}, \tilde{\rho}_*(\mathcal{N}_P))$ is an HN-pair for $(\mathbb{G}', \tilde{\rho}_*(\mathcal{N}))$. Moreover, the map (6.4) above is compatible with taking HN-vectors in an obvious way.

Next, the Galois group $\check{\Gamma}_0 = \mathrm{Gal}(\overline{K}_0/\check{K}_0)$ of the reflex field $\check{K}_0 = \check{K}_0(\mathbb{G}, \mathcal{N})$ acts on $\Theta(\mathbb{G}, \mathcal{N})$. Indeed, any $\gamma \in \check{\Gamma}_0$ leaves $\mathcal{N}$ stable. Hence for any HN-pair $(\mathbb{P}, \mathcal{N}_P)$, the pair $(\mathbb{P}, {}^{\gamma}\mathcal{N}_P)$ satisfies condition (i) in Definition 9.6.4. Furthermore, since the push-forward map $X_*(P)_{\mathbb{Q}}/P \longrightarrow X_*(A'_{\mathbb{P}})_{\mathbb{Q}}$ is Γ_0-equivariant, we have $\deg(\mathbb{P}, {}^{\gamma}\mathcal{N}_P) = \deg(\mathbb{P}, \mathcal{N}_P)$, hence condition (ii) of Definition 9.6.4 is also satisfied. We get an action

$$\begin{array}{rcl} \breve{\Gamma}_0 \times \Theta(\mathbb{G}, \mathcal{N}) & \longrightarrow & \Theta(\mathbb{G}, \mathcal{N}) \\ (\gamma, \theta) & \longmapsto & {}^{\gamma}\theta. \end{array}$$

We write $[\theta]$ for the Galois orbit of θ and we note that for any $\theta' \in [\theta]$, we have $H_{\theta'} = H_{\theta}$, thus we may simply write $H_{[\theta]}$.

Now, let K be an extension field of $\breve{K}_0$. Given a filtration $\mathcal{F} \in \breve{\mathcal{F}}(G, \mathcal{N})(K)$, we would like to define its HN-type $\theta_{\mathcal{F}}$. The first ingredient is its HN-psgp

$$\mathbb{P}_{\mathcal{F}} = \mathbb{P}_{\bullet_{N^{\mathbb{G}}_{\mathcal{F}}}} ,$$

and the second ingredient is the filtration

$$\mathcal{F}_{P_{\mathcal{F}}} = \mathcal{F}_{P_{\bullet_{N^{\mathbb{G}}_{\mathcal{F}}}}} \in \mathrm{Fil}_K(\omega^{P_{\mathcal{F}}}) ,$$

given by Proposition 4.2.17. However, in general there is no well-defined map

$$\mathrm{Fil}_K(\omega^{P_{\mathcal{F}}}) \longrightarrow X_*(P_{\mathcal{F}})_{\mathbb{Q}}/P_{\mathcal{F}}.$$

Indeed, by the definition of the RHS, in order to define such a map we need to choose an embedding of the algebraic closure of K_0 in K into $\overline{K}_0$. Such a choice in general matters† and we only get a map

$$\mathrm{Fil}_K(\omega^{P_{\mathcal{F}}}) \longrightarrow (X_*(P_{\mathcal{F}})_{\mathbb{Q}}/P_{\mathcal{F}})/\Gamma_0.$$

To get something better, we fix a finite Galois extension $\tilde{K}_0$ of K_0 which splits the group G. Thus in particular, the action of $\mathrm{Gal}(\overline{K}_0/\tilde{K}_0)$ on $X_*(P_{\mathcal{F}})_{\mathbb{Q}}/P_{\mathcal{F}}$ is trivial. It follows that for any extension K of $\tilde{K}_0$ we have a well-defined map

$$\mathrm{Fil}_K(\omega^{P_{\mathcal{F}}}) \longrightarrow X_*(P_{\mathcal{F}})_{\mathbb{Q}}/P_{\mathcal{F}}$$

and we may consider the conjugacy class $\mathcal{N}_{P_{\mathcal{F}}}$ of $\mathcal{F}_{P_{\mathcal{F}}}$. The pair $(\mathbb{P}_{\mathcal{F}}, \mathcal{N}_{P_{\mathcal{F}}})$ obviously satisfies axiom (i) in Definition 9.6.4. Moreover, Proposition 9.2.28 shows that axiom (ii) is also satisfied.

Definition 9.6.6 Let K be an extension field of $\tilde{K}_0$ and let $\mathcal{F} \in \mathrm{Fil}_K(\omega^{\mathbb{G}})$. The pair $(\mathbb{P}_{\mathcal{F}}, \mathcal{N}_{P_{\mathcal{F}}})$ defined above is called the HN-*pair of* $\mathcal{F}$. Its $J^{\mathbb{G}}(\mathbb{Q}_p)$-conjugacy class $\theta_{\mathcal{F}}$ is the HN-*type of* $\mathcal{F}$.

If K is only an extension of $\breve{K}_0$ then we can only define $[\theta]_{\mathcal{F}}$, its HN-*type modulo* $\breve{\Gamma}_0$.

† Since we deal with analytic spaces rather than with algebraic varieties, we cannot restrict our attention to the case when K is a subfield of $\overline{K}_0$, as we did in Chapter VI.

Using this construction we get a decomposition

$$\breve{\mathcal{F}}(G,\mathcal{N})(K) = \dot{\bigcup}_{[\theta]\in\Theta(\mathbb{G},\mathcal{N})/\Gamma_{\breve{K}_0}} \breve{\mathcal{F}}(\mathbb{G},\mathcal{N})(K)_{[\theta]}$$

for any extension K of $\breve{K}_0$, and a sharper decomposition

$$\breve{\mathcal{F}}(G,\mathcal{N})(K) = \dot{\bigcup}_{\theta\in\Theta(\mathbb{G},\mathcal{N})} \breve{\mathcal{F}}(\mathbb{G},\mathcal{N})(K)_{\theta}$$

for any extension K of $\tilde{K}_0$. Moreover, for any $\breve{K}_0$-linear morphism of fields $\iota : K \longrightarrow K'$ between extensions of $\tilde{K}_0$, the base change map $\breve{\mathcal{F}}(G,\mathcal{N})(K) \longrightarrow \breve{\mathcal{F}}(G,\mathcal{N})(K')$ takes $\breve{\mathcal{F}}(\mathbb{G},\mathcal{N})(K)_\theta$ into $\breve{\mathcal{F}}(\mathbb{G},\mathcal{N})(K')_{\gamma_\iota\theta}$ where $\gamma_\iota \in \mathrm{Gal}(\tilde{K}_0/\breve{K}_0)$ is the restriction of ι to $\tilde{K}_0$.

We note that the above decompositions are compatible with the HN-vector decomposition (6.1), page 277 in the sense that, for example,

$$\breve{\mathcal{F}}(\mathbb{G},\mathcal{N})(K)_H = \dot{\bigcup}_{\{\theta|H_\theta=H\}} \breve{\mathcal{F}}(\mathbb{G},\mathcal{N})(K)_\theta .$$

In particular, we have $\breve{\mathcal{F}}(\mathbb{G},\mathcal{N})(K)_{\theta_{\mathbb{G},\mathcal{N}}} = \breve{\mathcal{F}}(\mathbb{G},\mathcal{N})(K)^{ss}$.

In order to study the analytic properties of these decompositions, we will first work over $\tilde{K}_0$ and add the sign $\tilde{}$ to denote the base change of any object to $\tilde{K}_0$. For example, $\tilde{\mathcal{F}}(G,\mathcal{N})^{an}$ is the $\tilde{K}_0$-analytic space associated with the variety $\breve{\mathcal{F}}(G,\mathcal{N})_{\tilde{K}_0}$. According to the above discussion, we have a disjoint union

$$\tilde{\mathcal{F}}(G,\mathcal{N})^{an} = \dot{\bigcup}_{\theta\in\Theta(\mathbb{G},\mathcal{N})} \tilde{\mathcal{F}}(\mathbb{G},\mathcal{N})_\theta ,$$

where we put

$$\tilde{\mathcal{F}}(\mathbb{G},\mathcal{N})_\theta := \{x \in \tilde{\mathcal{F}}(G,\mathcal{N})^{an} \mid \theta_{\mathcal{F}_x} = \theta\}.$$

We also have a sharper decomposition according to HN-pairs

$$\tilde{\mathcal{F}}(\mathbb{G},\mathcal{N})_\theta = \dot{\bigcup}_{(\mathbb{P},\mathcal{N}_P)\in\theta} \tilde{\mathcal{F}}(\mathbb{P},\mathcal{N}_P)_\theta ,$$

where

$$\begin{aligned} \tilde{\mathcal{F}}(\mathbb{P},\mathcal{N}_P)_\theta &:= \{x \in \tilde{\mathcal{F}}(G,\mathcal{N})^{an} \mid (\mathbb{P}_{\mathcal{F}_x}, \mathcal{N}_{P_{\mathcal{F}_x}}) = (\mathbb{P},\mathcal{N}_P)\} \\ &= \tilde{\mathcal{F}}(\mathbb{G},\mathcal{N})_\theta \cap \tilde{\mathcal{F}}(P,\mathcal{N}_P)^{an} . \end{aligned}$$

Using the action of $J^{\mathbb{G}}(\mathbb{Q}_p)$, we may also rewrite the above decomposition as a bijection

$$J^{\mathbb{G}}(\mathbb{Q}_p) \times^{J^{\mathbb{P}}(\mathbb{Q}_p)} \tilde{\mathcal{F}}(\mathbb{P},\mathcal{N}_P)_\theta \xrightarrow{\sim} \tilde{\mathcal{F}}(\mathbb{G},\mathcal{N})_\theta , \tag{6.5}$$

where $J^{\mathbb{P}}(\mathbb{Q}_p) = \underline{\mathrm{Aut}}(N^{\mathbb{P}}) = P(K_0) \cap J^{\mathbb{G}}(\mathbb{Q}_p)$.

On the other hand, denote by $\mathbb{M}$ the Levi quotient of $\mathbb{P}$, by $\mathcal{N}_M$ the induced conjugacy class, and by π_P the reduction map $\tilde{\mathcal{F}}(P, \mathcal{N}_P)^{an} \longrightarrow \tilde{\mathcal{F}}(M, \mathcal{N}_M)^{an}$, which was already shown in Lemma 5.3.6 to be a locally trivial fibration in affine spaces. Then Proposition 9.2.28 shows that inside $\tilde{\mathcal{F}}(P, \mathcal{N}_P)^{an}$ we have

$$\tilde{\mathcal{F}}(\mathbb{P}, \mathcal{N}_P)_\theta = \pi_P^{-1}(\tilde{\mathcal{F}}(\mathbb{M}, \mathcal{N}_M)^{ss}).$$

For any weak morphism $\rho : \mathbb{G} \dashrightarrow \mathbb{G}'$, the very definition of the HN-filtration shows that

$$\tilde{\rho}^{an}(\tilde{\mathcal{F}}(\mathbb{P}, \mathcal{N}_P)_\theta) \subset \tilde{\mathcal{F}}(\mathbb{P}', \mathcal{N}_{P'})_{\theta'} , \tag{6.6}$$

where $(\mathbb{P}', \mathcal{N}_{P'})$ is the HN-pair defined below (6.4), page 279.

Example 9.6.7 Let $\mathbb{G} = \mathbb{GL}(V, \Phi_V)$ for a given isocrystal $N = (V, \Phi_V)$. We note that an HN-pair $(\mathbb{P}, \mathcal{N}_P)$ is the same as the specification of a pair $(\gamma, {}^\bullet N)$, where ${}^\bullet N$ is a flag of isocrystals and γ is an HN-type in the sense of Chapter VIII, i.e., a pair $(\underline{h}, \underline{g})$ of multiplicity functions. Here $\underline{h}$ is determined by ${}^\bullet N$, while $\underline{g}$ corresponds to $\mathcal{N}_P$ as explained in Example 6.3.3. Further, what is now denoted by $\tilde{\mathcal{F}}(\mathbb{P}, \mathcal{N}_P)_\theta$ is what was denoted by $\mathcal{F}(N, \nu)_{\gamma, {}^\bullet N}$ in Proposition 8.3.5.

Now we are able to prove the main result of this section.

Theorem 9.6.8 *Let $\theta \in \Theta(\mathbb{G}, \mathcal{N})$ and let $(\mathbb{P}, \mathcal{N}_P)$ be a representative* HN*-pair of θ.*

(i) *The set $\tilde{\mathcal{F}}(\mathbb{P}, \mathcal{N}_P)_\theta$ is a locally closed $\breve{K}_0$-analytic subset of $\tilde{\mathcal{F}}(G, \mathcal{N})^{an}$.*

(ii) *The map $\tilde{\mathcal{F}}(\mathbb{P}, \mathcal{N}_P)_\theta \xrightarrow{\pi_P} \tilde{\mathcal{F}}(\mathbb{M}, \mathcal{N}_M)^{ss}$ is an affine bundle.*

(iii) *We have*

$$\overline{\tilde{\mathcal{F}}(\mathbb{P}, \mathcal{N}_P)_\theta} \setminus \tilde{\mathcal{F}}(\mathbb{P}, \mathcal{N}_P)_\theta \subset \bigcup_{H > H_\theta} \tilde{\mathcal{F}}(\mathbb{G}, \mathcal{N})_H.$$

(iv) *The stratum $\tilde{\mathcal{F}}(\mathbb{G}, \mathcal{N})_\theta$ is locally closed in $\tilde{\mathcal{F}}(G, \mathcal{N})$ and the bijective map*

$$J^{\mathbb{G}}(\mathbb{Q}_p) \times^{J^{\mathbb{P}}(\mathbb{Q}_p)} \tilde{\mathcal{F}}(\mathbb{P}, \mathcal{N}_P)_\theta \xrightarrow{\sim} \tilde{\mathcal{F}}(\mathbb{G}, \mathcal{N})_\theta$$

is a homeomorphism. Moreover, for any extension K of $\breve{K}_0$ we have

$$\tilde{\mathcal{F}}(\mathbb{G}, \mathcal{N})(K)_\theta = \tilde{\mathcal{F}}(\mathbb{G}, \mathcal{N})_\theta(K).$$

(v) *Assume further that the automorphism group $J^{\mathbb{G}}$ of $N^{\mathbb{G}}$ is reductive and connected. Then the stratum $\tilde{\mathcal{F}}(\mathbb{G}, \mathcal{N})_\theta$ is closed in $\tilde{\mathcal{F}}(\mathbb{G}, \mathcal{N})_{H_\theta}$.*

Proof Statement (ii) was already explained above. Statement (i) follows as in the proof of Proposition 8.3.5.

Let us prove statement (iii) by reduction to the GL_n-case. First of all, the closure of the generalized Bruhat cell $\tilde{\mathcal{F}}(P, \mathcal{N}_P)^{an}$ in $\tilde{\mathcal{F}}(G, \mathcal{N})^{an}$, a Schubert variety, is a union of generalized Bruhat cells

$$\overline{\tilde{\mathcal{F}}(P, \mathcal{N}_P)^{an}} = \dot{\bigcup}_{\mathcal{N}'_P \leq \mathcal{N}_P} \tilde{\mathcal{F}}(P, \mathcal{N}'_P)^{an}$$

where $\mathcal{N}'_P$ runs through the conjugacy classes of $\mathbb{Q}$-1-PS of P which induce $\mathcal{N}$ on G, and where $\leq$ stands for the Bruhat order, as in Definition 9.4.6. We will first prove that

$$\overline{\tilde{\mathcal{F}}(P, \mathcal{N}_P)^{an}} \subset \bigcup_{H \geq H_\theta} \tilde{\mathcal{F}}(\mathbb{G}, \mathcal{N})_H. \tag{6.7}$$

Indeed, given a morphism $\rho : \mathbb{G} \longrightarrow \mathbb{G}'$, we have defined in (6.4), page 279 the push-forward $(\mathbb{P}', \mathcal{N}_{P'})$ of $(\mathbb{P}, \mathcal{N}_P)$ and we clearly have $\tilde{\rho}_*^{an}(\overline{\tilde{\mathcal{F}}(P, \mathcal{N}_P)^{an}}) \subset \overline{\tilde{\mathcal{F}}(P', \mathcal{N}_{P'})^{an}}$. If $\mathbb{G}'$ is (a product of) $\mathbb{GL}(V, \Phi_V)$, the inclusion (6.7) above for $\mathbb{G}'$ was obtained in the course of the proof of statement (iii) of Proposition 8.3.5. Therefore, taking $\mathbb{G}' = \prod \mathbb{GL}(V, \Phi_V)$ and $\rho = \prod \rho_V$ for (V, Φ_V, ρ_V) running over the set $\mathcal{I}_{\mathbb{Q}_p}$ of Proposition 9.4.2 (iii), and arguing as in the proof of Theorem 9.6.2, we get (6.7) for $\mathbb{G}$.

Now it remains to show that

$$\overline{\tilde{\mathcal{F}}(P, \mathcal{N}_P)^{an}} \cap \tilde{\mathcal{F}}(\mathbb{G}, \mathcal{N})_{H_\theta} = \tilde{\mathcal{F}}(\mathbb{P}, \mathcal{N}_P)_\theta. \tag{6.8}$$

Let ${}^\bullet N^{\mathbb{G}}_{\mathbb{P}, \mathcal{N}_P}$ be the filtration of $N^{\mathbb{G}}$ defined by any lifting λ as in point (ii) of Definition 9.6.4. We claim that

$${}^\bullet N^{\mathbb{G}}_{\mathcal{F}_x} = {}^\bullet N^{\mathbb{G}}_{\mathbb{P}, \mathcal{N}_P}, \quad \forall x \in \overline{\tilde{\mathcal{F}}(P, \mathcal{N}_P)^{an}} \cap \tilde{\mathcal{F}}(\mathbb{G}, \mathcal{N})_{H_\theta}.$$

Indeed, this is true if $\mathbb{G}$ is a product of $\mathbb{GL}(V, \Phi_V)$, as a byproduct of item (iii) of Proposition 8.3.5. Therefore, by definition of the HN-filtration in Theorem 9.2.18, and since for any morphism $\rho : \mathbb{G} \longrightarrow \mathbb{G}'$, we have ${}^\bullet N^{\mathbb{G}'}_{\mathbb{P}', \mathcal{N}_{P'}} = \rho_*({}^\bullet N^{\mathbb{G}}_{\mathbb{P}, \mathcal{N}_P})$, the statement follows for $\mathbb{G}$. In other words, for any x in the LHS of (6.8) above, the HN-psgp of $\mathcal{F}_x$ is $\mathbb{P}$ and the degree of $\mathcal{F}_{x,P}$ is $\deg(\mathbb{P}, \mathcal{N}_P)$. Thus there is $\mathcal{N}'_P \leq \mathcal{N}_P$ such that $\deg(\mathbb{P}, \mathcal{N}'_P) = \deg(\mathbb{P}, \mathcal{N}_P)$ and $x \in \tilde{\mathcal{F}}(\mathbb{P}, \mathcal{N}'_P)_{\theta'}$. Here θ' is the HN-type of the HN-pair $(\mathbb{P}, \mathcal{N}'_P)$, and has the same HN-vector $H_{\theta'} = H_\theta$ as θ. By Corollary 9.4.15, we conclude that $\mathcal{N}'_P = \mathcal{N}_P$. This proves (6.7) and statement (iii) of the Proposition.

We now turn to statement (v). Since $J^{\mathbb{G}}$ is assumed to be reductive and connected, we know by Proposition 9.3.6 that its subgroup $J^{\mathbb{P}}$ is a psgp over $\mathbb{Q}_p$. Therefore there is a compact open subgroup C such that $J^{\mathbb{G}}(\mathbb{Q}_p) = C.J^{\mathbb{P}}(\mathbb{Q}_p)$ and we may apply readily the argument used in Proposition 8.3.5 to derive statement (v) from statement (iii).

Let us turn to statement (iv). In this case there may not exist any compact open subgroup C of $J^{\mathbb{G}}(\mathbb{Q}_p)$ as above. Nevertheless, we may find an increasing sequence $(C_n)_{n \in \mathbb{N}}$ of open compact subgroups of $J^{\mathbb{G}}(\mathbb{Q}_p)$ such that $J^{\mathbb{G}}(\mathbb{Q}_p) =$

$\bigcup_{n\in\mathbb{N}} C_n.J^{\mathbb{P}}(\mathbb{Q}_p)$. This follows from the existence of a Levi decomposition of $J^{\mathbb{G}}$, cf. [23], Prop. 5.1.

For each $n \in \mathbb{N}$, denote by $\tilde{\mathcal{F}}(\mathbb{G},\mathcal{N})_{\theta,n}$ the subset $C_n{\cdot}\tilde{\mathcal{F}}(\mathbb{P},\mathcal{N}_P)_\theta$ of $\tilde{\mathcal{F}}(\mathbb{G},\mathcal{N})_\theta$. The argument of Proposition 8.3.5 shows that the bijection

$$C_n \times^{J^{\mathbb{P}}(\mathbb{Q}_p)\cap C_n} \tilde{\mathcal{F}}(\mathbb{P},\mathcal{N}_P)_\theta \longrightarrow \tilde{\mathcal{F}}(\mathbb{G},\mathcal{N})_{\theta,n}\,,$$

induced by the action map of $J^{\mathbb{G}}(\mathbb{Q}_p)$ on $\tilde{\mathcal{F}}(G,\mathcal{N})$, is a homeomorphism onto its image, and this image is a closed subset of $\tilde{\mathcal{F}}(\mathbb{G},\mathcal{N})_{H_\theta}$. Therefore, if we can prove that each $\tilde{\mathcal{F}}(\mathbb{G},\mathcal{N})_{\theta,n}$ is open in $\tilde{\mathcal{F}}(\mathbb{G},\mathcal{N})_\theta$, we deduce that (6.5), page 281 is a homeomorphism and its image is locally closed (we also note that the last statement of item (iv) is straightforward).

Let us assume that $\tilde{\mathcal{F}}(\mathbb{G},\mathcal{N})_{\theta,n}$ is not open in $\tilde{\mathcal{F}}(\mathbb{G},\mathcal{N})_\theta$ and let us choose an element

$$x \in \tilde{\mathcal{F}}(\mathbb{G},\mathcal{N})_{\theta,n} \cap \overline{\tilde{\mathcal{F}}(\mathbb{G},\mathcal{N})_\theta \setminus \tilde{\mathcal{F}}(\mathbb{G},\mathcal{N})_{\theta,n}}.$$

Let $(x_k)_{k\in\mathbb{N}}$ be a sequence of points in $\tilde{\mathcal{F}}(\mathbb{G},\mathcal{N})_\theta \setminus \tilde{\mathcal{F}}(\mathbb{G},\mathcal{N})_{\theta,n}$ which converges to x and for each k choose a pair (j_k, x_k^0) with $x_k^0 \in \tilde{\mathcal{F}}(\mathbb{P},\mathcal{N}_P)_\theta$ and $j_k \in J^{\mathbb{G}}(\mathbb{Q}_p)$, such that $x_k = j_k x_k^0$. By what was just explained, $\tilde{\mathcal{F}}(G,\mathcal{N})_{\theta,n}$ is open and closed in $\tilde{\mathcal{F}}(G,\mathcal{N})_{\theta,n'}$ for any $n' > n$. Therefore the sequence $(j_k)_{k\in\mathbb{N}}$ is not bounded in $J^{\mathbb{G}}(\mathbb{Q}_p)/J^{\mathbb{P}}(\mathbb{Q}_p)$. However, we have an inclusion

$$J^{\mathbb{G}}(\mathbb{Q}_p)/J^{\mathbb{P}}(\mathbb{Q}_p) \subset (G(K_0)/P(K_0))^\varphi.$$

By the same argument as used to show the compactness of the subset $\mathbf{G}_d(V)$ considered in the proof of Proposition 8.2.1, one proves that $(G(K_0)/P(K_0))^\varphi$ is a compact subset of $G(K_0)/P(K_0)$. Therefore we may assume that the sequence $(j_k P(K_0))_{k\in\mathbb{N}}$ has a limit, say $gP(K_0)$, in the compact set $(G(K_0)/P(K_0))^\varphi$. This limit does not lie in $J^{\mathbb{G}}(\mathbb{Q}_p)/J^{\mathbb{P}}(\mathbb{Q}_p)$.

Note that the K_0-psgp ${}^gP$ is φ-stable. We will abuse notation and denote by ${}^g\mathbb{P}$ the corresponding a.g.s. (this should not suggest that conjugation by g induces an isomorphism $\mathbb{P} \xrightarrow{\sim} {}^g\mathbb{P}$). Moreover the pair $({}^g\mathbb{P}, {}^g\mathcal{N}_P)$ is an HN-pair with HN-vector H_θ. Its associated HN-type ${}^g\theta$ is distinct from θ since $gP(K_0)$ does not belong to $J^{\mathbb{G}}(\mathbb{Q}_p)/J^{\mathbb{P}}(\mathbb{Q}_p)$. Now by continuity of the action of $G(K_0)$ on $\tilde{\mathcal{F}}(G,\mathcal{N})$, our point x lies in the closure of the generalized Bruhat cell $\tilde{\mathcal{F}}({}^gP, {}^g\mathcal{N}_P)$. By assumption it also lies in $\tilde{\mathcal{F}}(\mathbb{G},\mathcal{N})_{H_\theta}$, hence by (6.8), page 283 it lies in $\tilde{\mathcal{F}}({}^g\mathbb{P}, {}^g\mathcal{N}_P)_{{}^g\theta}$. However, this last subset has empty intersection with $\tilde{\mathcal{F}}(\mathbb{G},\mathcal{N})_{\theta,n}$, leading to a contradiction. □

Without the reductive assumption in point (v), the HN-type stratum $\tilde{\mathcal{F}}(\mathbb{G},\mathcal{N})_\theta$ need not be closed in $\tilde{\mathcal{F}}(\mathbb{G},\mathcal{N})_{H_\theta}$, as the following example shows.

Example 9.6.9 Let $\mathbb{G} = \mathbb{GL}(V,\Phi_V)$ for an isocrystal (V,Φ_V) over $\mathbb{F}_p$ with

V of dimension 3. Further, let Φ_V be a non-trivial and non-regular unipotent element. Then the set of 1-dimensional sub-isocrystals is in natural bijection with $\mathbb{P}^1(\mathbb{Q}_p)$. However, the group $J^{\mathbb{G}}$ (the centralizer of Φ_V) acts with two orbits : a point, resp. an affine line $\mathbb{A}^1(\mathbb{Q}_p)$, corresponding to the case where the 2-dimensional quotient isocrystal is trivial, resp. unipotent. Hence, if $\tilde{\mathcal{F}}(\mathbb{G}, \mathcal{N}) = \mathbb{P}(V)$ is the projective space, we see that the HN-type stratum corresponding to $\mathbb{A}^1(\mathbb{Q}_p)$ is not closed in its HN-vector stratum.

The above example and the proof of statement (iv) of Theorem 9.6.8 suggest a way to gather HN-type strata so as to obtain closed subsets of HN-vector strata. This will also clarify the relationship to our former result, Proposition 8.3.5, in the GL_n-case. In the following definition, we denote by $\tilde{\nu} := \nu \circ sl$ the "slope $\mathbb{Q}$-1-PS" of $\mathbb{G}$, where sl was defined in (1.3), page 225.

Definition 9.6.10 Two HN-pairs $(\mathbb{P}, \mathcal{N}_P)$ and $(\mathbb{P}', \mathcal{N}_{P'})$ are called *slope-equivalent* if there is $g \in G_{\tilde{\nu}}(K_0)$ such that $P' = {}^gP$ and $\mathcal{N}_{P'} = {}^g\mathcal{N}_P$.

We note that, although conjugation by an element g as in the definition generally does *not* induce an isomorphism $\mathbb{P} \longrightarrow \mathbb{P}'$, it always induces an isomorphism $A'_{\mathbb{P}} \xrightarrow{\sim} A'_{\mathbb{P}'}$. Moreover, through the latter isomorphism, we have $\deg(\mathbb{P}', \mathcal{N}_{P'}) = {}^g\deg(\mathbb{P}, \mathcal{N}_P)$. The slope-equivalence relation descends to the set of HN-types and in fact a slope-equivalence class of HN-pairs is a union of HN-types (considered as equivalence classes, i.e., $J_{\mathbb{G}}(\mathbb{Q}_p)$-conjugacy classes of HN-pairs). By definition, the slope-equivalence class of an HN-pair $(\mathbb{P}, \mathcal{N}_P)$ can be identified with $(G_{\tilde{\nu}}(K_0)/P_{\tilde{\nu}}(K_0))^{\varphi}$, while its equivalence class (HN-type) inside can be identified with $J^{\mathbb{G}}(\mathbb{Q}_p)/J^{\mathbb{P}}(\mathbb{Q}_p) = G_{\tilde{\nu}}(K_0)^{\varphi}/P_{\tilde{\nu}}(K_0)^{\varphi}$.

Lemma 9.6.11 *When $N^{\mathbb{G}}$ is split semi-simple, slope-equivalence and equivalence of HN-pairs coincide.*

Proof By Proposition 9.1.31 we may identify $\tilde{\nu}$ and ν. Then, as explained above Definition 9.3.1, φ induces an effective descent datum on $G_{\tilde{\nu}}$ whose descent is $J^{\mathbb{G}}$. Similarly, φ induces an effective descent datum on $P_{\tilde{\nu}} = P \cap G_{\tilde{\nu}}$, with descended group $J^{\mathbb{P}}$, and on the K_0-variety $G_{\tilde{\nu}}/P_{\tilde{\nu}}$, with descended $\mathbb{Q}_p$-variety $J^{\mathbb{G}}/J^{\mathbb{P}}$. Now, let $(P, \mathcal{N}_P)$ and $(P', \mathcal{N}_{P'})$ be conjugate by some element $g \in G_{\tilde{\nu}}(K_0)$. Then both $P_{\tilde{\nu}}$ and $P'_{\tilde{\nu}}$ descend to $\mathbb{Q}_p$-psgps of $J^{\mathbb{G}}$ which are conjugate under $J^{\mathbb{G}}(K_0)$, hence by Borel–Tits theory they are also conjugate by some element $j \in J^{\mathbb{G}}(\mathbb{Q}_p)$. Since $gj^{-1} \in P_{\tilde{\nu}}(K_0)$, we also have $\mathcal{N}_{P'} = {}^j\mathcal{N}_P$. Therefore $(P, \mathcal{N}_P)$ is equivalent to $(P', \mathcal{N}_{P'})$. □

Now, the argument of the proof of statement (iv) of Theorem 9.6.8, joined with the observation that $(G(K_0)/P(K_0))^{\varphi}$ contains $(G_{\tilde{\nu}}(K_0)/P_{\tilde{\nu}}(K_0))^{\varphi}$ as *closed* subset, shows the following.

Corollary 9.6.12 *Let $\langle\theta\rangle$ be a slope-equivalence class of HN-types. The set*

$$\tilde{\mathcal{F}}(\mathbb{G},\mathcal{N})_{\langle\theta\rangle} := \dot{\bigcup}_{\theta\in\langle\theta\rangle}\tilde{\mathcal{F}}(\mathbb{G},\mathcal{N})_{\theta}$$

is a closed subset of $\tilde{\mathcal{F}}(\mathbb{G},\mathcal{N})_{H_\theta}$. □

We now compare these HN-stratifications with those of Chapter VIII.

Example 9.6.13 Let $\mathbb{G} = \mathbb{GL}(V,\Phi_V)$ for some isocrystal $N = (V,\Phi_V)$, and let $\mathcal{N}$ correspond to ν or the multiplicity function g in the notation of Chapter VIII. We have already noted a dictionary $(\mathbb{P},\mathcal{N}_{\mathbb{P}}) \longleftrightarrow (\gamma,{}^{\bullet}N)$ right above Theorem 9.6.8. However the notion of HN-type from Definition 9.6.5 is sharper than that of Chapter VIII. In fact, an HN-type $\gamma = (\underline{h},\underline{g})$ in the sense of Definition 8.3.1, and for which there is at least one flag ${}^{\bullet}\overline{N}$ whose slope filtration has multiplicity function $\underline{h}$, is the same as a slope-equivalence class in the sense of Definition 9.6.10. In particular, the previous Corollary shows that the HN-type strata considered in Proposition 8.3.5 are always closed in their corresponding HN-vector strata.

We now turn to finiteness properties. Our first remark will be somewhat negative. Indeed, the following example shows that *the set $\Theta(\mathbb{G},\mathcal{N})$ need not be finite.*

Example 9.6.14 Let $L = \mathbb{F}_p$. We know that a reductive a.g.s. $\mathbb{G}$ over $\mathrm{Isoc}(\mathbb{F}_p)$ is the same as a pair (G,g), with G a reductive group over $\mathbb{Q}_p$ and $g \in G(\mathbb{Q}_p)$. The automorphism φ is simply the conjugation by g, and $J^{\mathbb{G}} = G_g$ is the centralizer of g in G. For simplicity let us assume that G is split, and let us choose $\mathcal{N}$ to be the conjugacy class of some $\mathbb{Q}$-1-PS ν of G defined over $\mathbb{Q}_p$ and such that $B := P_\nu$ is a Borel subgroup of G. Let us further assume that g lies in the unipotent radical of P_ν. For psychological reasons, let us write u instead of g.

By definition, a psgp $\mathbb{P}$ of $\mathbb{G}$ is just a $\mathbb{Q}_p$-psgp P of G which contains u. If this psgp is a Borel subgroup, then it can be extended uniquely to an HN-pair $(\mathbb{P},\mathcal{N}_{\mathbb{P}})$, which is conjugate under $G(\mathbb{Q}_p)$ to the pair (B,ν). Therefore, the *slope-equivalence* class of (B,ν) can be identified with the set of Borel subgroups of G which contain u, i.e., the set $S_u(\mathbb{Q}_p)$ of $\mathbb{Q}_p$-points of the Springer fiber of u. Moreover, the HN-types contained in this slope-equivalence class correspond to the set $G_u(\mathbb{Q}_p)\backslash S_u(\mathbb{Q}_p)$ of $G_u(\mathbb{Q}_p)$-orbits inside this Springer fiber. Now let O be the $G(\mathbb{Q}_p)$-conjugacy class of u, and put

$$S_O := \{(B',u') \mid u' \in O \cap B'(\mathbb{Q}_p)\}.$$

The map $G(\mathbb{Q}_p) \times S_u(\mathbb{Q}_p) \longrightarrow S_O$, $(g, B') \mapsto ({}^gB', {}^gu)$ identifies the induced $G(\mathbb{Q}_p)$-set $G(\mathbb{Q}_p) \times^{G_u(\mathbb{Q}_p)} S_u(\mathbb{Q}_p)$ with S_O, and therefore induces a bijection

$$G_u(\mathbb{Q}_p)\backslash S_u(\mathbb{Q}_p) \xrightarrow{\sim} G(\mathbb{Q}_p)\backslash S_O.$$

On the other hand, the map $G(\mathbb{Q}_p) \times (B(\mathbb{Q}_p) \cap O) \longrightarrow S_O$, $(g, u') \mapsto ({}^gB, {}^gu')$ induces a bijection $G(\mathbb{Q}_p) \times^{B(\mathbb{Q}_p)} (B(\mathbb{Q}_p) \cap O) \longrightarrow S_O$ and in turn a bijection

$$B(\mathbb{Q}_p)\backslash(B(\mathbb{Q}_p) \cap O) \xrightarrow{\sim} G(\mathbb{Q}_p)\backslash S_O.$$

Therefore, the set of HN-types which are slope-equivalent to (B, u) can be identified with the set of $B(\mathbb{Q}_p)$-orbits in $O \cap B(\mathbb{Q}_p)$. Such sets have been investigated in [128] (over an algebraically closed field rather than $\mathbb{Q}_p$ but this does not matter here). It is shown there that, as soon as G has a factor of type different from A_1, A_2, A_3, A_4 or C_2, the number of unipotent conjugacy classes in B is *infinite*. Therefore, in such a situation, one can find u such that $B(\mathbb{Q}_p)$ has infinitely many orbits in $B(\mathbb{Q}_p) \cap O$. In this case, the slope-equivalence class of the pair (B, u) contains infinitely many HN-types.

After this bad news, let us turn to a positive result.

Proposition 9.6.15 *The set of slope-equivalence classes of HN-pairs for* $(\mathbb{G}, \mathcal{N})$ *is finite. In particular, the set of HN-vectors which actually contribute to the HN-vector stratification of* $\tilde{\mathcal{F}}(\mathbb{G}, \mathcal{N})$ *in Theorem 9.6.2 is finite.*

By Lemma 9.6.11, this proposition implies that the set of HN-types is finite when $N^{\mathbb{G}}$ is split semi-simple.

Proof Let $\mathcal{P}_{\tilde{\nu}}$ be the set of all K_0-psgps of G through which $\tilde{\nu}$ factors, and let $\mathcal{P}_{\tilde{\nu}}^{\mathcal{N}}$ be the set of pairs $(P, \mathcal{N}_P)$, with $P \in \mathcal{P}_{\tilde{\nu}}$ and $\mathcal{N}_P$ a conjugacy class for P mapping to $\mathcal{N}$. By definition, the set of slope-equivalence classes for $(\mathbb{G}, \mathcal{N})$ is a subset of the set of $G_{\tilde{\nu}}(K_0)$-conjugacy classes in $\mathcal{P}_{\tilde{\nu}}^{\mathcal{N}}$. Forgetting $\mathcal{N}_P$ induces a map $\mathcal{P}_{\tilde{\nu}}^{\mathcal{N}} \longrightarrow \mathcal{P}_{\tilde{\nu}}$ with finite fibers. Therefore we are left to prove that $G_{\tilde{\nu}}(K_0)$ has finitely many orbits in $\mathcal{P}_{\tilde{\nu}}$. To prove this, observe that, on the one hand, any $P \in \mathcal{P}_{\tilde{\nu}}$ contains a maximal K_0-split torus through which $\tilde{\nu}$ factors, on the other hand, any two maximal K_0-split tori through which $\tilde{\nu}$ factors are contained in $G_{\tilde{\nu}}$ and therefore are conjugate under $G_{\tilde{\nu}}(K_0)$. In this way, one gets an injection of $G_{\tilde{\nu}}(K_0)\backslash\mathcal{P}_{\tilde{\nu}}$ into the finite set of K_0-psgps of G which contain a given maximal K_0-split torus. □

In Theorem 9.6.8, we have studied HN-strata over the base field $\tilde{K}_0$. Now using the action of the Galois group $\mathrm{Gal}(\tilde{K}_0/\breve{K}_0)$, we can descend the results to $\breve{K}_0$.

Corollary 9.6.16 *Let* $[\theta] \in \Theta(\mathbb{G}, \mathcal{N})/\check{\Gamma}_0$. *The set* $\check{\mathcal{F}}(\mathbb{G}, \mathcal{N})_{[\theta]}$ *is a locally closed subset of* $\check{\mathcal{F}}(G, \mathcal{N})^{an}$ *such that for any field extension* K *of* $\check{K}_0$, *we have*

$$\check{\mathcal{F}}(\mathbb{G}, \mathcal{N})(K)_{[\theta]} = \check{\mathcal{F}}(\mathbb{G}, \mathcal{N})_{[\theta]}(K).$$

Its preimage by the morphism $\tilde{\mathcal{F}}(G, \mathcal{N})^{an} \longrightarrow \check{\mathcal{F}}(G, \mathcal{N})^{an}$ *of analytic spaces over* $\check{K}_0$ *is the union*

$$\tilde{\mathcal{F}}(\mathbb{G}, \mathcal{N})_{[\theta]} = \dot{\bigcup}_{\theta \in [\theta]} \tilde{\mathcal{F}}(\mathbb{G}, \mathcal{N})_\theta.$$

Moreover, if $N^{\mathbb{G}}$ *is split semi-simple, then* $\check{\mathcal{F}}(\mathbb{G}, \mathcal{N})(K)_{[\theta]}$ *is a closed subset of* $\check{\mathcal{F}}(\mathbb{G}, \mathcal{N})_{H_{[\theta]}}$. □

We now look at the behavior of the HN-type stratification when extending the perfect field L to another one L'. As in Construction 9.1.29, we denote by $\mathbb{G}'$ the a.g.s. over $\mathrm{Isoc}_{L'}$ obtained from $\mathbb{G}$. After choosing an algebraic closure for K_0' and an embedding $\overline{K_0} \hookrightarrow \overline{K_0'}$, the element $\mathcal{N} \in X_*(G)_{\mathbb{Q}}/G$ defines an element $\mathcal{N}'$ in $X_*(G')_{\mathbb{Q}}/G'$. We denote by $\check{K}_0'$ the compositum of $\check{K}_0$ and K_0' in $\overline{K_0'}$. By the same process, any HN-pair $(\mathbb{P}, \mathcal{N}_P)$ for $(\mathbb{G}, \mathcal{N})$ gives rise to an HN-pair $(\mathbb{P}', \mathcal{N}_{P'})$ of $\mathbb{G}'$ having the same HN-vector H_θ. We denote by θ' the HN-type of $(\mathbb{P}', \mathcal{N}_{P'})$.

Proposition 9.6.17 *With the above notation, we have*

$$\tilde{\mathcal{F}}(\mathbb{P}', \mathcal{N}_{P'})_{\theta'} \subseteq \tilde{\mathcal{F}}(\mathbb{P}, \mathcal{N}_P)_\theta \hat{\otimes}_{\check{K}_0} \check{K}_0' \subseteq \tilde{\mathcal{F}}(\mathbb{P}', \mathcal{N}_{P'})_{\theta'} \cup \bigcup_{H > H_\theta} \tilde{\mathcal{F}}(\mathbb{G}', \mathcal{N}')_H.$$

Moreover, if $N^{\mathbb{G}}$ *is split semi-simple, then the first inclusion is an equality.*

Proof The first inclusion and the equality in the split semi-simple case follow from Proposition 9.5.3 (v) and Theorem 9.6.8 (ii). The second inclusion follows from the fact that its LHS sits in the Bruhat cell $\tilde{\mathcal{F}}(P', \mathcal{N}_{P'})$ which is contained in the RHS by (the proof of) statement (iii) in Theorem 9.6.8. □

Before proceeding, let us make some remarks.

Remarks 9.6.18 (i) Let $(\mathbb{G}', \mathcal{N}') := (\mathbb{G}_{\mathrm{ad}}, \mathcal{N}_{\mathrm{ad}})$ be the adjoint pair of $(\mathbb{G}, \mathcal{N})$ and let $\rho : \mathbb{G} \longrightarrow \mathbb{G}_{\mathrm{ad}}$ be the canonical projection. Then the map (6.4), page 279 is a bijection and the isomorphism $\tilde{\mathcal{F}}(G, \mathcal{N}) \xrightarrow{\sim} \tilde{\mathcal{F}}(G_{\mathrm{ad}}, \mathcal{N}_{\mathrm{ad}})$ identifies the HN-stratifications of $(\mathbb{G}, \mathcal{N})$ and $(\mathbb{G}_{\mathrm{ad}}, \mathcal{N}_{\mathrm{ad}})$, according to (6.6), page 282.

(ii) If $\rho : \mathbb{G} \overset{\sim}{\dashrightarrow} \mathbb{G}'$ is a weak isomorphism, then obviously the map (6.4), page 279 is a bijection and the isomorphism $\tilde{\mathcal{F}}(G, \mathcal{N}) \xrightarrow{\sim} \tilde{\mathcal{F}}(G', \tilde{\rho}_* \mathcal{N})$ identifies the HN-stratification of $(\mathbb{G}, \mathcal{N})$ and $(\mathbb{G}', \tilde{\rho}_* \mathcal{N})$. Moreover, ρ induces an isomorphism $J^{\mathbb{G}} \xrightarrow{\sim} J^{\mathbb{G}'}$ and the above identification is compatible with the respective actions of $J^{\mathbb{G}}(\mathbb{Q}_p)$ and $J^{\mathbb{G}'}(\mathbb{Q}_p)$.

Before closing this section, we investigate a little further the notion of HN-types in the case of a filtered G-isocrystal, i.e., a pair $(\mathbb{G}_b, \mathcal{N})$ for a reductive group G over $\mathbb{Q}_p$ and an element $b \in G(K_0)$. Then a K_0-psgp P of G_{K_0} is φ-stable iff it is "$b\sigma$-stable," i.e., $P = {}^{b\sigma}P$, meaning stable under "the natural descent datum composed with conjugation by b." As we already noticed in Remark 9.1.28, this does not imply that P is defined over $\mathbb{Q}_p$ (i.e., is σ-stable), and also not that it contains b (i.e., is b-stable). However, this certainly implies that the conjugacy class of P is σ-stable. Thus if G is quasi-split, then this conjugacy class contains a rational element.

Henceforth, we assume that G is quasi-split. We note that, thanks to Remark 9.6.18 and to Lemma 9.3.11, this is not really a restrictive hypothesis, as soon as L is large enough. In fact, we will also assume that L is algebraically closed, so that $\mathcal{N}^{\mathbb{G}}$ is split semi-simple, or equivalently, the σ-conjugacy class of b is decent (see Lemma 9.1.33).

In order to get a more convenient description of HN-types we have to (and may) modify b within its σ-conjugacy class.

Lemma 9.6.19 *In the given σ-conjugacy class, there exists an element b with the following properties.*

(i) *b is s-decent, for some integer s.*
(ii) *ν_b is defined over $\mathbb{Q}_p$ (and therefore, so is G_{ν_b}, and $b \in G_{\nu_b}(\mathbb{Q}_{p^s})$).*
(iii) *There is a minimal $\mathbb{Q}_p$-psgp P_0^J in J_b such that $b \in P_0^J(\mathbb{Q}_{p^s}) \subset G_{\nu_b}(\mathbb{Q}_{p^s})$.*

Proof That we can σ-conjugate our initial b so that the first two properties are fulfilled was already shown in the proof of Lemma 9.3.8. In this situation $b \in G_{\nu_b}(\mathbb{Q}_{p^s})$ is a basic element for the centralizer G_{ν_b}, which is a $\mathbb{Q}_p$-Levi subgroup of G and thus is also quasi-split.

Now let us start with an arbitrary minimal $\mathbb{Q}_p$-psgp Q_0^J in J_b. Its base change $Q^J_{0,\mathbb{Q}_{p^s}}$ in $G_{\nu_b,\mathbb{Q}_{p^s}}$ is $b\sigma$-stable. Thus the $G_{\nu_b}(\mathbb{Q}_{p^s})$-conjugacy class of this base change is σ-stable and, since G_{ν_b} is quasi-split, it contains a member ${}^gQ^J_{0,\mathbb{Q}_{p^s}}$ which is σ-stable. Here g is some element in $G_{\nu_b}(\mathbb{Q}_{p^s})$. Therefore, changing b to $gb\sigma(g)^{-1}$ (which is again s-decent and has slope morphism $\nu_{gb\sigma(g)^{-1}} = g\nu_b g^{-1} = \nu_b$), the latter psgp ${}^gQ^J_{0,\mathbb{Q}_{p^s}}$ is both σ-stable and $b\sigma$-stable, hence is the base change of a $\mathbb{Q}_p$-psgp P_0^J of J_b, which has the required properties. □

Let us assume that b is chosen as in the foregoing Lemma, and let us call *standard* the $\mathbb{Q}_p$-psgps of J_b which contain P_0^J. Since $P^J_{0,\mathbb{Q}_{p^s}} \subset G_{\nu_b,\mathbb{Q}_{p^s}}$ is $b\sigma$-stable and Int(b)-stable, it is also σ-stable, hence it descends to a $\mathbb{Q}_p$-psgp of G_{ν_b}, which will be denoted by P_0^ν. Hence we have $P^J_{0,\mathbb{Q}_{p^s}} = P^\nu_{0,\mathbb{Q}_{p^s}}$ inside $G_{\nu_b,\mathbb{Q}_{p^s}}$.

Lemma 9.6.20 *The map $P \mapsto P_{K_0}$ induces a bijection*

$$\{\mathbb{Q}_p\text{-psgps of } G \text{ containing } P_0^\nu\} \xrightarrow{\sim} \{b\sigma\text{-stable } K_0\text{-psgps of } G \text{ up to } J_b(\mathbb{Q}_p)\text{-conjugacy}\}$$

Proof We note first that if P is a $\mathbb{Q}_p$-psgp of G containing P_0^ν, then in particular $P(K_0)$ contains the element b and therefore the base change P_{K_0} is $b\sigma$-stable. Hence the map is well defined. Let us construct an inverse map. Start with a $b\sigma$-stable psgp P of G_{K_0}. Then $P \cap G_\nu$ descends to a $\mathbb{Q}_p$-psgp of J_b, call it P^J. There is a unique standard $\mathbb{Q}_p$-psgp of J_b to which P^J is conjugate by some element $j \in J_b(\mathbb{Q}_p)$. Then ${}^jP$ is $b\sigma$-stable and $b \in P_0^J(K_0) \subset {}^jP(K_0)$, so P is also σ-stable, i.e., $\mathbb{Q}_p$-rational. Its descent to $\mathbb{Q}_p$ necessarily contains that of P^J_{0,K_0}, namely P_0^ν. This provides us with the desired inverse map. □

Building on this lemma, we get "standard HN-pairs" representatives for HN-types, i.e., a bijection

$$\Theta(\mathbb{G}_b, \mathcal{N}) \xrightarrow{\sim} \left\{\text{pairs } (P, \mathcal{N}_P) \mid P \text{ a } \mathbb{Q}_p\text{-psgp of } G \text{ containing } P_0^\nu\right\}.$$

Here $\mathcal{N}_P$ denotes, as usual, the filtration induced by $\mathcal{N}$. The advantage of a standard HN-pair $(P, \mathcal{N}_P)$ as above is that $b \in P(K_0)$, so that the associated a.g.s. is $\mathbb{P} = \mathbb{P}_b$. Hence the stratum $\tilde{\mathcal{F}}(\mathbb{P}_b, \mathcal{N}_P)_{(\mathbb{P},\mathcal{N}_P)}$ is an affine bundle over the period domain $\tilde{\mathcal{F}}(\mathbb{M}_b, \mathcal{N}_M)$ associated to the Levi quotient of P. In other words, the stratification remains in the world of G-isocrystals and does not involve more general C-isocrystals.

In order to have a more concrete view of $\mathcal{N}_P$, we need to fix a maximal torus.

Lemma 9.6.21 *In the given σ-conjugacy class there exists b satisfying the requirements of Lemma 9.6.19 and the following additional one.*

(iv) *P_0^J contains a maximal maximally split $\mathbb{Q}_p$-torus T^J whose base change $T^J_{\mathbb{Q}_{p^s}}$ is also σ-stable (besides being $b\sigma$-stable).*

Proof Start with b as in Lemma 9.6.19. We first show that, after σ-conjugating b, we may find a Levi component M_0^J of P_0^J whose base change to $\mathbb{Q}_{p^s}$ is both $b\sigma$-stable and σ-stable (or equivalently, such that $b \in M_0^J(\mathbb{Q}_{p^s})$). Indeed, if we start with an arbitrary Levi component M^J of P_0^J, then $M^J_{\mathbb{Q}_{p^s}}$ is $b\sigma$-stable and is conjugate to a σ-stable Levi component of $P^J_{0,\mathbb{Q}_{p^s}}$ by some element $p \in P_0^J(\mathbb{Q}_{p^s})$. The element $b' := pb\sigma(p)^{-1}$ again fulfils the requirements of Lemma 9.6.19, with ${}^pP_0^J$ a psgp as in the third item of that lemma, and ${}^pM^J$ a Levi component whose base change to $\mathbb{Q}_{p^s}$ is both $b'\sigma$-stable and σ-stable, as desired.

Now let us replace b by b'. Let us fix a Levi component M_0^J of P_0^J as above.

By construction, the base change $M^J_{0,\mathbb{Q}_{p^s}}$ descends to a Levi component M^ν_0 of P^ν_0. Then, start with an arbitrary maximal maximally split $\mathbb{Q}_p$-torus T in M^J_0. The base change $T_{\mathbb{Q}_{p^s}} \subset M^J_{0,\mathbb{Q}_{p^s}}$ is $b\sigma$-stable. In particular, the conjugacy class of $T_{\mathbb{Q}_{p^s}}$ is σ-stable, and since M^ν_0 is quasi-split over $\mathbb{Q}_p$, this conjugacy class contains a σ-stable (i.e., $\mathbb{Q}_p$-rational) member ${}^mT$ with $m \in M^J_0(\mathbb{Q}_{p^s})$. Now the element $b' = mb\sigma(m)^{-1}$ fulfils the requirements of Lemma 9.6.19 with ${}^mP^J_0$ a psgp as in point (iii) of that lemma, and ${}^mM^J_0$ a Levi component as in the above paragraph. But the torus ${}^mT$ is a maximal maximally split torus of ${}^mP^J_0$ whose base change to $\mathbb{Q}_{p^s}$ is σ-stable, as required. □

Henceforth we assume that b fulfils the requirements of the foregoing lemma, and we fix a torus T^J as in this requirement. Then $T^J_{\mathbb{Q}_{p^s}}$ descends to a maximal $\mathbb{Q}_p$-torus T in P^ν_0 such that $b \in N_G(T)(\mathbb{Q}_{p^s})$. Also ν_b factors through T.

Note that T is also a maximal torus in G. Let W be its absolute Weyl group and let $\nu_G \in X_*(T)/W$ correspond to $\mathcal{N}$.

Proposition 9.6.22 *Let G be quasi-split over $\mathbb{Q}_p$ and assume that L is algebraically closed. Let $b \in G(K_0)$ be chosen as in Lemma 9.6.21 above.*

There is a natural bijection between $\Theta(\mathbb{G}_b, \mathcal{N})$ and the set of pairs (P, ν_P) consisting of a $\mathbb{Q}_p$-psgp $P \subset G$ containing P^ν_0 and an element $\nu_P \in X_*(T)/W_P$ satisfying the two conditions:

(a) $\nu_P \equiv \nu_G \mod W$;

(b) *let $\mu(P, \nu_P)$ be the push-forward of $\nu_P - \nu_b$ by the composition of morphisms $X_*(T)_\mathbb{Q} \longrightarrow X_*(A'_P)_\mathbb{Q} \xrightarrow{\sim} X_*(A_P)_\mathbb{Q}$. Then $\langle \mu(P, \nu_P), \alpha \rangle > 0$ for all $\alpha \in \Delta_P$.*

Here, as usual, A_P is the split component of the unique Levi subgroup of P which contains T and W_P is the Weyl group of T in P.

Proof We have already seen how $\Theta(\mathbb{G}_b, \mathcal{N})$ may be identified with the set of pairs $(P, \mathcal{N}_P)$ consisting of a $\mathbb{Q}_p$-psgp of G containing P^ν_0 and a conjugacy class $\mathcal{N}_P$ mapping to $\mathcal{N}$ satisfying condition (ii) in Definition 9.6.4. Now, using the bijection $X_*(P)/P \xrightarrow{\sim} X_*(T)/W_P$, we may replace $\mathcal{N}_P$ by its image ν_P, and then one has to work out condition (ii) of Definition 9.6.4 in terms of ν_P. To do this, note that $A'_P = A'_{\mathbb{P}_b}$ and that $\mu(P, \nu_P)$ is the unique lifting of $\deg(\mathbb{P}_b, \mathcal{N}_P)$ to a cocharacter of P factoring through T. Now, as in Chapter VI, §3, the positivity condition on $\mu(P, \nu_P)$ says exactly that P is the psgp of G attached to $\mu(P, \nu_P)$. □

Notes and References The question (that came up in Example 9.6.14) of when the number of orbits of a parabolic on its unipotent radical (over an algebraically closed field of characteristic 0) is finite has been studied intensely after the paper of Kashin [128] that concerned Borel subgroups. For instance, Röhrle [189] shows that this number is infinite as soon as the length of the descending central series of the unipotent radical of the parabolic is ≥ 6.

7 Relation to GIT

When dealing with the GIT criterion in Chapter VIII, §4, we only considered *split semi-simple* isocrystals. Without this hypothesis, this criterion generally fails. In the present context, again the GIT criterion only works for split semi-simple isocrystals $N^{\mathbb{G}}$. However, to avoid complications with "invariant inner products," we will write this section in the more classical context of *decent G-isocrystals*.

So let $(\mathbb{G}_b, \mathcal{N})$ be a PD-pair, with G a reductive group over $\mathbb{Q}_p$ and $b \in G(K_0)$ a *decent* element. Let $J = J_b$. Let $\mathcal{F} = \mathcal{F}(G, \mathcal{N})$ be the corresponding partial flag variety over the reflex field $E = E(G, \mathcal{N})$. We fix an invariant inner product $(\,,\,)$ on G, cf. Definition 6.2.1. In Chapter VI we have constructed an ample $\mathbb{Q}$-line bundle $\mathcal{L}_N$ on $\mathcal{F}$. It is given by

$$\mathcal{L}_N = G \times^P \mathbb{G}_{a, -\nu^*} .$$

Here $\nu \in \mathcal{N}$ is a $\mathbb{Q}$-1-PS which gives rise to a point $x \in \mathcal{F}(\overline{E})$. Further, $P = P_x$ is the corresponding psgp. Applying the above machinery to the inverse $\lambda_b := -\nu_b$ of our slope homomorphism, we get an ample $\mathbb{Q}$-line bundle $\mathcal{L}_b := \mathcal{L}_{-\lambda_b^*}$ on the flag variety

$$\mathcal{F}^b := \mathcal{F}(G_{K_0}, \{\lambda_b\}) = G_{K_0}/P_{\lambda_b}.$$

Consider the closed embedding

$$i : \mathcal{F}(G, \mathcal{N})_{\breve{K}_0} \hookrightarrow \mathcal{F}(G, \mathcal{N})_{\breve{K}_0} \times \mathcal{F}^b_{\breve{K}_0},$$

given by the identity on the first factor and by the $\breve{K}_0$-rational point λ_b of $\mathcal{F}^b$ on the second factor. Let $\mathcal{L} := i^*(\mathcal{L}_N \times \mathcal{L}_b)$. We consider $\mathcal{L}$ as a $J_{\breve{K}_0}$-equivariant line bundle.

Remark 9.7.1 Let $\rho : G \hookrightarrow G'$ be a closed embedding, and let $\mathcal{N}' = \rho_*(\mathcal{N})$, $b' = \rho(b)$. Assume that the invariant inner product on G is the restriction of the invariant inner product on G'. Then the homogeneous $\mathbb{Q}$-line bundle $\mathcal{L}$ on $\mathcal{F}(G, \mathcal{N})$ is isomorphic to the restriction of the corresponding homogeneous $\mathbb{Q}$-line bundle $\mathcal{L}'$ under the closed embedding

$$\mathcal{F}(G, \mathcal{N})_{\breve{K}_0} \hookrightarrow \mathcal{F}(G', \mathcal{N}') \otimes_{\breve{K}_0'} \breve{K}_0 ,$$

cf. Remark 6.2.6.

Example 9.7.2 Let $G = \mathrm{GL}_n$ and let $\mathcal{N}$ be given by the element $\nu \in (\mathbb{Q}^n)_+ \subset X_*(T)_\mathbb{Q}$. Here T denotes the diagonal torus in G. In Example 6.2.7 we saw that $\mathcal{L}_\mathcal{N}$ coincides with the line bundle $\mathcal{L}(\nu)$ considered in Chapter II, §2. By the same reason it follows that the line bundle $\mathcal{L}_b$ coincides with $\mathcal{L}(\nu_0)$, cf. Chapter VIII (4.1), page 214. Thus we see that the line bundle $\mathcal{L}$ coincides as well with the one in Chapter VIII (4.1), page 214.

Theorem 9.7.3 *Let $(G, b, \mathcal{N})$ be as above and fix an invariant inner product on G. Let $\mathcal{L}$ denote the corresponding ample homogeneous $\mathbb{Q}$-line bundle on $\mathcal{F}_{\breve{K}_0} = \mathcal{F}(G, \mathcal{N})_{\breve{K}_0}$ over $\breve{K}_0$. Let K be a field extension of $\breve{K}_0$, and let $x \in \mathcal{F}(K)$, with corresponding $\mathbb{Q}$-filtration $\mathcal{F}_x$ of ω^G over K. Then*

(i) *$(b, \mathcal{F}_x)$ is semi-stable if and only if, for all* 1-PS *λ of J defined over $\mathbb{Q}_p$ which factor through G_{der}, the Hilbert–Mumford inequality $\mu^\mathcal{L}(x, \lambda) \geq 0$ holds,*

(ii) *$(b, \mathcal{F}_x)$ is weakly admissible if and only if, for all* 1-PS *λ of J defined over $\mathbb{Q}_p$, the inequality $\mu^\mathcal{L}(x, \lambda) \geq 0$ holds.*

Proof The proof of part (i) of this theorem is – with the corresponding modifications – the same as in Theorem 6.2.8.

Thanks to the isogeny $G \to G_{\mathrm{ab}} \times G_{\mathrm{ad}}$ and part (i) applied to G_{ad}, it suffices to treat part (ii) in the case of a torus $G = T$. Let $\mathcal{F}_x$ be given by a $\mathbb{Q}$-1-PS λ_x of T. By Proposition 5.1.4, it follows easily that $(b, \mathcal{F}_x)$ is weakly admissible if and only if $\lambda_x + \nu_b$ is orthogonal to all $\mathbb{Q}_p$-rational characters of T. Since T is commutative, we have $J = T$. Let $(\,,\,)$ be any invariant inner product on T. Then $\mu^\mathcal{L}(x, \lambda) = -((\lambda, \lambda_x) + (\lambda, \nu))$ for all 1-PS λ of T. Thus the Hilbert–Mumford inequality reads $(\lambda, \lambda_x) + (\lambda, \nu) \leq 0$ for all $\mathbb{Q}_p$-rational 1-PS λ of T. Replacing λ by $-\lambda$ we see that this condition is satisfied if and only if $\lambda_x + \nu$ is orthogonal to all $\mathbb{Q}_p$-rational 1-PS of T. But the $\mathbb{Q}_p$-rational characters of T correspond to the $\mathbb{Q}_p$-rational 1-PS of T since the inner product is $\mathrm{Gal}(\overline{\mathbb{Q}}_p/\mathbb{Q}_p)$-invariant. The statement follows. □

Notes and References The statement of (ii) of Theorem 9.7.3 in the case of a torus appears in [183], Prop. 1.21. The proof here is taken from [219].

8 *G*-isocrystals with coefficients

In this section we briefly discuss the variant of G-isocrystals "with coefficients in F," where F is a finite extension of $\mathbb{Q}_p$. We use the definitions and notation of Chapter VIII, §5. In particular we have the unramified extension K_F of F with its relative Frobenius automorphism σ_F and the mutually equivalent F-linear tannakian categories $\mathrm{Isoc}(L)_F$ (of isocrystals with coefficients in F) and

σ_F-K_F-Sp (of σ_F-K_F-spaces). Starting from there, all definitions of Section 1 carry over to this setting almost verbatim. Let us give some examples.

Let G be an algebraic group over F. Here is the analogue of a G-isocrystal.

Definition 9.8.1 A *G-σ_F-K_F-space* is an exact faithful $\otimes$-functor

$$N : \mathrm{Rep}_F(G) \longrightarrow \sigma_F\text{-}K_F\text{-Sp}$$

which commutes with the natural tensor functors over K_F.

The set of G-σ_F-K_F-spaces yields a groupoid G-σ_F-K_F-Sp. On the other hand, let $\mathcal{B}(G, K_F)$ be the groupoid consisting of elements b in $G(K_F)$ with morphisms given by σ_F-conjugacy. As in the case of G-isocrystals, there is an equivalence of categories

$$\begin{array}{ccc} \mathcal{B}(G, K_F) & \longrightarrow & G\text{-}\sigma_F\text{-}K_F\text{-Sp} \\ b & \mapsto & N_b\,. \end{array}$$

Using the slope decomposition of σ_F-K_F-spaces, we then get a $\mathbb{Q}$-1-PS

$$\tilde{\nu}_b : \mathbb{D}_{K_F} \longrightarrow G_{K_F},$$

and all basic definitions and results on G-isocrystals have obvious analogues in this setting. For example, an element $b \in G(K_F)$ is *s-decent* if $\mathbb{F}_{q^s} \subset L$, the morphism $s\tilde{\nu}_b$ factors through $\mathbb{G}_{m,K_F}$ and we have the decency equation

$$(b\sigma_F)^s = s\tilde{\nu}_b(\varpi_F)\sigma_F^s.$$

In this case, the role of $\mathbb{Q}_{p^s}$ is played by "the" unramified extension F_s of degree s of F.

As in the case of G-isocrystals, it is more convenient to enlarge the setting and consider more general tensor functors $C \longrightarrow \sigma_F$-$K_F$-Sp, where C is a tannakian category over F. According to Remark 9.1.21, such tensor functors are equivalent to *augmented group schemes over σ_F-K_F-Sp* in the following sense.

Definition 9.8.2 An *a.g.s. over σ_F-K_F*-Sp is a triple $\mathbb{G} = (G, \varphi, \nu)$, where

- G is an affine group scheme over K_F,
- $\varphi : \sigma_{F,*}G \xrightarrow{\sim} G$ is an isomorphism of groups over K_F such that O_G is a union of finite-dimensional φ-stable subspaces,
- $\nu : D_{L,F} \longrightarrow G$ is an augmentation of the affine group scheme (G, φ) over σ_F-K_F-Sp, as in Definition 9.1.15.

Here $D_{L,F} = \underline{\mathrm{Aut}}^{\otimes}(\omega_F)$ is the automorphism group scheme over K_F of the fiber functor $\omega_F : \sigma_F$-K_F-Sp $\longrightarrow \mathrm{Vec}_{K_F}$.

The group scheme $D_{L,F}$ comes with a "Weil descent datum" $\varphi_{L,F}$, so that $(D_{L,F}, \varphi_{L,F})$ is the fundamental group scheme of the tannakian category σ_F-K_F-Sp. Moreover, the slope decomposition of σ_F-K_F-spaces yields a slope morphism

$$sl_F : (\mathbb{D}_{K_F}, \mathrm{id} \times \varphi_{\sigma_F}) \longrightarrow (D_{L,F}, \varphi_{L,F})$$

which is an analogue of (1.3), page 225 in Section 1.

Now, it is clear how to adapt the definitions and results of the preceding sections to this new setting, e.g. the degree of a filtration, semi-stability, HN-filtration, etc. In this context a PD-pair is a pair $(\mathbb{G}, \mathcal{N})$ consisting of a reductive a.g.s. over σ_F-K_F-Sp and a conjugacy class $\mathcal{N} \in X_*(G)_{\mathbb{Q}}/G$. To such a PD-pair, we associate an upper semi-continuous map with locally closed fibers

$$\tilde{\mathcal{F}}(G, \mathcal{N})^{an} \longrightarrow \Theta(\mathbb{G}, \mathcal{N}).$$

Here $\tilde{\mathcal{F}}(G, \mathcal{N})^{an}$ is the analytic flag variety over some finite Galois extension $\tilde{K}_F$ of K_F which splits the quasi-split inner form of G. Further, $\Theta(\mathbb{G}, \mathcal{N})$ is the set of HN-types defined in the same way as for isocrystals, preordered by dominance of the associated HN-vectors.

However, we are now going to see that these objects are not more general than those of the previous sections. As a first step, we note that the tensor equivalences (5.2), page 216 and (5.3), page 217 of Chapter VIII between σ_F-K_F-Sp and $\mathrm{Isoc}(L)_F$ extend to equivalences between (augmented) group schemes over each of these two Tannaka categories.

Definition 9.8.3 An *augmented group scheme over* $\mathrm{Isoc}(L)_F$ is a triple $\mathbb{G}' = (G', \varphi', \nu')$, where

- G' is an affine group scheme over $K_0 \otimes_{\mathbb{Q}_p} F$,
- $\varphi' : (\sigma \otimes \mathrm{id})_* G \xrightarrow{\sim} G$ is an isomorphism of group schemes over $K_0 \otimes F$ such that O_G is a union of finite-dimensional φ-stable subspaces,
- $\nu' : D'_{L,F} \longrightarrow G'$ is an augmentation of the affine group scheme (G', φ') over $\mathrm{Isoc}(L)_F$, as in Definition 9.1.15.

Here $D'_{L,F} = \underline{\mathrm{Aut}}^{\otimes}(\omega_0)$ is the automorphism group scheme over $K_0 \otimes_{\mathbb{Q}_p} F$ of the forgetful functor ω_0 from $\mathrm{Isoc}(L)_F$ to the category of locally free modules over $K_0 \otimes_{\mathbb{Q}_p} F$.

The group scheme $D'_{L,F}$ is endowed with a "Weil descent datum" $\varphi'_{L,F}$, so that $(D'_{L,F}, \varphi'_{L,F})$ is the fundamental group scheme of the tannakian category $\mathrm{Isoc}(L)_F$. Explicitly, since $\mathrm{Isoc}(L)_F$ is simply the base change of $\mathrm{Isoc}(L)$ from $\mathbb{Q}_p$ to F, we have by Chapter IV (1.6), page 114

$$D'_{L,F} = D_L \times_{\mathbb{Q}_p} F \text{ and } \varphi'_{L,F} = \varphi_L \times \mathrm{id},$$

with the notation introduced above (1.2), page 225.

Recall that the construction of the equivalences between σ_F-K_F-Sp and $\mathrm{Isoc}(L)_F$ in Chapter V, §5 involves the idempotent ε_0 such that $\varepsilon_0(K_0 \otimes_{\mathbb{Q}_p} F) = K_F$ and the "diagonal" embedding $\tau : K_F \hookrightarrow K_0 \otimes_{\mathbb{Q}_p} F = \prod_i K_F^{(i)}$. Then, formula (5.2), page 216 in Chapter VIII implies that

$$D_{L,F} = \varepsilon_0^* D'_{L,F} \text{ and } \varphi_{L,F} = \varepsilon_0^*((\varphi'_{L,F})^f),$$

while formula (5.3), page 217 says that

$$D'_{L,F} = \tau^* D_{L,F} \text{ and } \varphi'_{L,F} = (\varphi_{L,F}, \mathrm{id}, \ldots, \mathrm{id}).$$

We can also compare slope morphisms. Let $sl \times \mathrm{id}_F : \mathbb{D}_{K_0 \otimes F} \longrightarrow D'_{L,F}$ be the base change of the usual slope morphism $sl : \mathbb{D}_{K_0} \longrightarrow D_L$. Then the computation in Chapter VIII, §5, shows that

$$\tau^*(sl_F) = \frac{1}{[F : \mathbb{Q}_p]} \cdot (sl \times \mathrm{id}_F),$$

or equivalently,

$$\varepsilon_0^*(sl \times \mathrm{id}_F) = [F : \mathbb{Q}_p] \cdot sl_F.$$

More generally, the equivalences of Chapter VIII, §5, induce equivalences $\mathbb{G} \longleftrightarrow \mathbb{G}'$ which are abstractly characterized by diagrams of the form:

$$\begin{array}{ccc} \mathrm{Rep}_F(\mathbb{G}) & \overset{\sim}{\longleftrightarrow} & \mathrm{Rep}_F(\mathbb{G}') \\ {\scriptstyle N^{\mathbb{G}}}\downarrow & & \downarrow{\scriptstyle N^{\mathbb{G}'}} \\ \sigma_F\text{-}K_F\text{-Sp} & \overset{\sim}{\longleftrightarrow} & \mathrm{Isoc}(L)_F\,, \end{array}$$

Concretely, the relation between $\mathbb{G}$ and $\mathbb{G}'$ is as follows.

- We have $G = \varepsilon_0^* G'$ and $G' = \tau^* G$.
- We have $\varphi = \varepsilon_0^*(\varphi'^f)$ and $\varphi' = (\varphi, \mathrm{id}, \ldots, \mathrm{id})$.
- Finally, $\nu = \varepsilon_0^* \nu'$ and $\nu' = \tau^* \nu$.

We note that the automorphism group $J^{\mathbb{G}}$ of $N^{\mathbb{G}}$ and $N^{\mathbb{G}'}$ is a group scheme over F in this context. Also, we have a bijection $^\bullet N^{\mathbb{G}} \leftrightarrow {}^\bullet N^{\mathbb{G}'}$ between filtrations of $N^{\mathbb{G}}$ and filtrations of $N^{\mathbb{G}'}$. Concretely, if $^\bullet N^{\mathbb{G}}$ is split by a $\lambda : \mathbb{D}_{K_F} \longrightarrow G$, then $^\bullet N^{\mathbb{G}'}$ is split by $\tau^*\lambda : \mathbb{D}_{K_0 \otimes F} \longrightarrow G'$ and, conversely, if $^\bullet N^{\mathbb{G}'}$ is split by a $\lambda' : \mathbb{D}_{K_0 \otimes F} \longrightarrow G'$, then $^\bullet N^{\mathbb{G}}$ is split by $\varepsilon_0^* \lambda' : \mathbb{D}_{K_F} \longrightarrow G$.

Definition 9.8.4 Let $\mathbb{G} = (G, \varphi, \nu)$ be an a.g.s. over σ_F-K_F-Sp. We define its *associated* a.g.s. over Isoc(L) (without coefficients) $\mathbb{G}_0 = (G_0, \varphi_0, \nu_0)$ by setting

$$\begin{aligned} G_0 &:= \mathrm{Res}_{K_0\otimes F|K_0}(G') \\ \varphi_0 &:= \mathrm{Res}_{K_0\otimes F|K_0}(\varphi') \\ \nu_0 &: D_L \xrightarrow{\text{can}} \mathrm{Res}_{K_0\otimes F|K_0}(D'_{L,F}) \xrightarrow{\mathrm{Res}(\nu')} \mathrm{Res}_{K_0\otimes F|K_0}(G'). \end{aligned}$$

In other words, $\mathbb{G}_0$ is deduced from $\mathbb{G}'$ by merely forgetting the additional action of F. Here are some relations between $\mathbb{G}$ and $\mathbb{G}_0$.

(i) We have a "restriction of scalars" functor

$$\mathrm{Res}_{F|\mathbb{Q}_p} : \mathrm{Rep}_F(\mathbb{G}) \longrightarrow \mathrm{Rep}_{\mathbb{Q}_p}(\mathbb{G}_0).$$

Concretely, to $(V_F, \Phi_F, \rho) \in \mathrm{Rep}_F(\mathbb{G})$ we associate (V_0, Φ_0, ρ_0) where (V_0, Φ_0) is the RHS of Chapter VIII (5.3), page 217, and ρ_0 is the restriction to G_0 of the action $\rho' : G' \longrightarrow \mathrm{Aut}_{K_0\otimes F}(V_0)$.

(ii) We have a natural isomorphism

$$X_*(A'_{\mathbb{G}})_{\mathbb{Q}} \xrightarrow{\sim} X_*(A'_{\mathbb{G}_0})_{\mathbb{Q}}.$$

There are two ways to see this, internal or external. For the internal approach, let us fix an embedding $K_F \hookrightarrow \overline{K}_0$. We use the notation $\widetilde{\Gamma}_0$ of Section 4 and similarly we define $\widetilde{\Gamma}_F := \{\gamma \in \mathrm{Aut}(\overline{K}_0) \mid \gamma_{|K_F} \in \sigma_F^{\mathbb{Z}}\}$. Thus $\widetilde{\Gamma}_F$ is in a natural way a subgroup of index $d = [F : \mathbb{Q}_p]$ of $\widetilde{\Gamma}_0$. Now consider the sequence $G \twoheadrightarrow T_G \twoheadrightarrow A'_{\mathbb{G}}$ where T_G is the maximal quotient torus of G. As in Section 4, $X^*(T_G)$ is endowed with an action of the group $\widetilde{\Gamma}_F$ and by definition $A'_{\mathbb{G}}$ is characterized by the identity $X^*(A'_{\mathbb{G}}) = X^*(T_G)^{\widetilde{\Gamma}_F}$. Further, $(T_G)_0$ is the maximal quotient torus of G_0, so we also have $X^*(A'_{\mathbb{G}_0}) = X^*((T_G)_0)^{\widetilde{\Gamma}_0}$. However, we have the relation $X^*((T_G)_0) = \mathrm{Ind}_{\widetilde{\Gamma}_F}^{\widetilde{\Gamma}_0} X^*(T_G)$, which implies $X^*((T_G)_0)^{\widetilde{\Gamma}_0} = X^*(T_G)^{\widetilde{\Gamma}_F}$.

For the external approach, recall from Proposition 9.2.11 (i), that the $\mathbb{Q}$-vector space associated to $X^*(A'_{\mathbb{G}})$ can be identified with the $\mathbb{Q}$-vector space associated to the Picard group of $\mathrm{Rep}_F(\mathbb{G})$. Now if $N_F := (V_F, \Phi_F, \rho)$ is a one-dimensional object in $\mathrm{Rep}_F(\mathbb{G})$, its restriction of scalars $N_0 := (V_0, \Phi_0, \rho_0)$ is a d-dimensional object of $\mathrm{Rep}_{\mathbb{Q}_p}(\mathbb{G}_0)$. Then it can be checked that the map $N_F \mapsto \bigwedge^d N_0$ is an isomorphism of the Picard groups which coincides with the previous one.

(iii) We have a bijection

$$^\bullet N^{\mathbb{G}} \leftrightarrow {}^\bullet N^{\mathbb{G}_0}$$

between filtrations of $N^{\mathbb{G}}$ and filtrations of $N^{\mathbb{G}_0}$. Concretely, if $^\bullet N^{\mathbb{G}}$ is split by $\lambda : \mathbb{D}_{K_F} \longrightarrow G$, then $^\bullet N^{\mathbb{G}_0}$ is split by the adjoint $\lambda_0 : \mathbb{D}_{K_0} \longrightarrow G_0$ to $\tau^*\lambda : \mathbb{D}_{K_0\otimes F} \longrightarrow G'$. Conversely if $N^{\mathbb{G}_0}$ is split by λ_0, then $N^{\mathbb{G}}$ is split by $\varepsilon_0^*\lambda'$ where

$\lambda' : \mathbb{D}_{K_0 \otimes F} \longrightarrow G'$ is the adjoint to λ_0. *However, it is convenient to renormalize this bijection by letting λ correspond to $\frac{1}{d}\lambda_0$.* With this renormalization, the following diagram is commutative:

$$\begin{array}{ccc} \mathrm{Fil}(N^{\mathbb{G}}) & \stackrel{\sim}{\longleftrightarrow} & \mathrm{Fil}(N^{\mathbb{G}_0}) \\ \downarrow & & \downarrow \\ X_*(A'_{\mathbb{G}})_{\mathbb{Q}} & \stackrel{\sim}{\longleftrightarrow} & X_*(A'_{\mathbb{G}_0})_{\mathbb{Q}} \,. \end{array}$$

Here the vertical arrows are the push-forward maps, i.e., the degree maps. Indeed, it is enough to check this for $\mathbb{G} = \mathbb{T}$ a torus, in which case the above diagram is identified with the following:

$$\begin{array}{ccc} X_*(\mathbb{T})_{\mathbb{Q}}^{\tilde{\Gamma}_F} & \stackrel{\sim}{\longleftrightarrow} & \left(\mathrm{Ind}_{\tilde{\Gamma}_F}^{\tilde{\Gamma}_0}(X_*(\mathbb{T})_{\mathbb{Q}})\right)^{\tilde{\Gamma}_0} \\ \downarrow & & \downarrow \\ X_*(\mathbb{T})_{\mathbb{Q},\tilde{\Gamma}_F} & \stackrel{\sim}{\longleftrightarrow} & \left(\mathrm{Ind}_{\tilde{\Gamma}_F}^{\tilde{\Gamma}_0}(X_*(\mathbb{T})_{\mathbb{Q}})\right)_{\tilde{\Gamma}_0} \,. \end{array}$$

Here the vertical maps are the natural projections. The horizontal maps are deduced from the natural inclusion $X_*(\mathbb{T})_{\mathbb{Q}} \subset \mathrm{Ind}_{\tilde{\Gamma}_F}^{\tilde{\Gamma}_0}(X_*(\mathbb{T})_{\mathbb{Q}})$. The bottom map is the natural projection, while the top map is the $\tilde{\Gamma}_0/\tilde{\Gamma}_F$-averaging map, thanks to the extra factor $\frac{1}{d}$. This makes the diagram commute.

(iv) The automorphism groups satisfy the following relation

$$J^{\mathbb{G}_0} = \mathrm{Res}_{F|\mathbb{Q}_p}(J^{\mathbb{G}}).$$

(v) $\mathbb{G}$ is connected reductive if and only if $\mathbb{G}_0$ is connected reductive. In this case, if $\mathbb{P}$ is a psgp of $\mathbb{G}$, then $\mathbb{P}_0$ is a psgp of $\mathbb{G}_0$, and this defines a bijection between both sets of psgps. This bijection is compatible with item (ii) in the following sense. If $\mathbb{P} = \mathbb{P}_\lambda$ for some λ par-equivalent to ${}^{\varphi_F}\lambda$, then $\mathbb{P}_0 = \mathbb{P}_{\lambda_0}$ with the notation of item (ii).

Example 9.8.5 We start with a group G over F and an element $b \in G(K_F)$, whence with an a.g.s. $\mathbb{G}_b$ over σ_F-K_F-Sp, associated to the G-σ_F-K_F-space N_b. Define $G_{|\mathbb{Q}_p} := \mathrm{Res}_{F|\mathbb{Q}_p}(G)$ and put $b_0 := (b, 1, \ldots, 1) \in G_{|\mathbb{Q}_p}(K_0) = \prod_i G(K_F^{(i)})$. Then the a.g.s. over $\mathrm{Isoc}(L)$, associated by Definition 9.8.4 to $\mathbb{G}_b$, is the one defined by the $G_{|\mathbb{Q}_p}$-isocrystal N_{b_0}. In fact, the map $b \mapsto b_0$ underlies an equivalence of groupoids $\mathcal{B}(G, K_F) \xrightarrow{\sim} \mathcal{B}(G_0, K_0)$ which on the level of morphisms is given by $g \in G(K_F) \mapsto g^0 := (g, \sigma(g), \ldots, \sigma^{f-1}(g)) \in G_{|\mathbb{Q}_p}(K_0)$.

Example 9.8.6 Let $\mathbb{G} = \mathbb{GL}(V_F, \Phi_F)$ for some σ_F-K_F-space $N_F := (V_F, \Phi_F)$.

Then a filtration of $N^{\mathbb{G}}$ of course is the same as a filtration of N_F. Let (V_0, Φ_0, κ) be the associated isocrystal with coefficients in F as in Chapter VIII (5.3), page 217. Then we have a canonical embedding $\mathbb{G}_0 \hookrightarrow \mathbb{GL}(V_0, \Phi_0)$ which, on the level of K_0-groups, is given by the embedding $\mathrm{Res}_{K_0 \otimes F|K_0}(\mathrm{GL}_{K_0 \otimes F}(V_0)) \longrightarrow \mathrm{GL}_{K_0}(V_0)$. In this case a filtration of $N^{\mathbb{G}_0}$ is the same as an F-stable filtration of the isocrystal (V_0, Φ_0).

We now let filtrations enter the game. Let K be an extension field of K_F and let $\mathcal{F}$ be a filtration on ω^G over K. We associate to $\mathcal{F}$ a filtration $\mathcal{F}_0$ on ω^{G_0} over K in the same way as in Chapter VIII, §5. More precisely, we have the following expression for G_0,

$$
\begin{aligned}
G_0 &= \mathrm{Res}_{K_0 \otimes F|K_0}(G \times_{K_F, \tau} (K_0 \otimes F)) \\
&= \prod_{i=0}^{f-1} \mathrm{Res}_{K_F^{(i)}|K_0}(G \times_{K_F, \sigma^i} K_F^{(i)}) ,
\end{aligned}
$$

which shows that $\mathrm{Res}_{K_F|K_0}(G)$ is canonically a direct factor of G_0. Therefore, G is also canonically a direct factor of G_{0,K_F}, and more generally, G_K is canonically a direct factor of $G_{0,K}$. Now, if the filtration $\mathcal{F}$ is split by some 1-PS $\lambda : \mathbb{D}_K \longrightarrow G_K$, we let $\mathcal{F}_0$ be split by the 1-PS $\lambda_0 : \mathbb{D}_K \longrightarrow G_{0,K}$ which coincides with λ on the component G_K and is trivial on the other component.

Lemma 9.8.7 *With the above notation,* $\deg(\mathbb{G}, \mathcal{F})$ *corresponds to* $\deg(\mathbb{G}_0, \mathcal{F}_0)$ *under the isomorphism of item* (ii) *above. Assume further that* $\mathbb{G}$*, or equivalently* $\mathbb{G}_0$*, is reductive. Then the* HN*-filtrations* $^\bullet N_{\mathcal{F}}^{\mathbb{G}}$ *and* $^\bullet N_{\mathcal{F}_0}^{\mathbb{G}_0}$ *match under the renormalized bijection of item* (iii) *above.*

Proof For the first assertion, we use the external characterization of degrees of Proposition 9.2.11, and the external approach to the isomorphism of item (ii). Then all we have to do is to check that for any (rank 1) object (V_F, Φ_F, ρ) in $\mathrm{Rep}_F(\mathbb{G})$ with restriction of scalars (V_0, Φ_0, ρ_0) we have $\deg(V_F, \Phi_F, \rho_* \mathcal{F}) = \deg(V_0, \Phi_0, \rho_{0,*} \mathcal{F}_0)$. This was already used to get identity (5.5), page 219 of Chapter VIII.

For the second assertion, again by the external definition of the HN-filtration, we may assume $\mathbb{G} = \mathbb{GL}(V_F, \Phi_F)$ for some σ_F-K_F-space $N_F := (V_F, \Phi_F)$. In this case we have to compare the HN-filtrations on the filtered σ_F-K_F-space $(N_F, \mathcal{F}_F)$ and on the associated filtered isocrystal $(N_0, \mathcal{F}_0)$ as in Example 9.8.6. By devissage, it will suffice to prove that $(N_F, \mathcal{F}_F)$ is semi-stable of slope μ if and only if $(N_0, \mathcal{F}_0)$ is semi-stable of slope $\frac{1}{d}\mu$. But this is the content of Lemma 8.5.5 and identity (5.5), page 219 of Chapter VIII. □

For the rest of this section we assume that $\mathbb{G}$ is reductive and connected. In order to compare conjugacy classes of $\mathbb{Q}$-1-PS of G and G_0 we need to fix a K_0-linear embedding of fields $K_F \hookrightarrow \overline{K}_0$. We denote by Γ_F the corresponding Galois group. The canonical embedding of G as a direct factor of G_{0,K_F} explained above induces a Γ_F-equivariant map

$$\begin{array}{ccc} X_*(G)_{\mathbb{Q}}/G & \longrightarrow & X_*(G_0)_{\mathbb{Q}}/G_0 \\ \mathcal{N} & \mapsto & \mathcal{N}_0 \,. \end{array}$$

In particular, both $\mathcal{N}$ and $\mathcal{N}_0$ have the same reflex field, say $\breve{K}_F$. Moreover, the same embedding induces also an isomorphism of $\breve{K}_F$-varieties

$$\breve{\mathcal{F}}(G,\mathcal{N}) \xrightarrow{\sim} \breve{\mathcal{F}}(G_0,\mathcal{N}_0).$$

We note that the map $X_*(G)_{\mathbb{Q}}/G \longrightarrow X_*(G_0)_{\mathbb{Q}}/G_0$ is functorial in $\mathbb{G}$. In particular, for any psgp $\mathbb{P}$ of $\mathbb{G}$, we have a Γ_F-equivariant commutative square:

$$\begin{array}{ccc} X_*(P)_{\mathbb{Q}}/P & \xrightarrow{\mathcal{N}_P \mapsto \mathcal{N}_{P_0}} & X_*(P_0)_{\mathbb{Q}}/P_0 \\ \downarrow & & \downarrow \\ X_*(G)_{\mathbb{Q}}/G & \xrightarrow[\mathcal{N} \mapsto \mathcal{N}_0]{} & X_*(G_0)_{\mathbb{Q}}/G_0 \,. \end{array}$$

In fact this diagram is even *cartesian*, since conjugacy classes of the form $\mathcal{N}_0$ are characterized by the property of being trivial on the complementary factor of $G_{\overline{K}_0}$ in $G_{0,\overline{K}_0}$.

Now we fix a PD-pair $(\mathbb{G},\mathcal{N})$ over F. By the foregoing discussion, we have associated to it a PD-pair $(\mathbb{G}_0,\mathcal{N}_0)$ over $\mathbb{Q}_p$. Let us choose further a finite Galois subextension $\tilde{K}_F$ of K_F in $\overline{K}_0$ which splits G and, as usual now, let us denote with a $\tilde{}$ any base change to the compositum field $\tilde{K}_F$.

Theorem 9.8.8 *The map $(\mathbb{P},\mathcal{N}_P) \mapsto (\mathbb{P}_0,\mathcal{N}_{P_0})$ induces a bijection from $\Theta(\mathbb{G},\mathcal{N})$ to $\Theta(\mathbb{G}_0,\mathcal{N}_0)$ and the diagram*

$$\begin{array}{ccc} \tilde{\mathcal{F}}(G,\mathcal{N})^{an} & \xrightarrow{\sim} & \tilde{\mathcal{F}}(G_0,\mathcal{N}_0)^{an} \\ \downarrow & & \downarrow \\ \Theta(\mathbb{G},\mathcal{N}) & \xrightarrow{\sim} & \Theta(\mathbb{G}_0,\mathcal{N}_0) \end{array}$$

is commutative and equivariant under the action of $\mathrm{Gal}(\tilde{K}_F/\breve{K}_F)$.

Proof From item (v) below Definition 9.8.4 and the cartesian diagram above, the map $(\mathbb{P},\mathcal{N}_P) \mapsto (\mathbb{P}_0,\mathcal{N}_{P_0})$ induces a bijection between pairs consisting of a psgp and a conjugacy class above $\mathcal{N}$, resp. $\mathcal{N}_0$. It remains to check that this bijection preserves the property (ii) of Definition 9.6.4 of HN-pairs. So, let us start with a lifting $\lambda : \mathbb{D}_{K_F} \longrightarrow P$ of $\deg(\mathbb{P}, \mathcal{N}_P)$ and assume that $P = P_\lambda$, so that $(\mathbb{P}, \mathcal{N}_P)$ is an HN-pair. Then

$$P_0 = P_{\lambda_0} = P_{\frac{1}{d}\lambda_0}$$

with the notation of items (v) and (iii). But the commutative diagram of item (iii) says that $\frac{1}{d}\lambda_0$ is a lifting of $\deg(\mathbb{P}_0, \mathcal{N}_{P_0})$, hence $(\mathbb{P}_0, \mathcal{N}_{P_0})$ is also an HN-pair. Conversely, for the same reason, if $(\mathbb{P}_0, \mathcal{N}_{P_0})$ is a HN-pair then so is $(\mathbb{P}, \mathcal{N}_P)$. Whence the first assertion of the theorem.

Now the remaining assertions follow from the matching property of HN-filtrations in Lemma 9.8.7. □

We end this section with the Drinfeld space over F.

Example 9.8.9 Let $\mathbb{G} = \mathbb{G}_1$ be the a.g.s. over σ_F-K_F-Sp given by $G = \mathrm{GL}_{n,F}$ and the identity element $1 \in G(K_F)$, cf. Example 9.8.5. Let ν be the 1-PS of G given by the Hodge vector $\nu = (n-1, (-1)^{(n-1)}) \in (\mathbb{Z}^n)_+$ with respect to the diagonal torus in G, cf. Examples 8.2.7. Let $\mathcal{N} = \{\nu\} \in X_*(G)/G$ be the induced conjugacy class. Then going through the above definitions, one sees that $\breve{\mathcal{F}}(\mathbb{G}, \mathcal{N})^{\mathrm{wa}}$ can be identified with the Drinfeld space relative to F, i.e., $\breve{\mathcal{F}}(\mathbb{G}, \mathcal{N})^{\mathrm{wa}} = \Omega_F^n = \mathbb{P}_F^{n-1} \setminus \bigcup_{H \subsetneq F^n} \mathbb{P}(H)$.

Finally, we mention that also Example 9.5.7 (realizing the dual Drinfeld space by starting with a central division algebra of invariant $-\frac{1}{n}$) has a counterpart relative to F. In fact, in Example 9.5.7 one merely has to replace D by a central division algebra over F with the same invariant. The remaining data are the same, as well.

X. Cohomology of Period Domains over p-adic Fields

In this chapter we consider the compactly supported étale ℓ-adic cohomology of period domains. We will write down in the basic case a recursive formula for the Euler–Poincaré characteristic in a Grothendieck group of $J^{\mathbb{G}}(\mathbb{Q}_p)$-representations (possibly infinite-dimensional ones). Once again, we use the Langlands Lemma to resolve the corresponding recursion formula. The general argument is essentially the same as for finite fields, but the representation theory of p-adic reductive groups is somewhat different from that of finite reductive groups. We start with a review of the necessary background on smooth representations of p-adic groups.

1 Generalized Steinberg representations

In this section, we let G denote a reductive group over $\mathbb{Q}_p$. In contrast to the first two parts of this book, where representations of finite groups were considered, we are interested here in *smooth* representations of $G(\mathbb{Q}_p)$ on vector spaces over the field $\overline{\mathbb{Q}}_\ell$ [42]. Here *smooth* means that the stabilizer of any vector has to be open in $G(\mathbb{Q}_p)$. Such representations are generally infinite-dimensional, but we do not need to consider any topology on the underlying vector space. They form in an obvious way an abelian category $\mathrm{Rep}^\infty(G(\mathbb{Q}_p))$.

We note that the subcategory of finitely generated objects is much bigger than that of finite length objects (unless G is anisotropic). Moreover, both subcategories are known to be abelian [16], but the canonical map between their Grothendieck groups is far from being injective, see [49]. In our computation of Euler–Poincaré characteristics, we will be therefore concerned with the Grothendieck group of *finite length* representations. We note that it is not a ring, since tensor products of irreducible representations do not need to have finite length, not even to be finitely generated. If H is any locally pro-finite group, we denote by $K_0^{\mathrm{fl}}(H)$ the Grothendieck group of smooth finite length H-representations.

An important way of constructing smooth representations is via *smooth parabolic induction*. Let P be a rational psgp of G with Levi quotient M. If (W, σ) is a smooth representation of $M(\mathbb{Q}_p)$, we put

$$\mathrm{Ind}_P^G(\sigma) := \left\{ f : G(\mathbb{Q}_p) \longrightarrow W \mid f \text{ smooth}, \ f(gp) = \sigma(m(p))f(g), \ \forall p \in P(\mathbb{Q}_p) \right\}.$$

The left regular action of $G(\mathbb{Q}_p)$ is obviously smooth, so we get a functor

$$\mathrm{Ind}_P^G : \mathrm{Rep}^\infty(M(\mathbb{Q}_p)) \to \mathrm{Rep}^\infty(G(\mathbb{Q}_p))$$

from the category of smooth representations of $M(\mathbb{Q}_p)$ to the category of smooth representations $G(\mathbb{Q}_p)$. Moreover, this functor is exact, [17], Prop. 1.9 (a), and is known to carry finite length representations to finite length representations. In particular, we get a homomorphism

$$\mathrm{Ind}_P^G : K_0^{\mathrm{fl}}(M(\mathbb{Q}_p)) \to K_0^{\mathrm{fl}}(G(\mathbb{Q}_p)).$$

As in the case of finite groups of Lie type, we will be mostly interested in the case where σ is trivial, and we define again

$$i_P^G := \mathrm{Ind}_P^G(\mathbf{1}) = \overline{\mathbb{Q}}_\ell[G(\mathbb{Q}_p)/P(\mathbb{Q}_p)]^\infty. \tag{1.1}$$

Here the expression $\overline{\mathbb{Q}}_\ell[X]^\infty$ denotes the set of locally constant $\overline{\mathbb{Q}}_\ell$-valued functions on the topological space X. If Q is another rational psgp which contains P, we have a natural inclusion $i_Q^G \subset i_P^G$, so we may again define the *generalized Steinberg representation* in this context,

$$v_P^G := i_P^G / \sum_{Q \supsetneq P} i_Q^G \,. \tag{1.2}$$

In some regards, the behavior of these representations in the p-adic setting is simpler than in the finite field case.

Theorem 10.1.1 (i) *v_P^G is the unique irreducible quotient of i_P^G, for any $\mathbb{Q}_p$-rational psgp P of G.*

(ii) *For any $\mathbb{Q}_p$-rational psgp P of G, the map $Q \mapsto v_Q^G$ is a bijection from the set of $\mathbb{Q}_p$-rational psgps of G which contain P to the set of Jordan–Hölder constituents of i_P^G.*

(iii) *$v_P^G \simeq v_{P'}^G$ if and only if P' is conjugate to P.*

(iv) *Let Δ_0 be the set of simple roots of the relative root system of G with respect to a fixed minimal $\mathbb{Q}_p$-psgp. Let $I \subset \Delta_0$ and let P_I be the corresponding std psgp. Then*

$$v_{P_I}^G = (-1)^{|I|} \sum_{J \supset I} (-1)^{|J|} \, i_{P_J}^G$$

in $K_0^{\mathrm{fl}}(G)$. More precisely, the following complex is an acyclic resolution of $v_{P_I}^G$ by smooth G-modules,

$$0 \to i_G^G \to \bigoplus_{\substack{I\subset K\subset\Delta_0\\ |\Delta_0\setminus K|=1}} i_{P_K}^G \to \bigoplus_{\substack{I\subset K\subset\Delta_0\\ |\Delta_0\setminus K|=2}} i_{P_K}^G \to \cdots \to \bigoplus_{\substack{I\subset K\subset\Delta_0\\ |K\setminus I|=1}} i_{P_K}^G \to i_{P_I}^G \to v_{P_I}^G \to 0.$$

Here the differentials $d_{K',K} : i_{P_{K'}}^G \to i_{P_K}^G$ are defined as in Theorem 7.1.9.

(v) *We have*

$$i_P^G = \sum_{Q\supset P} v_Q^G$$

in $K_0^{\mathrm{fl}}(G)$. In particular, the generalized Steinberg representations v_P^G, as P ranges over std psgps, form a basis of the subspace of $K_0^{\mathrm{fl}}(G)$ formed by those virtual representations which appear in $i_{P_0}^G$.

Proof For points (i)–(iii) and (v), see [24], Ch. X 4.6-4.11 or [41]. For point (iv), we use the following distributivity property [24], Lemma X, 4.5: suppose $V_0, V_1, \ldots, V_m$ are G-submodules of i_P^G; then

$$V_0 \cap (V_1 + \cdots + V_m) = (V_0 \cap V_1) + \cdots + (V_0 \cap V_m). \qquad (1.3)$$

Therefore the exactness of the complex in (iv) follows from Lemma 3.2.8. □

Remark 10.1.2 We thus see that the nature of the representations i_P^G in the p-adic case is quite different from the finite field case, and the same applies to the generalized Steinberg representations v_P^G. In the p-adic case, the representations i_P^G are indecomposable, multiplicity-free, and their irreducible constitutents are precisely the representations v_Q^G, as Q runs through the $\mathbb{Q}_p$-rational psgps containing P. Also, the generalized Steinberg representations v_P^G are always irreducible, and non-isomorphic for non-conjugate $\mathbb{Q}_p$-psgps (recall from Theorem 7.1.3 that in the finite field case *associate* psgps yield isomorphic generalized Steinberg representations). Also, note that the distributivity property (1.3) above is easy to prove in the p-adic case, whereas in the finite field case it only holds for G-submodules of the form i_Q^G, and is quite tricky, cf. Proposition 3.2.9.

Notes and References The irreducibility of the generalized Steinberg representations is due to Casselman [40] (with proofs appearing in [41]). Casselman also proved that the generalized Steinberg representations are precisely the irreducible *admissible* G-representations V with non-vanishing group cohomology $H^*(G, V)$. He also showed that the only unitarizable representations among generalized Steinberg representations are the full Steinberg representation (corresponding to a minimal psgp) and the trivial representation, and used these results to give a new proof of (a strengthening of) the p-adic vanishing theorem of Garland [88]. For more information, see [24].

2 Compactly supported ℓ-adic cohomology of strata

In this section, we briefly recall some aspects of étale cohomology for non-Archimedean spaces, as developed by Berkovich [13]. We will use *compactly supported ℓ-adic cohomology*. Indeed, we are not aware of any reasonable ℓ-adic cohomology without support for analytic spaces. We note that most of Berkovich's published work on étale cohomology of analytic spaces is for *torsion* coefficients. The passage to ℓ-adic coefficients is explained in [77] and [48] (it is more delicate than for algebraic varieties).

For simplicity, we work over an algebraically closed complete non-Archimedean field K. Following Berkovich, a *germ of K-analytic spaces*, also called simply a *K-germ*, is a pair (X, S) where X is a K-analytic space and $S \subset X$ is a subset of the underlying topological space. The subset S, with the topology induced by X, should be considered as the underlying topological space of the K-germ (X, S). A *morphism* $(X', S') \xrightarrow{\varphi} (X, S)$ of K-germs is a morphism $X \xrightarrow{\varphi} X'$ of K-analytic spaces such that $\varphi(S') \subset \varphi(S)$. Any K-analytic space X can be seen as a K-germ with $S = X$. In this case we simply write X rather than (X, X). Berkovich associates in [13], 4.2, to a K-germ (X, S) a topos $\widetilde{(X, S)}_{et}$ (i.e., a category of "étale sheaves on (X, S)") and, to any morphism of K-germs, a morphism of the associated topoi. Then he studies the ordinary cohomology $H^*((X, S), \mathcal{F})$ and the compactly supported one $H^*_c((X, S), \mathcal{F})$, for any abelian étale sheaf $\mathcal{F}$. However, most of the deep properties are proved (and valid) only for torsion sheaves. The ℓ-adic compactly supported cohomology of a K-germ is defined using the formalism explained in [77], 4.1 and [48], B.2. In particular, it is the hypercohomology of the functor $\Gamma_!((X, S), -)$ of global sections with compact supports (in the naive sense) evaluated on some suitable complex of $\mathbb{Z}_\ell$-sheaves, and tensored with $\overline{\mathbb{Q}}_\ell$.

We now give a list of properties enjoyed by the series of functors $H^i_c(-) = H^i_c(-, \overline{\mathbb{Q}}_\ell)$, $i \in \mathbb{N}$, from the category of K-germs to that of $\overline{\mathbb{Q}}_\ell$-vector spaces. All these properties are due to Berkovich when $\overline{\mathbb{Q}}_\ell$ is replaced by $\mathbb{Z}/\ell^n\mathbb{Z}$. Using the techniques of [77] and [48], one can carry them over to the ℓ-adic case.

(I) For $i > 2\dim(X) + 1$, we have $H^i_c((X, S)) = 0$. This follows from [13], Thm. 4.2.6 for torsion coefficients and from [77], Cor. 4.1.9 (b) in the ℓ-adic case.

(II) If T is an open subset of S with complement R, then there is a long exact sequence

$$\begin{aligned}\cdots \longrightarrow H^i_c((X,S)) \longrightarrow \quad & H^i_c((X,R)) \quad \longrightarrow \\ & H^{i+1}_c((X,T)) \quad \longrightarrow H^{i+1}_c((X,S)) \longrightarrow \cdots\end{aligned}$$

This comes from [13], Prop. 5.2.6 (ii) in both the torsion and the ℓ-adic cases.

(III) If $(X',S') \xrightarrow{\varphi} (X,S)$ induces a homeomorphism $S' \xrightarrow{\sim} S$ and isomorphisms $\mathscr{H}(s) \xrightarrow{\sim} \mathscr{H}(\varphi^{-1}(s))$ for all $s \in S$, then φ induces isomorphisms

$$H^i_c((X',S')) \xrightarrow{\sim} H^i_c((X,S)).$$

This follows from [13], Prop. 4.3.4 in both the torsion and the ℓ-adic cases.

(IV) As a particular case, if S is a K-analytic locally closed subset of X, then

$$H^i_c((X,S)) = H^i_c(S).$$

(V) If $S \subset X(K)$, then $H^i_c((X,S)) = 0$ for $i > 0$ and

$$H^0_c((X,S)) = \overline{\mathbb{Q}}_\ell[S]^\infty_c,$$

where the RHS is the set of compactly supported smooth functions on the totally disconnected set S. Indeed, the spectral sequence at the bottom of page 96 of [13] shows that the morphism of topoi (denoted therein by π_*) from $\widetilde{(X,S)}_{et}$ to the topos of sheaves on the set S with the induced topology, is an equivalence. The claim then follows from the fact that S is totally disconnected, hence has no higher cohomology.

(VI) More generally, let (X,S) be as in (V) and let $(Y,R) \xrightarrow{f} (X,S)$ be a morphism of K-germs. It induces a functor $f_!$ of "direct image with proper supports" (see [13], 5.1, 5.2) from the étale sheaves on (Y,R) to those on (X,S), and we have $H^i_c((Y,R),\mathcal{F}) = \Gamma_c((X,S),R^if_!(\mathcal{F}))$ for any abelian sheaf. Indeed, this follows from the Leray spectral sequence [13], Thm. 5.2.2 and the vanishing of higher cohomology on (X,S) as above. In fact, thanks to this property, the passage from ℓ-torsion to ℓ-adic sheaves is unusually easy and, therefore, we have sheaves of $\overline{\mathbb{Q}}_\ell$-vector spaces $R^if_!\overline{\mathbb{Q}}_\ell$ on S such that

$$H^i_c((Y,R)) = \Gamma_c(S,R^if_!\overline{\mathbb{Q}}_\ell).$$

Moreover, these sheaves satisfy the following base change property. Let

$$\begin{array}{ccc}(Y',R') & \xrightarrow{g'} & (Y,R)\\ {\scriptstyle f'}\downarrow & & \downarrow{\scriptstyle f}\\ (X',S') & \xrightarrow[g]{} & (X,S)\end{array}$$

be a cartesian diagram with $S' \subset X'(K)$, then we have canonical isomorphisms:

$$g^* R^i f_! \overline{\mathbb{Q}}_\ell \xrightarrow{\sim} R^i f'_! \overline{\mathbb{Q}}_\ell.$$

Indeed this follows from the formula $(R^i f_! \overline{\mathbb{Q}}_\ell)_s = H^i_c((Y_s, R \cap Y_s))$ for the stalk of $R^i f_! \overline{\mathbb{Q}}_\ell$ at $s \in S$, which in turn is proved as [13], Thm. 5.3.1.

(VII) In the special case where $(Y, R) = (X, S) \times (Z, T)$ and f is the projection onto the first factor, the sheaf $R^i f_! \overline{\mathbb{Q}}_\ell$ is the constant sheaf with fibers $H^i_c((Z, T))$.

(VIII) If X is the analytification of some algebraic variety X^{alg} over K, then $H^i_c(X) = H^i_c(X^{\mathrm{alg}})$. This is proved in [13], Thm. 7.1.1 in the torsion case; the ℓ-adic case follows.

(IX) If (X, S) is defined over a complete subfield K', then the group $G_{K|K'}$ of continuous field automorphisms of K which fix K' acts continuously on $H^i_c((X, S))$ (for the finest locally convex topology on this $\overline{\mathbb{Q}}_\ell$-vector space).

(X) Let $Y \xrightarrow{\pi} X$ be a locally trivial affine bundle of dimension d defined over a complete subfield K'. Denote by t the Tate character of $G_{K|K'}$ on $\overline{\mathbb{Q}}_\ell$. Then, for any $S \subset X$, we have isomorphisms

$$H^i_c((Y, \pi^{-1}(S))) \simeq H^{i-2d}_c((X, S)) \otimes t^d$$

as representations of $G_{K|K'}$. This follows from the Leray spectral sequence [13], Thm. 5.2.2.

(XI) If an analytic group G acts on (X, S), then the action of $G(K)$ on $H^i_c((X, S))$ is smooth (i.e., locally constant). This is proved in [14], Cor. 7.8 in the torsion case, and in [77], Cor. 4.1.19 in the ℓ-adic case.

We shall now apply this formalism to period domains. Let L be a perfect field of characteristic p, and let $(\mathbb{G}, \mathcal{N})$ be a PD-pair as in Definition 9.5.1, consisting of a reductive a.g.s. $\mathbb{G} = (G, \varphi, \nu)$ over Isoc(L), and a conjugacy class $\mathcal{N}$ of $\mathbb{Q}$-1-PS of G. As right before Definition 9.6.6, we fix a finite Galois extension $\tilde{K}_0$ of K_0 which splits G. We abbreviate the objects obtained by base change to $\tilde{K}_0$ by $\tilde{\mathcal{F}} := \tilde{\mathcal{F}}(G, \mathcal{N})^{an}$, $\tilde{\mathcal{F}}^{ss} := \tilde{\mathcal{F}}(\mathbb{G}, \mathcal{N})^{ss}$ and more generally $\tilde{\mathcal{F}}_\theta := \tilde{\mathcal{F}}(\mathbb{G}, \mathcal{N})_\theta$ for any HN-type $\theta \in \Theta(\mathbb{G}, \mathcal{N})$. Also we replace $\tilde{}$ by $\bar{}$ to denote base change to the completion of an algebraic closure $\overline{K}_0$. Finally we denote by $\tilde{t}_0$ the Tate twist representation of $\tilde{\Gamma}_0$.

Proposition 10.2.1 *Let $\theta \in \Theta(\mathbb{G}, \mathcal{N})$ and fix an HN-pair $(\mathbb{P}, \mathcal{N}_P) \in \theta$. Denote by $(\mathbb{M}, \mathcal{N}_M)$ the Levi quotient PD-pair and put $d_\theta :=$ $\dim(\tilde{\mathcal{F}}(\mathbb{P}, \mathcal{N}_P)) -$ $\dim(\tilde{\mathcal{F}}(\mathbb{M}, \mathcal{N}_M))$. Then there is a $J^{\mathbb{G}}(\mathbb{Q}_p) \times \tilde{\Gamma}_0$-equivariant isomorphism of smooth representations,*

$$H^i_c((\overline{\mathcal{F}}, \overline{\mathcal{F}}_\theta)) \simeq \mathrm{ind}^{J^{\mathbb{G}}(\mathbb{Q}_p)}_{J^{\mathbb{P}}(\mathbb{Q}_p)} \left(H^{i-2d_\theta}_c(\overline{\mathcal{F}}(\mathbb{M}, \mathcal{N}_M)^{ss}) \right) \otimes \tilde{t}_0^{d_\theta}.$$

Here ind denotes "compact" induction. Explicitly,

$$\mathrm{ind}_{J^{\mathbb{P}}(\mathbb{Q}_p)}^{J^{\mathbb{G}}(\mathbb{Q}_p)}(V) = \Big\{\varphi : J^{\mathbb{G}}(\mathbb{Q}_p) \longrightarrow V \mid \varphi \text{ smooth}, \varphi(gp) = p \cdot \varphi(g), \forall p \in J^{\mathbb{P}}(\mathbb{Q}_p),$$
$$\mathrm{supp}(\varphi) \text{ compact modulo } J^{\mathbb{P}}(\mathbb{Q}_p)\Big\}.$$

If $J^{\mathbb{G}}$ is a reductive group, then we know by Proposition 9.3.6 that $J^{\mathbb{P}}$ is a parabolic subgroup. Hence $J^{\mathbb{G}}(\mathbb{Q}_p)/J^{\mathbb{P}}(\mathbb{Q}_p)$ is compact, so that $\mathrm{ind}_{J^{\mathbb{P}}(\mathbb{Q}_p)}^{J^{\mathbb{G}}(\mathbb{Q}_p)} = \mathrm{Ind}_{J^{\mathbb{P}}(\mathbb{Q}_p)}^{J^{\mathbb{G}}(\mathbb{Q}_p)}$.

Proof Consider the action map

$$G(K_0) \times^{P(K_0)} \tilde{\mathcal{F}}(P, N_P)^{an} \stackrel{\psi}{\longrightarrow} \tilde{\mathcal{F}}(G, N)^{an}.$$

This analytic morphism induces the bijection of Chapter IX (6.5), page 281. Moreover, since $J^{\mathbb{G}}(\mathbb{Q}_p) \subset G(K_0)$ and since $\tilde{\mathcal{F}}(\mathbb{P}, N_P)_\theta$ is an analytic subset of $\tilde{\mathcal{F}}(G, N)^{an}$, the second condition on residue fields in the cohomological property (III) above is satisfied. Therefore ψ induces isomorphisms

$$H_c^i((G \times^P \tilde{\mathcal{F}}(P, N_P)^{an}, J^{\mathbb{G}} \times^{J^{\mathbb{P}}} \tilde{\mathcal{F}}(\mathbb{P}, N_P)_\theta)) \stackrel{\sim}{\longrightarrow} H_c^i((\tilde{\mathcal{F}}, \tilde{\mathcal{F}}_\theta)).$$

Here we have abbreviated $G(K_0)$ to G, $J^{\mathbb{G}}(\mathbb{Q}_p)$ to $J^{\mathbb{G}}$, etc., to save space. Now consider the Cartesian square of germs of analytic spaces:

$$\begin{array}{ccc} \Big(G \times \tilde{\mathcal{F}}(P, N_P), J^{\mathbb{G}} \times \tilde{\mathcal{F}}(\mathbb{P}, N_P)_\theta\Big) & \stackrel{g'}{\longrightarrow} & \Big(G \times^P \tilde{\mathcal{F}}(P, N_P), J^{\mathbb{G}} \times^{J^{\mathbb{P}}} \tilde{\mathcal{F}}(\mathbb{P}, N_P)_\theta\Big) \\ {\scriptstyle f'}\downarrow & & \downarrow{\scriptstyle f} \\ \Big(G, J^{\mathbb{G}}\Big) & \stackrel{g}{\longrightarrow} & \Big(G/P, J^{\mathbb{G}}/J^{\mathbb{P}}\Big) . \end{array}$$

By property (VI) we thus have $H_c^i((\tilde{\mathcal{F}}, \tilde{\mathcal{F}}_\theta)) \simeq \Gamma_c(J^{\mathbb{G}}/J^{\mathbb{P}}, R^i f_! \overline{\mathbb{Q}}_\ell)$ and also $g^* R^i f_! \overline{\mathbb{Q}}_\ell \simeq R^i f'_! \overline{\mathbb{Q}}_\ell$. We note that for any sheaf F on $J^{\mathbb{G}}/J^{\mathbb{P}}$, the set $\Gamma(J^{\mathbb{G}}, g^*F)$ carries an action of $J^{\mathbb{P}}$. Moreover, since the projection map $J^{\mathbb{G}} \stackrel{g}{\longrightarrow} J^{\mathbb{G}}/J^{\mathbb{P}}$ is open, the natural map $\Gamma(J^{\mathbb{G}}/J^{\mathbb{P}}, F) \longrightarrow \Gamma(J^{\mathbb{G}}, g^*F)$ is a bijection from the LHS onto the $J^{\mathbb{P}}$-invariants in the RHS. This bijection yields by restriction a bijection

$$\Gamma_c(J^{\mathbb{G}}/J^{\mathbb{P}}, F) \stackrel{\sim}{\longrightarrow} \Gamma_!(J^{\mathbb{G}}, g^*F)^{J^{\mathbb{P}}},$$

where the sign $\Gamma_!$ on the RHS denotes sections with compact support modulo $J^{\mathbb{P}}$. On the other hand, by property (VII), $R^i f'_! \overline{\mathbb{Q}}_\ell$ is the constant sheaf on $J^{\mathbb{G}}$ with stalks $H_c^i(\tilde{\mathcal{F}}(\mathbb{P}, N_P)_\theta)$. Therefore we have

$$\Gamma_!(J^{\mathbb{G}}, R^i f'_! \overline{\mathbb{Q}}_\ell) =$$
$$\Big\{\varphi : J^{\mathbb{G}} \longrightarrow H_c^i(\tilde{\mathcal{F}}(\mathbb{P}, N_P)_\theta) \mid \varphi \text{ smooth}, \mathrm{supp}(\varphi) \text{ compact mod } J^{\mathbb{P}}\Big\}.$$

Gathering all pieces, we get an isomorphism

$$\Gamma_c(J^{\mathbb{G}}/J^{\mathbb{P}}, R^i f_! \overline{\mathbb{Q}}_\ell) \simeq \mathrm{ind}_{J^{\mathbb{P}}}^{J^{\mathbb{G}}} \left(H_c^i(\overline{\mathcal{F}}(\mathbb{P}, \mathcal{N}_P)_\theta) \right).$$

Thus the proposition now follows from property (X) above and statement (ii) of Theorem 9.6.8. □

In the next corollary we consider the stratum $\breve{\mathcal{F}}(\mathbb{G}, \mathcal{N})_{[\theta]}$ in $\breve{\mathcal{F}}(G, \mathcal{N})$ which is associated to the Galois orbit $[\theta]$ of θ in $\Theta(\mathbb{G}, \mathcal{N})$. We denote by Γ_θ the stabilizer of θ in $\breve{\Gamma}_0$ and by $\breve{t}_0$ the *Tate twist* representation of $\breve{\Gamma}_0$.

Corollary 10.2.2 *With the foregoing notation, there is a $J^{\mathbb{G}}(\mathbb{Q}_p)\times\breve{\Gamma}_0$-equivariant isomorphism*

$$H_c^i((\overline{\mathcal{F}}, \overline{\mathcal{F}}_{[\theta]})) \simeq \mathrm{Ind}_{\Gamma_\theta}^{\breve{\Gamma}_0} \left(\mathrm{ind}_{J^{\mathbb{P}}(\mathbb{Q}_p)}^{J^{\mathbb{G}}(\mathbb{Q}_p)} \left(H^{i-2d_\theta}(\overline{\mathcal{F}}(\mathbb{M}, \mathcal{N}_M)^{\mathrm{ss}}) \right) \right) \otimes \breve{t}_0^{d_\theta}. \qquad \square$$

In general very little is known about the smooth induction $\mathrm{ind}_{J^{\mathbb{P}}(\mathbb{Q}_p)}^{J^{\mathbb{G}}(\mathbb{Q}_p)}$ unless $J^{\mathbb{G}}$ is a reductive group.

Corollary 10.2.3 *Assume that $J^{\mathbb{G}}$ is a reductive group. Then the cohomology spaces $H_c^i(\overline{\mathcal{F}}(\mathbb{G}, \mathcal{N})^{\mathrm{ss}})$ are finite length representations of $J^{\mathbb{G}}(\mathbb{Q}_p)$. Moreover, in the Grothendieck group $K_0^{\mathrm{fl}}(J^{\mathbb{G}}(\mathbb{Q}_p) \times \breve{\Gamma}_0)$ of finite length representations of $J^{\mathbb{G}}(\mathbb{Q}_p) \times \breve{\Gamma}_0$, we have the following expression :*

$$\chi_c(\breve{\mathcal{F}}(G, \mathcal{N})) = \sum_{[\theta]\in\theta(\mathbb{G},\mathcal{N})/\breve{\Gamma}_0} \mathrm{Ind}_{\Gamma_\theta}^{\breve{\Gamma}_0} \left(\mathrm{Ind}_{J^{\mathbb{P}}(\mathbb{Q}_p)}^{J^{\mathbb{G}}(\mathbb{Q}_p)} \left(\chi_c(\breve{\mathcal{F}}(\mathbb{M}, \mathcal{N}_M)^{\mathrm{ss}}) \right) \right) \otimes \breve{t}_0^{d_\theta}.$$

Proof The finite length property follows inductively on the rank of $\mathbb{G}$ from the spectral sequence associated to the HN-stratification (or successive applications of property (II) listed above) and the fact that parabolic induction takes finite length objects to finite length objects. Also the expression follows from successive applications of property (II) above and the previous corollary. □

Notes and References It would also have been possible to use Huber's theory of adic spaces [121] to obtain these results. In this setting the HN-strata are *pseudo-adic* subspaces of the adic space $\breve{\mathcal{F}}(G, \mathcal{N})^{ad}$. The étale cohomology formalism of Huber [121] applies also to such subsets. This approach is taken in [171] to determine the individual cohomology groups of period spaces.

3 The Euler–Poincaré characteristic in the basic case

Let $\mathbb{G} = (G, \varphi, \nu_G)$ be a reductive augmented group scheme over Isoc(L). Let $\mathcal{N}$ be a conjugacy class of 1-PS of G. In this section we resolve the recursion relation of Corollary 10.2.3, in the special case where the isocrystal $N^{\mathbb{G}}$

is *basic*, in the sense of Definition 9.3.1. This assumption is a drastic restriction, which allows us to apply the machinery of Chapter VII. Indeed, by the comment below Definition 9.3.1, we have a weak isomorphism

$$\psi : \mathbb{G} \overset{\sim}{\dashrightarrow} \mathbb{J}_1^{\mathbb{G}},$$

where $\mathbb{J}_1^{\mathbb{G}} = (J_{K_0}^{\mathbb{G}}, \mathrm{id}_{J^{\mathbb{G}}} \times \varphi_\sigma, 1)$ is the a.g.s. over $\mathrm{Isoc}(L)$ associated to the trivial $J^{\mathbb{G}}$-isocrystal. Hence, by Proposition 9.5.3 (iii) and Remark 9.6.18 (ii), we have an equality in $K_0^{\mathrm{fl}}(J^{\mathbb{G}}(\mathbb{Q}_p) \times \check{\Gamma}_0)$,

$$\chi_c(\check{\mathcal{F}}(\mathbb{J}_1^{\mathbb{G}}, \tilde{\psi}_*(\mathcal{N}))^{\mathrm{ss}}) = \chi_c(\check{\mathcal{F}}(\mathbb{G}, \mathcal{N})^{\mathrm{ss}}).$$

We simplify notation by putting $J := J^{\mathbb{G}}$. The LHS falls into the realm of filtered vector spaces of Part 2 and we may denote it simply by $\chi_c(\check{\mathcal{F}}(J, \tilde{\psi}_*(\mathcal{N}))^{\mathrm{ss}})$. Therefore, all we have to do is to adapt Chapter VII to the group J. Besides the notation (G is now J and k is $\mathbb{Q}_p$), the only new feature is that J does not need to be quasi-split.

We fix a minimal $\mathbb{Q}_p$-psgp P_0 of J and a Levi subgroup M_0 of P_0. We denote by $\mathcal{P}_J$ the set of $\mathbb{Q}_p$-psgps of J which contain P_0. Then we use the same notation $A_{P_0} = S, \Delta_{P_0}, \tau_P^J, \hat{\tau}_P^J$, etc., as in the Introduction and in Chapter VII, §1. We also fix a maximal torus T in M_0 defined over $\mathbb{Q}_p$ and choose a Borel subgroup B of J over $\overline{\mathbb{Q}}_p$, containing T and which is contained in P_0. We denote by $\nu \in \overline{C}_{T,B} \subset X_*(T)_{\mathbb{Q}}$ the unique representative of $\mathcal{N} \in X_*(G)_{\mathbb{Q}}/G = X_*(J)_{\mathbb{Q}}/J$. Thus, we see by Chapter IX (2.4), page 241 that

$$\nu^{\mathbb{G}} := \nu - \nu_G \circ sl$$

is a representative of $\tilde{\psi}_*(\mathcal{N})$. We denote by $\rho : X_*(T)_{\mathbb{Q}} \longrightarrow X_*(S)_{\mathbb{Q}}$ the Galois-averaging map. Then we have the same identification as Chapter VII (2.5), page 172 :

$$\Theta(\mathbb{G}, \mathcal{N}) = \Theta(J, \tilde{\psi}_*(\mathcal{N})) = \left\{(Q, w) \mid Q \in \mathcal{P}_J, w \in W_Q \backslash W / W_\nu, \tau_Q^J(\rho(w\nu^{\mathbb{G}})) = 1\right\}.$$

Here we recall that all Weyl groups occurring are *absolute* Weyl groups of T. The corresponding action of W on $\nu_G \circ sl$ is trivial since $N^{\mathbb{G}}$ is basic. Hence

$$w\nu^{\mathbb{G}} = w\nu - \nu_G \circ sl\,, \ \forall w \in W.$$

In particular, we get $W_\nu = W_{\nu^{\mathbb{G}}}$. We also note that the action of $\check{\Gamma}_0$ on $\Theta(\mathbb{G}, \mathcal{N})$ corresponds to the L-action of $\check{\Gamma}_0$ on W. By the definition of $\check{K}_0$, this L-action preserves W_ν, possibly in contrast to the ordinary action when J is not quasi-split. Moreover, if θ corresponds to (Q, w) then we have $d_\theta = \ell(w)$, as in Chapter VII. In this setting we may rewrite the recursive formula of Corollary 10.2.3

to get an analogue of Chapter VII (2.6), page 172, with the same notation:

$$\chi_c(\breve{\mathcal{F}}(J,\tilde{\psi}_*(\mathcal{N}))) \;=\; \sum_{Q\in\mathcal{P}_J}\ \sum_{[w]\in[W_Q\backslash W/W_\nu]_0} \tau_Q^J(\rho(w\nu^{\mathbb{G}}))\cdot t_0^{\ell([w])}\cdot$$
$$\mathrm{Ind}_{\breve{\Gamma}_{[w]}}^{\breve{\Gamma}_0}\,(\,\mathrm{Ind}_{Q(\mathbb{Q}_p)}^{J(\mathbb{Q}_p)}\,(\chi_c(\breve{\mathcal{F}}(M_Q,w\nu^{\mathbb{G}})^{\mathrm{ss}})))\,. \qquad (3.1)$$

Here $[w]$ runs over the set $[W_Q\setminus W/W_\nu]_0$ of $\breve{\Gamma}_0$-orbits in $W_Q\backslash W/W_\nu$ and $\breve{\Gamma}_{[w]}$ is the stabilizer of $[w]$. Also, we have denoted by M_Q the Levi quotient of Q, which is identified with the a.g.s. $\mathbb{M}_{Q,1}$ over $\mathrm{Isoc}(L)$ attached to the trivial isocrystal.

Theorem 10.3.1 *In the Grothendieck group $K_0^{\mathrm{fl}}(J(\mathbb{Q}_p)\times\breve{\Gamma}_0)$, there is the equality*

$$\chi_c(\breve{\mathcal{F}}(\mathbb{G},\mathcal{N})^{\mathrm{ss}}) = \sum_{P\in\mathcal{P}_J}(-1)^{a_P^J}\cdot \sum_{[w]\in[W/W_\nu]_0} \mathrm{ind}_{\breve{\Gamma}_{[w]}}^{\breve{\Gamma}_0}\cdot t_0^{\ell(w)}\cdot\widetilde{\tau}_P^J(w\nu^{\mathbb{G}})\cdot i_P^J\,.$$

Proof The proof is formally the same as that of Theorem 7.2.2. It consists of inverting the recursive formula in Corollary 10.2.3 with the help of the Langlands Lemma for J and the following formula for the Euler–Poincaré characteristic of the flag variety $\breve{\mathcal{F}}(J,\tilde{\psi}_*(\mathcal{N}))$:

$$\chi_c(\breve{\mathcal{F}}(J,\tilde{\psi}_*(\mathcal{N}))) = \sum_{[w]\in[W/W_\nu]_0} \mathrm{ind}_{\breve{\Gamma}_{[w]}}^{\breve{\Gamma}_0}\cdot t_0^{\ell([w])}\,.$$

In Chapter VII, this formula was obtained for a quasi-split group, using the Bruhat decomposition. So it applies in particular to the quasi-split inner form J' of J. However the $\breve{\Gamma}_0$-action on $\overline{\mathcal{F}}(J,\tilde{\psi}_*(\mathcal{N})) = \overline{\mathcal{F}}(J',\tilde{\psi}_*(\mathcal{N}))$ associated to the $\breve{K}_0$-rational model $\breve{\mathcal{F}}(J,\tilde{\psi}_*(\mathcal{N}))$ is deduced from that associated to the $\breve{K}_0$-rational model $\breve{\mathcal{F}}(J',\tilde{\psi}_*(\mathcal{N}))$ by a 1-cocycle $\breve{\Gamma}_0 \longrightarrow J_{\mathrm{ad}}(\overline{K}_0)$. Since the group $J_{\mathrm{ad}}(\overline{K}_0)$ acts trivially on the ℓ-adic cohomology spaces of $\breve{\mathcal{F}}(J,\tilde{\psi}_*(\mathcal{N}))$, the above formula still holds for J. □

Again, as in Chapter VII, thanks to the transition matrix (cf. Theorem 10.1.1) between the families of the i_P^Js and v_P^Js, for $P\in\mathcal{P}_J$, we may rewrite this result as follows. We set for $[w]\in[W/W_\nu]_0$,

$$\Delta(w\nu) := \{\alpha\in\Delta_0 \mid \langle\rho(w\nu),\varpi_\alpha\rangle > 0\}\,.$$

Here $(\varpi_\alpha)_\alpha \in X^*(S)_{\mathbb{Q}}$ are the relative fundamental weights of J. This set is well-defined, i.e., independent of the choice of the representative of $[w]$. Also, as in Chapter VII, we introduce the std psgp $P(w\nu)$ with set of simple roots occurring in its unipotent radical equal to $\Delta(w\nu)$. Using the notation introduced above Theorem 7.2.3 we get the following expansion.

Theorem 10.3.2 *In the Grothendieck group $K_0^{\mathrm{fl}}(J(\mathbb{Q}_p)\times\breve{\Gamma}_0)$, there is the equality*

$$\chi_c(\breve{\mathcal{F}}(\mathbb{G},\mathcal{N})^{ss}) = \sum_{w\in[W/W_\nu]_0} (-1)^{a^G_{P(w\nu)}} \operatorname{ind}^{\breve{\Gamma}_0}_{\breve{\Gamma}_{[w]}} \cdot \mathfrak{t}_0^{\ell(w)} \cdot v^J_{P(w\nu)}. \qquad \square$$

As in the finite field case, the individual cohomology spaces are known in this context. The main result of [171] is the following formula. Again we will not prove this formula here.

Theorem 10.3.3

$$H_c^*(\breve{\mathcal{F}}(\mathbb{G},\mathcal{N})^{ss}) = \bigoplus_{[w]\in[W/W_\nu]_0} \operatorname{ind}^{\breve{\Gamma}_0}_{\breve{\Gamma}_{[w]}} \cdot \mathfrak{t}_0^{\ell(w)} \cdot v^J_{P(w\nu)}[-2\ell(w) - |\Delta(w\nu)|].$$

Here again the symbol $[-m], m \in \mathbb{N}$, indicates that the corresponding module is shifted into degree m of the graded cohomology ring.

Example 10.3.4 If $\mathbb{G} = (G, \varphi, \nu_G)$ is an a.g.s. attached to the trivial isocrystal (hence $J = G$) and if G is quasi-split, then the formula has the same shape as the one in Theorem 7.2.4.

Example 10.3.5 Reconsider Example 8.2.7 (iv). Thus $G = \mathrm{GL}_4(\mathbb{Q}_p)$, $\nu = (1,1,0,0) \in \mathbb{Q}^4$ and $\nu_G \circ sl = (\frac{1}{2},\frac{1}{2},\frac{1}{2},\frac{1}{2}) \in \mathbb{Q}^4$, $\mathcal{F} = \mathrm{Gr}_2(\mathbb{Q}_p^4)$. Hence $J = \mathrm{GL}_2(D)$ where $D = D_{\frac{1}{2}}$ and $S = \{\mathrm{diag}(t\cdot 1_D, t^{-1}\cdot 1_D) \mid t \in \mathbb{G}_m\}$. Since G is split, the action of $\breve{\Gamma}_0$ on W is trivial.

Let $s_1 = s_{\alpha_1}$, $s_2 = s_{\alpha_2}$, $s_3 = s_{\alpha_3}$ be the reflections of GL_4 corresponding to the std simple roots $\alpha_1, \alpha_2, \alpha_3$. Then we have

- $\Delta_0 = \{\alpha_0\}$, where $\alpha_0 = \alpha_2|S$
- $\omega_{\alpha_0} = (\frac{1}{2},\frac{1}{2},-\frac{1}{2},-\frac{1}{2}) \in X_*(S)_{\mathbb{Q}}$
- $W_\nu = S_2 \times S_2$
- $W/W_\nu = \{1, s_2, s_1s_2, s_3s_2, s_1s_3s_2, s_2s_1s_3s_2\}$.

The set of std psgps is $\mathcal{P} = \{P_0, J\}$.

The sets $\Delta(w\nu)$, $w \in W/W_\nu$, are determined by the following table.

w	1	s_2	s_1s_2	s_3s_2
$w\nu$	$(1,1,0,0)$	$(1,0,1,0)$	$(0,1,1,0)$	$(1,0,0,1)$
$\Delta(w\nu)$	Δ_0	$\emptyset$	$\emptyset$	$\emptyset$

w	$s_1s_3s_2$	$s_2s_1s_3s_2$
$w\nu$	$(0,1,0,1)$	$(0,0,1,1)$
$\Delta(w\nu)$	$\emptyset$	$\emptyset$

Theorem 10.3.3 gives

$$H_c^0(\breve{\mathcal{F}}(\mathbb{G},\mathcal{N})^{\mathrm{ss}}) = 0 \qquad H_c^5(\breve{\mathcal{F}}(\mathbb{G},\mathcal{N})^{\mathrm{ss}}) = 0$$

$$H_c^1(\breve{\mathcal{F}}(\mathbb{G},\mathcal{N})^{\mathrm{ss}}) = v_{P_0}^J \qquad H_c^6(\breve{\mathcal{F}}(\mathbb{G},\mathcal{N})^{\mathrm{ss}}) = \breve{t}_0^3 \cdot i_J^J$$

$$H_c^2(\breve{\mathcal{F}}(\mathbb{G},\mathcal{N})^{\mathrm{ss}}) = \breve{t}_0 \cdot i_J^J \qquad H_c^7(\breve{\mathcal{F}}(\mathbb{G},\mathcal{N})^{\mathrm{ss}}) = 0$$

$$H_c^3(\breve{\mathcal{F}}(\mathbb{G},\mathcal{N})^{\mathrm{ss}}) = 0 \qquad H_c^8(\breve{\mathcal{F}}(\mathbb{G},\mathcal{N})^{\mathrm{ss}}) = \breve{t}_0^4 \cdot i_J^J$$

$$H_c^4(\breve{\mathcal{F}}(\mathbb{G},\mathcal{N})^{\mathrm{ss}}) = (\breve{t}_0^2 \cdot i_J^J)^{\oplus 2} \qquad H_c^i(\breve{\mathcal{F}}(\mathbb{G},\mathcal{N})^{\mathrm{ss}}) = 0, \quad \forall i > 8.$$

Remark 10.3.6 We note that also Remark 3.3.11 (regarding the object in the derived category of $J(\mathbb{Q}_p) \times \breve{\Gamma}_0$-representations defined by the cohomology complex $R\Gamma_c(\breve{\mathcal{F}}(\mathbb{G},\mathcal{N})^{\mathrm{ss}}) = R\Gamma_c(\breve{\mathcal{F}}(\mathbb{G},\mathcal{N})^{\mathrm{ss}} \otimes_{\breve{K}_0} \overline{K}_0, \bar{\mathbb{Q}}_\ell))$ has an analogue in this situation. In fact, consider for each $w \in [W/W_\nu]_0$ the complex

$$C_{[w]}^\bullet : 0 \to \mathrm{ind}_{\breve{\Gamma}_{[w]}}^{\breve{\Gamma}_0} \otimes i_J^J \to \bigoplus_{\substack{\Delta_0\setminus\Delta(w\nu)\subset K\subset\Delta_0\\ |\Delta_0\setminus K|=1}} \mathrm{ind}_{\breve{\Gamma}_{[w]}}^{\breve{\Gamma}_0} \otimes i_{P_K}^J \to \bigoplus_{\substack{\Delta_0\setminus\Delta(w\nu)\subset K\subset\Delta_0\\ |\Delta_0\setminus K|=2}} \mathrm{ind}_{\breve{\Gamma}_{[w]}}^{\breve{\Gamma}_0} \otimes i_{P_K}^J \to$$
$$\cdots \to \bigoplus_{\substack{\Delta_0\setminus\Delta(w\nu)\subset K\subset\Delta_0\\ |K\setminus(\Delta_0\setminus\Delta(w\nu))|=1}} \mathrm{ind}_{\breve{\Gamma}_{[w]}}^{\breve{\Gamma}_0} \otimes i_{P_K}^J \to \mathrm{ind}_{\breve{\Gamma}_{[w]}}^{\breve{\Gamma}_0} \otimes i_{P(w\nu)}^J. \quad (3.2)$$

Again, the representation i_J^J is in degree 0 and $P_{\Delta_0\setminus\Delta(w\nu)} = P(w\nu)$. By Theorem 10.1.1 the cohomology of this complex is concentrated in degree $|\Delta(w\nu)|$ and yields the twisted generalized Steinberg representation $\mathrm{ind}_{\breve{\Gamma}_{[w]}}^{\breve{\Gamma}_0} \otimes v_{P(w\nu)}^J$. Furthermore, the twisted and shifted complex $\breve{t}_0^{\ell(w)} \cdot C_{[w]}^\bullet[-2\ell(w)]$ gives exactly the w-contribution in the cohomology of $H_c^*(\breve{\mathcal{F}}(\mathbb{G},\mathcal{N})^{\mathrm{ss}})$.

By Proposition 11.5.10 below, the object $R\Gamma_c(\breve{\mathcal{F}}(\mathbb{G},\mathcal{N})^{\mathrm{ss}})$ splits in the derived category of $J(\mathbb{Q}_p)$-representations, i.e.,

$$R\Gamma_c(\breve{\mathcal{F}}(\mathbb{G},\mathcal{N})^{\mathrm{ss}}) = \bigoplus_{w\in[W/W_\nu]_0} \breve{t}_0^{\ell(w)} \cdot C_{[w]}^\bullet[-2\ell(w)]\,, \quad (3.3)$$

where we consider the trivial $J(\mathbb{Q}_p)$-action on $\breve{t}_0^{\ell(w)} \cdot \mathrm{ind}_{\breve{\Gamma}_{[w]}}^{\breve{\Gamma}_0}$. This splitting is, however, no longer true if we also consider the $\breve{\Gamma}_0$-action, because of the inertia subgroup. In the Drinfeld case one can recover the missing monodromy operator induced by the inertia group on the complex $R\Gamma_c(\Omega^n)$, cf. [48], as was already pointed out in Remark 3.3.11. We will return to this issue in more detail in Part 4 of this monograph.

By Proposition 9.5.3 (iv), the period domain $\breve{\mathcal{F}}(\mathbb{G},\mathcal{N})^{\mathrm{ss}}$ depends only on the adjoint a.g.s. $\mathbb{G}_{\mathrm{ad}}$. Also, if $\mathbb{G} = \prod_i \mathbb{G}_i$ is a direct product of a.g.s. over Isoc(L), the corresponding period domain $\breve{\mathcal{F}}(\mathbb{G},\mathcal{N})^{\mathrm{ss}} = \prod_i \breve{\mathcal{F}}(\mathbb{G}_i,\mathcal{N}_i)^{\mathrm{ss}}$ splits into a product of corresponding period domains. When $\mathbb{G}$ is adjoint and $N^{\mathbb{G}}$ is basic, we can split $\mathbb{G}$ as a product of "simple" $\mathbb{G}_i$, meaning that each $J^{\mathbb{G}_i}$ is a simple $\mathbb{Q}_p$-group. Indeed, such a decomposition follows from the usual decomposition of $J^{\mathbb{G}}$ as a product of simple factors, together with the existence of a weak isomorphism $J^{\mathbb{G}} \dashrightarrow \mathbb{G}$, cf. Chapter IX (3.3), page 252.

Now let us assume that J is $\mathbb{Q}_p$-simple and adjoint. Then J is of the form $J = \mathrm{Res}_{F|\mathbb{Q}_p}(J')$ for an absolutely simple group J' over some finite extension F of $\mathbb{Q}_p$. Then $\mathcal{N}$ is of the form $(\mathcal{N}_1,\dots,\mathcal{N}_n)$ corresponding to the $\mathbb{Q}_p$-embeddings of F into $\overline{K}_0$. Here $n = [F:\mathbb{Q}_p]$ and $\mathcal{N}_1,\dots,\mathcal{N}_n$ are conjugacy classes of J'.

The same proof as that of Corollary 7.2.7 now gives the following vanishing/non-vanishing property.

Corollary 10.3.7 *Let $\mathbb{G} = (G,\varphi,\nu_G)$ be an a.g.s. over* Isoc(L) *such that $J^{\mathbb{G}}$ is $\mathbb{Q}_p$-simple and adjoint (this implies that $N^{\mathbb{G}}$ is basic). Let $\mathcal{N} = (\mathcal{N}_1,\dots,\mathcal{N}_n)$ be non-trivial, i.e., there is at least one index $1 \le i \le n$ such that $\mathcal{N}_i$ is not defined by a trivial* $\mathbb{Q}$-1-PS. *Let $d = |\Delta_0|$ be the semi-simple $\mathbb{Q}_p$-rank of J. Then*

$$H_c^i(\breve{\mathcal{F}}(\mathbb{G},\mathcal{N})^{\mathrm{ss}}) = 0 \ \ \textit{for } i \textit{ with } 0 \le i \le d-1$$

and

$$H_c^d(\breve{\mathcal{F}}(\mathbb{G},\mathcal{N})^{\mathrm{ss}}) = v_{P_0}^J.$$

Furthermore, the top cohomology is

$$H_c^{2\ell(w_0)}(\breve{\mathcal{F}}(\mathbb{G},\mathcal{N})^{\mathrm{ss}}) = t_E^{\ell(w_0)} \cdot i_J^J,$$

where w_0 is of maximal length in W/W_ν, i.e., $\dim \breve{\mathcal{F}}(G,\mathcal{N}) = \ell(w_0)$.

Notes and References Theorem 10.3.2 is due to Kottwitz and the third author, and is stated without proof in [185]. A sketch of the proof, based on suggestions of R. Huber, is given in the unpublished manuscript [184]. At the time some foundational questions regarding the ℓ-adic cohomology with compact supports of analytic spaces were still unsolved.

Theorem 10.3.3 is proved by the second author in [174]. The analogue for torsion coefficients is treated in [171]. Strictly speaking, in [174] resp. [171] only the case of a quasi-split group is considered, which, by the Kottwitz Lemma, allows an easier description of the complement of period domains using Bruhat cells, cf. also Chapter XI, §1.

Part 4

Complements

XI. Further Aspects of Period Domains

This final chapter has a more informal character. On the one hand, we want to give some guidance for approaching recent publications on period domains (some of which of the present authors). On the other hand, we want to flesh out the remarks in the Introduction on some of the open problems in this area.

In the first section we explain the *fundamental complex* and the geometry around it. This is the crucial tool in the proof of Theorems 3.3.10, 7.2.4, 10.3.3 in the text, which are only stated in this book, but not proved. In the second section, we compare period domains over a finite field with Deligne-Lusztig varieties. The third section describes special features of the Drinfeld space Ω_K^n for a local non-Archimedean field K, which are either not shared by other period domains, or are unknown in general. The fourth section describes the conjectural *admissible* analytic subset of period domains with its local system over it, and Hartl's candidate for it in a special case. In the fifth section we discuss the problem of determining the cohomology *complex* of period domains, and explain what is known in the Lubin–Tate and the Drinfeld cases.

1 The fundamental complex

In this section we give the construction of an acyclic resolution of an étale sheaf on the complement of a period domain. This complex appears in [170, 171], and is the main ingredient for the determination of the individual cohomology groups of period domains (as opposed to the Euler–Poincaré characteristic). We first consider period domains over finite fields. Afterwards, we address the p-adic situation.

Let $(G, \mathcal{N})$ be a PD-pair over a finite field k, cf. Definition 6.1.2. Let $\mathcal{F} = \mathcal{F}(G, \mathcal{N})$ be the corresponding flag variety and $\mathcal{F}^{ss} = \mathcal{F}(G, \mathcal{N})^{ss}$ the associated period domain inside $\mathcal{F}$, cf. Definition 6.1.11. Consider the complement $Y := \mathcal{F} \setminus \mathcal{F}^{ss}$, with its reduced scheme structure. This is a closed subvariety which is defined over the reflex field E of $(G, \mathcal{N})$. In the following, we will use the same notation concerning root systems, etc., as in Part 2. So, let $\Delta_0 = \{\alpha_1, \ldots, \alpha_d\}$ be the set of simple relative roots with respect to a fixed maximal split torus S in G and a Borel subgroup B containing S. Let $(\omega_\alpha)_\alpha$ be the set of relative fundamental co-weights. Hence $\omega_\alpha \in X_*(S_{\mathrm{der}})$ and $\langle \omega_\alpha, \beta \rangle = \delta_{\alpha\beta}, \forall \alpha, \beta \in \Delta_0$. After choosing an invariant inner product on G, we have an ample $\mathbb{Q}$-line bundle $\mathcal{L}$ on $\mathcal{F}$ corresponding to $\mathcal{N}$, cf. Lemma 6.2.5. For any subset $I \subset \Delta_0$, we set

$$Y_I(\bar{k}) := \{x \in \mathcal{F}(\bar{k}) \mid \mu^{\mathcal{L}}(x, \omega_\alpha) < 0, \ \forall \alpha \notin I\}.$$

Again, as in the introduction to Chapter VII, let $P_I \subset G$ be the std psgp generated by B and the simple reflections $s_\alpha \in W_0$, for $\alpha \in I$.

Lemma 11.1.1 *Let $I \subset \Delta_0$. Then $Y_I(\bar{k})$ is the set of $\bar{k}$-valued points of a closed subvariety Y_I of Y defined over E. The natural action of G on $\mathcal{F}$ restricts to an action of P_I on Y_I.*

Proof Obviously $Y_{I\cap J}(\bar{k}) = Y_I(\bar{k}) \cap Y_J(\bar{k})$, for all subsets $I, J \subset \Delta_0$. Hence it suffices to prove the first statement for a maximal proper subset which is of the form $I = \Delta_0 \setminus \{\alpha\}$. Let $\mathcal{F} \hookrightarrow \mathbb{P}(V)$ be a G-linearized embedding into some projective space which is defined over E and such that $\mathcal{O}(1)|\mathcal{F} = \mathcal{L}$, cf. [164], Prop. 1.7. Let $X_{\omega_\alpha}(\bar{k}) = \{x \in \mathbb{P}(V)(\bar{k}) \mid \mu^{\mathcal{O}(1)}(x, \omega_\alpha) < 0\}$. It suffices to prove that the set $X_{\omega_\alpha}(\bar{k})$ is induced by a closed subvariety of $\mathbb{P}(V)$. Let $V = \oplus_{i\in\mathbb{Q}} V_i$ be the $\mathbb{Q}$-grading induced by ω_α. Let $\hat{x} \in \mathbb{A}(V)(\bar{k})$ be a representative of x. Write $\hat{x} = \sum_i x_i$ with $x_i \in V_i$. Then the slope $\mu^{\mathcal{O}(1)}(x, \omega_\alpha)$ is given by $\max\{-i \mid x_i \neq 0\}$, cf. Chapter II (2.3), page 37. Thus the set $X_{\omega_\alpha}(\bar{k})$ is induced by the closed subvariety $\mathbb{P}(\oplus_{i>0} V_i) \subset \mathbb{P}(V)$ and the first assertion follows.

The second statement results immediately from the identity (cf. [164], Prop. 2.7)

$$\mu^{\mathcal{L}}(px, \omega_\alpha) = \mu^{\mathcal{L}}(x, \omega_\alpha), \quad \forall p \in P_{\omega_\alpha}(\bar{k}). \qquad \square$$

Let F be an étale sheaf on Y. Let $g \in G(k)$ and let $I \subset \Delta_0$. We denote by

$$\phi_{g,I} : gY_I \hookrightarrow Y$$

the corresponding closed embedding. Now let $I \subset J$ be two subsets of Δ_0 with $|J \setminus I| = 1$. Let further $g \in (G/P_I)(k), h \in (G/P_J)(k)$ be two elements, such that g is mapped to h under the canonical projection $(G/P_I)(k) \longrightarrow (G/P_J)(k)$. In this case, we define

$$p_{I,J}^{g,h} : (\phi_{h,J})_*(\phi_{h,J})^* F \longrightarrow (\phi_{g,I})_*(\phi_{g,I})^* F$$

to be the natural morphism of étale sheaves on Y which is induced by the closed embedding $gY_I \hookrightarrow hY_J$. If g is not mapped to h, then we set $p_{I,J}^{g,h} = 0$. Finally, we define

$$p_{I,J} = \bigoplus_{(g,h)\in(G/P_I)(k)\times(G/P_J)(k)} p_{I,J}^{g,h} : \bigoplus_{h\in(G/P_J)(k)} (\phi_{h,J})_*(\phi_{h,J})^* F \longrightarrow \bigoplus_{g\in(G/P_I)(k)} (\phi_{g,I})_*(\phi_{g,I})^* F.$$

For two arbitrary subsets $I, J \subset \Delta_0$ with $|J| - |I| = 1$, we set

$$d_{I,J} = \begin{cases} (-1)^i p_{I,J} & \text{if } J = I \cup \{\alpha_i\} \\ \\ 0 & \text{if } I \not\subset J \end{cases}.$$

We obtain a complex of étale sheaves on Y, called the *fundamental complex* of F,

$$0 \to F \to \bigoplus_{\substack{I\subset\Delta_0 \\ |\Delta_0 - I|=1}} \bigoplus_{g\in(G/P_I)(k)} (\phi_{g,I})_*(\phi_{g,I})^* F \to \bigoplus_{\substack{I\subset\Delta_0 \\ |\Delta_0 - I|=2}} \bigoplus_{g\in(G/P_I)(k)} (\phi_{g,I})_*(\phi_{g,I})^* F \to$$
$$\cdots \to \bigoplus_{\substack{I\subset\Delta_0 \\ |\Delta_0 - I|=d-1}} \bigoplus_{g\in(G/P_I)(k)} (\phi_{g,I})_*(\phi_{g,I})^* F \to \bigoplus_{g\in(G/B)(k)} (\phi_{g,\emptyset})_*(\phi_{g,\emptyset})^* F \to 0. \quad (1.1)$$

Theorem 11.1.2 ([170]) *The complex (1.1) is acyclic.*

We will not prove this theorem here, but we will indicate a crucial step in the proof.

Lemma 11.1.3 *Let λ be a $\mathbb{Q}$-1-PS of G and let $x \in \mathcal{F}(\bar{k})$. Let $v_x \in \mathcal{N}$ be a representative of x factoring over the same maximal torus T as λ. Then*

$$\mu^{\mathcal{L}}(x, \lambda) = -(v_x, \lambda).$$

(Here on the right appears the fixed invariant inner product on $X_*(T)_{\mathbb{R}}$.)

Proof By choosing a faithful representation $G \longrightarrow \mathrm{GL}(V)$ we are reduced to the case $G = \mathrm{GL}(V)$, cf. the proof of Theorem 6.2.8. But then the statement follows from the identity (2.6), page 40 in Chapter II, and Lemma 1.2.4. □

The following proposition says that the slope function is affine-linear on each chamber in the rational flag complex. Its proof uses the previous lemma.

Proposition 11.1.4 *Let $x \in \mathcal{F}(\bar{k})$ be any point. Then*

$$\mu^{\mathcal{L}}(x, \sum\nolimits_{\alpha\in\Delta_0} r_\alpha \omega_\alpha) = \sum\nolimits_{\alpha\in\Delta_0} r_\alpha \mu^{\mathcal{L}}(x, \omega_\alpha), \quad \forall r_a \in \mathbb{R}, 0 \le r_\alpha \le 1. \quad \square$$

This proposition is then used for the proof of the acyclicity of the complex (1.1), page 321. More precisely, let $y \in Y$. Then the fiber in y of the fundamental complex can be identified with the chain complex with coefficients in F_y of a subcomplex of the combinatorial Tits complex of $G(k)$. Proposition 11.1.4 is used to construct a topological retract of this subcomplex by convex geometry, cf. [170].

The following corollary of Proposition 11.1.4 will be used in the determination of the cohomology, but also has other applications, cf. next section.

Corollary 11.1.5 *Let $x \in \mathcal{F}(\bar{k})$. Then $x \in Y(\bar{k})$ (i.e. x is not semi-stable) iff there are $g \in G(k)$ and $\alpha \in \Delta_0$ such that $\mu^{\mathcal{L}}(x, \mathrm{Int}(g) \circ \omega_\alpha) < 0$.* □

Hence we may write Y as the union

$$Y = \bigcup_{\alpha \in \Delta_0} \bigcup_{g \in G(k)} gY_{\Delta_0 \setminus \{\alpha\}}. \tag{1.2}$$

It turns out that the subvarieties Y_I have a simple description: the proposition below describes them as a union of Schubert cells. Recall from Chapter VII, §2 the notation $[W/W_\nu]_E$ for the set of Γ_E-orbits on W/W_ν. For a subset $I \subset \Delta_0$, let

$$\Omega_I = \{[w] \in [W/W_\nu]_E \mid (w\nu, \omega_\alpha) > 0 \ \ \forall \alpha \notin I\}.$$

Here w is any element of the orbit $[w]$. Let $\nu \in \mathcal{N}$ be a representative lying in the closed Weyl chamber and write $\mathcal{F} = G/P_\nu$.

Proposition 11.1.6 *The closed subvarieties Y_I have the following description in terms of the generalized Bruhat cells of G with respect to P_ν,*

$$Y_I = \bigcup_{[w] \in \Omega_I} P_I w P_\nu / P_\nu = \bigcup_{[w] \in \Omega_I} BwP_\nu / P_\nu.$$

Proof By Lemma 11.1.1 we know that the subvarieties are stable under the action of P_I. In particular, they are union of P_I-orbits. On the other hand, let $x_\nu \in \mathcal{F}(E)$ be the point induced by the 1-PS ν. Then by Lemma 11.1.3 we deduce that

$$\mu^{\mathcal{L}}(wx_\nu, \omega_\alpha) = -(w\nu, \omega_\alpha).$$

The statements follow. □

Now we are going to indicate how Theorem 11.1.2 is used to calculate the cohomology groups of $\mathcal{F}^{ss}$. The fundamental complex (1.1), page 321 induces a spectral sequence

$$E_1^{p,q} = H^q_{\acute{e}t}\Big(Y \times_k \bar{k}, \bigoplus_{\substack{I \subset \Delta_0 \\ |\Delta_0 \setminus I| = p+1}} \bigoplus_{g \in (G/P_I)(k)} (\phi_{g,I})_*(\phi_{g,I})^* F \times_k \bar{k}\Big) \Longrightarrow H^{p+q}_{\acute{e}t}(Y \times_k \bar{k}, F \times_k \bar{k}).$$

Let $F = \mathbb{Q}_\ell$. Then

$$E_1^{p,q} = \bigoplus_{\substack{I\subset\Delta_0 \\ |\Delta_0\setminus I|=p+1}} \bigoplus_{(G/P_I)(k)} H^q_{ét}(Y_I \times_k \bar{k}, \mathbb{Q}_\ell).$$

By Proposition 11.1.6 the varieties Y_I are Schubert varieties. Their cohomology is given by

$$H^*_{ét}(Y_I \times_k \bar{k}, \mathbb{Q}_\ell) = \bigoplus_{w\in\Omega_I} t^{\ell(w)}[-2\ell(w)].$$

In [170, 171] it is proved that the spectral sequence degenerates in E_2 and that it computes the ℓ-adic cohomology of Y. Then, by considering the long exact cohomology sequence for cohomology with compact supports with respect to $Y \subset \mathcal{F} \supset \mathcal{F}^{ss}$, one obtains the formula for the cohomology of $\mathcal{F}^{ss}$ in Theorem 7.2.4.

Now we want to discuss how to modify the fundamental complex (1.1), page 321 in the p-adic setting. Since the construction of this complex rests on the GIT criterion Theorem 9.7.3, we can deal only with PD-pairs $(\mathbb{G}_b, \mathcal{N})$ coming from a triple $(G, b, \mathcal{N})$ with G reductive over $\mathbb{Q}_p$ and b a decent element in $G(K_0)$. In fact, we will also assume that b is *basic*. Let $\mathcal{F}^{ss} = \mathcal{F}(G, b, \mathcal{N})^{ss}$ be the corresponding period domain in $\mathcal{F}$, and let $Y := \mathcal{F}^{an} \setminus \mathcal{F}^{ss}$ be the closed complement. Two difficulties arise. The first is that Y is not an analytic subspace of $\mathcal{F}^{an}$. The second is that, if we naively imitate the above construction of the complex for finite fields, we must deal, on the face of it, with infinitely many summands, indexed by a compact totally disconnected set of the form $((G/P)(K_0))^{b\sigma}$.

These difficulties are addressed in [171]. In accordance with our general treatment of period domains in the previous chapters, we will give a version of the construction in [171] in the setting of analytic spaces. We consider the germ $\mathcal{Y} := (\mathcal{F}^{an}, Y)$ of analytic spaces, as in Chapter X, §2. Moreover, we will assume that $b = 1$, and therefore $J = G$. This is not a serious restriction, as is explained at the beginning of Chapter X, §3. It will simplify the exposition.

We use the same notation as right before Lemma 11.1.1 for finite fields, except that we have to replace the Borel subgroup B by a minimal psgp P_0, since G need not be quasi-split. Hence, to any subset $I \subset \Delta_0$ we associate a closed subvariety Y_I of $\mathcal{F}$, which is invariant under the psgp P_I of G. We define a germ of analytic spaces

$$\mathcal{Y}_I := \left(G^{an} \times^{P_I^{an}} Y_I^{an}, G(\mathbb{Q}_p) \times^{P_I(\mathbb{Q}_p)} Y_I^{an}\right).$$

It is endowed with a natural continuous action of $G(\mathbb{Q}_p)$, and the inclusion $Y_I^{an} \hookrightarrow \mathcal{F}^{an}$ induces a $G(\mathbb{Q}_p)$-equivariant morphism of germs $\phi_I : \mathcal{Y}_I \longrightarrow \mathcal{Y}$.

For any subset $J \subset \Delta_0$ containing I, the inclusion $Y_I \hookrightarrow Y_J$ induces a $G(\mathbb{Q}_p)$-equivariant morphism of germs $\mathcal{Y}_I \longrightarrow \mathcal{Y}_J$. Therefore we get a strict simplicial G-equivariant germ of $\breve{K}_0$-analytic spaces:

$$\cdots \Rrightarrow \coprod_{|\Delta_0 - I|=2} \mathcal{Y}_I \rightrightarrows \coprod_{|\Delta_0 - I|=1} \mathcal{Y}_I \longrightarrow \mathcal{Y} \longrightarrow 0\ . \qquad (1.3)$$

Theorem 11.1.7 *Let y be any geometric point of $\mathcal{F}^{an}$ with image in Y. Then the fiber of (1.3) over y is a homotopically trivial strict simplical t.d. space.*

This is the content of the proof of [171], Thm 3.3. As a corollary, for any étale sheaf F on $\mathcal{Y}$, we can consider the complex of étale sheaves on Y, the *fundamental complex* in the p-adic context,

$$0 \to F \to \bigoplus_{\substack{I\subset\Delta_0 \\ |\Delta_0 - I|=1}} (\phi_I)_*(\phi_I)^* F \to \bigoplus_{\substack{I\subset\Delta_0 \\ |\Delta_0 - I|=2}} (\phi_I)_*(\phi_I)^* F \to$$
$$\cdots \to \bigoplus_{\substack{I\subset\Delta_0 \\ |\Delta_0 - I|=d-1}} (\phi_I)_*(\phi_I)^* F \to (\phi_\emptyset)_*(\phi_\emptyset)^* F \to 0. \qquad (1.4)$$

The last theorem is then used to establish the p-adic analogue of Theorem 11.1.2.

Theorem 11.1.8 *The complex (1.4) is acyclic.*

This fact can then be used to prove Theorem 10.3.3.

Notes and References The proof of the acyclicity of the fundamental complex (1.1), page 321 uses the fact that in GIT the set of 1-PS of G which violates the semi-stability condition of a fixed point in a G-variety is a convex subset inside the rational flag complex, cf. [164], Cor. 2.16. The observation that in our case the slope function is affine-linear, and not only convex (as is typical in GIT), is due to B. Totaro.

In [171], the fundamental complex in the p-adic case is constructed in the context of Huber's adic spaces. In Huber's setting, Y is a *pseudo-adic* subset of $\mathcal{F}^{ad}$ and is endowed with an étale site. Furthermore, in order to deal with the infinitely many summands in the naive analogue of the finite field case, in [171] a mixture is used between the direct product and the direct sum, which makes use of the topology of the index set. It is an exercise to check that the complex (1.4) above is the same as the one in [171], up to translating between adic and analytic spaces. In [171] it is shown that Theorem 11.1.8 holds for arbitrary overconvergent sheaves on the flag variety. A variant of this construction, leading to the determination of the *continuous* ℓ-adic cohomology, is given in [174].

2 The relation to Deligne–Lusztig varieties

Besides period domains, there is another class of varieties attached to a reductive group over a finite field, Deligne–Lusztig varieties (DL-varieties, for short). These varieties were introduced by Deligne and Lusztig in [58] with the aim of constructing representations of the corresponding finite group of Lie type. Let us recall their definition. In this section we denote the finite ground field by k, and we have fixed an algebraic closure $\bar{k}$. We adopt the convention of [58], by which a variety is identified with its set of points over $\bar{k}$.

Let $k = \mathbb{F}_q$ be a finite field with q elements, where q is a power of a prime number p. Let G be a reductive group over $\mathbb{F}_q$. Let $T \subset G$ be a maximal torus contained in a Borel subgroup, and fix such a Borel subgroup B. We denote by W the Weyl group of G and by $X = X_G = G/B$ the corresponding complete flag variety. There is a bijection $G\backslash(X \times X) \cong W$ which gives rise to *the relative position map* (associate to an element of $X \times X$ the G-orbit containing it)

$$\mathrm{inv} : X \times X \to W.$$

Definition 11.2.1 Let $w \in W$. The *Deligne–Lusztig variety* associated to (G, w) is the locally closed subset of X given by

$$X(w) = X_G(w) = \{x \in X \mid \mathrm{inv}(x, Fx) = w\}\,.$$

Here $F : X \to X$ denotes the Frobenius map over $\mathbb{F}_q$.

It is known ([58], 1.4) that $X(w)$ is a smooth quasi-projective variety of dimension $\ell(w)$, which is equipped with an action of $G(\mathbb{F}_q)$. If F^e is the minimal power of F with $F^e(w) = w$, then $X(w)$ is defined over $\mathbb{F}_{q^e}$. However, in this section, all varieties will be considered over the algebraic closure $\bar{k}$ of $\mathbb{F}_q$.

Recall the Drinfeld space $\Omega^n_k = \Omega(k^n)$ from Example 2.1.7. It turns out that the Drinfeld space $\Omega^n_{\mathbb{F}_q}$ is isomorphic to a DL-variety.

Example 11.2.2 Let $G = \mathrm{GL}_n$. Then $W = S_n$. Let $w = s_1 s_2 \cdots s_{n-1} = (1, 2, \ldots, n) \in S_n$ be the *standard Coxeter element*. Let $V = \bar{k}^n$. Then

$$\begin{aligned} X_G(w) \;\; = \;\; & \{(0) \subset \mathcal{V}_1 \subset \mathcal{V}_2 \subset \cdots \subset \mathcal{V}_{n-1} \subset \mathcal{V}_n = V \mid \\ & \qquad F(\mathcal{V}_i) \subset \mathcal{V}_{i+1},\ F(\mathcal{V}_i) \neq \mathcal{V}_i,\ \forall i\}. \end{aligned}$$

Thus, via the projection map $X \longrightarrow \mathbb{P}^{n-1}$ which maps a complete flag $\mathcal{V}_\bullet$ to $\mathcal{V}_1$, the DL-variety $X_G(w)$ can be identified with $\Omega^n_{\mathbb{F}_q}$, cf. [58], §2. The Drinfeld space $\Omega^n_{\mathbb{F}_q}$, being the open complement in $\mathbb{A}^{n-1}$ of all hyperplanes defined over $\mathbb{F}_q$, can be identified with the affine variety, cf. [58], (2.2.2),

$$\Omega^n_{\mathbb{F}_q} = \mathrm{Spec}\left[\bar{k}[X_1, \ldots, X_n]/\left(\left(\det(X_i^{q^{j-1}})_{i,j=1,\ldots,n}\right)^{q-1} - (-1)^{n-1}\right)\right]^{\mathbb{F}^\times_{q^n}}.$$

For any Coxeter element w for GL_n, the corresponding DL-variety $X_G(w)$ is universally homeomorphic to $\Omega^n_{\mathbb{F}_q}$, cf. [155], Prop. 1.10.

Here is another example which brings out the seeming similarity between DL-varieties and period domains.

Example 11.2.3 Let $G = \mathrm{GL}_3$. Let $V = \bar{k}^3$. The DL-variety attached to the longest element $w = w_0$ in S_3 is given as follows.

$$X(w) = \{(0) \subset \mathcal{V}_1 \subset \mathcal{V}_2 \subset V \mid F\mathcal{V}_1 + \mathcal{V}_2 = V, \mathcal{V}_1 + F\mathcal{V}_2 = V\}.$$

For a period domain attached to GL_3, we take up Example 2.1.9, but we rewrite it in terms of the Frobenius map. So let $\nu = (x_1 > x_2 > x_3)$. Then

$$\mathcal{F}(\nu)^{\mathrm{ss}} =$$

$$\left\{(0) \subset \mathcal{V}_1 \subset \mathcal{V}_2 \subset V \;\middle|\; \begin{array}{l} x_1 - x_2 > x_2 - x_3 : \mathcal{V}_1 + F\mathcal{V}_1 + F^2\mathcal{V}_1 = V \\ x_1 - x_2 < x_2 - x_3 : \mathcal{V}_2 \cap F\mathcal{V}_2 \cap F^2\mathcal{V}_2 = (0) \\ x_1 - x_2 = x_2 - x_3 : \mathcal{V}_1 \cap F\mathcal{V}_1 = (0), \mathcal{V}_2 + F\mathcal{V}_2 = V \end{array}\right\}.$$

It turns out that this superficial similarity is quite misleading. In fact, it turns out that the Drinfeld space is essentially the only period domain that is at the same time a DL-variety.

Let us formulate this precisely. Recall that DL-varieties only depend on the adjoint group and are compatible with products, just as is the case for period domains. Now let $G = \mathrm{Res}_{\mathbb{F}_{q'}|\mathbb{F}_q}(G')$ be $\mathbb{F}_q$-simple and adjoint. Here G' is an absolutely simple adjoint group over $\mathbb{F}_{q'}$. Any conjugacy class $\mathcal{N}$ of $\mathbb{Q}$-1-PS of G over $\bar{k}$ is of the form $(\mathcal{N}_1, \ldots, \mathcal{N}_t)$ where $t = [\mathbb{F}_{q'} : \mathbb{F}_q]$ and $\mathcal{N}_1, \ldots, \mathcal{N}_t$ are conjugacy classes of $\mathbb{Q}$-1-PS G'. In the case of PGL_n, we identify as usual the set of conjugacy classes $\mathcal{N}$ with elements in

$$(\mathbb{Q}^n)^0_+ = \left\{(x_1, \ldots, x_n) \in \mathbb{Q}^n \mid x_1 \geq x_2 \geq \cdots \geq x_n, \sum_i x_i = 0\right\}.$$

Proposition 11.2.4 ([168]) *Let G be absolutely simple of adjoint type over $\mathbb{F}_q$. A Deligne–Lusztig variety $X_G(w)$ is never universally homeomorphic to a period domain of the form $\mathcal{F}(G, \mathcal{N})^{\mathrm{ss}}$, unless $G = \mathrm{PGL}_n$, and w is a Coxeter element and $\mathcal{N}$ corresponds to $\nu \in (\mathbb{Q}^n)^0_+$ of the form $\nu = (x, y^{(n-1)})$ or $\nu = (x^{(n-1)}, y)$ with $x > y$, in which case $X_G(w)$ and $\mathcal{F}(G, \mathcal{N})^{\mathrm{ss}}$ are both universally homeomorphic to $\Omega^n_{\mathbb{F}_q}$.*

More generally, let $G = \mathrm{Res}_{\mathbb{F}_{q'}|\mathbb{F}_q}(G')$ be simple of adjoint type, and let $t = [\mathbb{F}_{q'} : \mathbb{F}_q]$. Then a Deligne–Lusztig variety $X_G(w)$ is never universally homeomorphic to a period domain $\mathcal{F}(G, \mathcal{N})^{\mathrm{ss}}$, unless $G' = \mathrm{PGL}_n$, and w is a Coxeter element in the sense of [155], 1.7, *and $\mathcal{N}$ is of the form $(\nu_1, \ldots, \nu_t) \in ((\mathbb{Q}^n)^0_+)^t$ with $\nu_i = 0$ for all indices $i = 1, \ldots, t$, except one index, where the entry is of* *the form* $(x, y^{(n-1)})$ *or* $(x^{(n-1)}, y)$ *with* $x > y$. *In this case* $X_G(w)$ *and* $\mathcal{F}(G, \mathcal{N})^{ss}$ *are both universally homeomorphic to* $\Omega^n_{\mathbb{F}_{q'}}$. □

This proposition comes about by comparing a cohomology vanishing theorem for DL-varieties with the cohomology non-vanishing theorem Corollary 7.2.7 for period domains. Recall that this non-vanishing says that if G is a k-simple group of adjoint type and $\mathcal{N}$ is non-trivial, then $H^d_c(\mathcal{F}(G, \mathcal{N})^{ss}) \neq 0$, where $d = |\Delta_0|$ is the semi-simple k-rank of G. On the other hand, there is the following vanishing theorem for DL-varieties.

Proposition 11.2.5 *Let* $w \in W$. *Then*

$$H^i_c(X_G(w)) = 0 \text{ for } 0 \leq i < \ell(w).$$

Remark 11.2.6 This vanishing theorem is due to Digne, Michel and Rouquier [69], Cor. 3.3.22. When $q \geq h$ (where h denotes the Coxeter number of G) then all DL-varieties $X_G(w)$ are affine, cf. [58], Thm. 9.7. In this case, the vanishing statement follows by Poincaré duality from a general vanishing theorem for the étale cohomology of affine varieties [7]. In fact, in [168], it is proved in general that

$$H^i_c(X_G(w), \mathcal{F}) = 0 \quad \text{for } 0 \leq i < \ell(w),$$

for any smooth *prime-to-p* $\mathbb{Q}_\ell$-sheaf $\mathcal{F}$ on $X_G(w)$. This applies in particular to the Deligne–Lusztig sheaves $\mathcal{F}_\theta$ attached to an ℓ-adic character θ of $T(\bar{k})^{wF}$.

In order to play out the vanishing theorem for DL-varieties against the non-vanishing theorem for period domains, the following observation ([168]) is used.

Proposition 11.2.7 *Let G be a simple group of adjoint type over an algebraically closed field k. For any proper parabolic subgroup P, the following inequality holds,*

$$\mathrm{rk}(G) \leq \dim G/P,$$

with strict inequality, except when $G = \mathrm{PGL}_n$ *and P is a parabolic subgroup of type* $(n-1, 1)$ *or* $(1, n-1)$. □

Remark 11.2.6 leads to the question of which DL-varieties or period domains are affine. For period domains, it turns out that only the Drinfeld space has this property. The precise statement is the following proposition.

Proposition 11.2.8 *Let* $k = \mathbb{F}_q$, *and let G be a k-simple adjoint group, of the form* $G = \mathrm{Res}_{\mathbb{F}_{q'}|\mathbb{F}_q}(G')$ *for an absolutely simple* $\mathbb{F}_{q'}$*-group* G'. *Let* $\mathcal{N}$ *be* *non-trivial. Then* $\mathcal{F}(G,\mathcal{N})^{ss}$ *is affine if and only if* $G' = \mathrm{PGL}_n$ *and* $\mathcal{N} = (\mathcal{N}_1,\ldots,\mathcal{N}_t)$*, where all entries* $\mathcal{N}_i$ *are trivial except one, which is of Drinfeld type, i.e., of the form* $(x^{(n-1)},y)$ *or* $(x,y^{(n-1)})$*, for* $x > y$*. In this last case* $\mathcal{F}(G,\mathcal{N})^{ss} \simeq \Omega^n_{\mathbb{F}_{q'}}$.

Proof The last assertion is easy. It implies the "if" direction, since the Drinfeld space is affine, cf. Example 11.2.2.

Now let us prove the converse. If $\mathcal{F}(G,\mathcal{N})^{ss}$ is affine, then we deduce by the general vanishing property for the étale cohomology of affine varieties and Corollary 7.2.7 that $\dim \mathcal{F}(G,\mathcal{N})^{ss} \leq d = |\Delta_0|$. Further $\mathcal{F}(G,\mathcal{N})^{ss}$ is open in $\mathcal{F}(G,\mathcal{N})$, hence $\dim \mathcal{F}(G,\mathcal{N})^{ss} = \dim \mathcal{F}(G,\mathcal{N})$.

First case: G absolutely simple adjoint Let $\nu \in \mathcal{N}$ be a representative. Then we have $\dim \mathcal{F}(G,\mathcal{N}) = \dim G/P_\nu$ and $d \leq r = \mathrm{rk}(G)$ (= absolute rank of G). We get by Proposition 11.2.7 a chain of inequalities

$$\dim \mathcal{F}(G,\mathcal{N})^{ss} \leq d \leq r \leq \dim \mathcal{F}(G,\mathcal{N})\,. \tag{2.1}$$

Hence $d = r = \dim G/P_\nu$, and $G = \mathrm{PGL}_n$ and P_ν is a parabolic subgroup of type $(n-1,1)$ or $(1,n-1)$. This then means that $\mathcal{N}$ is of Drinfeld type, and the result follows.

Second case: $G = \mathrm{Res}_{\mathbb{F}_{q'}|\mathbb{F}_q}(G')$ *with* G' *absolutely simple adjoint* Let $r = \mathrm{rk}(G')$ be the absolute rank of G'. Let $t_1 \geq 1$ be the number of indices i, where $\mathcal{N}_i$ is nontrivial. The inequality (2.1) above is now replaced by

$$\dim \mathcal{F}(G,\mathcal{N})^{ss} \leq d \leq rt_1 \leq \dim \mathcal{F}(G,\mathcal{N})\,. \tag{2.2}$$

Since $d \leq r$, we deduce from the fact all inequalities in (2.2) above are equalities, that $d = r$ and $t_1 = 1$. As in the absolutely simple case we deduce that $G' = \mathrm{PGL}_n$, and that for the one index i with non-trivial $\mathcal{N}_i$ this conjugacy class of PGL_n corresponds to $(x,y^{(n-1)})$ or $(x^{(n-1)},y)$ with $x > y$. The corresponding period domain is $\Omega^n_{\mathbb{F}_{q'}}$. □

The question "which DL-varieties are affine?" has been open for more than 30 years. Let us list a few results in this direction.

(i) For $q \geq h - 1$, any DL-variety $X_G(w)$ is affine (Haastert [101]). This is a strengthening of the original Deligne–Lusztig result mentioned in Remark 11.2.6 above.
(ii) Any DL-variety is quasi-affine (i.e., open in an affine variety) (Haastert [102]).
(iii) If w is of minimal length in its F-conjugacy class, then $X_G(w)$ is affine [19, 112, 168].

(iv) Assume that there exists an F-stable subset I of the set S of simple roots, and that w lies in the corresponding parabolic subgroup W_I. Assume further that there exists a non-negative integer d such that there is an identity in the braid monoid associated to (W, S) of the form

$$\underline{w}F(\underline{w}) \cdot \cdots \cdot F^d(\underline{w}) = \underline{w}_I \cdot a$$

for a suitable element a. Then $X_G(w)$ is affine. Here w_I denotes the longest element of W_I. This condition implies the condition in (iii), see [19].

(v) So far not a single DL-variety is known which is definitely not affine, but many potential candidates exist over $\mathbb{F}_2$: see the Notes and References.

DL-varieties have interesting étale coverings which give rise to the smooth ℓ-adic sheaves $\mathcal{F}_\theta$ mentioned above, and which in turn serve to construct interesting representations of $G(\mathbb{F}_q)$. One may ask whether period domains can be used in a similar way. The following proposition shows that in this respect also, period domains are quite different from DL-varieties.

Proposition 11.2.9 ([175]) *Let G be absolutely simple adjoint over k. Then $\mathcal{F}(G, \mathcal{N})^{ss}$ is simply connected, unless $G = \mathrm{PGL}_n$ and $\nu = (x_1 \geq x_2 \geq \cdots \geq x_n) \in (\mathbb{Q}^n)^0_+$ with $x_2 < 0$ or $x_{n-1} > 0$. In the latter case we have $\pi_1(\mathcal{F}(G, \mathcal{N})^{ss}) = \pi_1(\Omega^n_k)$. More precisely, in the latter case, there is a morphism*

$$\mathcal{F}(G, \mathcal{N})^{ss} \longrightarrow \Omega^n_k$$

with fibers in partial flag varieties. This morphism is an isomorphism if and only if $\mathcal{N}$ is of Drinfeld type, i.e., $\nu = (x^{(n-1)}, y)$, or $\nu = (x, y^{(n-1)})$ for $x > y$.

More generally, let $G = \mathrm{Res}_{k'|k} G'$ be k-simple adjoint. Then $\mathcal{F}(G, \mathcal{N})^{ss}$ is simply connected, unless $G' = \mathrm{PGL}_n$ and the following two conditions are satisfied. Write $\nu_i = (x_1^{[i]} \geq x_2^{[i]} \geq \cdots \geq x_n^{[i]}) \in (\mathbb{Q}^n)^0_+$, $i = 1, \ldots, t$. Then there is a unique j with $1 \leq j \leq t$, such that

(i) $x_2^{[j]} < 0$ *or* $x_{n-1}^{[j]} > 0$,

(ii) $\sum_{i \neq j} x_1^{[i]} < -x_2^{[j]}$ *if* $x_2^{[j]} < 0$, resp. $\sum_{i \neq j} x_n^{[i]} > -x_{n-1}^{[j]}$ *if* $x_{n-1}^{[j]} > 0$.

In the latter case we have $\pi_1(\mathcal{F}(G, \mathcal{N})^{ss}) = \pi_1(\Omega^n_{k'})$. More precisely, in the latter case, there is a morphism

$$\mathcal{F}(G, \mathcal{N})^{ss} \longrightarrow \Omega^n_{k'}$$

with fibers in products of partial flag varieties. This morphism is an isomorphism if and only if $\mathcal{N}_j$ is of Drinfeld type, i.e., $\nu_j = (x^{(n-1)}, y)$, or $\nu_j = (x, y^{(n-1)})$ for $x > y$ and $\mathcal{N}_i$ is trivial for all $i \neq j$.

The proof of this proposition is based on the following observation: $\mathcal{F}(G,\mathcal{N})$ is simply connected and hence $\mathcal{F}(G,\mathcal{N})^{ss}$ is simply-connected as well, provided that $\operatorname{codim} Y \geq 2$, cf. [1], Ch. XI, Cor. 1.2. So we are led to investigate the codimension of Y. We sketch the method.

By identity (1.2), page 322, we know that

$$Y = \bigcup\nolimits_{\alpha\in\Delta_0} \bigcup\nolimits_{g\in G(k)} gY_{\Delta_0\setminus\{\alpha\}}.$$

We conclude that

$$\dim Y = \max\nolimits_{\alpha\in\Delta_0} \dim Y_{\Delta_0\setminus\{\alpha\}}.$$

On the other hand, each subvariety $Y_{\Delta\setminus\{\alpha\}}$ is by Proposition 11.1.6 a union of the Schubert cells BwP/P, where $P = P_\nu$. Here w ranges over those elements in the set of shortest length representatives W^P of cosets in W/W_P with $(\omega_\alpha, w\nu) > 0$. The dimension of BwP/P is $\ell(w)$, cf. [25]. Hence

$$\dim Y_{\Delta_0\setminus\{\alpha\}} = \max\{\ell(w) \mid w \in W^P,\ (\omega_\alpha, w\nu) > 0\}. \tag{2.3}$$

Let w_0 resp. w_0^P be the longest element of the Weyl group W resp. of W^P. Then $w_0 = w_0^P \cdot w_P$, where w_P is the longest element in W_P. Then $w_0\nu = w_0^P\nu$ and $\dim \mathcal{F}(G,\mathcal{N}) = \ell(w_0^P)$. Now one examines by explicit computation in root systems when it happens that $\dim Y = \dim \mathcal{F}(G,\mathcal{N}) - 1$, i.e., when there are pairs of simple roots $\alpha, \beta \in \Delta_0$ such that $(\omega_\alpha, s_\beta \cdot w_0\nu) > 0$. This leads to the condition in Proposition 11.2.9: in all cases, except the indicated exceptions, $\operatorname{codim} Y \geq 2$, and hence $\mathcal{F}(G,\mathcal{N})^{ss}$ is simply connected.

Now let us discuss the cases when $\mathcal{F}(G,\mathcal{N})^{ss}$ is not simply connected. Let us first suppose that G is absolutely simple and adjoint. Then $G = \mathrm{PGL}_n$ and $\nu = (x_1, x_2, \ldots, x_n) \in (\mathbb{Q}^n)^0_+$ with $x_2 < 0$ or $x_{n-1} > 0$. We rewrite ν in the form $\nu = (y_1^{(n_1)}, \ldots, y_r^{(n_r)})$, with $y_1 > y_2 > \cdots > y_r$ and $n_i \geq 1$, $i = 1, \ldots, r$. Let $V = \bar{k}^n$. Then $\mathcal{F} = \mathcal{F}(G,\mathcal{N})$ is the variety of filtrations

$$(0) \subset \mathcal{F}^{y_1} \subset \mathcal{F}^{y_2} \subset \cdots \subset \mathcal{F}^{y_r} = V$$

with

$$\dim \mathcal{F}^{y_i} = n_1 + \cdots + n_i.$$

If $x_2 < 0$ then $n_1 = 1$ resp. if $x_{n-1} > 0$ then $n_r = 1$. Then one computes easily that

$$\begin{aligned} \mathcal{F}(G,\mathcal{N})^{ss} &= \{\mathcal{F}^\bullet \in \mathcal{F}(G,\mathcal{N}) \mid \mathcal{F}^{y_1} \in \Omega_k^n\} && \text{if } x_2 < 0 \\ \mathcal{F}(G,\mathcal{N})^{ss} &= \{\mathcal{F}^\bullet \in \mathcal{F}(G,\mathcal{N}) \mid \mathcal{F}^{y_{r-1}} \in \check{\Omega}_k^n\} && \text{if } x_{n-1} > 0\,. \end{aligned}$$

Hence the projections

$$\begin{array}{ccc} \mathcal{F} \longrightarrow \mathbb{P}^{n-1} & \text{resp.} & \mathcal{F} \longrightarrow \check{\mathbb{P}}^{n-1} \\ \mathcal{F}^\bullet \mapsto \mathcal{F}^{y_1} & & \mathcal{F}^\bullet \mapsto \mathcal{F}^{y_{r-1}} \end{array}$$

induce surjective proper maps

$$\mathcal{F}(G,\mathcal{N})^{\mathrm{ss}} \longrightarrow \Omega^n_k \quad \text{resp.} \quad \mathcal{F}(G,\mathcal{N})^{\mathrm{ss}} \longrightarrow \check{\Omega}^n_k, \tag{2.4}$$

in which the fibers are generalized flag varieties. Furthermore, these projections are isomorphisms precisely if $\mathcal{F}^{y_2} = V$, resp. if $\mathcal{F}^{y_{r-2}} = (0)$. These two cases correspond to the Drinfeld cases.

Now let $G = \mathrm{Res}_{\mathbb{F}_{q'}|\mathbb{F}_q}(G')$, with $G' = \mathrm{PGL}_n$ and $t = [\mathbb{F}_{q'} : \mathbb{F}_q]$ and $\nu = (\nu_1, \ldots, \nu_t)$ such that $\pi_1(\mathcal{F}(G,\mathcal{N})^{\mathrm{ss}}) \neq \{1\}$. After reindexing we may suppose that $j = 1$, i.e., $\nu_1 \in (\mathbb{Q}^n)^0_+$ is the vector with $\sum_{i\neq 1} x_1^{[i]} < -x_2^{[1]}$ or $\sum_{i\neq 1} x_n^{[i]} > -x_{n-1}^{[1]}$, depending on whether $x_2^{[1]} < 0$ or $x_{n-1}^{[1]} > 0$. The flag variety $\mathcal{F}(G,\mathcal{N})$ is the product

$$\mathcal{F}(G,\mathcal{N}) = \prod_{j=1}^{t} \mathcal{F}(\mathrm{PGL}_n, \mathcal{N}_j),$$

where $\mathcal{N}_j$ is the PGL_n-conjugacy class of ν_j. Let $\nu_1 = (y_1^{(n_1)}, \ldots, y_r^{(n_r)})$ with $y_1 > y_2 > \cdots > y_r$ and $n_i \geq 1$, $i = 1, \ldots, r$. The corresponding period domain is then (cf. [175])

$$\mathcal{F}(G,\mathcal{N})^{\mathrm{ss}} = \mathcal{F}(\mathrm{PGL}_n, \mathcal{N}_1)^{\mathrm{ss}} \times \prod_{j\geq 2} \mathcal{F}(\mathrm{PGL}_n, \mathcal{N}_j).$$

In the case $\sum_{i\neq 1} x_1^{[i]} < -x_2^{[1]}$, we have

$$\mathcal{F}(\mathrm{PGL}_n, \mathcal{N}_1)^{\mathrm{ss}} = \{\mathcal{F}^\bullet \in \mathcal{F}(\mathrm{PGL}_n, \mathcal{N}_1) \mid \mathcal{F}^{y_1} \in \Omega^n_{k'}\}.$$

For $\sum_{i\neq 1} x_n^{[i]} > -x_{n-1}^{[1]}$, we have

$$\mathcal{F}(\mathrm{PGL}_{n,k'}, \mathcal{N}_1)^{\mathrm{ss}} = \{\mathcal{F}^\bullet \in \mathcal{F}(\mathrm{PGL}_n, \mathcal{N}_1) \mid \mathcal{F}^{y_{r-1}} \in \check{\Omega}^n_{k'}\}.$$

Now the rest of the argument is as in the absolutely simple case.

Remark 11.2.10 The above proposition again shows that period domains are never affine outside the case of the Drinfeld space, cf. Proposition 11.2.8. Indeed, if $\mathcal{F}^{\mathrm{ss}}$ is affine, then for any open dense embedding of $\mathcal{F}^{\mathrm{ss}}$ into a scheme, the codimension of its complement is at most 1, cf. EGA IV, 21.12.7. It follows that we are in the exceptional cases of Proposition 11.2.9. If $\mathcal{F}^{\mathrm{ss}}$ is affine, it does not contain any positive-dimensional projective subvariety. Hence the morphism from $\mathcal{F}^{\mathrm{ss}}$ to the Drinfeld space in Proposition 11.2.9 is an isomorphism.

Proposition 11.2.9 also gives a variant of Proposition 11.2.4. We first state the following result.

Lemma 11.2.11 *Let $w \in W$. Then $|T(\bar{k})^{wF}| > 1$, unless $q = 2$ and the torus over k, obtained by twisting T with w, is split.*

Proof We use the notation of the proof of Proposition 7.1.4. Thus $Y = X_*(T)$ and the action of F on Y is written as $F = q \cdot \sigma$, where σ is an automorphism of finite order. Then $T(\bar{k})^{wF}$ can be identified with $Y/(w^{-1}F - 1)Y$, cf. [58]. So, we have to investigate when $w^{-1}F - 1$ is an automorphism of Y. Let $\alpha = w^{-1}\sigma$. Then α is of finite order. If $q\alpha - 1$ is an automorphism of Y, then all its eigenvalues have absolute value 1. An eigenvector equation $(q\alpha - 1)v = \lambda v$ gives

$$(\lambda + 1)v = q\alpha(v).$$

Since α is of finite order, we obtain $|\lambda + 1| = q$. Together with $|\lambda| = 1$, this implies $q = 2$ and $\lambda = 1$. Hence $w^{-1}F = q$, and this implies the assertion. □

Let $w \in W$. Let $\tilde{G}$ be the simply-connected cover of the adjoint group G. Then $X_{\tilde{G}}(w) = X(w) = X_G(w)$. Let $\tilde{T}$ be the inverse image of T in $\tilde{G}$. Recall the $\tilde{T}(\bar{k})^{wF}$-torsor $\pi : \tilde{X}(\dot{w}) \longrightarrow X(w)$ defined by Deligne and Lusztig [58]. Up to isomorphism it is independent of the representative $\dot{w}$ of w. It is equipped with an action of $\tilde{G}(k) \times \tilde{T}(\bar{k})^{wF}$ such that the map π is $\tilde{G}(k)$-equivariant. The following proposition is due to Bonnafé and Rouquier [20].

Proposition 11.2.12 *Let $X(w)$ be connected. Then $\tilde{X}(\dot{w})$ is connected.* □

Suppose that $X_G(w)$ is a period domain. Hence $X_G(w)$ is connected, and hence, by Proposition 11.2.12, so is $\tilde{X}(\dot{w})$. On the other hand, since $X_G(w)$ is connected, w is an elliptic element of W, cf. [21, 94]. Hence by Lemma 11.2.11 (applied to $\tilde{T}$), the étale covering $\tilde{X}(\dot{w})$ is non-trivial, and hence $X_G(w)$ is not simply connected. Let us apply Proposition 11.2.9. We are in the exceptional case of this proposition. But $X_G(w)$ is quasi-affine [102], and hence contains no projective subvarieties of positive dimension. This implies that the morphism to the Drinfeld space in Proposition 11.2.9 is an isomorphism. Hence $X_G(w)$ is isomorphic to the Drinfeld space.

So, the upshot is that in the conclusion of Proposition 11.2.4, if $X_G(w)$ is isomorphic to a period domain $\mathcal{F}(G, \mathcal{N})^{ss}$, then it is isomorphic to $\Omega^n_{\mathbb{F}_{q'}}$.

Notes and References The property of a DL-variety $X_G(w)$ to be affine, only depends on the *cyclic shift* class of w, cf. [168]. The statement in item (iii) right before Proposition 11.2.9 was proved in [168] for split classical groups, checking case by case that an element of minimal length in its F-conjugacy class is a cyclic shift of an element which satisfies the combinatorial Deligne–Lusztig criterion for affineness [58], 9.6. This method was simplified and extended by He [112] to all cases. Statement (iv) is due to Bonnafé/ Rouquier [19] who also show that it implies (iii). However, this deduction is implicitly also based on case distinctions, and even computer calculations. The condition in (iv) is natural from the point of view of Broué's *abelian defect conjecture*, cf. the Introduction of [19]. Calculations of C. Bonnafé and X. He seem to indicate that not only the elements satisfying the hypothesis in item (iii), but also those that satisfy the hypothesis in item (iv) are all cyclic shifts of elements which satisfy the DL-criterion. The following DL-data (G, w) (for $q = 2$) are candidates for non-affine DL-varieties (He [113]).

(i) G of type G_2, $w = s_1s_2s_1$ or $w = s_2s_1s_2$.
(ii) G of type A_5 or ${}^2\mathrm{A}_5$, $w = s_1s_3s_5s_2s_4s_1s_3s_5$.
(iii) G of type B_3, $w = s_1s_3s_2s_1s_3$.
(iv) G of type D_4 or ${}^2\mathrm{D}_4$ or ${}^3\mathrm{D}_4$, $w = s_1s_3s_4s_2s_1s_3s_4$.

Also, for G of type A_n, or ${}^2\mathrm{A}_n$ for $n \geq 5$, G of type B_n or C_n for $n \geq 3$, G of type D_n or ${}^2\mathrm{D}_n$ for $n \geq 4$, G of type E_6 or ${}^2\mathrm{E}_6$, and G of type E_7, E_8, F_4 there are always DL-data (G, w) (for $q = 2$) for which $X_G(w)$ are candidates for non-affine DL-varieties. This list of candidates for non-affine DL-varieties given above are elements such that no cyclic shift of them satisfies the DL-criterion. It is remarkable that the ground field is $\mathbb{F}_2$ in all these candidates for non-affine DL-varieties.

We note that the vanishing theorem for DL-local systems, mentioned in Remark 11.2.6, substantially simplifies Lusztig's classification of representations of finite groups of Lie type, cf. Introduction of [156] (modulo the computer calculations mentioned above).

Proposition 11.2.9 does not give a complete answer to the problem of determining the fundamental groups of period domains over finite fields, since the fundamental group of $\Omega^n_{\mathbb{F}_q}$ seems to be unknown. A natural guess would be that the Deligne–Lusztig étale Galois cover with Galois group $\mathbb{F}^\times_{q^n}$ is the universal *tame* Galois covering of $\Omega^n_{\mathbb{F}_q}$.

Over a p-adic field, some of the above statements still make sense or admit natural analogues. For example, one might ask for a classification of those period domains that are Stein spaces. Is the Drinfeld space the only example? Also one might want to compute the π_1 of a period domain. But here the first crucial difference is that flag varieties are not simply connected any longer. For example, the Gross–Hopkins period map defines a surjective homomorphism $\pi_1(\mathbb{P}^{n-1}_{K_0(\overline{\mathbb{F}}_p)}) \twoheadrightarrow \mathrm{SL}_n(\mathbb{Q}_p)$, see [50], Prop. 7.4. On the other hand, the algebraic part π_1^{alg} (see [50]) of flag varieties is trivial and one may hope for an explicit list of period domains with non-vanishing π_1^{alg}, similar to that of Proposition 11.2.9.

3 The Drinfeld space for a p-adic field

In this section K will denote a p-adic field, i.e., a finite extension of $\mathbb{Q}_p$. Our aim is to describe some noteworthy features of the Drinfeld space $\Omega_K^n = \Omega(K^n)$ relative to K, cf. Example 9.8.9. These features are either not shared by other period domains over K, or unknown for them.

One striking feature is the connection of Ω_K^n to the Bruhat–Tits building of $\mathrm{PGL}_n(K)$. Let us recall briefly the definition of the Bruhat–Tits building of $G(K) = \mathrm{PGL}_n(K)$. Let $O = O_K$ be the ring of integers and denote by $\pi \in O_K$ a uniformizing element. Fix a norm $|\ | : K \longrightarrow \mathbb{R}_{\geq 0}$ on K such that $q = |\pi^{-1}|$ is a power of p. The Bruhat–Tits building $\mathcal{B}$ of $G(K)$ is the simplicial complex whose vertices are given by similarity classes $[L]$ of O-lattices in K^n. Here two lattices L and L' are called *similar* if there is $a \in K^\times$ with $a \cdot L_1 = L_2$. The m-simplices of $\mathcal{B}$ are defined to be sets $\{[L_0], \ldots, [L_m]\}$ of $m+1$ similarity classes of lattices satisfying

$$L_0 \subsetneq L_1 \subsetneq \cdots \subsetneq L_m \subsetneq \pi^{-1} L_0.$$

This simplicial complex is equipped with a natural action of $G(K) = \mathrm{PGL}_n(K)$. The stabilizers of finite subsets in $\mathcal{B}$ are compact open subgroups of $G(K)$, cf. [216], 3.2.

Another approach to the Bruhat–Tits building is by homothety classes of norms on the K-vector space K^n. Recall that a *norm* on a K-vector space V is a map

$$\varsigma : V \longrightarrow \mathbb{R}_{\geq 0}$$

satisfying the following conditions:

(i) $\varsigma(\lambda \cdot v) = |\lambda| \cdot \varsigma(v)$ for all $\lambda \in K$ and $v \in V$
(ii) $\varsigma(v + w) \leq \sup\{\varsigma(v), \varsigma(w)\}$
(iii) $\varsigma(v) > 0$ for all $v \in V \setminus \{0\}$.

We denote by $\mathcal{N}$ the topological space of norms on K^n. Here the topology is defined by the metric on $\mathcal{N}$,

$$d(\varsigma_1, \varsigma_2) = \log_q \Big(\sup_{K^n} \frac{\varsigma_1}{\varsigma_2} \cdot \sup_{K^n} \frac{\varsigma_2}{\varsigma_1} \Big).$$

In [92] Goldmann and Iwahori show that the geometric realization $|\mathcal{B}|$ of $\mathcal{B}$ can be identified G-equivariantly with the topological space $\mathcal{N}/\mathbb{R}_{>0}$ of homothety classes of norms on K^n. The latter space is equipped with the quotient topology.

Berkovich [15] defines a continuous $G(K)$-equivariant map

$$\tau : \Omega_K^n \longrightarrow |\mathcal{B}|.$$

Let us recall its definition. Let $R = K[T_1, \ldots, T_n]$ be the polynomial ring over K in n indeterminates. We identify $\mathcal{N}$ with the space of norms on the K-subspace $R_1 = K \cdot T_1 \oplus \cdots \oplus K \cdot T_d$ of linear forms on K^n. Let $\tilde{\Omega}^n_K$ be the preimage of Ω^n_K under the projection map $(\mathbb{A}^n_K)^{an} \setminus \{0\} \longrightarrow (\mathbb{P}^{n-1}_K)^{an}$. Recall [12], §1.5, that $(\mathbb{A}^n_K)^{an}$ is by definition the set of multiplicative semi-norms on R which extend the fixed norm on K. Then we may identify $\tilde{\Omega}^n_K$ with the open analytic subspace

$$\tilde{\Omega}^n_K = \{f \in (\mathbb{A}^n_K)^{an} \mid f_{|R_1} \text{ is a norm }\}.$$

Thus we get a natural map

$$\tilde{\Omega}^n_K \longrightarrow \mathcal{N},$$

which sends a semi-norm $x \in \tilde{\Omega}^n_K$ to its restriction to R_1. This map induces, via the identification $|\mathcal{B}| = \mathcal{N}/\mathbb{R}_{>0}$, a map on the quotients

$$\tau : \Omega^n_K \longrightarrow |\mathcal{B}|.$$

This map is continuous, proper and surjective [15]. On (classical) points $x = [x_1 : \cdots : x_n] \in \Omega^n_K(K')$, where K' is a finite extension of K, it is given by

$$\tau(x)(z) = \left|\sum\nolimits_i x_i z_i\right|, \qquad z = (z_1, \ldots, z_n) \in K^n.$$

Berkovich [15] constructs as follows a section j of τ. Let $\varsigma \in \mathcal{N}$. Let $e_1, \ldots, e_n$ be an orthogonal basis of R_1 relative to ς, i.e., $\varsigma(\sum_{i=1}^n a_i e_i) = \max |a_i|\varsigma(e_i)$. Then $\tilde{j}(\varsigma) \in \tilde{\Omega}^n_K$ is defined by

$$\left|\left(\sum\nolimits_\nu a_\nu \mathbf{e}^\nu\right)(\tilde{j}(\varsigma))\right| = \max\nolimits_\nu |a_\nu| \varsigma(\mathbf{e})^\nu.$$

The section $j : |\mathcal{B}| \longrightarrow \Omega^n_K$ is then obtained from $\tilde{j}$ by dividing out scalars. Berkovich proves that $j \circ \tau$ is a $\mathrm{PGL}_n(K)$-equivariant proper retraction map, and deduces that Ω^n_K is contractible, [12], §6.1.

Using this retraction map and some function-theoretic arguments, Berkovich [15] then proves a version of the main theorem of projective geometry.

Proposition 11.3.1 *For any non-Archimedean extension K' of K*

$$\mathrm{Aut}(\Omega^n_K \widehat{\otimes}_K K') = \mathrm{PGL}_n(K).$$

We next address the question of whether there exists a quotient of Ω^n_K by a co-compact discrete subgroup of $\mathrm{PGL}_n(K)$.

Proposition 11.3.2 *Let $\Gamma \subset G(K)$ be a discrete subgroup. Then the following conditions are equivalent:*

(i) *the K-analytic space $\Gamma\backslash\Omega^n_K$ is proper;*
(ii) *the topological space $\Gamma\backslash|\mathcal{B}|$ is compact;*

(iii) Γ *is co-compact in* $\mathrm{PGL}_n(K)$.

In this case, the quotient $\Gamma\backslash\Omega^n_K$ *is the analytic space associated to a projective algebraic variety over* Spec K.

The equivalence of the three conditions comes from the properness of the map τ, cf. Berkovich [15]. The algebraization of the quotient is proved in the context of rigid-analytic varieties by Mustafin [165] and by Kurihara [140]. More precisely, $\Gamma\backslash\Omega^n_K$ has a π-adic formal scheme model which is proper and such that the relative dualizing sheaf is a relatively ample line bundle, cf. also Theorem 11.3.5 below. Therefore the assertion follows from Grothendieck's formal GAGA.

The quotient by Γ essentially determines Γ in the following sense.

Proposition 11.3.3 *Let* $\Gamma_1, \Gamma_2 \subset G(K)$ *be discrete co-compact subgroups. Then* $\Gamma_1\backslash\Omega^n_K$ *is isomorphic to* $\Gamma_2\backslash\Omega^n_K$ *as K-analytic spaces if and only if* Γ_1 *is conjugate to* Γ_2. *In this case the set of isomorphisms* $\Gamma_1\backslash\Omega^n_K \xrightarrow{\sim} \Gamma_2\backslash\Omega^n_K$ *is in bijection with*

$$C(\Gamma_1, \Gamma_2)/\Gamma_1,$$

where $C(\Gamma_1, \Gamma_2) = \{g \in \mathrm{PGL}_n(K) \mid g\Gamma_1 g^{-1} = \Gamma_2\}$.

Berkovich's [15] proof applies even to the equicharacteristic case; another proof is due to M. Ishida and F. Kato [123].

One may consider $(\mathbb{P}^{n-1}_K)^{an}$ as a compactification of Ω^n_K as $G(K)$-space. The points on the boundary are radically different from those in the interior, in the following sense.

Proposition 11.3.4 *Let* $x \in (\mathbb{P}^{n-1}_K)^{an}$. *Then the stabilizer* $G(K)_x$ *of* x *in* $G(K)$ *is compact if and only if* $x \in \Omega^n_K$.

Proof If $x \in \Omega^n_K$, then $G(K)_x$ is a closed subgroup of the stabilizer of $\tau(x) \in \mathcal{B}$. But the latter are compact subgroups of $\mathrm{PGL}_n(K)$.

Now suppose that $G(K)_x$ is compact, but that $x \notin \Omega^n_K$. Let $C = \widehat{\overline{\mathcal{H}(x)}}$ be the completion of an algebraic closure of the residue field $\mathcal{H}(x)$ of x. Then we may identify x with a line in the C-vector space C^n. Since $x \notin \Omega^n_K$, there exists a proper K-rational subspace V_x of C^n containing x. Let $P = P(V_x)$ be the maximal K-psgp stabilizing V_x. Its unipotent radical U acts via the identity on V_x. Thus it fixes the point x as well. Hence we see that $U(K) \subset G(K)_x$. But $U(K)$ is certainly not compact, a contradiction. □

Finally, we mention the existence of formal models of Ω^n_K, in the following sense.

Theorem 11.3.5 *There exists a π-adic strictly semi-stable formal scheme $\hat{\Omega}^n_{O_K}$ locally of finite type over* Spf O_K *whose associated analytic space is Ω^n_K. It is equipped with an action of* $\mathrm{GL}_n(K)$ *which induces the one on Ω^n_K.*

In fact, the formal model $\hat{\Omega}^n_{O_K}$ we have in mind is the Deligne–Drinfeld model [71]. There are a number of other constructions of it [165, 140, 163, 220], but they all yield the same model.

This concludes our list of special features of the Drinfeld space. What about other period domains? As the previous section has amply demonstrated, the Drinfeld space occupies a very special position among all period domains, and therefore caution has to be exercised when trying to extend some of the above results from the Drinfeld space to other period domains. We mention only one result in this direction.

Let G be a split simply connected simple group G over K, and let $\mathcal{N}$ be a conjugacy class of $\mathbb{Q}$-1-PS of G. Let $\mathbb{G} = \mathbb{G}_1 = (G, 1)$ be the associated a.g.s. over the tannakian category σ_K-K_K-Sp, cf. Chapter IX, §8. Just as in the case of the Drinfeld space (Example 9.8.9) we associate the period domain $\mathcal{F}(\mathbb{G}, \mathcal{N})^{ss}$ to the pair $(\mathbb{G}, \mathcal{N})$. Van der Put and Voskuil [220] consider such period domains $(\mathcal{F}(\mathbb{G}, \mathcal{N})^{ss})^{rig}$ in the rigid-analytic sense. In [220] the question is raised of when it happens that the semi-stable points in $\mathcal{F}(G, \mathcal{N})$ are all stable, i.e. $\mathcal{F}(G, \mathcal{N})^{ss} = \mathcal{F}(G, \mathcal{N})^{s}$. Here a point is *stable* if in the Definition 1.1.13 of semi-stability for GL_n, the symbol $\leq$ is replaced by $<$ for non-trivial subobjects, in the usual way. The same modification applies to define the notion of stability for arbitrary reductive groups. Then it follows by the same proof as in Chapter II, §2, resp. Chapter VI, §2, that a point is stable if and only if the Hilbert–Mumford criterion is satisfied for all maximal K-tori of G (where $<$ is replaced by $\leq$ for all nontrivial $\mathbb{Q}$-1-PS). In fact, van der Put and Voskuil use the Hilbert–Mumford inequality as the *definition* of a stable point.

Example 11.3.6 Let $G = \mathrm{GL}_n$ and $\mathcal{N}$ be as in Example 9.8.9. Hence $\mathcal{F}(G, \mathcal{N})^{ss} = \Omega^d_K$. Then $\mathcal{F}(G, \mathcal{N})^{s} = \mathcal{F}(G, \mathcal{N})^{ss}$.

Proposition 11.3.7 ([220]) *Let G be a split simply connected simple group over K and let $\mathcal{N}$ be a conjugacy class of* $\mathbb{Q}$-1-PS *of G.*

(i) *Consider the root system with respect to a maximal torus $T \subset G$. Let $\nu \in \mathcal{N}$ be a representative with $\nu \in X_*(T)_{\mathbb{Q}}$. Then $\mathcal{F}(G, \mathcal{N})^{ss} = \mathcal{F}(G, \mathcal{N})^{s}$ if and only if ν is not contained in any hyperplane spanned by co-roots.*

(ii) *Let $\Gamma \subset G(K)$ be a discrete co-compact subgroup and assume $\mathcal{F}(G, \mathcal{N})^{s} = \mathcal{F}(G, \mathcal{N})^{ss}$. Then $\Gamma\backslash(\mathcal{F}(G, \mathcal{N})^{ss})^{rig}$ is a proper rigid-analytic variety.*

(iii) *In the case $\mathcal{F}(G, \mathcal{N})^{s} = \mathcal{F}(G, \mathcal{N})^{ss}$ there is a π-adic formal scheme locally of finite type over O_K whose generic fibre is $(\mathcal{F}(G, \mathcal{N})^{ss})^{rig}$.*

Remarks 11.3.8 (i) For proving part (i), van der Put and Voskuil use the approach via hypersimplices as in Proposition 2.4.11. If G is not of type A_n, then it can be shown that the hyperplane given by the orthogonal complement $\alpha^\perp \subset X_*(T)_\mathbb{Q}$ of a root α is spanned by co-roots. This implies that, if the identity $\mathcal{F}(G,\mathcal{N})^{ss} = \mathcal{F}(G,\mathcal{N})^{s}$ holds true, then the conjugacy class of psgps associated to $\mathcal{N}$ is that of Borel subgroups. If G is of type A_n, the situation is more subtle, but van der Put and Voskuil give a necessary and sufficient condition on ν for the validity of the above identity.

(ii) In contrast to the Drinfeld half-space, the quotient $\Gamma\backslash(\mathcal{F}(G,\mathcal{N})^{ss})^{rig}$ is in general not algebraic. In fact, if G is different from SL_n, none of these quotients is algebraic, cf. [220], Thm. 4.14. In the case $G = \mathrm{SL}_n$, van der Put and Voskuil give a complete list of those period domains for which the quotient is algebraic. Curiously, this list coincides precisely with the list of period domains over finite fields which are not simply connected, cf. Proposition 11.2.9.

(iii) It is not clear if the formal models constructed by van der Put and Voskuil have semi-stable reduction.

Notes and References On the level of rigid-analytic spaces the map τ was introduced by Drinfeld [71]. He uses this map for constructing admissible open affinoid coverings of $(\Omega^n_K)^{rig}$. Werner [221] constructs a compactification $\overline{|\mathcal{B}|}$ of $|\mathcal{B}|$ by means of semi-norms. She proves that the map τ extends to a $G(K)$-equivariant map $(\mathbb{P}^{n-1}_K)^{an} \longrightarrow \overline{|\mathcal{B}|}$. An interesting question in light of Proposition 11.3.4 is if there is a compactification of Ω^n_K such that the stabilizers of boundary points are compact. One may regard such a compactification as a smooth compactification in the world of $G(K)$-spaces, whereas the obvious compactification $(\mathbb{P}^{n-1}_K)^{an}$ is singular from this point of view. From the proof of Proposition 11.3.4 it would seem that the correct analogue of this question over a finite field k would be to find a compactification such that all points at the boundary have trivial stabilizers in $U(k)$.

For other period domains besides the Drinfeld space, this sharp dichotomy between the stabilizers of points inside the period domain, or on its boundary fails. In some unpublished notes, Voskuil (cf. also [220]) constructs, for any (rigid-analytic) period domain, a $G(K)$-equivariant map $I : (\mathcal{F}(G,\mathcal{N})^{ss})^{rig} \longrightarrow \mathcal{P}(\mathcal{B})$ to the power set $\mathcal{P}(\mathcal{B})$ of the Bruhat–Tits building. If $\mathcal{F}(G,\mathcal{N})^{ss} = \mathcal{F}(G,\mathcal{N})^{s}$, then $I(x)$ consists of a single point and I yields a generalization of τ.

We point out that we do not know if the analogue of Proposition 11.3.1 is true over finite fields.

The content of this section is in part discussed in [186].

4 Local systems on p-adic period domains

In this section we discuss period domains in the p-adic setting and the conjectural existence of interesting local systems on them. These local systems arise from Fontaine's theory of Grothendieck's "mysterious functor."

We begin with a brief review of Fontaine's theory. Let L be a perfect field of characteristic $p > 0$ and set $K_0 = W(L)_{\mathbb{Q}}$. Let K be a complete non-Archimedean extension of K_0, with corresponding valuation ring O_K (the ring of elements of norm ≤ 1). We fix an algebraic closure $\bar{K}$ of K and denote by C its completion. Let $\mathcal{R}(O_{\bar{K}})$ be the projective limit of copies of $O_{\bar{K}}/pO_{\bar{K}}$ enumerated by the non-negative integers, with the p-power map as transition morphisms. This is a perfect ring of characteristic p which has a $\bar{L}$-algebra structure, cf. [75]. Let $A_{\text{inf}}(O_{\bar{K}}) = W(\mathcal{R}(O_{\bar{K}}))$ be the ring of Witt vectors of $\mathcal{R}(O_{\bar{K}})$. We denote by

$$\theta : A_{\text{inf}}(O_{\bar{K}}) \longrightarrow O_{\bar{K}}$$

the natural surjective ring homomorphism. Its kernel is a principal ideal. Let $A_{\text{cris}}(O_{\bar{K}})$ be the p-adic completion of the divided power hull of this kernel. The Frobenius automorphism of $A_{\text{inf}}(O_{\bar{K}})$ induces a Frobenius endomorphism ϕ of $A_{\text{cris}}(O_{\bar{K}})$. Also, $A_{\text{cris}}(O_{\bar{K}})$ is equipped with the divided power filtration $F^{\bullet}(A_{\text{cris}}(O_{\bar{K}}))$. In $A_{\text{cris}}(O_{\bar{K}})$ there is the characteristic element $t \in F^1(A_{\text{cris}}(O_{\bar{K}}))$ satisfying $\phi(t) = pt$. Then $B_{\text{cris}}(K)$ is obtained from $A_{\text{cris}}(O_{\bar{K}})$ by inverting t and p. This is a filtered K_0-algebra which is equipped with a σ-linear Frobenius automorphism ϕ. Furthermore, there is a natural surjective ring homomorphism

$$F^0(B_{\text{cris}}(K))/F^1(B_{\text{cris}}(K)) \longrightarrow C. \tag{4.1}$$

Now let K be a *finite* extension of K_0. Then also the residue field k_K of K is a finite extension of L, and K is a finite totally ramified extension of $K_0(k_K)$. Denote by Γ_K the Galois group of $\bar{K}$ over K. Then Γ_K acts on $B_{\text{cris}}(K)$ in a ϕ-semi-linear way and one has $B_{\text{cris}}(K)^{\Gamma_K} = K_0(k_K)$. Recall that a *p-adic representation* of Γ_K is a finite-dimensional $\mathbb{Q}_p$-vector space U equipped with a continuous linear action of Γ_K. There is always the inequality [81], 5.1,

$$\dim_{K_0(k_K)} (U \otimes_{\mathbb{Q}_p} B_{\text{cris}}(K))^{\Gamma_K} \leq \dim U. \tag{4.2}$$

The representation U is called *crystalline* if equality holds. The category $\text{Rep}_{\text{cris}}(\Gamma_K)$ of all crystalline representations is tannakian [80].

The $K_0(k_K)$-vector space $D_{\text{cris}}(U) = (U \otimes_{\mathbb{Q}_p} B_{\text{cris}}(K))^{\Gamma_K}$ inherits the structure of an isocrystal over k_K which is equipped with a filtration over K. The functor

$$D_{\text{cris}} : \text{Rep}_{\text{cris}}(\Gamma_K) \quad \longrightarrow \quad \text{FilIsoc}_{k_K}^{K}$$

is a fully faithful $\mathbb{Q}_p$-linear tensor functor. A filtered isocrystal over K is called *admissible* if it lies in the essential image of this functor. Any admissible filtered isocrystal is weakly admissible, cf. [80]. The converse is due to Colmez and Fontaine [44].

Theorem 11.4.1 *Let K be a finite extension of K_0. Then every weakly admissible filtered isocrystal in* $\mathrm{FilIsoc}_{k_K}^K$ *is admissible.*

Hence the functor D_{cris} induces an equivalence of categories

$$D_{\mathrm{cris}} : \mathrm{Rep}_{\mathrm{cris}}(\Gamma_K) \xrightarrow{\sim} (\mathrm{FilIsoc}_{k_K}^K)^{\mathrm{wa}},$$

with obvious notation. Its inverse is given by

$$V_{\mathrm{cris}} : (\mathrm{FilIsoc}_{k_K}^K)^{\mathrm{wa}} \xrightarrow{\sim} \mathrm{Rep}_{\mathrm{cris}}(\Gamma_K), \qquad (N, \mathcal{F}) \mapsto \mathcal{F}^0(N \otimes_{K_0(k_K)} B_{\mathrm{cris}}(K))^{\Phi}. \tag{4.3}$$

Example 11.4.2 Let X be a p-divisible group over O_K and let $T_p(X)$ be its Tate module. Then $V_p(X) = T_p(X) \otimes_{\mathbb{Z}_p} \mathbb{Q}_p$ is a crystalline representation. The dual $D_{\mathrm{cris}}(V_p(X))^* = (N, \mathcal{F})$ of the filtered isocrystal $D_{\mathrm{cris}}(V_p(X))$ has the following description. The isocrystal N is the rational Dieudonné module of the special fibre X_k of X, in the sense of [63] (contravariant Dieudonné module). The filtration $\mathcal{F}^\bullet$ has the form

$$\mathcal{F}^i = \begin{cases} N_K & i \leq 0 \\ (\mathrm{Lie}\, X)_K^\vee & i = 1 \\ (0) & i \geq 2. \end{cases}$$

Here $(\mathrm{Lie}\, X)_K^\vee$ is a submodule of N_K, via the identification of N_K with the dual of the Lie algebra of the universal vector extension of X [162].

Let $(\mathbb{G}, \mathcal{N})$ be a PD-pair relative to a perfect field L of characteristic p, with reflex field $\breve{K}_0 = \breve{K}_0(\mathbb{G}, \mathcal{N})$, cf. Definition 9.5.1. Let $\breve{\mathcal{F}}^{\mathrm{wa}} = \breve{\mathcal{F}}(\mathbb{G}, \mathcal{N})^{\mathrm{wa}}$ be the associated period domain over $\breve{K}_0$, an open analytic subset of the partial flag variety $\breve{\mathcal{F}}(\mathbb{G}, \mathcal{N})^{an} = \breve{\mathcal{F}}^{an}$. In what follows we will always assume that this period domain is non-empty, cf. Theorem 9.5.10.

For any representation (V, Φ_V, ρ) in $\mathrm{Rep}_{\mathbb{Q}_p}(\mathbb{G})$ and any $\mathcal{F} \in \breve{\mathcal{F}}(K)$, where $K/\breve{K}_0$ is finite, we may consider the filtered isocrystal $((V, \Phi_V)_{k_K}, \rho_*(\mathcal{F}))$. Here $(V, \Phi_V)_{k_K}$ is the isocrystal over k_K obtained from (V, Φ) via base change from L to k_K, cf. Chapter VIII (1.2), page 190. If $((V, \Phi_V)_{k_K}, \rho_*(\mathcal{F}))$ is admissible, then $(V, \Phi_V, \rho_*(\mathcal{F}))$ is weakly admissible, cf. Proposition 8.2.1 (iii). Hence, if $((V, \Phi_V)_{k_K}, \rho_*(\mathcal{F}))$ is admissible for every representation (V, Φ_V, ρ) in $\mathrm{Rep}_{\mathbb{Q}_p}(\mathbb{G})$ (one faithful representation is enough), then $\mathcal{F} \in \breve{\mathcal{F}}(\mathbb{G}, \mathcal{N})^{\mathrm{wa}}(K)$. The converse holds if the $\mathrm{Rep}_{\mathbb{Q}_p}(\mathbb{G})$-isocrystal $\mathcal{N}^{\mathbb{G}}$ is split semi-simple. In order to avoid confusion, we therefore call a point $\mathcal{F} \in \breve{\mathcal{F}}(K)$ *admissible* if $((V, \Phi_V)_{k_K}, \rho_*(\mathcal{F}))$ is weakly admissible (or equivalently, by Theorem 11.4.1, admissible), for every representation (V, Φ_V, ρ) in $\mathrm{Rep}_{\mathbb{Q}_p}(\mathbb{G})$. Note that if the K-valued point $\mathcal{F}$ is admissible, then for every finite extension K' of K, the point $\mathcal{F}$, considered as a K'-valued point of $\breve{\mathcal{F}}$ is admissible, and the corresponding Galois representation of $\Gamma_{K'}$ is simply the restriction of the Galois representation of Γ_K corresponding to $\mathcal{F}$, considered as K-valued point.

If the $\mathrm{Rep}_{\mathbb{Q}_p}(\mathbb{G})$-isocrystal $N^{\mathbb{G}}$ is split semi-simple, then a point of $\breve{\mathcal{F}}^{an}$ with values in a finite extension K of $\breve{K}_0$ (i.e., a *classical point*) is admissible if and only if it factors through $\breve{\mathcal{F}}^{\mathrm{wa}}$. This applies in particular to the case when $L = \bar{L}$ is algebraically closed.

Assume $\mathcal{F} \in \breve{\mathcal{F}}(K)$ is admissible. Then using the Fontaine functor (4.3), page 340, and denoting by $\omega_0 : \mathrm{Rep}_{\mathrm{cris}}(\Gamma_K) \longrightarrow \mathrm{Vec}_{\mathbb{Q}_p}$ the natural forgetful functor, we get a fiber functor on $\mathrm{Rep}_{\mathbb{Q}_p}(\mathbb{G})$

$$\begin{aligned} \omega_{\mathrm{cris},\mathcal{F}} : \mathrm{Rep}_{\mathbb{Q}_p}(\mathbb{G}) &\longrightarrow \mathrm{Vec}_{\mathbb{Q}_p}, \\ (V, \Phi_V, \rho) &\mapsto \omega_0(V_{\mathrm{cris}}((V, \Phi_V)_{k_K}, \rho_* \mathcal{F})). \end{aligned} \tag{4.4}$$

Let

$$G_{\mathcal{F}} := \underline{\mathrm{Aut}}^{\otimes}(\omega_{\mathrm{cris},\mathcal{F}}).$$

This is a $\mathbb{Q}_p$-rational form of the underlying group scheme of $\mathbb{G}$. We also obtain a continuous homomorphism

$$\Gamma_K \longrightarrow G_{\mathcal{F}}(\mathbb{Q}_p). \tag{4.5}$$

It is natural to expect that the isomorphism class of the fiber functor $\omega_{\mathrm{cris},\mathcal{F}}$ is independent of the admissible filtration $\mathcal{F} \in \breve{\mathcal{F}}^{\mathrm{wa}}(K)$. We mention the following result, which pertains to the case when $L = \bar{L}$ is algebraically closed, and when $\mathbb{G} = \mathbb{G}_b$, for a reductive group G over $\mathbb{Q}_p$ and an element $b \in G(K_0)$ (recall from Example 9.1.22 that, in this situation, the natural functor $\mathrm{Rep}_{\mathbb{Q}_p}(G) \longrightarrow \mathrm{Rep}_{\mathbb{Q}_p}(\mathbb{G})$ is a tensor equivalence). We denote by $\hat{G}$ the Langlands dual group of G, and by $Z(\hat{G})$ its center. Then $\Gamma_{\mathbb{Q}_p}$ acts on $Z(\hat{G})$.

Proposition 11.4.3 *Let $L = \bar{L}$ be algebraically closed and let K be a finite extension of K_0. Identify $H^1(\mathbb{Q}_p, G)$ with the torsion subgroup $X^*(Z(\hat{G})^{\Gamma_{\mathbb{Q}_p}})_{\mathrm{tors}}$ of $X^*(Z(\hat{G})^{\Gamma_{\mathbb{Q}_p}})$, cf. [137]. Then the difference $\mathrm{cl}(\omega^G, \omega_{\mathrm{cris},\mathcal{F}})$ between ω^G and $\omega_{\mathrm{cris},\mathcal{F}}$, as an element of $H^1(\mathbb{Q}_p, G)$, is independent of $\mathcal{F} \in \breve{\mathcal{F}}(\mathbb{G}_b, \mathcal{N})^{\mathrm{wa}}(K)$ and is given by the following expression,*

$$\mathrm{cl}(\omega^G, \omega_{\mathrm{cris},\mathcal{F}}) = \kappa(b) - \mathcal{N}^{\sharp}. \qquad \square$$

Here $\kappa(b)$ is the image of b under the Kottwitz homomorphism [136]

$$G(K_0) \longrightarrow X^*(Z(\hat{G})^{\Gamma_{\mathbb{Q}_p}})$$

and $\mathcal{N}^\sharp \in X^*(Z(\hat{G})^{\Gamma_{\mathbb{Q}_p}})$ is the restriction of the character of $Z(\hat{G})$ defined by $\mathcal{N}$.

Let us assume that all $\omega_{\mathrm{cris},\mathcal{F}}$ are indeed isomorphic. Then one would like to glue them together to some "local system" on $\breve{\mathcal{F}}^{\mathrm{wa}}$. However, examples show that this hope is too naive. The main reason is that these fiber functors make only sense for *admissible classical* points, and such points, even though dense in $\breve{\mathcal{F}}^{\mathrm{wa}}$, are still too scarce to distinguish open subsets. The following conjecture takes this objection into account.

Conjecture 11.4.4 *There exists a dense open subset $\breve{\mathcal{F}}^{\mathrm{a}}$ of $\breve{\mathcal{F}}^{\mathrm{wa}}$, invariant under the automorphism group $J(\mathbb{Q}_p)$ of the a.g.s. $\mathbb{G}$ (cf. Chapter IX (3.2), page 252), and a tensor functor $\mathcal{L}_{\mathbb{G},\mathcal{N}} : \mathrm{Rep}_{\mathbb{Q}_p}(\mathbb{G}) \longrightarrow \mathbb{Q}_p\text{-}\mathrm{Loc}_{\breve{\mathcal{F}}^{\mathrm{a}}}$ into the category of local systems of $\mathbb{Q}_p$-vector spaces on $\breve{\mathcal{F}}^{\mathrm{a}}$ with the following property.*

For any finite extension K of $\breve{K}_0$, a point $x \in \breve{\mathcal{F}}(K)$ factors through $\breve{\mathcal{F}}^{\mathrm{a}}$ if and only if x is admissible, and in this case the fiber functor $\omega_{\mathrm{cris},\mathcal{F}_x}$ is isomorphic to the composition of $\mathcal{L}_{\mathbb{G},\mathcal{N}}$ with the functor "fiber at x."

We refer to [50], §4 for the definition of local systems of $\mathbb{Q}_p$-vector spaces on Berkovich spaces. In fact, the tensor functor $\mathcal{L}_{\mathbb{G},\mathcal{N}}$ should factor through the tannakian category $\mathbb{Q}_p\text{-}\mathrm{Loc}_{\breve{\mathcal{F}}^{a}}/J(\mathbb{Q}_p)$ of local systems of $\mathbb{Q}_p$-vector spaces on $\breve{\mathcal{F}}^{a}$, to which the action of $J(\mathbb{Q}_p)$ is lifted in a smooth way, in the sense of Fargues [78], IV.8.2.

Remark 11.4.5 Given a local system on a dense open subset of $\breve{\mathcal{F}}^{\mathrm{wa}}$, there exists a unique maximal open subset to which this local system extends [108]. We may always assume when discussing Conjecture 11.4.4 that $\breve{\mathcal{F}}^{\mathrm{a}}$ has this maximal property. Then one may ask whether the property stated in the conjecture relative to the fiber functors $\omega_{\mathrm{cris},\mathcal{F}_x}$ for classical admissible points x characterizes $\mathcal{L}_{\mathbb{G},\mathcal{N}}$ and $\breve{\mathcal{F}}^{\mathrm{a}}$ uniquely.

Remark 11.4.6 It is also possible to restate the conjecture in terms of fundamental groups. In [50], de Jong develops a good theory of étale coverings of an analytic space X. Such coverings form a Galois category, and any geometric point $\bar{x}$ defines a "fiber functor," whose associated fundamental group $\pi_1(X, \bar{x})$ is a pro-discrete group. Further, when X is connected, the category $\mathbb{Q}_p\text{-}\mathrm{Loc}_X$ is equivalent to the category of continuous finite-dimensional $\mathbb{Q}_p$-representations of $\pi_1(X, \bar{x})$, [50], Thm. 4.2. Therefore the question arises whether the conjectured space $\breve{\mathcal{F}}^{\mathrm{a}}$ is connected. If so, then in these terms, the tensor functor $\mathcal{L}_{\mathbb{G},\mathcal{N}}$ is given by a continuous morphism

$$\pi_1(\breve{\mathcal{F}}^{\mathrm{a}}, \bar{x}) \longrightarrow G_{\mathcal{F}_{\bar{x}}}(\mathbb{Q}_p)$$

which, any time the geometric point $\bar{x}$ is above a classical admissible point

x, when composed with the natural morphism $\Gamma_{K_x} \longrightarrow \pi_1(\breve{\mathcal{F}}^{\mathrm{a}}, \bar{x})$ yields the morphism (4.5), page 341, and which is compatible with changes of base points in a sense that we will not make precise here. To incorporate the expected lifted action of $J(\mathbb{Q}_p)$ one would have to look at the fundamental group of the stack quotient of $\breve{\mathcal{F}}^{\mathrm{a}}$ by $J(\mathbb{Q}_p)$ (if such an object has a right to existence).

Remark 11.4.7 As noted above, if the $\mathrm{Rep}_{\mathbb{Q}_p}(\mathbb{G})$-isocrystal $N^{\mathbb{G}}$ is split semi-simple, then a classical point of $\breve{\mathcal{F}}$ is admissible iff it lies in $\breve{\mathcal{F}}^{\mathrm{wa}}$. Hence in this case, the conjectural open set $\breve{\mathcal{F}}^{\mathrm{a}}$ contains all classical points of $\breve{\mathcal{F}}^{\mathrm{wa}}$, and each such point defines a fiber functor on $\mathrm{Rep}_{\mathbb{Q}_p}(\mathbb{G})$. This applies in particular to the case when L is algebraically closed.

Remark 11.4.8 In [183], 1.37, Conjecture 11.4.4 is formulated, in the case when L is algebraically closed, in terms of rigid-analytic varieties. In fact, in this case, passing to associated rigid-analytic varieties, the open embedding $\breve{\mathcal{F}}^{\mathrm{a}} \longrightarrow \breve{\mathcal{F}}^{\mathrm{wa}}$ yields a bijective étale morphism

$$(\breve{\mathcal{F}}^{\mathrm{a}})^{rig} \longrightarrow (\breve{\mathcal{F}}^{\mathrm{wa}})^{rig}$$

of rigid-analytic varieties over $\breve{K}_0$. It is in terms of this morphism of rigid-analytic varieties, that the conjecture is formulated in [183]. We refer to [50], Prop. 5.1, for the correspondence between local systems of $\mathbb{Q}_p$-vector spaces on analytic spaces and on their associated rigid-analytic varieties. The formulation of Conjecture 11.4.4 in terms of analytic spaces comes from [108].

Remark 11.4.9 An interesting duality phenomenon arises as follows (in order to bring out more clearly the characteristic features of the situation, we are making simplifying assumptions). Let us assume that L is algebraically closed and let us start with a PD-pair of the form $\mathbb{G} = \mathbb{G}_b$, where G is a reductive group over $\mathbb{Q}_p$, and where $b \in G(K_0)$ is *basic*, cf. Definition 9.3.1. Then the automorphism group $J = J_b$ is an inner form of G, which we will denote by $\check{G}$. Then b can also be considered as an element of $J(K_0) = G(K_0)$. We set $\check{b} = b^{-1}$, and let $\check{\mathcal{N}}$ be defined by the inverse of the $\mathbb{Q}$-1-PS defining $\mathcal{N}$. We therefore obtain a flag variety $\mathcal{F}(\check{G}, \check{\mathcal{N}})$ over the same reflex field as $\mathcal{F}(G, \mathcal{N})$, and a period domain $\breve{\mathcal{F}}(\check{G}, \check{\mathcal{N}})^{\mathrm{wa}}$. We call $\breve{\mathcal{F}}(\check{G}, \check{\mathcal{N}})^{\mathrm{wa}}$ the *dual* period domain to $\breve{\mathcal{F}}(\check{G}, \check{\mathcal{N}})^{\mathrm{wa}}$. It is clear using Lemma 9.3.12 that, starting conversely from $(\check{G}, \check{b}, \check{\mathcal{N}})$, we get back to $(G, b, \mathcal{N})$.

Now let us make the additional assumption that the inner form $G_{\mathcal{F}}$ of Proposition 11.4.3 is isomorphic to G. Then the same is true for the dual pair as well. Note that this assumption is automatic when the derived group of G is simply connected. In this case there should be a close relation between the postulated tensor functors $\mathcal{L}_{\mathbb{G},\mathcal{N}}$ and $\mathcal{L}_{\check{\mathbb{G}},\check{\mathcal{N}}}$, cf. [183], 5.54. In fact, there should be an equivalence of topoi between the topos of étale $\check{G}(\mathbb{Q}_p)$-sheaves on $\check{\mathcal{F}}(G,\mathcal{N})^{\mathrm{a}}$ where the action of $\check{G}(\mathbb{Q}_p)$ is smooth and the topos of $G(\mathbb{Q}_p)$-sheaves on $\check{\mathcal{F}}(\check{G},\check{\mathcal{N}})^{\mathrm{a}}$ where the action of $G(\mathbb{Q}_p)$ is smooth.

The pair obtained from $G = \mathrm{GL}_n$ and the minuscule co-weight $(1, 0^{(n-1)})$ and the basic element of slope $1/n$ leads to the period domain $\mathbb{P}^{n-1}_{K_0}$. Its dual is the dual Drinfeld space $\check{\Omega}^n_{\mathbb{Q}_p} \widehat{\otimes}_{\mathbb{Q}_p} K_0$, cf. Example 9.5.7. In this case, the equivalence of topoi mentioned above is due to Faltings [74] and to Fargues [78], Thm. IV.13.1.

Let us now assume that the PD-pair is of the form $(\mathbb{G},\mathcal{N})$, with $\mathbb{G} = \mathbb{G}_b$, where G is a reductive group over $\mathbb{Q}_p$, and where $b \in G(K_0)$. We furthermore make the following assumption.

$$\begin{array}{c}\textit{There exists a faithful representation } (V,\rho) \textit{ of } G \\ \textit{such that all weights of } \nu \textit{ on } V \textit{ are } 0 \textit{ or } 1.\end{array} \tag{4.6}$$

Here ν is any representative 1-PS corresponding to $\mathcal{N}$.

Remark 11.4.10 The assumption (4.6) above implies that ν is a minuscule co-weight, cf. [198], §3. Furthermore, assume that G does not possess a proper normal subgroup through which ν factors. Then by Serre [198] all weights of ν in V form one orbit under the Weyl group, the group G has no simply connected factor group, and all irreducible components of G_{der} are of type A_n, B_n, C_n or D_n. Furthermore, in case V is an irreducible representation, $\dim V$ is even, except possibly when all irreducible factors of G_{der} are of type A.

Under the assumption (4.6) above, Hartl [108] constructs an open $\breve{K}_0$-analytic subspace $\check{\mathcal{F}}^{\mathrm{H}}$ of $\check{\mathcal{F}}^{\mathrm{wa}}$ that is a candidate for the subset $\check{\mathcal{F}}^{\mathrm{a}}$ in Conjecture 11.4.4. A different definition of this subset was subsequently given by Faltings [75], which, however, leads to the same subset. Let us briefly indicate these definitions.

We start with Faltings' definition, because it is based on the Fontaine rings introduced at the beginning of this section. Namely, let K be a complete non-Archimedean extension field of K_0. Let $N = (V, \Phi)$ be an isocrystal over L and let $\mathcal{F}^\bullet$ be a filtration of $V \otimes_{K_0} K$ with support in $\{0, 1\}$, given by a subspace $\mathcal{F}$ of $V \otimes_{K_0} K$. We define a $\mathbb{Z}$-filtration on $V \otimes_{K_0} B_{\mathrm{cris}}(K)$ as follows. Let

$$\begin{aligned}\mathcal{F}^0 &= V \otimes_{K_0} F^0 B_{\mathrm{cris}}(K)\\ \mathcal{F}^1 &= \text{the preimage of } \mathcal{F} \otimes_K C \text{ under the composition of surjective maps}\\ &\quad V \otimes_{K_0} F^0(B_{\mathrm{cris}}(K)) \longrightarrow V \otimes_{K_0} F^0(B_{\mathrm{cris}}(K))/F^1(B_{\mathrm{cris}}(K)) \longrightarrow V \otimes_{K_0} C.\end{aligned}$$

Finally the other filtration steps are obtained by shifting with powers of the characteristic element t, i.e., $\mathcal{F}^{i+2} = t\mathcal{F}^i$, $\forall i \in \mathbb{Z}$.

Then the object $(N, \mathcal{F}^\bullet)$ of $\mathrm{FilIsoc}_L^K$ is called *admissible* in the sense of Faltings [75], if there exists an isomorphism

$$V \otimes_{K_0} B_{\mathrm{cris}}(K) \simeq (B_{\mathrm{cris}}(K))^n$$

which respects the filtrations and the Frobenius on both sides. Here the Frobenius on the LHS is induced by the Frobenius on both factors.

If K is a finite extension of K_0, then this definition of admissibility coincides with the definition of Fontaine.

Now we describe the definition of Hartl. To do so, we have to recall some facts from the p-adic differential equations approach to Fontaine theory, cf. [11] and [108].

Again let K be a complete non-Archimedean extension of K_0. We take up the notation at the beginning of this section. In particular, we have the perfect ring $\mathcal{R}(O_{\bar{K}})$. Let $\tilde{E}$ be the fraction field of $\mathcal{R}(O_{\bar{K}})$, and set $\tilde{A} = W(\tilde{E})$. For every positive real number r, Hartl defines a subring $\tilde{A}^{(0,r]}$ of $\tilde{A}$, and introduces a certain completion $\tilde{B}^{]0,r]}$ of $\tilde{A}^{(0,r]}[\frac{1}{p}]$. Finally, let $\tilde{B}^{\dagger}_{rig} = \tilde{B}^{\dagger}_{rig}(K) = \bigcup_{r>0} \tilde{B}^{]0,r]}$. Then $\tilde{B}^{\dagger}_{rig}$ is a K_0-algebra, equipped with a σ-linear automorphism ϕ, which restricts to isomorphisms $\phi : \tilde{B}^{]0,r]} \longrightarrow \tilde{B}^{]0,r/p]}$, for every $r > 0$.

The characteristic element t introduced above lies in $\tilde{B}^{]0,1]}$. Furthermore, there is a surjective homomorphism $\theta : \tilde{B}^{]0,1]} \longrightarrow C$ which contains t in its kernel.

Now let $N = (V, \Phi)$ be an isocrystal over L and let $\mathcal{F}^\bullet$ be a filtration of $V \otimes_{K_0} K$ with support in $\{0, 1\}$, given by a subspace $\mathcal{F}$ of $V \otimes_{K_0} K$. Define a module $\overline{M}_{\mathcal{F}}^{]0,1]}$ over $\tilde{B}^{]0,1]}$ by taking the inverse image of $\mathcal{F} \otimes_K C$ under the surjection induced by θ,

$$V \otimes_{K_0} \tilde{B}^{]0,1]} \longrightarrow V \otimes_{K_0} C,$$

and set for $n \geq 1$, $\overline{M}_{\mathcal{F}}^{]0,p^{-n}]} = \phi^n(\overline{M}_{\mathcal{F}}^{]0,1]})$, where

$$\phi : V \otimes_{K_0} \tilde{B}^{[0,p^{-i}]} \longrightarrow V \otimes_{K_0} \tilde{B}^{[0,p^{-(i+1)}]}$$

is the induced Frobenius. Then

$$M_{\mathcal{F}} = \Big(\bigcap_n \overline{M}_{\mathcal{F}}^{]0,p^{-n}]}\Big) \otimes_{\tilde{B}^{]0,1]}} \tilde{B}^{\dagger}_{rig}$$

is a ϕ-submodule of $V \otimes_{K_0} \tilde{B}^{\dagger}_{rig}$.

The object $(N, \mathcal{F}^\bullet)$ of $\mathrm{FilIsoc}_L^K$ is called *admissible* in the sense of Hartl [108], if there exists an isomorphism

$$M_{\mathcal{F}} \simeq (\tilde{B}^{\dagger}_{rig}(K))^n$$

which respects the Frobenius on both sides. Again it can be proved [10], Thm.

V.2 that, when K is a finite extension of K_0, this definition of admissibility coincides with the definition of Fontaine.

Theorem 11.4.11 *Let $L = \bar{L}$.*

(i) *Both definitions of admissibility coincide.*

(ii) *Assume that the PD-pair $(\mathbb{G}_b, \mathcal{N})$ satisfies the assumption* (4.6), *page 344. There is an open $\breve{K}_0$-analytic subset $\breve{\mathcal{F}}(\mathbb{G}, \mathcal{N})^{\mathrm{H}}$ of $\breve{\mathcal{F}}(\mathbb{G}, \mathcal{N})^{\mathrm{wa}}$ such that for any non-Archimedean field K containing $\breve{K}_0$, and any representation (V, ρ) as in* (4.6), *page 344, the set $\breve{\mathcal{F}}(\mathbb{G}, \mathcal{N})^{\mathrm{H}}(K)$ coincides with the set of points $x \in \breve{\mathcal{F}}(\mathbb{G}, \mathcal{N})(K)$ such that the object $(N_b(V), \rho(\mathcal{F}_x))$ of* $\mathrm{FilIsoc}_L^K$ *is admissible.*

(iii) *For any representation (V, ρ) as in* (4.6), *page 344, there is a local system $\mathcal{L}_V$ of $\mathbb{Q}_p$-vector spaces over $\breve{\mathcal{F}}(\mathbb{G}, \mathcal{N})^{\mathrm{H}}$ such that the fiber of $\mathcal{L}_V$ at a point x with values in a finite extension K of $\breve{K}_0$ is equal to $V_{\mathrm{cris}}(N_b(V)_{k_K}, \rho_*(\mathcal{F}_x))$.*

Proof (i) is contained in the proof of [109], Thm. 3.5, see also [108], Thm. 6.2. The statement (ii) is contained in [108], Prop. 5.2, and the statement (iii) in [108], Thm. 5.5. □

Remarks 11.4.12 (i) The statement of Theorem 11.4.11 is weaker than Conjecture 11.4.4 in two respects (even under the assumption of the existence of a faithful representation V of G satisfying (4.6), page 344). First of all, in Conjecture 11.4.4 there is built in a functoriality statement in $\mathrm{Rep}_{\mathbb{Q}_p}(G)$, and this functoriality is missing in (iii) of Theorem 11.4.11. However, assume that $G = \mathrm{GL}(V)$, where V satisfies the assumption (4.6), page 344. Note that in this case $G = G_{\mathcal{F}}$ since GL_n has trivial cohomology. In this case the desired functor $\mathcal{L}_{\mathbb{G},\mathcal{N}} : \mathrm{Rep}_{\mathbb{Q}_p}(G) \longrightarrow \mathbb{Q}_p\text{-}\mathrm{Loc}_{\breve{\mathcal{F}}^{\mathrm{H}}}$ can be constructed by using the point of view of representations of fundamental groups, cf. Remark 11.4.6. Indeed, $\mathcal{L}_V$ defines for every connected component of $\breve{\mathcal{F}}(\mathbb{G}, \mathcal{N})^{\mathrm{H}}$ a representation of the fundamental group with values in $\mathrm{GL}(V)$ which, by Remark 11.4.6, then defines the desired tensor functor $\mathcal{L}_{\mathbb{G},\mathcal{N}}$. Something similar is true if G is the group of symplectic similitudes for a symplectic form on V. For details, see [108], Thm. 6.2, b).

Second, it is not clear that $\breve{\mathcal{F}}(\mathbb{G}, \mathcal{N})^{\mathrm{H}}$ is the maximal open $\breve{K}_0$-analytic subset to which any of the local systems in Theorem 11.4.11 extends, cf. Remark 11.4.5. Hartl [108] conjectures that this is indeed the case.

(ii) The assertions (i)–(iii) of Theorem 11.4.11 should continue to hold for any perfect field L. In addition, the following assertion seems reasonable.

(iv) *Let L' be a perfect field extension of L, and let $(\mathbb{G}', \mathcal{N}')$ be the corresponding PD pair relative to L', cf. Proposition 9.2.31 (i.e., $\mathbb{G}' = (G, b)$, where b is* *now considered as an element of* $G(K_0'))$. *Then the inclusion*

$$\breve{\mathcal{F}}(\mathbb{G}',\mathcal{N}')^{\mathrm{wa}} \subset \breve{\mathcal{F}}(\mathbb{G},\mathcal{N})^{\mathrm{wa}}\widehat{\otimes}_{\breve{K}_0}\breve{K}_0'$$

induces an isomorphism

$$\breve{\mathcal{F}}(\mathbb{G}',\mathcal{N}')^{\mathrm{H}} \simeq \breve{\mathcal{F}}(\mathbb{G},\mathcal{N})^{\mathrm{H}}\widehat{\otimes}_{\breve{K}_0}\breve{K}_0'.$$

Remark 11.4.13 Let us sketch the relation of Theorem 11.4.11 to period morphisms. To simplify this explanation, let us assume that $G = \mathrm{GL}(V)$ (in addition to the assumption that all weights occurring in V are 0 or 1). In fact, let us assume that $\mathcal{N}$ corresponds to the minuscule vector $\nu = (1^{(d)}, 0^{(n-d)})$. Let $\mathbb{X}$ be a p-divisible group over L with rational Dieudonné module $(V,\Phi) = (V \otimes K_0, b\cdot(\mathrm{Id}\otimes\sigma))$. Let $\mathcal{M}$ be the formal moduli space of p-divisible groups X equipped with a quasi-isogeny ρ from the reduction modulo p of X to the constant p-divisible group defined by $\mathbb{X}$, cf. [183], Ch. 2. Then $\mathcal{M}$ is a formal scheme, locally formally of finite type over $\mathrm{Spf}(W(L))$. When L is algebraically closed, and more generally when $\mathbb{X}$ is *decent*, this is proved in [183], Thm. 2.16. The general case is due to de Jong and Lau [146]. Let $\mathcal{M}^{rig}$ be the associated rigid-analytic variety over K_0 and let $\mathcal{M}^{an}$ be the associated K_0-analytic space. Then there is the *period morphism* [183], Ch. 5,

$$\pi : \mathcal{M}^{an} \longrightarrow \breve{\mathcal{F}}(\mathbb{G},\mathcal{N}), \tag{4.7}$$

which is an étale morphism [183], even in the sense of Berkovich [77]. Then by Faltings [75], the image of π is the Hartl space $\breve{\mathcal{F}}(\mathbb{G},\mathcal{N})^{\mathrm{H}}$. The local system $\mathcal{L}_V$ is then defined by the rational Tate module of the universal p-divisible group over $\mathcal{M}^{an}$.

Hence $\mathcal{M}^{an}$ is an étale covering space of $\breve{\mathcal{F}}(\mathbb{G},\mathcal{N})^{\mathrm{H}}$. By Hartl [108], this étale covering may be reconstructed directly from $\breve{\mathcal{F}}(\mathbb{G},\mathcal{N})^{\mathrm{H}}$ and the local system $\mathcal{L} = \mathcal{L}_V$ on $\breve{\mathcal{F}}(\mathbb{G},\mathcal{N})^{\mathrm{H}}$: in fact, $\mathcal{M}^{an}$ is the space of $\mathbb{Z}_p$-lattices in $\mathcal{L}$, in the sense of de Jong [50], §4.

In general, the inclusion of $\breve{\mathcal{F}}(\mathbb{G},\mathcal{N})^{\mathrm{a}}$ in $\breve{\mathcal{F}}(\mathbb{G},\mathcal{N})^{\mathrm{wa}}$ is rarely an equality. The following proposition [108] supports this statement.

Proposition 11.4.14 *Let* $L = \bar{L}$. *Assume that* $G = \mathrm{GL}_n$ *and* $b \in G(K_0)$ *and let* $\mathcal{N}$ *correspond to a minuscule co-weight of the form* $(1^{(d)}, 0^{(n-d)})$ *and assume (as always) that* $\breve{\mathcal{F}}(\mathbb{G},\mathcal{N})^{\mathrm{wa}}$ *is non-empty. Then* $\breve{\mathcal{F}}(\mathbb{G},\mathcal{N})^{\mathrm{H}} = \breve{\mathcal{F}}(\mathbb{G},\mathcal{N})^{\mathrm{wa}}$ *in precisely the following cases (we give the Newton vector of b, i.e., the opposite type of the slope filtration):*

$$-\nu_0 = \begin{cases} (1^{(n_1)}, (1/m)^{(m)}, 0^{(n_0)}) & n_0+m+n_1 = n, n_0 \geq 0, m \geq 0, n_1 \geq 0 \\ (1^{(n_1)}, ((m-1)/m)^{(m)}, 0^{(n_0)}) & n_0+m+n_1 = n, n_0 \geq 0, m \geq 0, n_1 \geq 0 \\ (1^{(n_1)}, (1/2)^{(4)}, 0^{(n_0)}) & n_0+m+n_1 = n, n_0 \geq 0, m \geq 0, n_1 \geq 0. \end{cases}$$

(Here in the cases when $m = 0$ the middle part is supposed to be absent. Case by case, the integer d entering into the definition of $\mathcal{N}$ is equal to $n_1 + \inf(1, m)$, resp. $n_1 + \sup(0, m-1)$, resp. $n_1 + 2$.)

Remark 11.4.15 For simplicity of the following discussion, assume that $n_0 = n_1 = 0$. Then the first alternative of the previous proposition corresponds to the Lubin–Tate case, when $\breve{\mathcal{F}}(\mathbb{G}, \mathcal{N})^{\mathrm{wa}} = \mathbb{P}^{n-1}$ (the space of lines), and the second alternative corresponds to $\check{\mathbb{P}}^{n-1}$ (the space of hyperplanes). In these cases, it is well-known [116] that the corresponding period morphism is surjective. The rational Tate module of the Lubin–Tate family defines then a $\mathbb{Q}_p$-local system on $\breve{\mathcal{F}}(\mathbb{G}, \mathcal{N})^{\mathrm{wa}}$, so that in this case $\breve{\mathcal{F}}(\mathbb{G}, \mathcal{N})^{\mathrm{H}} = \breve{\mathcal{F}}(\mathbb{G}, \mathcal{N})^{\mathrm{a}} = \breve{\mathcal{F}}(\mathbb{G}, \mathcal{N})^{\mathrm{wa}}$.

On the other hand, the third alternative relative to GL_4 is a total surprise, and still has to be "explained."

Notes and References Theorem 11.4.1 was conjectured by Fontaine [80]. After the proof of Colmez and Fontaine [44], there have been other proofs by Berger [10], by Colmez [43], and by Kisin [130].

Proposition 11.4.3 is due to Colmez and Fontaine [44]. More precisely, Wintenberger [222] proved that the formula in this proposition follows from Theorem 11.4.1, which was at the time still a conjecture. The formula itself was conjectured in [183] and proved there in the case when G_{der} is simply connected. It would be interesting to generalize Proposition 11.4.3 to the case of a general PD-pair $(\mathbb{G}, \mathcal{N})$.

Theorem 11.4.11 is due to Faltings [75] and Hartl [108]. The description by Hartl of the open subset $\breve{\mathcal{F}}(\mathbb{G}, \mathcal{N})^{\mathrm{H}}$ as the image of the period morphism, cf. Remark 11.4.13, is related to an old question of Grothendieck who asked which filtrations of the rational Dieudonné module are attained by deforming a fixed p-divisible group over L, cf. [109].

The duality observation in Remark 11.4.9 originates with Gross and Hopkins [116] relative to the Drinfeld/Lubin–Tate pair, which then was proved by Faltings [74] and Fargues [78]. The speculation in general is in [183], 5.54, where, however, it is formulated as an existence statement of a common covering space of the period domains in duality. It was subsequently pointed out by G. Faltings and by L. Fargues that the concept of "covering space" needs clarification in this context.

It would be interesting to extend Proposition 11.4.14 to groups other than GL_n.

5 The cohomology complex of p-adic period domains

This section is a continuation of Chapter X, where we discussed cohomological aspects of period domains over p-adic fields. Because of their relation to Shimura varieties, we may expect a close relation between period domains and the local Langlands program. However, Theorem 10.3.3 shows that the Galois action on the cohomology of period domains is often very poor. For instance, the middle degree cohomology group of the Drinfeld period domain Ω^n_K for a p-adic field K is isomorphic to the Steinberg representation v^G_B as a $\mathrm{GL}_n(K)$-representation, whereas the Galois group only acts through some unramified character. On the other hand, the local Langlands correspondence [139] associates to v^G_B the *special representation* sp_n of the Weil group of K. Thus the ℓ-adic cohomology of Ω^n_K does not realize the local Langlands correspondence (for generalized Steinberg representations). In this section we will explain (following [48]) that, at least in the case of the Drinfeld space Ω^n_K, such a realization can be obtained by considering the cohomology *complex* of period domains in a suitable derived category.

We first recall the local Langlands correspondence for GL_n. Let K be a local non-Archimedean field with residue field k of characteristic p. We normalize as in Section 3 the absolute value on K by $|\pi| = q^{-1}$. Let $W = W_K$ be the *Weil group* of K. Recall that it appears in a commutative diagram

$$\begin{array}{ccccccccc} 1 & \longrightarrow & I & \longrightarrow & \Gamma_K & \longrightarrow & \Gamma_k & \longrightarrow & 1 \\ & & \| & & \cup & & \cup & & \\ 1 & \longrightarrow & I & \longrightarrow & W_K & \longrightarrow & \mathbb{Z} & \longrightarrow & 1\,, \end{array}$$

where $\Gamma_K \longrightarrow \Gamma_k$ is the natural homomorphism and $\mathbb{Z}$ is identified with the subgroup $\{\mathrm{Fr}^{\mathbb{Z}}\} \subset \Gamma_k$. Here $\mathrm{Fr} \in \Gamma_k$ denotes the geometric Frobenius. The Weil group is topologized such that the inertia subgroup $I \subset W$ is open and such that multiplication by a lift of Fr induces a homeomorphism on W. By local class field theory, we may identify the (hausdorff) abelianized group W^{ab} topologically with the group of units $K^\times$. Thus it makes sense to write $|w|$ for any element $w \in W$. Fix a prime $\ell \neq p$. In the following all representations are in $\bar{\mathbb{Q}}_\ell$-vector spaces.

The *local Langlands correspondence* for $G = \mathrm{GL}_n(K)$ is a compatible system of bijections for varying n

$$\begin{array}{ccc} |\,\mathrm{Rep}^\infty_{\mathrm{irr}}\,|(\,\mathrm{GL}_n(K)) & \longrightarrow & |\,\mathrm{Rep}^{\mathrm{F\text{-}ss}}_n\,|(W_K) \\ \pi & \mapsto & \sigma_n(\pi) \end{array} \tag{5.1}$$

between the set of isomorphism classes $|\,\mathrm{Rep}^\infty_{\mathrm{irr}}\,|(\,\mathrm{GL}_n(K))$ of smooth irreducible $\mathrm{GL}_n(K)$-representations and the set of isomorphism classes $|\,\mathrm{Rep}^{\mathrm{F\text{-}ss}}_n\,|(W_K)$ of continuous *F-semi-simple* W_K-representations of dimension n. Here a W_K-representation is called *F-semi-simple* if any lift of the Frobenius Fr in Γ_K acts semi-simply on the underlying vector space. We refer to [139] for the properties of this correspondence, which in fact characterize it uniquely. Under this bijection the supercuspidal $\mathrm{GL}_n(K)$-representations correspond to the irreducible W_K-representations in $|\operatorname{Rep}_n^{\text{F-ss}}|(W_K)$, and the discrete series representations of $\operatorname{GL}_n(K)$ to the indecomposable W_K-representations in $|\operatorname{Rep}_n^{\text{F-ss}}|(W_K)$.

Since we will also want to consider representations in $|\operatorname{Rep}_n^{\text{F-ss}}|(W_K)$ which are not irreducible, it will be convenient to consider the category of representations of the *Weil–Deligne* group $\mathrm{WD} = \mathrm{WD}_K$. We will not need the Weil–Deligne group, but only its representation category, which we recall now [214].

A *representation of* WD_K is given by a tuple (ρ, N) consisting of a continuous W_K-representation in a finite-dimensional vector space V (which is equipped with the discrete topology!), and a nilpotent endomorphism N of V such that, for all $w \in W$,

$$\rho(w)N\rho(w)^{-1} = |w| \cdot N.$$

Such a representation is called *semi-simple* if ρ is a semi-simple W_K-representation. Denote by $|\operatorname{Rep}_n^{\text{ss}}|(\mathrm{WD}_K)$ the set of isomorphism classes of semi-simple WD-representations. The relevance of WD-representations comes from Grothendieck's theorem that the tame inertia subgroup acts potentially unipotently on any continuous ℓ-adic representation of W. Indeed, there is a natural bijection

$$|\operatorname{Rep}_n^{\text{F-ss}}|(W_K) = |\operatorname{Rep}_n^{\text{ss}}|(\mathrm{WD}_K).$$

Under this correspondence, an irreducible object in $|\operatorname{Rep}_n^{\text{F-ss}}|(W_K)$ corresponds to a pair (ρ, N) in $|\operatorname{Rep}_n^{\text{ss}}|(\mathrm{WD}_K)$, where ρ is irreducible and where $N = 0$.

Example 11.5.1 Set

$$V = \bigoplus_{i=0}^{n-1} \bar{\mathbb{Q}}_\ell(i) \in |\operatorname{Rep}_n^{\text{F-ss}}|(W_K).$$

Then the W-representation $V \otimes |\;|^{\frac{1-n}{2}}$ corresponds under (5.1), page 349 to the trivial G-representation $\mathbf{1}$, i.e., $\sigma_n(\mathbf{1}) = V \otimes |\;|^{\frac{1-n}{2}}$. Let Δ be the set of std simple roots of G. For a subset $I \subset \Delta = \{\alpha_1, \ldots, \alpha_{n-1}\}$, let

$$N_I = \sum_{\alpha_{n-i} \in \Delta \setminus I} E_{i,i+1} \in M_n(\bar{\mathbb{Q}}_\ell)$$

be the nilpotent matrix, where $E_{i,i+1}$ is the matrix with entry 1 at position $(i, i+1)$ and zero otherwise. Then N_I defines a morphism $V \longrightarrow V(-1)$. Let

$$\tau_I = (V, N_I) \in |\operatorname{Rep}_n^{\text{ss}}|(\mathrm{WD}_K)$$

be the WD-representation with monodromy operator N_I. Then the generalized Steinberg representation $v_{P_I}^G$ of Chapter X (1.2), page 304 corresponds to the WD-representation

$$\sigma_n(v_{P_I}^G) = \tau_I \otimes |\;|^{\frac{1-n}{2}}.$$

In particular, for $I = \emptyset$, we get the special representation $\tau_\emptyset \otimes |\,|^{\frac{1-n}{2}} \cong \mathrm{sp}_n$, and for $I = \Delta$ we get $V \otimes |\,|^{\frac{1-n}{2}} = \bigoplus_{i=-\frac{n-1}{2}}^{\frac{n-1}{2}} \bar{\mathbb{Q}}_\ell(i)$.

We now return to the cohomology of period domains, especially of Ω_K^n. Instead of considering the individual cohomology groups of Ω_K^n, we want to investigate its cohomology complex

$$R\Gamma_c(\Omega_K^n) = R\Gamma_c(\Omega_K^n \widehat{\otimes}_K \widehat{\overline{K}}, \overline{\mathbb{Q}}_\ell). \tag{5.2}$$

Since the action of $\mathrm{GL}_n(K)$ on Ω_K^n is continuous in the sense of Berkovich [14], this can be considered as an object of the bounded derived category $D^b(G)$ of the abelian category of smooth $\overline{\mathbb{Q}}_\ell$-representations of $\mathrm{GL}_n(K)$. We refer to [48], App. B for the technical details that the definition of this complex involves. The cohomology groups of this complex are the cohomology groups $H_c^*(\Omega_K^n)$ that were considered in Chapter X. Furthermore, the action of the Weil group W of K on $\Omega_K^n \widehat{\otimes}_K \widehat{\overline{K}}$ induces an action of W on $R\Gamma_c(\Omega_K^n)$, given by a morphism

$$\gamma : W \to \mathrm{End}_{D^b(G)}\,(R\Gamma_c(\Omega_K^n)). \tag{5.3}$$

Furthermore, for any $\pi \in |\,\mathrm{Rep}_{\mathrm{irr}}^\infty|(G)$, the complex of $\bar{\mathbb{Q}}_\ell$-vector spaces

$$\mathrm{RHom}_{D^b(G)}(R\Gamma_c(\Omega_K^n), \pi) \in D^b(\bar{\mathbb{Q}}_\ell) \tag{5.4}$$

is equipped with an action of W.

Now the bounded derived category $D^b(\bar{\mathbb{Q}}_\ell)$ of $\bar{\mathbb{Q}}_\ell$-vector spaces is equivalent to the triangulated category of $\mathbb{Z}$-graded $\bar{\mathbb{Q}}_\ell$-vector spaces with finite support (with respect to the grading). This equivalence is given by the cohomology functor

$$\mathcal{H}^* : C^\bullet \mapsto \bigoplus_{i\in\mathbb{Z}} \mathcal{H}^i(C^\bullet).$$

Hence applying this functor to (5.4) above, we obtain a graded $\bar{\mathbb{Q}}_\ell$-vector space

$$\mathcal{H}^*\left(\mathrm{RHom}_{D^b(G)}(R\Gamma_c(\Omega_K^n), \pi)\right)$$

equipped with an action of W. The following theorem is proved in [47].

Theorem 11.5.2 *Let π be a smooth irreducible representation of G. Unless π is a generalized Steinberg representation,*

$$\mathrm{RHom}_{D^b(G)}\,(R\Gamma_c(\Omega_K^n), \pi) = 0.$$

Let $\pi = v_{P_I}^G$ for some subset $I \subset \Delta$. Then as W-representations

$$\mathcal{H}^*\left(\mathrm{RHom}_{D^b(G)}(R\Gamma_c(\Omega_K^n), \pi)\right) \simeq \sigma_n(\pi) \otimes |\,|^{\frac{n-1}{2}}.$$

Remark 11.5.3 In this statement the grading on the LHS is disregarded. But in fact, the result proved in [48] is more precise. Let $P = P_I$, and write $I = \{i_1, \ldots, i_r\}$ with $i_1 < i_2 < \ldots < i_r$. For $k = 0, \ldots, r$, set $n_k = i_{k+1} - i_k$, with $i_0 = 0, i_{r+1} = n$, so that $\sum_k n_k = n$. Then τ_I can be written as follows as a sum of indecomposable representations of the Weil–Deligne group,

$$\tau_I \simeq \bigoplus_{k=0}^{r} \mathrm{sp}_{n_k}(i_k).$$

With these notations, the result above can be strengthened into an isomorphism of *graded* W-representations

$$\mathcal{H}^* \left(\mathrm{RHom}_{D^b(G)}(R\Gamma_c(\Omega^n_K), \pi)\right) \simeq \bigoplus_{k=0}^{r} \mathrm{sp}_{n_k}(i_k)[n-1-r+2k]. \qquad (5.5)$$

In particular, we see that the Steinberg representation v^G_B only contributes in the middle degree $n-1$; all other generalized Steinberg representations contribute also outside the middle degree, symmetrically around $n-1$.

Remarks 11.5.4 There are three features of this result which are remarkable. First, the result shows that the cohomology complex of Ω^n_K indeed gives a realization of part of the local Langlands correspondence for K. Second, whereas by Theorem 10.3.3 the cohomology groups $H^*_c(\Omega^n_K)$ involve only the n generalized Steinberg representations $v^G_{P_I}$ for I of the form $I = \{1, \ldots, i\}$ for $i = 0, \ldots, n-1$, here all 2^n generalized Steinberg representations appear. Third, whereas the cohomology groups $H^*_c(\Omega^n_K)$ are very unsymmetric around its middle degree (recall that $H^i_c(\Omega^n_K) = 0$ for i strictly below the middle degree, cf. Corollary 10.3.7), the above formula is arranged symmetrically around the middle degree.

Let us explain some ingredients of the proof of Theorem 11.5.2, and in particular clarify the relation to Theorem 10.3.3. Recall that a complex $C^\bullet$ in a derived category is called *splittable* if it is quasi-isomorphic to the complex with trivial differentials,

$$C^\bullet \simeq \bigoplus_i \mathcal{H}^i(C^\bullet)[-i].$$

An important ingredient of the proof of Theorem 11.5.2 is the following fact.

Proposition 11.5.5 *The object $R\Gamma_c(\Omega^n_K)$ of $D^b(G)$ is splittable, i.e., there is an isomorphism of G-modules,*

$$R\Gamma_c(\Omega^n_K) \xrightarrow{\sim} \bigoplus_i H^i_c(\Omega^n_K)[-i].$$

The proof follows from the following two statements.

Lemma 11.5.6 *Let $C^\bullet$ be a cohomological bounded complex in a derived category. Then $C^\bullet$ is splittable if*

$$\mathrm{Ext}^{i-j-1}(H^i(C^\bullet), H^j(C^\bullet)) = 0, \quad \forall (i, j) \in \mathbb{Z}^2. \qquad \square$$

This lemma naturally leads to the question of the determination of the Ext-groups of generalized Steinberg representations. This can in fact be answered for any reductive group. We first state the result for *semi-simple* groups and then comment on how it is extended to general reductive groups.

Proposition 11.5.7 ([48, 172]) *Let G be a p-adic semi-simple Lie group. Let $I, J \subset \Delta$ be subsets of a fixed system of simple roots Δ with respect to a maximal split torus and a minimal parabolic subgroup of G containing it. Set $\Delta(I, J) := (I \cup J) \setminus (I \cap J)$ and let $\delta(I, J) := |\Delta(I, J)|$. Then*

$$\mathrm{Ext}^i_G(v^G_{P_I}, v^G_{P_J}) = \begin{cases} \bar{\mathbb{Q}}_\ell & : \quad i = \delta(I, J) \\ 0 & : \quad \text{otherwise.} \end{cases}$$

If G is reductive, but not necessarily semi-simple, let r be the dimension of the maximal split central torus A_G. Then

$$\mathrm{Ext}^*_G(v^G_{P_I}, v^G_{P_J}) \simeq \bigwedge{}^*(\bar{\mathbb{Q}}^r_\ell)[-\delta(I, J)].$$

Note that in the case at hand, the semi-simple case is enough, since the action of $\mathrm{GL}_n(K)$ on the generalized Steinberg representations factors through the projective group PG. Similarly, the action of G on Ω^n_K factors through PG.

Applying Proposition 11.5.5 and Proposition 11.5.7, we get an identification of algebras,

$$\begin{aligned} \mathrm{End}_{D^b(PG)}(R\Gamma_c(\Omega^n_K)) &\simeq \bigoplus_{n-1 \geq j \geq i \geq 0} \mathrm{Ext}^{j-i}_{PG}(H^j_c(\Omega^n_K), H^i_c(\Omega^n_K)) \\ &\simeq \bigoplus_{n-1 \geq j \geq i \geq 0} \mathrm{Ext}^{j-i}_{PG}(v^G_{P_{(j-n+2,1^{(2(n-1)-j)})}}, v^G_{P_{(i-n+2,1^{(2(n-1)-i)})}})(j-i) \\ &\simeq \bigoplus_{n-1 \geq j \geq i \geq 0} \bar{\mathbb{Q}}_\ell(j-i), \end{aligned} \tag{5.6}$$

where the product on the RHS in the first two lines is given by the $\cup$-product, cf. [48]. The algebra structure in the third line is that of the "twisted upper triangular matrices" of size n. This cup product can be described explicitly using representatives of the Ext-groups, as follows [48].

Proposition 11.5.8 *Let G be split semi-simple. Let I, J, K be subsets of Δ. Assume that $\delta(K, J) = \delta(K, I) + \delta(I, J)$. Then the cup-product*

$$- \cup - : \mathrm{Ext}^{\delta(I,J)}(v^G_{P_I}, v^G_{P_J}) \otimes \mathrm{Ext}^{\delta(K,I)}_G(v^G_{P_K}, v^G_{P_I}) \longrightarrow \mathrm{Ext}^{\delta(K,J)}_G(v^G_{P_K}, v^G_{P_J})$$

is an isomorphism.

Let $\mathcal{A} = \bigoplus_{n-1\geq j\geq i\geq 0} \bar{\mathbb{Q}}_\ell(j-i)$ be the algebra of twisted upper triangular matrices. We denote by $\sigma : W \to \mathcal{A}$ the homomorphism which is induced by the W-representation $V = \mathrm{sp}_n \otimes |\ |^{\frac{n-1}{2}}$. One of the major steps in the proof of Theorem 11.5.2 is the following theorem.

Theorem 11.5.9 *There is an isomorphism of algebras making the following diagram commutative:*

$$\begin{array}{ccc} \mathrm{End}_{D^b(PG)}(R\Gamma_c(\Omega^n_K)) & \xrightarrow{\ \sim\ } & \mathcal{A}\ . \\ \uparrow{\scriptstyle\gamma} & \nearrow{\scriptstyle\sigma} & \\ W & & \end{array}$$

The main point of this theorem is that the inertia acts as strongly as possible on $R\Gamma_c(\Omega^n_K)$, i.e., that the monodromy endomorphism N satisfies $N^{n-1} \neq 0$. This in turn is proved by considering the geometric monodromy operator for a projective scheme quotient Ω^n_K/Γ, by a discrete co-compact subgroup, cf. Proposition 11.3.2. This involves the Rapoport-Zink spectral sequence [187], and the argument in fact proves the weight monodromy conjecture for such quotients.

Finally, using this result one proves Theorem 11.5.2 by checking that the Weil group actions coincide on both sides. In fact, we have

$$\begin{aligned} &\mathrm{RHom}_{D^b(PG)}(R\Gamma_c(\Omega^n_K), v^G_{P_I}) \\ &\cong \bigoplus_{i=0}^{n-1} \mathrm{RHom}_{D^b(PG)}(v^G_{P_{(i+1,1^{(n-i)-1})}}(-i), v^G_{P_I})[n-1+i] \\ &\cong \bigoplus_{i=0}^{n-1} \mathrm{Ext}^{\delta(i,I)}_{PG}(v^G_{P_{(i+1,1^{(n-i)-1})}}, v^G_{P_I})(i)[n-1+i-\delta(i,I)]\,, \end{aligned}$$

where $\delta(i, I) = \delta(\{\alpha_1, \ldots, \alpha_i\}, I)$. The action of N on the RHS is given by the cup-product with an element in

$$\bigoplus_{i=0}^{n-2} \mathrm{Ext}^1_{PG}\left(v^G_{P_{(i+2,1^{(n-i-2)})}}, v^G_{P_{(i+1,1^{(n-i-1)})}}\right)$$

having a non-zero entry in each summand. Now one applies Proposition 11.5.8 to check that the RHS coincides with the W-representation $\sigma_n(v^G_{P_I})$. This completes the sketch of the proof of Theorem 11.5.2.

A natural question is whether the methods used to prove Theorem 11.5.2 may be applied to other period domains besides Ω^n_K. Let us suppose, as in Chapter X, that the period domain arises from a PD-pair $(\mathbb{G}, \mathcal{N})$, where $\mathbb{G}$ is of the form $\mathbb{G}_b$ for some reductive group G over $\mathbb{Q}_p$ and some basic element $b \in G(\mathbb{Q}_p)$. By Theorem 10.3.3 the cohomology of period domains involves only generalized Steinberg representations. It turns out that again we may apply Proposition 11.5.7 to verify the condition in Lemma 11.5.6 to show that the complex $R\Gamma_c(\breve{\mathcal{F}}(\mathbb{G}_b, \mathcal{N}))$ is splittable.

Proposition 11.5.10 *The object $R\Gamma_c(\breve{\mathcal{F}}(\mathbb{G}_b, \mathcal{N})^{\mathrm{wa}})$ in $D^b(PJ_b)$ is splittable, i.e.,*

$$R\Gamma_c(\breve{\mathcal{F}}(\mathbb{G}_b, \mathcal{N})^{\mathrm{wa}}) \xrightarrow{\sim} \bigoplus_i H^i_c(\breve{\mathcal{F}}(\mathbb{G}_b, \mathcal{N})^{\mathrm{wa}})[-i].$$

Proof All we have to show is that the condition of Lemma 11.5.6 is satisfied. Consider Theorem 10.3.3 describing the ℓ-adic cohomology of $\breve{\mathcal{F}}^{\mathrm{wa}}$. For $w \in W/W_\nu$ (beware of the double notation!), let $I_w = \Delta \setminus \Delta(w\nu)$. Then P_{I_w} is just the parabolic subgroup $P(w\nu)$ in Theorem 10.3.2. So, let $w, w' \in W/W_\nu$. They produce essentially (i.e., we may disregard the Galois representation) the contributions $v^G_{P_{I_w}}$ and $v^G_{P_{I_{w'}}}$ in degree $2\ell(w) + |\Delta(w\nu)|$, resp. $2\ell(w') + |\Delta(w'\nu)|$. By Proposition 11.5.7 we know that $\mathrm{Ext}^k_{PG}(v^G_{P_{I_w}}, v^G_{P_{I_{w'}}}) \neq 0$ if and only if $k = \delta(I_w, I_{w'}) = |(I_w \cup I_{w'}) \setminus (I_w \cap I_{w'})|$. But

$$\begin{aligned}
2\ell(w) + |\Delta(w\nu)| - (2\ell(w') + |\Delta(w'\nu)|) - 1 &= |I_w \cup I_{w'}| - |I_w \cap I_{w'}| \\
\Leftrightarrow 2(\ell(w) - \ell(w')) + |\Delta(w\nu)| - |\Delta(w'\nu)| - 1 &= |I_w| + |I_{w'}| - 2|I_w \cap I_{w'}| \\
\Leftrightarrow 2(\ell(w) - \ell(w')) + |\Delta| - |I_w| - (|\Delta| - |I_{w'}|) - 1 &= |I_w| + |I_{w'}| - 2|I_w \cap I_{w'}| \\
\Leftrightarrow 2(\ell(w) - \ell(w')) - |I_w| + |I_{w'}| - 1 &= |I_w| + |I_{w'}| - 2|I_w \cap I_{w'}| \\
\Leftrightarrow 2(\ell(w) - \ell(w')) - 1 &= -2|I_w| - 2|I_w \cap I_{w'}|.
\end{aligned}$$

Thus this equality cannot be satisfied. □

Of course, this is not even half-way. For instance, one of the main ingredients in the proof of Theorem 11.5.9 is the existence of a semi-stable formal model for Ω^n_K, and the existence of projective algebraic quotients Ω^n_K/Γ. As we saw in Section 3, formal models are known to exist only in rare cases, and further they are not known to be semi-stable, except in the Drinfeld example; and also the existence of projective algebraic quotients is very rare.

On the other hand, computing the algebra $\mathrm{End}_{D(PJ_b)}(R\Gamma_c(\breve{\mathcal{F}}(\mathbb{G}_b, \mathcal{N})^{\mathrm{wa}})$ should be feasible more generally. It follows from Proposition 11.5.10 and Propositions 11.5.7 and 11.5.8 that $\mathrm{End}_{D(PJ_b)}(R\Gamma_c(\breve{\mathcal{F}}(\mathbb{G}_b, \mathcal{N})^{\mathrm{wa}})$ is always the path algebra $\mathcal{P}(Q)$ of a finite quiver Q. For instance, the algebra $\mathcal{A}$ in Theorem 11.5.9 is nothing but the path algebra of the quiver of type A_n. Here is a more interesting example.

Example 11.5.11 Let $G = \mathrm{GL}_3$, $b = 1$ and $\nu = (x_1 > x_2 > x_3)$ with $x_1 - x_2 =$

$x_2 - x_3$. By Theorem 10.3.3, the ℓ-adic cohomology with compact supports of $\breve{\mathcal{F}}^{\mathrm{wa}} = \breve{\mathcal{F}}(G,\nu)^{\mathrm{wa}}$ is given by

$$H_c^*(\breve{\mathcal{F}}^{\mathrm{wa}}) = v_B^G[-2] \oplus v_{P_{(2,1)}}^G(-1)[-3] \oplus v_{P_{(1,2)}}^G(-1)[-3] \oplus v_G^G(-2)[-4]^2 \oplus v_G^G(-3)[-6].$$

In this case

$$\mathrm{End}_{D(PJ_b)}(R\Gamma_c(\breve{\mathcal{F}}(G,b,\mathcal{N})^{\mathrm{wa}})) \simeq \mathcal{P}(Q),$$

where Q is the following quiver.

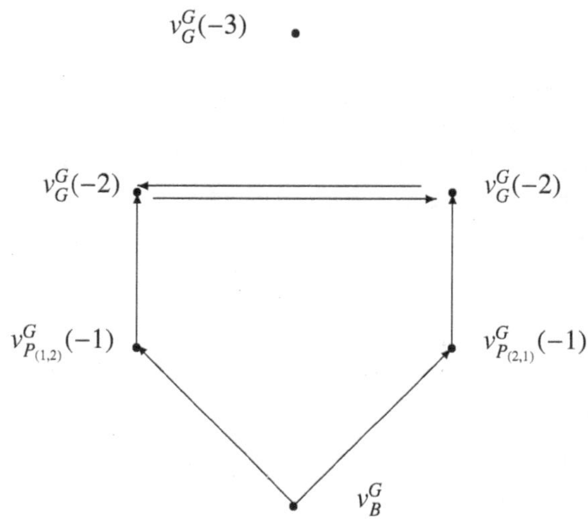

Notes and References The local Langlands correspondence for $G = \mathrm{GL}_n$ in the p-adic case is due to Harris and Taylor [107] and Henniart [114]. The Jacquet–Langlands correspondance is due to Rogawski [190] and to Deligne, Kazhdan and Vignéras [57].

The usual (naive) construction of $R\Gamma_c(\Omega_K^n)$ gives only an object in the bounded derived category of all G-modules instead of smooth G-modules. In an unpublished manuscript of Berkovich this problem is solved for étale finite coefficients. The extension to $\bar{\mathbb{Q}}_\ell$-coefficients is explained in [48], App. B. We remark that part of the inspiration for considering the cohomology complex instead of the individual cohomology groups comes from the analogy with the geometric version of Broué's abelian defect conjecture for GL_n over a finite field. The Ext-groups of Theorem 11.5.7 were determined by Casselman [40], [41], resp. Borel and Wallach [24], in the cases $I = \Delta$ or $J = \Delta$. The cases $I = \emptyset$ or $J = \emptyset$ are due to Schneider and Stuhler [196], who also conjectured the general result. This general result was proved in [48] and [172] simultaneously by substantially different methods. More precisely, the result in [172] applies to any semi-simple p-adic group, whereas the result in [48] is only for *split* groups; on the other hand, [48] also determines the algebra structure defined by the cup product, cf. Proposition 11.5.8.

As mentioned in the Introduction, considering the cohomology of period domains concerns only the bottom layer of a whole system of coverings. Indeed, it is the étale covering spaces of the *admissible subset* postulated in Conjecture 11.4.4 and constructed in many cases, that should involve more interesting representations in Langlands, resp. Jacquet–Langlands, correspondence. Here the Drinfeld cases and the Lubin–Tate cases are essentially solved (even in terms of cohomology complexes [47]), thanks to work of Boyer [31], Dat [47], Faltings [74], Fargues [78], Harris and Taylor [107] and Strauch [212]. More general cases are considered by Fargues [77] and Mantovan [159]. All these developments lie outside the scope of this book. However, the third alternative in Proposition 11.4.14 relative to GL_4 yields a period domain which should have more easily accessible étale coverings, to which one may hope to extend the methods developed here.

References

[1] *Revêtements étales et groupe fondamental (SGA 1).* Documents Mathématiques (Paris), 3. Société Mathématique de France, Paris, 2003. *Séminaire de Géométrie Algébrique du Bois Marie* 1960–61. Directed by A. Grothendieck, with two papers by M. Raynaud, updated and annotated. Lecture Notes in Math., 224, Springer, Berlin. Reprint of the 1971 original.

[2] J. F. Adams, *Lectures on Lie groups.* W. A. Benjamin, Inc., New York-Amsterdam, 1969.

[3] Y. André, Slope filtrations, arXiv:0812.3921, 2008.

[4] J. Arthur, A trace formula for reductive groups. I. Terms associated to classes in $G(\mathbf{Q})$. *Duke Math. J.*, 45(4):911–952, 1978.

[5] M. Artin, J. E. Bertin, M. Demazure, P. Gabriel, A. Grothendieck, M. Raynaud, J.-P. Serre, *Schémas en groupes. Fasc. 1: Exposés 1 à 4*, Volume 1963 of *Séminaire de Géométrie Algébrique de l'Institut des Hautes Études Scientifiques.* Institut des Hautes Études Scientifiques, Paris, 1963/1964.

[6] M. Artin, J. E. Bertin, M. Demazure, P. Gabriel, A. Grothendieck, M. Raynaud, J.-P. Serre, *Schémas en groupes. Fasc. 7: Exposés 23 à 26*, Volume 1963/64 of *Séminaire de Géométrie Algébrique de l'Institut des Hautes Études Scientifiques.* Institut des Hautes Études Scientifiques, Paris, 1965/1966.

[7] M. Artin, A. Grothendieck, J.-L. Verdier. Avec la collaboration de P. Deligne et B. Saint-Donat, *Théorie des topos et cohomologie étale des schémas.* Tome 3. *Séminaire de Géométrie Algébrique du Bois-Marie 1963–1964 (SGA 4).* Lecture Notes in Mathematics, Vol. 305. Springer-Verlag, Berlin-New York, 1973.

[8] M. F. Atiyah, Convexity and commuting Hamiltonians. *Bull. London Math. Soc.*, 14(1):1–15, 1982.

[9] M. F. Atiyah, R. Bott, The Yang–Mills equations over Riemann surfaces. *Philos. Trans. Roy. Soc. London Ser. A*, 308(1505):523–615, 1983.

[10] L. Berger, Équations différentielles p-adiques et (ϕ,N)-modules filtrés. Preprint, 2004.

[11] L. Berger, Représentations p-adiques et équations différentielles. *Invent. Math.* 148, no. 2:219–284, 2002.

[12] V. G. Berkovich, *Spectral Theory and Analytic Geometry over non-Archimedean Fields.* Volume 33 of *Mathematical Surveys and Monographs*, American Mathemati-

cal Society, Providence, RI, 1990.

[13] V. G. Berkovich, Étale cohomology for non-Archimedean analytic spaces. *Inst. Hautes Études Sci. Publ. Math.*, (78):5–161, 1994.

[14] V. G. Berkovich, Vanishing cycles for formal schemes. *Invent. Math.*, 115(3):539–571, 1994.

[15] V. G. Berkovich, The automorphism group of the Drinfeld half-plane. *C. R. Acad. Sci. Paris Sér. I Math.*, 321(9):1127–1132, 1995.

[16] J. N. Bernstein, Le "centre" de Bernstein. Edited by P. Deligne. In *Representations of Reductive Groups over a Local Field.* Travaux en Cours, pages 1–32, Hermann, Paris, 1984.

[17] J. N. Bernstein, A. V. Zelevinsky, Induced representations of reductive p-adic groups. I. *Ann. Sci. École Norm. Sup.* (4) 10, no. 4:441–472, 1977.

[18] G. D. Birkhoff, H. S. Vandiver, On the integral divisors of $a^n - b^n$. *Ann. of Math. (2)*, 5(4):173–180, 1904.

[19] C. Bonnafé, R. Rouquier, Affineness of Deligne–Lusztig varieties for minimal length elements. *J. Algebra*, 320(3):1200–1206, 2008.

[20] C. Bonnafé, R. Rouquier, Coxeter orbits and modular representations. *Nagoya Math. J.*, 183:1–34, 2006.

[21] C. Bonnafé, R. Rouquier, On the irreducibility of Deligne–Lusztig varieties. *C. R. Math. Acad. Sci. Paris*, 343(1):37–39, 2006.

[22] A. Borel, Automorphic *L*-functions. In *Automorphic Forms, Representations and L-functions (Proc. Sympos. Pure Math., Oregon State Univ., Corvallis, Ore., 1977), Part 2*, Proc. Sympos. Pure Math., XXXIII, pages 27–61. Amer. Math. Soc., Providence, R.I., 1979.

[23] A. Borel, J.-P. Serre, Théorèmes de finitude en cohomologie galoisienne. *Comment. Math. Helv.*, 39:111–164, 1964.

[24] A. Borel, N. Wallach, *Continuous Cohomology, Discrete Subgroups, and Representations of Reductive Groups.* Volume 67 of *Mathematical Surveys and Monographs*, American Mathematical Society, Providence, RI, second edition, 2000.

[25] A. Borel, *Linear Algebraic Groups. Graduate Texts in Mathematics*, Volume 126, Springer-Verlag, New York, second edition, 1991.

[26] A. Borel, J. Tits, Groupes réductifs. *Inst. Hautes Études Sci. Publ. Math.*, (27):55–150, 1965.

[27] N. Bourbaki, *Éléments de Mathématique. Fasc. XIV. Premiére partie: Les structures fondamentales de l'analyse. Livre II: Algèbre. Chapitre IV: Polynomes et fractions rationnelles. Chapitre V: Corps commutatifs.* Actualités Scientifiques et Industrielles, No. 1102. Hermann, Paris, 1950.

[28] N. Bourbaki, *Éléments de Mathématique. Fasc. XXXIV. Groupes et algèbres de Lie. Chapitre IV: Groupes de Coxeter et systèmes de Tits. Chapitre V: Groupes engendrés par des réflexions. Chapitre VI: systèmes de racines.* Actualités Scientifiques et Industrielles, No. 1337. Hermann, Paris, 1968.

[29] N. Bourbaki, *Éléments de Mathématique. Fasc. XXXVIII. Groupes et algèbres de Lie. Chapitre VII: Sous-algèbres de Cartan, éléments réguliers. Chapitre VIII: Algèbres de Lie semi-simples déployées.* Actualités Scientifiques et Industrielles, No. 1364. Hermann, Paris, 1975.

[30] N. Bourbaki, *Éléments de Mathématique. Algèbre commutative. Chapitres 8 et 9.* Springer, Berlin, 2006. Reprint of the 1983 original.

[31] P. Boyer, Monodromie du faisceau pervers des cycles évanescents de quelques

variétés de Shimura simples. *Invent. Math.*, 177, no. 2: 239–280, 2009.

[32] C. Breuil, Groupes p-divisibles, groupes finis et modules filtrés. *Ann. of Math. (2)*, 152(2):489–549, 2000.

[33] L. Breen, Tannakian categories. In *Motives (Seattle, WA, 1991)*, Proc. Sympos. Pure Math., 55, Part 1, pages 337–376, Amer. Math. Soc., Providence, RI, 1994.

[34] C. Breuil, P. Schneider, First steps towards p-adic Langlands functoriality. *J. Reine Angew. Math.*, 610:149–180, 2007.

[35] W. Brockman, M. Haiman, Nilpotent orbit varieties and the atomic decomposition of the q-Kostka polynomials. *Canad. J. Math.* 50, no. 3:525–537, 1998.

[36] F. Bruhat, J. Tits, Groupes algébriques sur un corps local. Chapitre III. Compléments et applications à la cohomologie galoisienne. *J. Fac. Sci. Univ. Tokyo Sect. IA Math.*, 34(3):671–698, 1987.

[37] H. Carayol, Nonabelian Lubin–Tate theory. In *Automorphic Forms, Shimura Varieties, and L-functions, Vol. II (Ann Arbor, MI, 1988)*, Volume 11 of *Perspect. Math.*, pages 15–39. Academic Press, Boston, MA, 1990.

[38] R. W. Carter, *Finite Groups of Lie Type*. Wiley Classics Library. John Wiley & Sons Ltd., Chichester, 1993. Reprint of the 1985 original.

[39] W. Casselman, Truncation exercises. In *Functional Analysis VIII*, Volume 47 of *Various Publ. Ser. (Aarhus)*, pages 84–104. Aarhus Univ., Aarhus, 2004.

[40] W. Casselman, On a p-adic vanishing theorem of Garland. *Bull. Amer. Math. Soc.*, 80:1001–1004, 1974.

[41] W. Casselman, A new non-unitarity argument for p-adic representations. *J. Fac. Sci. Univ. Tokyo Sect. IA Math.*, 28(3):907–928 (1982), 1981.

[42] W. Casselman, Introduction to the theory of admissible representations of p-adic reductive groups. http://www.math.ubc.ca/ cass/research/pdf/casselman.pdf

[43] P. Colmez, Espaces vectoriels de dimension finie. *J. Inst. Math. Jussieu* 1, no. 3, 331–439, 2002.

[44] P. Colmez, J.-M. Fontaine, Construction des représentations p-adiques semi-stables. *Invent. Math.*, 140(1):1–43, 2000.

[45] C. W. Curtis, G. I. Lehrer, A new proof of a theorem of Solomon–Tits. *Proc. Amer. Math. Soc.* 85, no. 2:154–156, 1982.

[46] C. W. Curtis, I. Reiner, *Methods of Representation Theory. Vol. II.* Pure and Applied Mathematics (New York). John Wiley & Sons Inc., New York, 1987.

[47] J.-F. Dat, Théorie de Lubin–Tate non-abélienne et représentations elliptiques. *Invent. Math.*, 169(1):75–152, 2007.

[48] J.-F. Dat, Espaces symétriques de Drinfeld et correspondance de Langlands locale. *Ann. Sci. École Norm. Sup. (4)*, 39(1):1–74, 2006.

[49] J.-F. Dat, On the K_0 of a p-adic group. *Invent. Math.* 140, no. 1:171–226, 2000.

[50] A. J. de Jong, Étale fundamental groups of non-Archimedean analytic spaces. *Compositio Math.*, 97(1-2):89–118, 1995.

[51] S. del Baño, On the Chow motive of some moduli spaces. *J. Reine Angew. Math.*, 532:105–132, 2001.

[52] P. Deligne, Travaux de Griffiths. In *Séminaire Bourbaki. Vol. 1969/70: Exposés 364–381*, Lecture Notes in Mathematics, Vol. 180, pages 213–237. Springer-Verlag, Berlin, 1971.

[53] P. Deligne, *Cohomologie étale. Séminaire de Géométrie Algébrique du Bois-Marie* (SGA $4\frac{1}{2}$). Avec la collaboration de J. F. Boutot, A. Grothendieck, L. Illusie

et J. L. Verdier. Lecture Notes in Mathematics, Volume 569. Springer-Verlag, Berlin, 1977.

[54] P. Deligne, Le groupe fondamental de la droite projective moins trois points. In *Galois groups over* **Q** *(Berkeley, CA, 1987)*, Volume 16 of *Math. Sci. Res. Inst. Publ.*, pages 79–297. Springer, New York, 1989.

[55] P. Deligne, Catégories tannakiennes. In *The Grothendieck Festschrift, Vol. II.* Volume 87 of *Progr. Math.*, pages 111–195. Birkhäuser Boston, Boston, MA, 1990.

[56] P. Deligne, Catégories tensorielles. *Mosc. Math. J.* 2, no. 2:227–248, 2002.

[57] P. Deligne, D. Kazhdan, M.-F. Vignéras, Représentations des algèbres centrales simples p-adiques. In *Representations of Reductive Groups over a Local Field*, Travaux en Cours, pages 33–117. Hermann, Paris, 1984.

[58] P. Deligne, G. Lusztig, Representations of reductive groups over finite fields. *Ann. of Math.* (2), 103(1):103–161, 1976.

[59] P. Deligne, G. Lusztig, Duality for representations of a reductive group over a finite field. *J. Algebra* 74, no. 1:284–291, 1982.

[60] P. Deligne, Théorie de Hodge. II. *Inst. Hautes Études Sci. Publ. Math.*, 40:5–57, 1971.

[61] P. Deligne, Travaux de Shimura. In *Séminaire Bourbaki, 23ème année (1970/71), Exp. No. 389*, pages 123–165. Lecture Notes in Math., Volume 244. Springer, Berlin, 1971.

[62] P. Deligne, J. S. Milne, A. Ogus, K.-Yen Shih, *Hodge Cycles, Motives, and Shimura Varieties. Lecture Notes in Mathematics*, Vol. 900. Springer-Verlag, Berlin, 1982.

[63] M. Demazure, *Lectures on p-divisible Groups, Lecture Notes in Mathematics*, Vol. 302. Springer-Verlag, Berlin, 1986. Reprint of the 1972 original.

[64] M. Demazure, P. Gabriel, *Groupes Algébriques. Tome I: Géométrie Algébrique, Généralités, Groupes Commutatifs*. Masson & Cie, Éditeur, Paris, 1970. Avec un appendice *Corps de classes local* par M. Hazewinkel.

[65] C. Deninger, A. Werner, Vector bundles on p-adic curves and parallel transport. *Ann. Sci. École Norm. Sup. (4)*, 38(4):553–597, 2005.

[66] V. V. Deodhar, On Bruhat ordering and weight-lattice ordering for a Weyl group. *Nederl. Akad. Wetensch. Indag. Math.*, 40(4):423–435, 1978.

[67] J. Dieudonné, Groupes de Lie et hyperalgèbres de Lie sur un corps de caractéristique $p > 0$. VII. *Math. Ann.*, 134:114–133, 1957.

[68] F. Digne, J. Michel, *Representations of Finite Groups of Lie Type*, Volume 21 of *London Mathematical Society Student Texts*. Cambridge University Press, Cambridge, 1991.

[69] F. Digne, J. Michel, R. Rouquier, Cohomologie des variétés de Deligne–Lusztig. *Adv. Math.*, 209(2):749–822, 2007.

[70] I. V. Dolgachev, Y. Hu, Variation of geometric invariant theory quotients. With an appendix by Nicolas Ressayre. *Inst. Hautes Etudes Sci. Publ. Math.*, 87:5–56, 1998.

[71] V. G. Drinfeld, Coverings of p-adic symmetric domains. *Funct. Anal. and Appl.*, 10:107–115, 1976.

[72] C. Ehresmann, Sur la topologie de certains espaces homogènes. *Ann. of Math. (2)*, 35(2):396–443, 1934.

[73] G. Faltings, Mumford-Stabilität in der algebraischen Geometrie. In *Proceedings of the International Congress of Mathematicians*, Volume 1, 2 (Zürich, 1994), pages 648–655. Birkhäuser, Basel, 1995.

[74] G. Faltings, A relation between two moduli spaces studied by V. G. Drinfeld. In

Algebraic Number Theory and Algebraic Geometry, Volume 300 of *Contemp. Math.*, pages 115–129. Amer. Math. Soc., Providence, RI, 2002.

[75] G. Faltings, Coverings of p-adic period domains, preprint 2007.

[76] G. Faltings, G. Wüstholz, Diophantine approximation on projective spaces. *Invent. Math.*, 116(1-3):109–138, 1994.

[77] L. Fargues, Cohomologie des espaces de modules de groupes p-divisibles et correspondances de Langlands locales. In *Variétés de Shimura, espaces de Rapoport–Zink et correspondances de Langlands locales. Astérisque*, 291, pages 1–199, 2004.

[78] L. Fargues, A. Genestier, V. Lafforgue, *L'isomorphisme entre les Tours de Lubin–Tate et de Drinfeld*, Volume 262 of *Progress in Mathematics*. Birkhäuser, 2008.

[79] S. Fomin, A. Zelevinsky, Recognizing Schubert cells. *J. Algebraic Combin.*, 12(1):37–57, 2000.

[80] J.-M. Fontaine, Modules galoisiens, modules filtrés et anneaux de Barsotti–Tate. *Astérisque*, 65:3–80, 1979.

[81] J.-M. Fontaine, Sur certains types de représentations p-adiques du groupe de Galois d'un corps local; construction d'un anneau de Barsotti–Tate. *Ann. of Math. (2)*, 115(3):529–577, 1982.

[82] J.-M. Fontaine, M. Rapoport, Existence de filtrations admissibles sur des isocristaux. *Bull. Soc. Math. France*, 133(133–134):73–86, 2005.

[83] R. Friedman, J. W. Morgan, On the converse to a theorem of Atiyah and Bott. *J. Algebraic Geom.*, 11(2):257–292, 2002.

[84] M. Fujimori, On systems of linear inequalities. *Bull. Soc. Math. France*, 131(1):41–57, 2003.

[85] M. Fujimori, Corrigenda: On systems of linear inequalities. *Bull. Soc. Math. France*, 132(4):613–616, 2004.

[86] W. Fulton, *Introduction to Toric Varieties*, Volume 131 of *Annals of Mathematics Studies*. Princeton University Press, Princeton, NJ, 1993.

[87] W. Fulton, J. Harris, *Representation Theory*, Volume 129 of *Graduate Texts in Mathematics*. Springer-Verlag, New York, 1991.

[88] H. Garland, p-adic curvature and the cohomology of discrete subgroups of p-adic groups. *Ann. of Math.* (2) 97:375–423, 1973.

[89] I. M. Gelfand, R. M. Goresky, R. D. MacPherson, V. V. Serganova, Combinatorial geometries, convex polyhedra, and Schubert cells. *Adv. in Math.*, 63(3):301–316, 1987.

[90] I. M. Gelfand, V. V. Serganova, Combinatorial geometries and the strata of a torus on homogeneous compact manifolds. *Uspekhi Mat. Nauk*, 42(2):107–134, 1987.

[91] A. Genestier, Espaces symétriques de Drinfeld. *Astérisque*, 234:ii+124, 1996.

[92] O. Goldman, N. Iwahori, The space of p-adic norms. *Acta Math.*, 109:137–177, 1963.

[93] M. Goresky, R. Kottwitz, R. MacPherson, Discrete series characters and the Lefschetz formula for Hecke operators. *Duke Math. J.* 89, no. 3:477–554, 1997.

[94] U. Görtz, On the connectedness of Deligne–Lusztig varieties. *Represent. Theory* 13:1–7, 2009.

[95] D. R. Grayson, Finite generation of K-groups of a curve over a finite field (after Daniel Quillen). In *Algebraic K-theory, Part I (Oberwolfach, 1980), Lecture Notes in Mathematics.* Vol. 966, pages 69–90. Springer, Berlin, 1982.

[96] P. A. Griffiths, W. Schmid, Recent developments in Hodge theory: a discussion of techniques and results. In *Discrete Subgroups of Lie Groups and Applications to*

Moduli (Internat. Colloq., Bombay, 1973), pages 31–127. Oxford Univ. Press, Bombay, 1975.

[97] P. A. Griffiths, Periods of integrals on algebraic manifolds. III. Some global differential-geometric properties of the period mapping. *Inst. Hautes Études Sci. Publ. Math.*, 38:125–180, 1970.

[98] A. Grothendieck, Groupes de Barsotti–Tate et cristaux. In *Actes du Congrès International des Mathématiciens (Nice, 1970), Tome 1*, pages 431–436. Gauthier-Villars, Paris, 1971.

[99] A. Grothendieck, Technique de descente et théorèmes d'existence en géométrie algébrique. I. Généralités. Descente par morphismes fidèlement plats. Séminaire Bourbaki 190 , Paris, 1959/60.

[100] V. Guillemin, S. Sternberg, Convexity properties of the moment mapping. *Invent. Math.*, 67(133-134):491–513, 1982.

[101] B. Haastert, Über die Deligne–Lusztig Varietäten reduktiver Gruppen, *Diplomarbeit Universität Bonn*, 1984.

[102] B. Haastert, Die Quasiaffinität der Deligne–Lusztig-Varietäten. *J. Algebra* 102 no. 1:186–193, 1986.

[103] W. J. Haboush, Reductive groups are geometrically reductive. *Ann. of Math. (2)*, 102(1):67–83, 1975.

[104] P. Hacking, S. Keel, J. Tevelev, Compactification of the moduli space of hyperplane arrangements. *J. Algebraic Geom.*, 15(4):657–680, 2006.

[105] G. Harder, M. S. Narasimhan, On the cohomology groups of moduli spaces of vector bundles on curves. *Math. Ann.*, 212:215–248, 1975.

[106] M. Harris, Local Langlands correspondences and vanishing cycles on Shimura varieties. In *European Congress of Mathematics, Vol. I (Barcelona, 2000)*, Volume 201 of *Progr. Math.*, pages 407–427. Birkhäuser, Basel, 2001.

[107] M. Harris, R. Taylor, *The Geometry and Cohomology of some Simple Shimura Varieties*, Volume 151 of *Annals of Mathematics Studies*. Princeton University Press, Princeton, NJ, 2001. With an appendix by Vladimir G. Berkovich.

[108] U. Hartl, On a conjecture of Rapoport and Zink. arXiv:math.NT/0605254.

[109] U. Hartl, On period spaces for p-divisible groups. *C. R. Math. Acad. Sci. Paris* 346, no. 21-22:1123–1128, 2008.

[110] U. Hartl, Period spaces for Hodge structures in equal characteristic. arXiv:math.NT/0511686.

[111] U. Hartl, R. Pink, Vector bundles with a Frobenius structure on the punctured unit disc. *Compos. Math.*, 140(3):689–716, 2004.

[112] X. He, On the affineness of Deligne–Lusztig varieties. *J. Algebra*, 320(3):1207–1219, 2008.

[113] X. He, *e-mail message to M. Rapoport*, 24/03/09.

[114] G. Henniart, Une preuve simple des conjectures de Langlands pour GL(n) sur un corps p-adique. *Invent. Math.*, 139(2):439–455, 2000.

[115] Nan-Kuo Ho, Chiu-Chu Melissa Liu, Yang–Mills connections on orientable and nonorientable surfaces, arXiv:0707.0258, to appear in *Memoirs of the American Mathematical Society*.

[116] M. J. Hopkins, B. H. Gross, Equivariant vector bundles on the Lubin–Tate moduli space. In *Topology and Representation Theory (Evanston, IL, 1992)*, Volume 158 of *Contemp. Math.*, pages 23–88. Amer. Math. Soc., Providence, RI, 1994.

[117] R. Howe, *Harish-Chandra Homomorphisms for p-Adic Groups. With the col-*

laboration of Allen Moy., Volume 59 of *CBMS Regional Conference Series in Mathematics*. American Mathematical Society, Providence, RI, 1985.

[118] R. E. Howe, C. C. Moore, Asymptotic properties of unitary representations. *J. Funct. Anal.*, 32(1):72–96, 1979.

[119] R. B. Howlett, G. I. Lehrer, On Harish-Chandra induction and restriction for modules of Levi subgroups. *J. Algebra*, 165(1):172–183, 1994.

[120] Y. Hu, (W,R)-matroids and thin Schubert-type cells attached to algebraic torus actions. *Proc. Amer. Math. Soc.*, 123(9):2607–2617, 1995.

[121] R. Huber, *Etale Cohomology of Rigid Analytic Varieties and Adic Spaces. Aspects of Mathematics*, E30. Friedr. Vieweg & Sohn, Braunschweig, 1996.

[122] J. E. Humphreys, *Reflection Groups and Coxeter Groups*, Volume 29 of *Cambridge Studies in Advanced Mathematics*. Cambridge University Press, Cambridge, 1990.

[123] M.-N. Ishida, F. Kato, The strong rigidity theorem for non-Archimedean uniformization. *Tohoku Math. J.* (2) 50, no. 4:537–555, 1998.

[124] T. Ito, Weight-monodromy conjecture for p-adically uniformized varieties. *Invent. Math.* 159, no. 3: 607–656, 2005.

[125] J. C. Jantzen, *Representations of Algebraic Groups*, Volume 131 of *Pure and Applied Mathematics*. Academic Press, Inc., Boston, MA, 1987.

[126] M. Kapranov, Chow quotients of Grassmannians. I. I. M. Gelfand Seminar, *Adv. Soviet Math., 16, Part 2*, pages 29–110, Amer. Math. Soc., Providence, RI, 1993.

[127] M. Kapranov, Veronese curves and Grothendieck–Knudsen moduli space $\overline{M}_{0,n}$. *J. Algebraic Geom.* 2, no. 2:239–262, 1993.

[128] V. Kashin, Orbits of an adjoint and co-adjoint action of Borel subgroups of a semisimple algebraic group. (Russian) Problems in group theory and homological algebra (Russian), *Matematika, Yaroslav. Gos. Univ., Yaroslavl*, 141–158, 1990.

[129] K. S. Kedlaya, Slope filtrations revisited. *Doc. Math.*, 10:447–525 (electronic), 2005.

[130] M. Kisin, Crystalline representations and F-crystals. In *Algebraic Geometry and Number Theory*, pages 459–496, vol. 253 of *Progr. Math.*, Birkhäuser Boston, Boston, MA, 2006.

[131] M. Kisin, G. I. Lehrer, Equivariant Poincaré polynomials and counting points over finite fields. *J. Algebra*, 247(2):435–451, 2002.

[132] M. Kneser, Galois-Kohomologie halbeinfacher algebraischer Gruppen über p-adischen Körpern. I. *Math. Z.*, 88:40–47, 1965.

[133] B. Kostant, On convexity, the Weyl group and the Iwasawa decomposition. *Ann. Sci. École Norm. Sup. (4)*, 6:413–455, 1974.

[134] R. E. Kottwitz, Shimura varieties and twisted orbital integrals. *Math. Ann.*, 269(3): 287–300, 1984.

[135] R. E. Kottwitz, Isocrystals with additional structure. *Compositio Math.*, 56(2): 201–220, 1985.

[136] R. E. Kottwitz, Isocrystals with additional structure. II. *Compositio Math.*, 109(3): 255–339, 1997.

[137] R. E. Kottwitz, Stable trace formula: cuspidal tempered terms. *Duke Math. J.* 51, no. 3:611–650, 1984.

[138] C. Ku, A new proof of the Solomon–Tits theorem. *Proc. Amer. Math. Soc.* 126, no. 7:1941–1944, 1989.

[139] S. S. Kudla, The local Langlands correspondence: the non-Archimedean case. In

Motives (Seattle, WA, 1991), Volume 55 of *Proc. Sympos. Pure Math.*, pages 365–391. Amer. Math. Soc., Providence, RI, 1994.

[140] A. Kurihara, Construction of p-adic unit balls and the Hirzebruch proportionality. *Amer. J. Math.*, 102(3):565–648, 1980.

[141] J.-P. Labesse, La formule des traces d'Arthur–Selberg, Séminaire Bourbaki, Volume 1984/85. *Astérisque*, 133-134:73–88, 1986.

[142] L. Lafforgue, *Chirurgie des Grassmanniennes*, Volume 19 of *CRM Monograph Series*. American Mathematical Society, Providence, RI, 2003.

[143] R. P. Langlands, Eisenstein series. In *Algebraic Groups and Discontinuous Subgroups – Boulder, Colo. 1965*, Volume 9 of *Proc. Sympos. Pure Math.*, pages 235–252. American Mathematical Society, Providence, RI, 1966.

[144] R. P. Langlands, M. Rapoport, Shimuravarietäten und Gerben. *J. Reine Angew. Math.*, 378:113–220, 1987.

[145] S. Lattermann, *Tannakakategorien.* Bonner Mathematische Schriften [Bonn Mathematical Publications], 196. Universität Bonn, Mathematisches Institut, Bonn, 1989.

[146] E. Lau, Rapoport–Zink spaces without decency. Preprint

[147] G. Laumon, M. Rapoport, U. Stuhler, $\mathscr{D}$-elliptic sheaves and the Langlands correspondence. *Invent. Math.*, 113(2):217–338, 1993.

[148] G. Laumon, Comparaison de caractéristiques d'Euler–Poincaré en cohomologie l-adique. *C. R. Acad. Sci. Paris Sér. I Math.*, 292(3):209–212, 1981.

[149] G. Laumon, *Cohomology of Drinfeld Modular Varieties. Part I. Geometry, counting of points and local harmonic analysis*, Volume 41 of *Cambridge Studies in Advanced Mathematics*. Cambridge University Press, Cambridge, 1996.

[150] G. Laumon, *Cohomology of Drinfeld modular varieties. Part II. Automorphic forms, trace formulas and Langlands correspondence.*, Volume 56 of *Cambridge Studies in Advanced Mathematics*. Cambridge University Press, Cambridge, 1997.

[151] G. Laumon, M. Rapoport, The Langlands lemma and the Betti numbers of stacks of G-bundles on a curve. *Internat. J. Math.*, 7(1):29–45, 1996.

[152] G. I. Lehrer, The spherical building and regular semisimple elements. *Bull. Austral. Math. Soc.*, 27(3):361–379, 1983.

[153] G. I. Lehrer, The cohomology of the regular semisimple variety. *J. Algebra*, 199(2):666–689, 1998.

[154] J. Lubin, J. Tate, Formal moduli for one-parameter formal Lie groups. *Bull. Soc. Math. France*, 94:49–59, 1966.

[155] G. Lusztig, Coxeter orbits and eigenspaces of Frobenius. *Invent. Math.*, 38(2):101–159, 1976/77.

[156] G. Lusztig, *Characters of Reductive Groups over a Finite Field.* Volume 107 of *Annals of Mathematics Studies.* Princeton University Press, Princeton, NJ, 1984.

[157] I. G. Macdonald, *Symmetric Functions and Hall Polynomials.* Oxford Mathematical Monographs. The Clarendon Press Oxford University Press, New York, second edition, 1995.

[158] J. I. Manin, Theory of commutative formal groups over fields of finite characteristic. *Uspehi Mat. Nauk*, 18(6 (114)):3–90, 1963.

[159] E. Mantovan, On certain unitary group Shimura varieties. In *Variétés de Shimura, Espaces de Rapoport–Zink et Correspondances de Langlands Locales. Astérisque*, 291, pages 201–331, 2004.

[160] E. Mantovan, On the cohomology of certain PEL-type Shimura varieties. *Duke Math. J.* 129, no. 3:573–610, 2005.

[161] M. Maruyama, The theorem of Grauert–Mülich–Spindler. *Math. Ann.* 255(3):317–333, 1981.

[162] W. Messing, *The Crystals Associated to Barsotti–Tate Groups: with Applications to Abelian Schemes*. Lecture Notes in Mathematics, Volume 264. Springer-Verlag, Berlin, 1972.

[163] D. Mumford, An algebraic surface with K ample, $(K^2) = 9$, $p_g = q = 0$. *Amer. J. Math.* 101, no. 1:233–244, 1979.

[164] D. Mumford, J. Fogarty, F. Kirwan, *Geometric Invariant Theory (third edition)*, Volume 34 of *Ergebnisse der Mathematik und ihrer Grenzgebiete (2)*. Springer-Verlag, Berlin, 1994.

[165] G. A. Mustafin, Non-Archimedean uniformization. *Mat. Sb. (N.S.)*, 105(147)(2):207–237, 1978.

[166] M. S. Narasimhan, C. S. Seshadri, Stable and unitary vector bundles on a compact Riemann surface. *Ann. of Math.*, 82(2):540–567, 1965.

[167] P. Orlik, H. Terao, *Arrangements of Hyperplanes*, Volume 300 of *Grundlehren der Mathematischen Wissenschaften*. Springer-Verlag, Berlin, 1992.

[168] S. Orlik, M. Rapoport, Deligne–Lusztig varieties and period domains over finite fields. *J. Algebra*, 320(3):1220–1234, 2008.

[169] S. Orlik, Kohomologie von Periodenbereichen über endlichen Körpern. *J. Reine Angew. Math.*, 528(1):201–233, 2000.

[170] S. Orlik, The cohomology of period domains for reductive groups over finite fields. *Ann. Sci. École Norm. Sup. (4)*, 34(1):63–77, 2001.

[171] S. Orlik, The cohomology of period domains for reductive groups over local fields. *Invent. Math.*, 162(3):523–549, 2005.

[172] S. Orlik, On extensions of generalized Steinberg representations. *J. Algebra*, 293(2):611–630, 2005.

[173] S. Orlik, On Harder–Narasimhan strata in flag manifolds. *Math. Z.*, 252(1):209–222, 2006.

[174] S. Orlik, The continuous cohomology of period domains over local fields. *Comment. Math. Helv.*, 82(3):665–681, 2007.

[175] S. Orlik, The fundamental group of period domains over finite fields. *Adv. Math.*, 219(2):728–741, 2008.

[176] G. Pappas, M. Rapoport, Φ-modules and coefficient spaces, *Moscow Math. J.*, 9:625–663, 2009.

[177] R. Pink, Hodge Structures over Function Fields.
Preprint, http://www.math.ethz.ch/ pink/publications.html.

[178] R. Pink, l-adic algebraic monodromy groups, cocharacters, and the Mumford–Tate conjecture. *J. Reine Angew. Math.*, 495:187–237, 1998.

[179] S. Ramanan, A. Ramanathan, Some remarks on the instability flag. *Tohoku Math. J. (2)*, 36(2):269–291, 1984.

[180] A. Ramanathan, Stable principal bundles on a compact Riemann surface. *Math. Ann.*, 213:129–152, 1975.

[181] M. Rapoport, Analogien zwischen den Modulräumen von Vektorbündeln und von Flaggen. *Jahresber. Deutsch. Math.-Verein.*, 99(4):164–180, 1997.

[182] M. Rapoport, M. Richartz, On the classification and specialization of F-isocrystals with additional structure. *Compositio Math.*, 103(2):153–181, 1996.

[183] M. Rapoport, T. Zink, *Period Spaces for p-divisible Groups*, Volume 141 of *Annals of Mathematics Studies*. Princeton University Press, Princeton, NJ, 1996.

[184] M. Rapoport, Two letters to B. Totaro, 1994.

[185] M. Rapoport, Non-Archimedean period domains. In *Proceedings of the Inter-*

national Congress of Mathematicians, Volume 1, 2 (Zürich, 1994), pages 423–434. Birkhäuser, Basel, 1995.

[186] M. Rapoport, Period domains over finite and local fields. In *Algebraic geometry–Santa Cruz 1995*, Volume 62 of *Proc. Sympos. Pure Math.*, pages 361–381. Amer. Math. Soc., Providence, RI, 1997.

[187] M. Rapoport, T. Zink, Über die lokale Zetafunktion von Shimuravarietäten. Monodromiefiltration und verschwindende Zyklen in ungleicher Charakteristik, *Invent. Math.*, 68(1):21–101, 1982.

[188] M. Reineke, The Harder–Narasimhan system in quantum groups and cohomology of quiver moduli. *Invent. Math.*, 152(2):349–368, 2003.

[189] G. Röhrle, Parabolic subgroups of positive modality. *Geom. Dedicata*, 60, no. 2:163–186, 1996.

[190] J. Rogawski, Representations of GL(n) and division algebras over a p-adic field. *Duke Math. J.* 50, no. 1:161–196, 1983.

[191] A. L. Rosenberg, The existence of fiber functors. In *The Gelfand Mathematical Seminars, 1996–1999*, Gelfand Math. Sem., pages 145–154. Birkhäuser Boston, Boston, MA, 2000.

[192] G. Rousseau, Immeubles sphériques et théorie des invariants. *C. R. Acad. Sci. Paris Sér. A-B*, 286(5):A247–A250, 1978.

[193] N. Saavedra Rivano, *Catégories Tannakiennes*. Lecture Notes in Mathematics, Volume 265. Springer-Verlag, Berlin, 1972.

[194] W. Schmid, Variation of Hodge structure: the singularities of the period mapping. *Invent. Math.*, 22:211–319, 1973.

[195] W. Schmid, L^2-cohomology and the discrete series. *Ann. of Math. (2)*, 103(2): 375–394, 1976.

[196] P. Schneider, U. Stuhler, The cohomology of p-adic symmetric spaces. *Invent. Math.*, 105(1):47–122, 1991.

[197] J.-P. Schneiders, Quasi-abelian categories and sheaves. *Mém. Soc. Math. Fr. (N.S.)*, (76):vi+134, 1999.

[198] J. P. Serre, Groupes algébriques associés aux modules de Hodge–Tate. In *Journées de Géométrie Algébrique de Rennes, Astérisque* 65, pages 155–188, 1979.

[199] E. de Shalit, The p-adic monodromy-weight conjecture for p-adically uniformized varieties. *Compos. Math.* 141, no. 1:101–120, 2005.

[200] S. S. Shatz, The decomposition and specialization of algebraic families of vector bundles. *Compositio Math.*, 35(2):163–187, 1977.

[201] V. P. Snaith, *Algebraic K-groups as Galois Modules*, Volume 206 of *Progress in Mathematics*. Birkhäuser Verlag, Basel, 2002.

[202] L. Solomon, The Steinberg character of a finite group with BN-pair. In *Theory of Finite Groups (Symposium, Harvard Univ., Cambridge, Mass., 1968)*, pages 213–221. Benjamin, New York, 1969.

[203] L. Solomon, A Mackey formula in the group ring of a Coxeter group. *J. Algebra*, 41(2):255–264, 1976.

[204] T. A. Springer, Regular elements of finite reflection groups. *Invent. Math.*, 25:159–198, 1974.

[205] T. A. Springer, A construction of representations of Weyl groups. *Invent. Math.*, 44(3):279–293, 1978.

[206] T. A. Springer, R. Steinberg, Conjugacy classes. In *Seminar on Algebraic Groups and Related Finite Groups (The Institute for Advanced Study, Princeton,*

N.J., 1968/69), Lecture Notes in Mathematics, Volume 131, pages 167–266. Springer, Berlin, 1970.

[207] R. P. Stanley, Some aspects of groups acting on finite posets. *J. Combin. Theory Ser. A*, 32(2):132–161, 1982.

[208] R. Steinberg, Regular elements of semisimple algebraic groups. *Inst. Hautes Études Sci. Publ. Math.*, (25):49–80, 1965.

[209] R. Steinberg, *Endomorphisms of Linear Algebraic Groups*. Memoirs of the American Mathematical Society, No. 80. American Mathematical Society, Providence, R.I., 1968.

[210] R. Steinberg, *Lectures on Chevalley groups*. Yale University, New Haven, Conn., 1968. Notes prepared by John Faulkner and Robert Wilson.

[211] J. R. Stembridge, Tight quotients and double quotients in the Bruhat order. *Electron. J. Combin.*, 11(2):Research Paper 14, 41 pp. (electronic), 2004/06.

[212] M. Strauch, Deformation spaces of one-dimensional formal modules and their cohomology. *Adv. Math.* 217, no. 3:889–951, 2008.

[213] D. B. Surowski, Permutation characters of finite groups of Lie type. *J. Austral. Math. Soc. Ser. A*, 27(3):378–384, 1979.

[214] J. Tate, Number theoretic background. In *Automorphic forms, Representations and L-functions (Proc. Sympos. Pure Math., Oregon State Univ., Corvallis, Ore., 1977), Part 2*, Proc. Sympos. Pure Math., XXXIII, pages 3–26. Amer. Math. Soc., Providence, R.I., 1979.

[215] J. Tits, Représentations linéaires irréductibles d'un groupe réductif sur un corps quelconque. *J. Reine Angew. Math.*, 247:196–220, 1971.

[216] J. Tits, Reductive groups over local fields. In *Automorphic Forms, Representations and L-functions (Proc. Sympos. Pure Math., Oregon State Univ., Corvallis, Ore., 1977), Part 1*, Proc. Sympos. Pure Math., XXXIII, pages 29–69. Amer. Math. Soc., Providence, R.I., 1979.

[217] A. N. Tjurin, The classification of vector bundles over an algebraic curve of arbitrary genus. *Izv. Akad. Nauk SSSR Ser. Mat.*, 29:657–688, 1965.

[218] B. Totaro, Tensor products of semistables are semistable. In *Geometry and Analysis on Complex Manifolds*, pages 242–250. World Sci. Publishing, River Edge, NJ, 1994.

[219] B. Totaro, Tensor products in p-adic Hodge theory. *Duke Math. J.*, 83(1):79–104, 1996.

[220] M. van der Put, H. Voskuil, Symmetric spaces associated to split algebraic groups over a local field. *J. Reine Angew. Math.*, 433:69–100, 1992.

[221] A. Werner, Compactification of the Bruhat–Tits building of PGL by seminorms. *Math. Z.*, 248(3):511–526, 2004.

[222] J.-P. Wintenberger, Propriétés du groupe tannakien des structures de Hodge p-adiques et torseur entre cohomologies cristalline et étale. *Ann. Inst. Fourier (Grenoble)* 47, no. 5:1289–1334, 1997.

[223] D. Zagier, Elementary aspects of the Verlinde formula and of the Harder–Narasimhan–Atiyah–Bott formula. In *Proceedings of the Hirzebruch 65 Conference on Algebraic Geometry (Ramat Gan, 1993)*, Volume 9 of *Israel Math. Conf. Proc.*, pages 445–462, Ramat Gan, 1996. Bar-Ilan University.

[224] T. Zink, *Cartiertheorie Kommutativer Formaler Gruppen. Teubner-Texte zur Mathematik*, 68. BSB B. G. Teubner Verlagsgesellschaft, Leipzig, 1984.

Index

For EU product safety concerns, contact us at Calle de José Abascal, 56–1°, 28003 Madrid, Spain or eugpsr@cambridge.org.

www.ingramcontent.com/pod-product-compliance
Lightning Source LLC
LaVergne TN
LVHW101456200926
848483LV00001B/139
* 9 7 8 0 5 2 1 1 9 7 6 9 4 *